STATISTICAL REASONING AND METHODS

STATISTICAL REASONING AND METHODS

RICHARD A. JOHNSON

KAM-WAH TSUI

University of Wisconsin at Madison

JOHN WILEY & SONS, INC.

NEW YORK • CHICHESTER • WEINHEIM • BRISBANE • SINGAPORE • TORONTO

Acquisitions Editor   *Brad Wiley II*
Marketing Manager   *Jay Kirsch*
Senior Production Editor   *Tracey Kuehn*
Cover Designer   *Carol C. Grobe*
Photo Editor   *Elaine Paoloni*
Illustration Coordinator   *Jaime Perea*
Illustration Studio   *Boris Starosta Technical Illustration*

Special thanks to FPG International, a New York City photo agency, for providing all of the photographs for the Statistical Reasoning cases.
Cover and chapter opening photos: Norman Myers/Bruce Coleman, Inc.

This book was set in 10/12 Berling by Progressive Information Technologies and printed and bound by Quebecor Printing, Fairfield. The cover was printed by Lehigh Press.

This book is printed on acid-free paper. ∞

The paper in this book was manufactured by a mill whose forest management programs include sustained yield harvesting of its timberlands. Sustained yield harvesting principles ensure that the numbers of trees cut each year does not exceed the amount of new growth.

ISBN 0-471-04205-6
Printed in the United States of America

10 9 8 7 6 5 4 3 2 1

To my students—for making statistics an interesting and exciting career

R. A. J.

To my parents, Chan-Wing Tsui and Wai-King Yeung

K. W. T.

Preface

Statistics plays an important role in our daily lives. It helps us understand the nature of situations that appear to be unpredictable and aid decision making when events are not certain to occur. Statistics concerns the collection, description, and making of generalizations from data. It plays an increasing role in almost all fields of activity and is used extensively in government reports and by professionals in many fields.

The aims of this book are to introduce basic statistical concepts and methods and to show how they enhance critical thinking and reasoning. The first chapter emphasizes critical thinking regarding data collection. Throughout the book, a number of cases are presented where statistics are improperly used and, consequently, the corresponding conclusions are invalid. Special material in each chapter, set out in boxes labeled *Statistical Reasoning*, presents and reinforces key ideas of statistical reasoning in story settings. Exercises help students learn to question the quality of data and think about what factors produce variation in the response.

Many intuitive and simple explanations motivate the key statistical concepts and terms. Numerous examples illustrate the methods in the context of situations most students experience in their everyday lives.

We present probability, a somewhat difficult subject, as it will be used to help make decisions and to make inferences. This novel treatment, including many examples, makes clear how the key ideas of probability underpin statistical concepts and inferences.

This book can be used for any first elementary one-quarter or one-semester statistics course. The mathematical level is elementary and suitable for students in two-year and four-year colleges as well as universities. The topics covered are the most important ones common to any first course in statistics for freshmen and sophomores. By concentrating on the main concepts and methods, we place more emphasis on reasoning than is currently available in most texts.

Responding to the current trend of having students perform experiments and collect data, we have included suggested class projects in most chapters. They help deepen students' understanding of how to collect good data to answer well posed questions and the role of randomization in comparative experiments.

Numerous computer outputs are included with the examples in the text. This makes it possible for an instructor to place less emphasis on the formulas if desired. Additional MINITAB commands and output are included in the many computer-based exercises.

To the students, we hope that you will become critical consumers of statistical information provided by others. Further, we hope that you will learn how to collect good data and assemble the information in a way that leads to better understanding of a product, service, or the environment.

DISTINGUISHING FEATURES

- Brief text focuses on extensive coverage of key topics. Extended intuitive explanations of the basic concepts and methods, beginning with sample mean, median, and variance, reinforce their importance.
- This book shows how basic statistical concepts and methods enhance critical thinking and reasoning. Chapter 1 helps students distinguish "good" samples from "bad" samples and criticize data. Statistical Reasoning boxes, throughout, describe reasoning in settings familiar to students.
- Chapter 4 develops probability from the point of view of its use in statistics. The laws of probability are applied to solve probability and inference problems of the kind encountered by students in their everyday activities.
- Early emphasis on designing the data collection process, in Chapters 1–3, allows the interested instructor to assign student projects early in the course. These can help students understand the key elements of statistical design, and students can refer to their own data when inference procedures are discussed.
- Special charts show the general themes connecting various inference procedures. These link statistical methods that students might otherwise assume are unrelated.

ORGANIZATION

The text can be divided into five parts.

1. Collecting and describing data (Chapters 1–3). Determining good and bad samples and understanding graphical and numerical summaries.
2. The ideas underlying inference (Chapters 4–7). Probability and its role; discrete and continuous distributions; the notions of sampling and sampling distributions.
3. Inferences about means (Chapters 8–10). Clear, simple explanations of confidence intervals and tests of hypotheses.
4. Inferences about proportions (Chapter 11). One-sample and two-sample inference procedures with some coverage of chi-square tests.
5. Regression (Chapter 12). Fitting a straight line to data; and associated inferences and model checking.

COURSE SEQUENCING

Most of the first nine chapters can be taught in sequence in a one-semester course. Chapter 3 could be skipped if student projects are not a priority. We typically finish with some inference about proportions and fitting a straight line.

SUPPLEMENTS

- An *Instructor's Manual* includes complete worked solutions to all of the exercises in the text.
- A *Test Bank* provides true–false, fill-in-the-blank, multiple-choice, and computational exercises for every chapter in the book. It is available in both hard copy and IBM-PC version formats.
- A *Data Disk* contains files for all large data sets in ASCII format. The data disk is available from the Wiley website at www.wiley.com/college. If unable to access the Internet, adopters of the text may also request a copy of the disk by contacting your local Wiley sales representative.

ACKNOWLEDGMENTS

We first want to express our thanks to long-time friend and colleague Gouri Bhattacharyya, coauthor with R. Johnson of a related book. His contributions are clearly evident in this text. Thanks to Alan Wong for his comments, David Tsui for his comments and assistance in typing some of the material, Ella Mae Matsumura for her editorial advice, and Erik Johnson for his help with the computer exercises. Thanks to Melissa Schultz for her careful proofreading and suggestions. Any errors that remain are, of course, our responsibility. We would appreciate having them brought to our attention.

We would like to thank Pat Buchanan (Pennsylvania State University), Gary Egan (Monroe Community College), Ruby Evans (Santa Fe Community College), Betsy Farber (Bucks County Community College), Gary Kulis (Mohawk Valley Community College), Nancy Lyons (University of Georgia), Larry Schubert (San Joaquin Delta College), and Vasant Waikar (Miami University) for their comments.

Contents

CHAPTER 1

About Statistics

Chapter Objectives

After reading this chapter, you should be able to

▶ Distinguish between populations and samples.
▶ Use a table of random digits to select a random sample.
▶ Begin to develop a critical viewpoint toward the collection of data.
▶ Give a clear statement of purpose for the collection of data.

1. WHY STUDY STATISTICS?

You are undoubtedly already familiar with the term *statistics* as it applies to the numbers of unemployed, college students in various disciplines, or the cost of living index, all of which are based on data collected by the government. In its earliest applications, government rulers numbered their subjects as the basis for collecting taxes and determining the size of armies they could raise. In fact, the word **statistics** is derived from the Latin word "status," meaning "state."

Gigantic strides made during the twentieth century have brought the discipline of statistics to prominence by providing ways for reaching reasoned conclusions based on data. Statistics now serves as an investigative tool and a guide to the unknown. Statistics helps provide

> **a systematic approach for obtaining reasoned answers together with some assessment of their reliability**

in situations where complete information is unobtainable or not available in a timely manner.

Answers provided by statistical approaches can provide the basis for making decisions or choosing actions. For example, city officials might examine whether the level of lead in the water supply is within safety standards. Because not all of the water can be checked, answers must be based on the partial information from samples of water that are collected for this purpose. As another example, a store wants to determine which days of the week and times of day they should expect the most shoppers in a typical month. Data on the number of customers at various times of day, collected daily for a few weeks, can provide a reference.

When information is sought, statistical ideas suggest a typical process with four crucial steps:

(a)   Set clearly defined goals for the investigation.

(b)   Make a plan of what data to collect and how to collect them.

(c)   Apply appropriate methods to extract information from the data.

(d)   Interpret the information and draw conclusions.

These indispensable steps will provide a frame of reference throughout as we develop the key ideas and methods of reasoning statistically.

In summary,

> **Statistics** as a subject provides a body of principles and methodology for designing the process of data collection, summarizing and interpreting the data, and drawing conclusions or generalities.

2. STATISTICS IN OUR LIVES

Fact finding through the collection and interpretation of data is not confined to professional researchers. In our attempts to understand the issues of air and water quality, improvement of the mass transit system we ride, or the performance of our favorite football team, numerical information and figures need to be reviewed and interpreted.

Our sources of information range from individual experience to reports in the news media, government records, and technical articles. As consumers of these reports, citizens need to be aware of the quality of the data and to have some understanding of statistical reasoning to properly interpret the data and evaluate the conclusions. Statistical criteria determine which conclusions are supported by the data and which are not.

The following examples illustrate the diversity of applications of statistics.

Repairing library books. Librarians need to estimate the number of their books that need repairing each year. Records from previous years provide one source of data. Librarians use their estimate to budget repair costs and any extra staff required.

Election time brings the pollsters into the limelight.

Gallup Poll. This, the best known of the national polls, produces estimates of the percentage of popular vote for each candidate based on interviews with a minimum of 1500 adults. Beginning several months before the presidential election, results are regularly published. These reports help predict winners and track changes in voter preferences.

Statistical approaches are also key to improving any type of manufacturing or service process.

Quality and productivity improvement. In the past 30 years, the United States has faced increasing competition in the world marketplace. An international revolution in quality and productivity improvement has heightened the pressure on the U.S. economy. The ideas and teaching of W. Edwards Deming helped rejuvenate Japan's industry in the late 1940s and 1950s. In the 1980s and 1990s, Deming stressed to American executives that, in order to survive, they must mobilize their work force to make a continuing commitment to quality improvement. His ideas have also been applied to government. The city of Madison, Wisconsin, has implemented quality improvement projects in the police department, and in bus repair and scheduling. In each case, the project goal was better service at less cost. Treating citizens as the customers of government services, the first step was to collect information from them to identify situations that needed improvement. One end result was the strategic placement of a new police substation and a subsequent increase in the number of foot patrolpersons to interact with the community.

Once a candidate project is selected for improvement, data must be collected to assess the current status and then more data collected on the effects of possible changes. At this stage, statistical skills in the collection and presentation of summaries are not only valuable but necessary for all participants.

In an industrial setting, statistical training for all employees—production line and office workers, supervisors, and managers—is vital to the quality transformation of American industry.

Monitoring advertising claims.   The public is constantly bombarded with commercials that claim the superiority of one product brand compared with others. When such comparisons are founded on sound experimental evidence, they serve to educate the consumer. Not infrequently however, misleading advertising claims are made due to insufficient experimentation, faulty analysis of data, or even blatant manipulation of experimental results. Government agencies and consumer groups must be prepared to verify the comparative quality of products by using adequate data collection procedures and proper methods of statistical analysis.

3. VARIATION AND THE ROLE OF STATISTICS IN LEARNING

Why do pine trees grow in one part of the country and not in another? Is it because of weather conditions, the chemical composition of the soil, or the existence of special live organisms? For our forests to remain renewable resources, these questions need to be answered. Experiments can be conducted to determine the major factors that influence the growth of pine trees. We can use statistical methods to efficiently search for the important factors. After careful considerations, one can formulate a tentative hypothesis about a factor that is influential for pine tree growth. Statistical procedures can guide the collection of appropriate data for supporting or refuting the hypothesis. Essentially, one carries out the four steps (a)–(d), presented in Section 1, for a typical statistical investigation. If this check confirms agreement between the data and what is suggested by the hypothesis, the hypothesis is verified to the extent it provides an explanation of the data. Learning has occurred.

If the check reveals disagreement, another hypothesis can be formulated and the investigative procedure repeated to examine its validity. The searching continues until, with good fortune, the important factors are identified.

In the context of a manufacturing process, what are the factors or materials that yield products that are insensitive to variation in the environment? A portable tape player is expected to work well in all temperatures as well as in very high or very low humidity conditions. If equipment is too sensitive to variation in the environment, its usefulness is questionable and it will not sell well over the long term.

STATISTICS REQUIRED TO EXPLAIN VARIATION

Whenever we wish to explain an observed pattern of variation, we are required to carefully examine the situation and then propose a potential theory or hypothesis. Statistics can play a key role in the verification step. It can provide guidance on how to design an experiment or survey to collect data efficiently. Then, following the methods described in later chapters, it provides the only methods for assessing the agreement between hypotheses and data that exhibit variation.

As another example, we notice that there is a wide variation of income among individuals. Naturally, we would like to search for explanations of this variation. Several hypotheses have been proposed. Many persons think that the variation is due to an age factor and educational levels. An older person has had more time to accumulate wealth or to attain a senior position.

As we learn more about a situation, how to complete a task, or play a sport, we typically get better results. The learning process can be made more efficient by applying a logical procedure for reasoning from available evidence. A conjecture is made and data are collected. Typically, an analysis of the data leads to a modified conjecture. New data are collected and the cycle continues until a satisfactory conjecture, theory, or explanation is obtained. This approach, called the **scientific method,** is similar to the investigative procedure described in Section 1. It is represented by the sequence of steps conjecture–experiment–verification. These steps are located on a circle in Figure 1 to emphasize the fact that one pass is usually not enough but that learning continues to take place as we cycle through the steps.

Figure 1. The conjecture–experiment–verification cycle for learning: The scientific method

Keep in mind that no conjecture is apt to be perfect so the verification step really asks only whether it provides an adequate explanation of the data currently available.

In any given application, the steps may run together and be supplemented by trial-and-error, good guesswork, and luck. Rarely, if ever, do we get it right the first time, so expect to repeat the cycle of steps. Do you remember your experiences when learning to swim or ride a bike?

4. TWO BASIC CONCEPTS—POPULATION AND SAMPLE

The examples above, where the evaluation of factual information is essential for acquiring new knowledge, motivate the development of statistical reasoning and tools taught in this text. These examples have some common characteristics. First, relevant data must be collected. Second, access to a complete set of data is often physically impossible or practically infeasible. When data are obtained from laboratory experiments, no matter how much experimentation has been performed, more can always be done. To collect an exhaustive set of data related to the damage sustained by all cars of a particular model under collision at a specified speed, every car of that model coming off the production lines would have to be subjected to a collision! In most situations, we must work with the partial information.

The distinction between the data actually acquired and the vast collection of all potential observations is a key to understanding statistics.

The source of each measurement is called a **unit.** It is usually a person or an object. To emphasize the population as the entire collection of units, we term it the **population of units.**

A **unit** is a single entity, usually a person or an object, whose characteristics are of interest.

The **population of units** is the complete collection of units about which information is sought.

There is another aspect to any population and that is the value, for each unit, of a characteristic or variable of interest. There can be several characteristics of interest for a given population of units. Some examples are given in Table 1.

TABLE 1   Examples of Populations, Units, and Variables

Population	Unit	Variables/Characteristics
All students currently enrolled in school	Student	GPA Number of credits Hours of work per week Major Right/Left-handed
All campus fast-food restaurants	Restaurant	Number of employees Seating capacity Hiring/Not hiring
All books in library	Book	Replacement cost Frequency of check-out Repairs needed

For a given variable or characteristic of interest, we call the collection of values, evaluated for every unit in the population, the **statistical population** or just the **population.** We refer to the collection of units as the **population of units** when there is a need to differentiate it from the collection of values.

> A statistical **population** is the set of all measurements (or record of some qualitative trait) corresponding to each unit in the entire collection of units about which information is sought.

The population represents the target of an investigation. We learn about the population by taking a sample from the population.

> A **sample** from a statistical population is the subset of measurements that are actually collected in the course of an investigation.

Factual information is crucial to any investigation. The sample needs to be representative of the situation and contain sufficient information to address the problems under investigation. Experiments are carried out to investigate major factors that can explain unwanted variation in the final product. For example, when making apple pies, factors such as the quality of the apples, oven temperature, time of baking, and mix of other ingredients can influence the taste.

The following examples clarify the meaning of population and sample and the distinction between good and bad samples.

Example 1   Illustrating population and sample
A newspaper headline reads, CAFFEINE CONSUMPTION CAN LEAD TO DEPENDENCE, and the article[1] describes a medical study in which ninety-nine subjects volunteered as having a problem, physically or psychologically, with caffeine dependence. Among these subjects, 11 were confirmed to have a dependence and agreed to take part in a trial. They alternated two weeks of taking pills containing caffeine with two weeks of taking look-alike pills without caffeine. The trial was a **double blind** trial because neither the subjects nor those administering the pills knew which set of pills had the active ingredient. Nine of the patients showed objective evidence of withdrawal.

Identify the population and sample.

[1]E. C. Strain, et al. "Caffeine Dependence Syndrome," *Journal of the American Medical Association* **272** (1994), pp. 1043–1048.

SOLUTION AND DISCUSSION   As the article implies, the information should apply to you and me. The population here could well be the withdrawal responses of all who consume caffeine in the United States or even in the world, although the cultural customs may vary the type of caffeine consumption from coffee breaks, to tea time, to kola nut chewing. The sample would consist of the withdrawal responses of the 11 persons who agreed to participate in the double blind trial.

Although this sample is the best available, we caution that volunteers may not mimic the population of caffeine-dependent persons.   ■

Example 2   A bad sampling practice

A host of a radio music show announced that she wanted to know which singer is the favorite among city residents. Listeners were then asked to call in and name their favorite singer.

Identify the population and sample. Comment on how to get a sample that is more representative of the city's population.

SOLUTION AND DISCUSSION   The population is the collection of singer preferences of all city residents and the purported goal was to learn who was the favorite singer. Because it would be nearly impossible to question all the residents in a large city, one must necessarily settle for taking a sample.

Having residents make a local call is certainly a low-cost method of getting a sample. The sample would then consist of the singers named by each person who calls the radio station. Unfortunately, with this selection procedure, the sample is not very representative of the responses from all city residents. Those who listen to the particular radio station are already a special subgroup with similar listening tastes. Furthermore, those listeners who take the time and effort to call are usually those who feel strongest about their opinions. The resulting responses could well be much stronger in favor of a particular country western or rock singer than is the case for preference among the total population of city residents or even those who listen to the station.

If the purpose of asking the question is really to determine the favorite singer of the city's residents, we have to proceed otherwise. One procedure commonly employed is a phone survey where the phone numbers are chosen at random. For instance, one can imagine that the numbers 0, 1, 2, 3, 4, 5, 6, 7, 8, and 9 are written on separate pieces of paper and placed in a hat. Slips are then drawn one at a time and replaced between drawings. Later, we will see that computers can mimic this selection quickly and easily. Four draws will produce a random telephone number within a three-digit exchange. Telephone numbers chosen in this manner will certainly produce a much more representative sample than the self-selected sample of persons who call the station.   ■

Self-selected samples consisting of responses to call-in or write-in requests will, in general, not be representative of the population. They arise primarily from subjects who feel strongly about the issue in question. To their credit, many TV news and entertainment programs now state that their call-in polls are nonscientific and merely reflect the opinions of those persons who responded.

5. ELEMENTS OF DATA COLLECTION

The next example illustrates the "what's good for a guinea pig is good for me" fallacy. Newspapers often feature stories of preliminary medical breakthroughs of new treatments or medicines that work well on guinea pigs or mice. The government now insists on a long careful process before new medicines can be tested on humans and their positive effects documented. In fact, many never fulfill their early promise and are not heard from again.

Example 3   Samples from a different population—or you and me?
A newspaper[2] reported a study in which a naturally occurring chemical that was injected into rats reduced their appetites. A substantial loss of weight was observed. The article states that there is hope it could lead to an appetite-suppressing pill for overweight persons. Comment.

SOLUTION AND DISCUSSION   This approach to weight loss seems to work well for laboratory rats even though they probably were not even concerned about losing weight. A human is much more complex biologically than a rat. Not everything that is good for a rat is good for a person and vice versa. The two populations of units are very different. Inferences valid for one will not necessarily hold for the other.   ■

Example 4   A nonrepresentative sample
Figure 2 (page 10) depicts eight kinds of whales swimming together. Hold your pencil above the figure and then close your eyes. Move your arm around in circles and place the pencil point on the figure. Repeat until the pencil point lands on a whale.
    Suppose we want to learn about size (the weight or length) for the population of eight swimming whales. Is the whale selected by this procedure representative?

SOLUTION AND DISCUSSION   According to the selection procedure, the pencil will have the same chance of hitting all areas of the same size. The first thing to notice is that the Blue whale is the largest. It will be selected the most often, whereas the Beluga whale, the smallest, will be selected least. In fact, the Blue whale will be selected more than 15 times as often as the Beluga whale because it is that much larger. The whale selected in this manner is much more likely to be one of the larger ones.   ■

Selecting units according to size can lead to misleading conclusions if this fact is not accounted for in the statistical analysis. In a less obvious situation, if a

[2]*The Wall Street Journal,* Jan. 4, 1996.

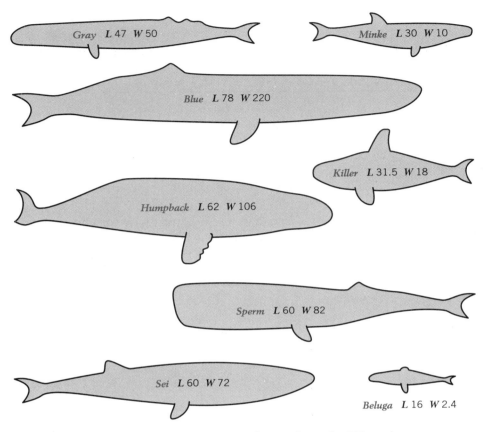

Figure 2. Eight whales swimming: Length (*L*, feet) and Weight (*W*, tons)

company checks the length of time persons have been on sick leave, among those absent on the first working day of the month, persons on long sick leaves will more likely be included than those who have been out for only one day during the month.

Data collected with a clear-cut purpose in mind are very different from **anecdotal data.** Most of us have heard people say they won money at a casino, but certainly most people cannot win most of the time as casinos are not in the business of giving away money. People tend to tell good things about themselves. In a similar vein, some drivers' lives are saved when they are thrown free of car wrecks because they were not wearing seat belts. Although such stories are told and retold, you must remember that there is really no opportunity to hear from those who would have lived if they had worn their seat belts. Anecdotal information is usually repeated because it has some striking feature that may not be representative of the mass of cases in the population. Consequently, it is not apt to provide reliable answers to questions.

USING A RANDOM NUMBER TABLE TO SELECT SAMPLES

The selection of a sample from a finite population must be done impartially and objectively. Putting slips of paper in a box and drawing them out is cumbersome and proper mixing may not be possible. However, the selection is easy to carry out using a chance mechanism called a **random number table**. Suppose ten balls numbered 0, 1, . . . , 9 are placed in an urn and shaken thoroughly. Then one is drawn and the digit recorded. It is then replaced, the balls shuffled, another one drawn, and the digit recorded. The digits in Table 1, Appendix B, were actually generated by a computer that closely simulates this procedure. A portion of this table is shown as Table 2.

TABLE 2   Random Digits: A Portion of Table 1, Appendix B

Row					25					
1	0695	7741	8254	4297	0000	5277	6563	9265	1023	5925
2	0437	5434	8503	3928	6979	9393	8936	9088	5744	4790
3	6242	2998	0205	5469	3365	7950	7256	3716	8385	0253
4	7090	4074	1257	7175	3310	0712	4748	4226	0604	3804
5	0683	6999	4828	7888	0087	9288	7855	2678	3315	6718
6	7013	4300	3768	2572	6473	2411	6285	0069	5422	6175
7	8808	2786	5369	9571	3412	2465	6419	3990	0294	0896
8	9876	3602	5812	0124	1997	6445	3176	2682	1259	1728
9	1873	1065	8976	1295	9434	3178	0602	0732	6616	7972
10	2581	3075	4622	2974	7069	5605	0420	2949	4387	7679
11	3785	6401	0540	5077	7132	4135	4646	3834	6753	1593
12	8626	4017	1544	4202	8986	1432	2810	2418	8052	2710
13	6253	0726	9483	6753	4732	2284	0421	3010	7885	8436
14	0113	4546	2212	9829	2351	1370	2707	3329	6574	7002
15	4646	6474	9983	8738	1603	8671	0489	9588	3309	5860

The chance mechanism that generated the random number table ensures that each of the single digits has the same chance of occurrence, that all pairs 00, 01, . . . , 99 have the same chance of occurrence, and so on. Further, each digit is unrelated to any other digit and this makes any collection of digits unrelated to any other collection of digits in the table. Because of these properties, the digits are called random.

Example 5   Using the table of random digits
Thirty different low-fat frozen dinners are available at the supermarket. Use Table 2 to select a sample of size $n = 4$ to study their fat content. Select the sample so that the same frozen dinner box does not appear twice in the sample.

SOLUTION AND DISCUSSION The first step is to number the boxes from 1 to 30, or to stack them in some order so they can be identified. The digits must be selected two at a time because the population size $N = 30$ is a two-digit number. We begin by arbitrarily selecting a row and column. We select row 9 and column 25. Reading the digits in columns 25 and 26, and proceeding downward, we obtain

$$06 \quad 04 \quad 46 \quad 28 \quad 04 \quad 27$$

We ignore the number 46 because it is greater than the population size 30. We also ignore any number when it appears a second time, as 4 does here. That is, we continue reading until four different numbers in the appropriate range are selected. Here the four boxes of frozen dinners numbered

$$6 \quad 4 \quad 28 \quad 27$$

will be tested for fat content. ■

For large sample size situations or frequent applications, it is more convenient to use computer software to choose the random numbers.

Example 6 Selecting a sample by random digit dialing
Suppose there is a single three-digit telephone exchange for the area where you wish to conduct a survey. Use the random digits in Table 2 to select five phone numbers.

SOLUTION AND DISCUSSION We arbitrarily decide to start at row 11 and column 17. Reading the digits in columns 17 through 20, and proceeding downward, we obtain

$$7132 \quad 8986 \quad 4732 \quad 2351 \quad 1603$$

These five numbers, together with the designated exchange, become the phone numbers to be called in the survey.

Every phone number, listed or unlisted, has the same chance of being selected. The same holds for every pair, every triplet, and so on. Commercial phones may have to be discarded and another number drawn from the table. If there are two exchanges in the area, separate selections could be done for each exchange. ■

ASSESSING THE QUALITY OF DATA

Data can be representative or nonrepresentative. The soundness of any conclusion or decision depends on the quality and the reliability of the data collected. Hospitals advertise their rates of success in open heart surgery. These success rates can be misleading if a hospital can choose its heart patients. In that case, a high survival rate may be due more to the initial condition of patients than the skillfulness of the operating team. In other words, the rate of success is based on a biased sample. Questions need to be asked so that the sampling process is understood and unwarranted conclusions about one hospital being better than another are avoided.

Example 7   Questioning the quality of data

There is fierce competition among cola manufacturers trying to get customers to switch to their brand and stick with it. The proportion of total sales due to a particular soft drink is called its market share. *The New York Times* recently reported the market shares, by type of beverage, in the carbonated soft-drink market.

Type of Beverage	Market Share (%)
Caffeinated cola	48.0
Caffeine-free cola	10.4
Lemon-lime	9.8
Dr Pepper	3.9
Other	27.9

Describe a main feature of these data. If this is to be a valid conclusion, the data must be sound. What questions should be asked concerning the data?

SOLUTION AND DISCUSSION   The data suggest that cola, either caffeinated or caffeine-free, dominates the soft-drink market. But, what are the source and quality of the data? Some relevant questions to ask are

On what basis are these percentage market shares calculated?

How reliable are these percentages?

That is, these data were likely based on a sample. Was the sample large enough so variation from sale to sale is overcome? Is the sample representative of the whole country? ∎

To summarize the main point of the example, all data should be questioned.

Insight to Statistical Reasoning

Always question data. Examine the sources of data to evaluate their representativeness and quality.

"Garbage in — Garbage out." If the source of information is not sound or reliable, how can any conclusions be valid?

Statistical Reasoning

Data Interpretation—Caution Is Necessary

A very large sample of men and women is collected to examine the relationship between the average weight of a person and the age group and the height of the person. The following table shows the average weights in six age groups, 18–24, 25–34, 35–44, 45–54, 55–64, and 65–74, for both men and women. This set of data was discussed in a Midwest newspaper under the headline, "Men Lose Weight Earlier."

FOR MEN

Age	18 to 24	25 to 34	35 to 44	45 to 54	55 to 64	65 to 74
5-foot-2	130	141	143	147	143	143
5-foot-3	135	145	148	152	147	147
5-foot-4	140	150	153	156	153	151
5-foot-5	145	156	158	160	158	156
5-foot-6	150	160	163	164	163	160
5-foot-7	154	165	169	169	168	164
5-foot-8	159	170	174	173	173	169
5-foot-9	164	174	179	177	178	173
5-foot-10	168	179	184	182	183	177
5-foot-11	173	184	190	187	189	182
6-foot-0	178	189	194	191	193	186
6-foot-1	183	194	200	196	197	190
6-foot-2	188	199	205	200	203	194

FOR WOMEN

Age	18 to 24	25 to 34	35 to 44	45 to 54	55 to 64	65 to 74
4-foot-9	114	118	125	129	132	130
4-foot-10	117	121	129	133	136	134
4-foot-11	120	125	133	136	140	137
5-foot-0	123	128	137	140	143	140
5-foot-1	126	132	141	143	147	144
5-foot-2	129	136	144	147	150	147
5-foot-3	132	139	148	150	153	151
5-foot-4	135	142	152	154	157	154
5-foot-5	138	146	156	158	160	158
5-foot-6	141	150	159	161	164	161
5-foot-7	144	153	163	165	167	165
5-foot-8	147	157	167	168	171	169

For example, let us look at the data for men in the group of height 5'8". It appears that the average weight of men first increases with age but then decreases in the later years. The peak occurs at a middle age group. It seems to be true also for men of other heights (especially the taller men). In contrast, the data for women of a given height show that the average weight does not seem to decrease until the women are in the highest age group, 65 to 74. It peaks for the age group 55 to 64.

Is the headline correct? What are some other possible explanations for such a phenomenon, other than the naive interpretation that men lose weight earlier?

One possible explanation is that the sample on which the table is based may not be a representative sample. However, there is a more intriguing and likely true explanation. These data show the weights only of people who are alive; the weights of those who have died are omitted. If those who die during a year are

Jim Cummins/FPG International

heavier on average than those who survive, the average weight of the survivors will decrease. For instance, take 45–54-year-olds. If all the factors remain the same for ten years, except some of the people died, the average weight of the 55–64-year-olds will appear to decrease. This will happen even if the weight of each individual survivor does not decrease.

Men in the 45–54 age group face a higher chance of heart attacks than younger men. Overweight men are at even higher risk, and hence may not survive heart attacks. Women have historically not faced a higher chance of heart attacks until they reach about 50 years of age. The data are consistent with the possibility that the apparent lower average weights simply reflect the greater mortality rate of heavier people.

To interpret the data properly and draw meaningful conclusions, one should not only pay attention to the numerical values or an apparent pattern in the data, but also understand the background of the data, how they were collected, and other factors that may give rise to these kinds of patterns in the data.

6. THE PURPOSEFUL COLLECTION OF DATA

The repair of library books, election polls, and other examples in Section 2 indicate how accurate data can answer important questions. Still, as described in Section 3, variation must be addressed. Many poor decisions are made both in business and everyday activities because of the failure to understand and account for variability. Certainly, the purchasing habits of one person may not represent the population. In one instance, a person spends $184 on groceries for the week while the next person spends $79. The amounts spent on groceries vary from person to person; that is, they are different. This state of disagreement is called **variation.** The good news is that, despite the presence of variation, we will learn how to combine the information in a large number of highly variable individual cases to obtain information about a population.

Just making the decision to collect data to answer a question, to provide the basis for taking action, or to improve a process is a key step. Once that decision has been made, an important next step is to develop a statement of purpose that is both specific and unambiguous. If the subject of the study is public transportation being behind schedule, you must carefully specify what is meant by late. Is it 1 minute, 5 minutes, or more than 10 minutes behind scheduled times that should result in calling a bus or commuter train late? Words like "soft" or "uncomfortable" in a statement are even harder to quantify. One common approach, for a quality such as comfort, is to ask passengers to rate the ride on public transportation on the five-point scale

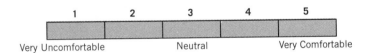

where the numbers 1 through 5 are attached to the scale, with 1 for very uncomfortable and so on through 5 for very comfortable.

A clearly specified statement of purpose will guide the choice of what data to collect and help ensure that they will be relevant to the purpose. Without a clearly specified purpose, or terms unambiguously defined, much effort can be wasted in collecting data that will not answer the question of interest.

The overall goal for a study can be quite general, but at each step a specific statement of purpose is needed. For instance, a primary health facility became aware that sometimes it was taking too long to return patients' phone calls. That is, patients would phone in with requests for information. These requests, in turn, had to be turned over to doctors or nurses who would collect the information and return the call. The overall objective was to understand the current procedure and then improve on it. As a good first step, it was decided to find how long it was taking to return calls under the current procedure. Variation in times from call to call is expected, so the purpose of the initial investigation is to benchmark the variability with the current procedure by collecting a sample of times.

PURPOSE: Obtain a reference or benchmark for the current procedure by collecting a sample of times to return a patient's call under
the current procedure.

For a sample of incoming calls collected during the week, the time received
was noted along with the request. When the return call was completed, the
elapsed time, in minutes, was recorded. Each of these times is represented as a
dot in Figure 3. Notice that over one-third of the calls took over 120 minutes, or
over two hours to return. This could be a long time to wait for information if it
concerns a child with a high fever or an adult with acute symptoms. If the purpose
was to determine what proportion of calls took too long to return, we would need
to agree on a more precise definition of "too long" in terms of number of minutes.
Instead, these data clearly indicate that the process needs improvement and the
next step is to proceed in that direction.

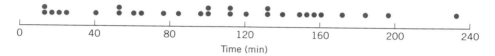

Figure 3. Time in minutes to return call

In any context, to pursue potential improvements of a process, one needs to
focus more closely on particulars. Three questions

When   Where   Who

should always be asked before gathering further data. More specifically, data
should be sought that will answer the following questions.

When do the difficulties arise? Is it during certain hours, certain days of the
week or month, or in coincidence with some other activities?

Where do the difficulties arise? Try to identify the locations of bottlenecks
and unnecessary delays.

Who was performing the activity and who was supervising? The idea is not
to pin blame, but to understand the roles of participants with the goal of
making improvements.

It is often helpful to construct a **cause-and-effect diagram** or **fishbone diagram**
to give yourself an illustrated guide to plan your analysis. The main centerline
represents the problem or the effect. A somewhat simplified fishbone chart is
shown in Figure 4 (page 18) for the *where* question regarding the location of delays
when returning patients' phone calls. The main centerline represents the problem:
Where are delays occurring? Calls come to the reception desk, but when these
lines are busy, the calls go directly to nurses on the third or fourth floor. The main
diagonal arms in Figure 4 represent the floors and the smaller horizontal lines

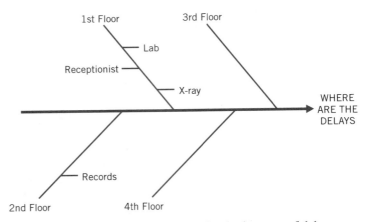

Figure 4. A cause-and-effect diagram for the location of delays

more specific locations on the floor where the delay could occur. For instance, the horizontal line representing a delay in retrieving a patient's medical record connects to the second floor diagonal line. The resulting figure resembles the skeleton of a fish. Consideration of the diagram can help guide the choice of what new data to collect.

Fortunately, the quality team conducting this study had already given preliminary consideration to the *When*, *Where*, and *Who* questions and recorded not only the time of day, but also day and person receiving the call. That is, their current data gave them a start on determining whether the time to return calls depends on when or where the call is received.

Although we go no further with this application here, the quality team next developed more detailed diagrams to study the flow of paper between the time the call is received and when it is returned. They then identified bottlenecks in the flow of information that were removed and the process was improved. In later chapters, you will learn how to compare and display data from two locations or old and new processes, but the key idea emphasized here is the purposeful collection of relevant data.

7. OBJECTIVES OF STATISTICS

The subject of statistics provides the methodology to make inferences about the population from the collection and analysis of sample data. These methods enable one to derive plausible generalizations and then assess the extent of uncertainty underlying these generalizations. Statistical concepts are also essential during the planning state of an investigation when decisions must be made as to the mode and extent of the sampling process.

The major objectives of statistics are:

1. To make **inferences** about a population from an analysis of information contained in sample data. This includes assessments of the extent of uncertainty involved in these inferences.
2. To **design the process and the extent of sampling** so that the observations form a basis for drawing valid inferences.

The design of the sampling process is an important step. A good design for the process of data collection permits efficient inferences to be made, often with a straightforward analysis. Unfortunately, even the most sophisticated methods of data analysis cannot, in themselves, salvage much information from data that are produced by a poorly planned experiment or survey.

The early use of statistics in the compilation and passive presentation of data has been largely superseded by the modern role of providing analytical tools with which data can be efficiently gathered, understood, and interpreted. Statistical concepts and methods make it possible to draw valid conclusions about the population, on the basis of a sample. Given its extended goal, the subject of statistics has penetrated all fields of human endeavor in which the evaluation of information must be grounded in data-based evidence.

The basic statistical concepts and methods, described in this book, form the core in all areas of application. We present examples drawn from a wide range of applications to help develop an appreciation of various statistical methods, their potential uses, and their vulnerabilities to misuse.

One goal of statistics is to understand and quantify the variation in the data and, if possible, to identify sources that contribute to this variation.

KEY IDEAS

A **population** is the complete set of measurements, or record of a qualitative trait, corresponding to the collection of all units about which information is desired.

A **sample** from the population is the subset of measurements that are actually collected.

Statistics is a body of principles for **designing the process of data collection** and making **inferences** about the population from the information in the sample.

A **statement of purpose** is a key step in designing the data collection process.

8. REVIEW EXERCISES

8.1   A newspaper headline reads,

U.S. TEENS TRUST, FEAR THEIR PEERS

and the article explains that a telephone poll was conducted of 1055 persons 13 to 17 years old. Identify a population and the sample.

8.2   A consumer magazine article asks,

HOW SAFE IS THE AIR IN AIRPLANES?

and then says that its study of air quality is based on measurements taken on 158 different flights of U.S.-based airlines. Identify a population and the sample.

8.3   A scientist[3] testifying before a congressional hearing on the effects of estrogenic pesticides told every man in the room that he "is half the man his grandfather was." The scientist was dramatizing a report on worldwide studies involving 14,847 men, over many different years and in many countries, that showed male semen concentrations have dropped approximately 50% over the past 50 years.

Identify a population and the sample.

8.4   A magazine that features the latest electronics and computer software for homes enclosed a short questionnaire on a postcard. Readers were asked to answer questions concerning their use and ownership of various software and hardware products, and to then send the card to the publisher. A summary of results appeared in a later issue of the magazine that used the data to make statements such as 40% of readers have purchased program X. Identify a population and sample and comment on the representativeness of the sample. Are readers who have not purchased any new products mentioned in the questionnaire as likely to respond as those who have purchased?

8.5   Each year a local weekly newspaper gives out "Best of the City" awards in categories such as restaurant, deli, pastry shop, and so on. Readers are asked to fill in their favorites on a form enclosed in this free weekly paper and then send it to the publisher. The establishment receiving the most votes is declared the winner in its category. Identify the population and sample and comment on the representativeness of the sample.

8.6   Which of the following are anecdotal and which are based on a sample?

(a)   Erik says that check-out is faster at a grocery store on Tuesday than any other night of the week.

---

[3]Louis Guillette. Hearing before the Subcommittee on Health and the Environment. Serial No. 103–87, Oct. 21, 1993.

(b)   Out of 58 cars clocked on a busy road, 22 were exceeding the speed limit.

(c)   During a tornado, a man wrapped himself in a mattress in his bedroom. He was sucked out of his home along with the mattress, but was unhurt.

8.7   Which of the following are anecdotal and which are based on a sample?

(a)   Joan says that adding a few drops of almond flavoring makes her cookies taste better.

(b)   Out of 102 customers at a quick-stop market, 31 made their purchases with a credit card.

(c)   Out of 80 persons who purchased catsup at the supermarket one night, 71 purchased the same brand as their previous purchase.

8.8   What is wrong with this statement of purpose?

Purpose: *Determine whether a newly designed recliner chair is comfortable.*
Give an improved statement of purpose.

8.9   What is wrong with this statement of purpose?

Purpose: *Determine whether a paper tissue product is soft.*
Give an improved statement of purpose.

8.10   Give a statement of purpose for determining how long it takes to wait in line and get cash from an automated teller machine during the lunch hour.

8.11   A consumer magazine article asked readers to respond to the question, "How do you like to spend your free time?," and from the responses of 100 persons it concluded that 74% consider shopping at least a hobby.
          Identify a population and the sample and comment on the representativeness of the sample.

8.12   According to the cause-and-effect diagram on page 18, where are the possible delays on the first floor?

8.13   Refer to the cause-and-effect diagram on page 18. The workers have now noticed that a delay could occur

(a)   On the fourth floor at the pharmacy

(b)   On the third floor at the practitioners' station

Redraw the diagram and include this added information

8.14   An article[4] describes an investigation that concluded that attractive lawyers, men and women, earn more than their less attractive peers. The finding was based on the earnings of 2500 lawyers and their photos. According to the article, initially it was thought that looks are less important in pro-

[4]*The Wall Street Journal*, Jan. 4, 1996, p. A1.

fessional positions. The findings, however, were consistent with earlier studies on the general work force.

(a) Write two questions that would be good to ask about this study.

(b) In your personal experience, are attractive persons favored?

(c) Would the conclusions of this study apply to all professions? Comment.

8.15 An article[5] describes a study on the impact of pesticides where a side effect was that the shaved areas on mice grew hair when treated with an estrogen blocker. The article headline says that the result provides a hint for curing baldness. Specify a population for the actual study and comment on the generalization suggested by the headline.

8.16 A campus political club has ninety active members listed on its membership roll. Use Table 1, Appendix B, of random digits to select 5 persons to be interviewed regarding the time they devote to club activities each week.

8.17 A city runs 60 buses daily. Use Table 1, Appendix B, of random digits to select 4 buses to inspect for cleanliness. (We started at row 71, columns 33 and 34, and read down.)

8.18 The following percentages from the National Highway Traffic Safety Administration show the relative speed (rounded to the nearest 5 mph) involved in U.S. automobile accidents in 1993:

20 mph or less	2.0%
25 or 30 mph	29.7%
35 or 40 mph	30.4%
45 or 50 mph	16.5%
55 mph	19.2%
60 or 65 mph	2.2%

(a) Can we conclude that it must be relatively safe to drive at a high speed? Why or why not?

(b) What type of sample information do we have here? Why do most accidents appear to occur around 30 mph or 55 mph?

ACTIVITIES TO IMPROVE UNDERSTANDING: STUDENT PROJECTS

Apply the ideas in this chapter to newspapers and magazines that you read. This is a good course-long activity to improve your statistical reasoning.

1. Find an article that contains an example of a nonrepresentative sample. Describe the sampling procedure and the weaknesses. How would you improve the sampling process? Attach a copy of the article to your report.

[5]*Wisconsin State Journal*, Oct. 29, 1996.

2. Find an example of an eye-catching headline or title for an article explaining the results of an investigation. Describe the sampling process. Give a few questions that should be asked before the conclusions should be accepted at face value. Attach a copy of the article to your report.

3. Think of an investigation to conduct as an individual or group project. At this point you can begin to plan.

   (a) Write a statement of purpose.

   (b) Specify the units, the population of units, and how to collect a sample.

   (c) Decide upon the characteristic or characteristics to include in your study.

   After reading the next chapter, you will want to review your plan and include some details on how you will summarize the data.

CHAPTER **2**

Organizing Data and Describing Patterns

Chapter Objectives

After reading this chapter, you should be able to

▶ Identify continuous and discrete data.
▶ Construct a frequency table.
▶ Create histograms and other graphical displays of data.
▶ Recognize the main features of the distribution of the data.
▶ Calculate the mean, variance, and standard deviation as summaries.
▶ Calculate the sample quartiles.

1. INTRODUCTION

In Chapter 1, we showed that numerical information is often required to acquire new knowledge or effectively improve a process. This chapter is concerned with methods for describing and summarizing data to highlight any important features or patterns they may contain. It is the goal of these methods to make the information in the data obvious.

Once data are collected, they should be organized and described by means of tables, graphs, and numerical summaries that we introduce below. The process of examining data by creating tables, plots, and summary numbers is called **exploratory data analysis**.

The choice of a particular display of data may be suitable for one type of data but not another.

2. VARIABLES AND DATA

We use the term **data** to mean numbers or features that represent characteristics of the sampling units. Grade point average is a characteristic, and if you are a sampling unit, your grade point average is a measurement or data. In discussing methods for summarizing the information in data, it is helpful to distinguish between two basic types of data:

1. **Quantitative** or **measurement data**
2. **Qualitative** or **categorical data**

Quantitative data arise when a characteristic is measured on a numerical scale. Examples include shoe size, intensity of an earthquake, the time in line at an automated teller, and the number of puppies in a litter.

Categorical data arise when a characteristic is a trait that is only classified in categories. Hair color {blonde, brown, red, black}, employment status {employed, unemployed}, gender {male, female}, and blood type {O, A, B, AB} are examples.

Numbers can be assigned to the categories and it is often convenient to do so.

> Nominal data result from assigning different numbers to categories as a code. The numbers themselves do not have a meaning.

The number 1 could be assigned to male and 2 to female. The resulting numerical data would be nominal. The numbers 1 and 2 could well be assigned in the reverse order.

If the categories have a natural ordering, an increasing or decreasing sequence of numbers could be assigned to the categories. The academic grade categories

{F, D, C, B, A} are ordered. There is extra information about categories above and below a given grade. Assigning the set of increasing numbers 0, 1, 2, 3, and 4 reflects the ordering and results in ordinal data.

> Ordinal data result from assigning an increasing (or decreasing) set of numbers to categories that preserves the intrinsic ordering of the categories.

The categories {none, slight, moderate, severe} for sunburn could be assigned 0, 1, 2, and 3, respectively, resulting in ordinal data.

The characteristics of a sampling unit are called variables. The term "variable" indicates that it varies from sampling unit to sampling unit. A quantitative variable is one that is naturally numerical, such as income or grade point average. On closer examination, there are two types of quantitative or measurement data.

> A discrete variable takes values that are distinct numbers with gaps between them.

Examples include shoe sizes such as 6, $6\frac{1}{2}$, 7, $7\frac{1}{2}$, . . . , which proceed in steps of $\frac{1}{2}$, and the number of puppies in a litter, which is an integer.

> A continuous variable takes any value in an interval.

To a reasonable approximation, time to wait in a check-out line, grade point average, and weight are continuous variables.

The data are of the same type as the variables. Discrete variables produce discrete data and continuous variables produce continuous data. Figure 1 (page 28) reviews the types of data and their relations.

A summary description of categorical data is discussed in Section 3. The remainder of this chapter is devoted to a descriptive study of measurement data, both discrete and continuous. As in the case of summarization and commentary on a long, wordy document, it is difficult to prescribe concrete steps for summary descriptions that work well for all types of measurement data. However, a few important aspects that deserve special attention are outlined here to provide general guidelines for this process.

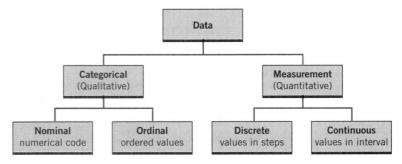

Figure 1. Types of data

Describing a Data Set of Measurements

1. **Summarization and description of the overall pattern**

    (a) Presentation of tables and graphs

    (b) Noting important features of the graphed data, including symmetry or departures from it

    (c) Scanning the graphed data to detect any observations that seem to stick far out from the major mass of the data—the outliers

2. **Computation of numerical measures**

    (a) A typical or representative value that indicates the center of the data

    (b) The amount of spread or variation present in the data

---

3. DESCRIBING DATA BY TABLES AND GRAPHS

CATEGORICAL DATA

When a qualitative trait is observed for a sample of units, each observation is recorded as a member of one of several categories. Our understanding of the variability in the trait data is enhanced by creating a summary in the form of a frequency table that shows the counts (**frequencies**) of individual categories. We also calculate the proportion (also called **relative frequency**) of observations in each category.

$$\text{Relative frequency of a category} = \frac{\text{Frequency in the category}}{\text{Total number of observations}}$$

Example 1   Relative frequencies from a student opinion poll
A campus press polled a sample of 280 undergraduate students to study student attitude toward a proposed change in the dormitory regulations. Each student was to respond as either support, oppose, or neutral in regard to the issue. The numbers were 152 support, 77 neutral, and 51 opposed. Tabulate the results and calculate the relative frequencies for the three response categories.

SOLUTION AND DISCUSSION   Table 1 records the frequencies in the second column, and the calculated relative frequencies in the third column.

TABLE 1   Summary Results of an Opinion Poll

Responses	Frequency	Relative Frequency
Support	152	$\dfrac{152}{280} = .543$
Neutral	77	$\dfrac{77}{280} = .275$
Oppose	51	$\dfrac{51}{280} = .182$
Total	280	1.000

The relative frequencies show that about 54% of the polled students supported the change, 18% opposed, and 28% were neutral.   ■

Remark: The relative frequencies provide the most relevant information about the pattern of the data. One should also state the sample size, which serves as an indicator of the credibility of the relative frequencies. (More on this in Chapter 11.)

Categorical data can be displayed in simple graphs. Computer software programs can easily construct the familiar bar charts and pie charts that regularly appear in newspapers and magazines. Amounts are indicated by heights of bars in a bar chart and by the areas of the slices in a pie chart.

Example 2   Constructing a bar chart and pie chart
There is fierce competition among cola manufacturers trying to get customers to switch to their brand and stick with it. The proportion of total sales of a particular soft drink is called its market share. *The New York Times* recently reported the market shares, by type of beverage, in the carbonated soft-drink market.

Type of Beverage	Market Share (%)
Caffeinated cola	48.0
Caffeine-free cola	10.4
Lemon-lime	9.8
Dr Pepper	3.9
Other	27.9

Graphically illustrate the market shares with a bar chart and a pie chart.

SOLUTION AND DISCUSSION To construct a bar chart, first draw a horizontal axis and locate the name of each type of soft drink according to the order listed in the table. Vertical bars of equal widths are drawn at each name. The height of a bar represents the percentage market share.

To construct a pie chart, you must calculate the angle of each slice. The size of the angle is equal to the proportion of market share times 360 degrees. For instance, $.48 \times 360 = 172.8°$ for caffeinated cola. With this choice, the area of the slice corresponds to the proportion.

The market share bar chart is shown in Figure 2(a) and the pie chart in Figure 2(b). Both charts reveal the extent to which cola, either caffeinated or caffeine-free, dominates the soft-drink market.

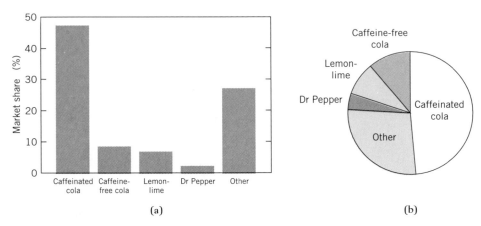

Figure 2. (a) Bar chart and (b) Pie chart for market share

Comparing the bar heights in Figure 2(a) or slice sizes in Figure 2(b), it is clear that caffeinated colas have the largest market share, followed by caffeine-free colas, which have only about one-fourth as much market share. ∎

Psychologists have suggested that most people are better able to compare bar heights than to compare pie slices. Pie slices are even harder to compare visually when the pie is tipped or given a thick edge, as in many newspaper graphic displays. Those displays should be avoided.

When questions arise that need answering but the decision makers lack precise knowledge of the state of nature or the full ramifications of their decisions, the best procedure is often to collect more data. In the context of quality improvement, if a problem is recognized, the first step is to collect data on the magnitude and possible causes. This information is most effectively communicated through graphical presentations.

A **Pareto diagram** is a special type of bar chart that is a powerful graphical technique for displaying events or categories according to their frequencies. According to Pareto's empirical law, any collection of events contains a small group of events that are the ones that occur most of the time. The Pareto diagram displays the categories ordered according to the heights of the bars, except for the catch-all category "other."

Figure 3 gives the Pareto diagram for the types of defects found in a day's production of facial tissue. The frequency for the first cause, tears, is 22 and the cumulative frequency for the first two causes combined is 22 + 15 = 37. This illustrates Pareto's law, with two of the causes being responsible for 37 out of 50 or 74% of the defects.

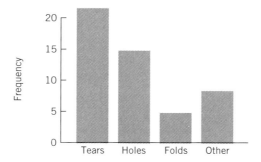

Figure 3. Pareto diagram of facial tissue defects

Statistical Reasoning

Graphic Display of Data Highlights Opportunities for Continuous Improvement

In the mid-1970s, a major camera manufacturer[1] decided to improve its products by finding out which advances would most benefit consumers. Instead of asking

[1]See S. Bisgaard, *Journal of Engineering Design* **3** (1992), pp. 31–47, for a description by N. Kano of Konica's approach to the development of its Picarri C35EF and Juspin C35AF cameras.

Ken Chernus/FPG International

the camera users directly what features they wanted in a camera, data were collected at several photo development labs on thousands of pictures that ordinary people had taken. The manufacturer classified the defective pictures into categories of causes, and drew a Pareto diagram similar to Figure 4.

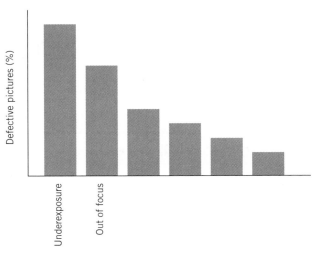

Figure 4. Pareto diagram for poor photos: First data

The diagram reveals that the primary cause of a poor photo is underexposure, which occurs when there is insufficient light. The second most frequent cause of defective pictures was an out-of-focus image.

In response to the finding that underexposure was the most frequent cause, the camera manufacturer pioneered the development of a built-in flash. No longer would people need to have a separate flash attachment and carry it everywhere. This advance was quickly copied by the other manufacturers.

Work then shifted to the out-of-focus problem. Three years later, the company was successful in developing an automatic focusing camera. This advance, too, was quickly copied by the competition.

Even after these two major advances, the camera manufacturer continued to seek improvements on the new cameras. When a further series of photos was analyzed, a Pareto diagram similar to Figure 5 revealed that the leading cause of defects was now blank film. That situation occurs when the film is not properly loaded. Can you imagine your feelings if the pictures you took at an important occasion turned out to be blank because the film was not loaded properly? These problems were overcome by new cameras that automatically loaded film. However, it was another manufacturer that was first to develop and produce them.

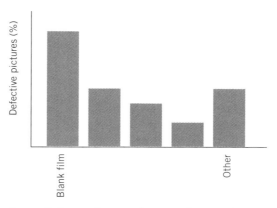

Figure 5. Pareto diagram for poor photos: The new cameras

We see that improvement is a continuous process and that the effective use of Pareto analysis, with relevant data, can be crucial to the product development. Who knows what features we will see in future cameras.

DISCRETE DATA

We next consider summary descriptions of measurement data and begin our discussion with discrete measurement scales. A data set is identified as discrete when the underlying scale is discrete and the distinct values observed are not too numerous.

Similar to our description of categorical data, the information in a discrete data set can be summarized in a frequency table that includes a calculation of the relative frequencies. In place of the qualitative categories, we now list the distinct numerical measurements that appear in the data set and then count their frequencies. Unlike the case with categorical data, where the categories can be displayed in any order, discrete data have a natural ordering following their distinct numerical values.

Example 3   Table giving the frequency distribution

The daily numbers of computer stoppages are observed over 30 days at a university computing center, and the data of Table 2 are obtained. Determine the frequency distribution.

TABLE 2   Daily Numbers of Computer Stoppages

1	3	1	1	0	1	0	1	1	0
2	2	0	0	0	1	2	1	2	0
0	1	6	4	3	3	1	2	4	0

SOLUTION AND DISCUSSION   The frequency distribution of this data set is presented in Table 3, where the last column shows the calculated relative frequencies. One stoppage per day is the most frequent. The frequency distribution has a long right tail.

TABLE 3   Frequency Distribution for Daily Number (x) of Computer Stoppages

Value x	Frequency	Relative Frequency
0	9	.300
1	10	.333
2	5	.167
3	3	.100
4	2	.067
5	0	.000
6	1	.033
Total	30	1.000

The frequency distribution of a discrete variable can be presented pictorially by drawing either lines or rectangles to represent the relative frequencies. First, the distinct values of the variable are located on the horizontal axis. For a **line diagram,** we draw a vertical line at each value and make the height of the line equal to the relative frequency. A **histogram** employs vertical rectangles instead of lines. These rectangles are centered at the values and their areas represent relative frequencies. Typically, the values proceed in equal steps so the rectangles are all of the same width and their heights are proportional to the relative frequencies as well as frequencies. Figure 6(a) shows the line diagram and (b) the histogram of the frequency distribution of Table 3.

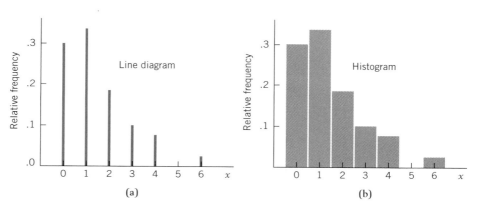

Figure 6. Graphic display of the frequency distribution of data in Table 3

DATA ON A CONTINUOUS VARIABLE

We now consider tabular and graphical presentations of data sets that contain numerical measurements on a virtually continuous scale. Of course, the recorded measurements are always rounded. In contrast with the discrete case, a data set of measurements on a continuous variable may contain many distinct values. Then, a table or plot of all distinct values and their frequencies will not provide a condensed or informative summary of the data.

The two main graphical methods for displaying a data set of measurements are the **dot diagram** and the **histogram.** Dot diagrams are employed when there are relatively few observations (say, less than 20 or 25); histograms are used with a larger number of observations.

Dot Diagram

When the data consist of a small set of numbers, they can be graphically represented by drawing a line with a scale covering the range of values of the measurements. Individual measurements are plotted above this line as prominent dots. The resulting diagram is called a **dot diagram** or **dot plot.**

Example 4   Dot diagram of survival times

The number of days the first six heart transplant patients at Stanford survived after their operations were 15, 3, 46, 623, 126, 64. Make a dot diagram.

SOLUTION AND DISCUSSION   These survival times extended from 3 to 623 days. Drawing a line segment from 0 to 700, we can plot the data as shown in Figure 7. This dot diagram shows a cluster of small survival times and a single rather large value.

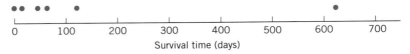

Figure 7. Dot diagram for the heart transplant data

Frequency Distribution on Intervals

When the data consist of a large number of measurements, a dot diagram may be quite tedious to construct. More seriously, overcrowding of the dots will mar the clarity of the diagram. In such cases, it is convenient to condense the data by grouping the observations according to intervals and recording the frequencies of the intervals. The main steps in this process are outlined as follows.

**Constructing a Frequency Distribution
for a Continuous Variable**

1. Find the minimum and the maximum values in the data set.

2. Choose intervals or cells of equal length that cover the range between the minimum and the maximum without overlapping. These are called class intervals, and their endpoints class boundaries.

3. Count the number of observations in the data that belong to each class interval. The count in each class is the class frequency or cell frequency.

4. Calculate the relative frequency of each class by dividing the class frequency by the total number of observations in the data:

$$\text{Relative frequency} = \frac{\text{Class frequency}}{\text{Total number of observations}}$$

The choice of the number and position of the class intervals is primarily a matter of judgment guided by the following considerations. The number of classes usually ranges from 5 to 15, depending on the number of observations in the data. Grouping the observations sacrifices information concerning how the observations are distributed within each cell. With too few cells, the loss of information is serious. On the other hand, if one chooses too many cells and the data set is relatively small, the frequencies from one cell to the next would jump up and down in a chaotic manner and no overall pattern would emerge. As an initial step, frequencies may be determined with a large number of intervals that can later be combined as desired to obtain a smooth pattern of the distribution.

Computers conveniently order data from smallest to largest so that the observations in any cell can easily be counted. The construction of a frequency distribution is illustrated in Example 5.

Example 5   Frequency distribution for bookstore sales

Most students purchase their books during the registration period. University bookstore receipts from 40 students provided the sales data of Table 4, where the values have been ordered from smallest to largest.

TABLE 4   The Data of Forty Cash Register Receipts (in Dollars) at a University Bookstore

3.20	11.70	13.64	15.60	15.89	28.44	29.07	37.34
41.81	43.35	43.94	49.51	49.82	51.20	51.43	52.47
53.72	53.92	54.03	56.89	63.80	66.40	68.64	70.15
70.98	74.52	76.68	77.84	80.91	84.04	85.70	86.48
88.92	89.28	91.36	91.62	98.79	102.39	104.21	124.27

Construct a frequency distribution of the sales data.

SOLUTION AND DISCUSSION   To construct a frequency distribution, we first notice that the minimum sale is $3.20 and the maximum sale $124.27. We choose class intervals of length 25 as a matter of convenience.

The selection of class boundaries is a bit of fussy work. Because the data have two decimal places, we could add a third decimal figure to avoid the possibility of any observation falling exactly on a boundary. For example, we could end the first class interval at 24.995. Alternatively, and more neatly, we could write 0–25 and make the **endpoint convention** that the left-hand limit is included but not the right. See Table 5 on page 38.

The first interval contains 5 observations, so its frequency is 5 and its relative frequency $\frac{5}{40} = .125$. The relative frequencies add to 1, as they should in any frequency distribution.   ■

In this book, we make the endpoint convention that the left-hand endpoint is included in the interval but the right-hand endpoint is not.

HISTOGRAM

A frequency distribution can be graphically presented as a histogram. To draw a histogram, we first mark the class intervals on the horizontal axis. When the class intervals all have the same length, we draw on each interval a vertical rectangle whose height is equal to the relative frequency.

TABLE 5   Frequency Distribution for Bookstore
Sales Data (Left endpoints included,
but right endpoints excluded)

Class Interval	Frequency	Relative Frequency
$ 0–25	5	$\frac{5}{40} = .125$
25–50	8	$\frac{8}{40} = .200$
50–75	13	$\frac{13}{40} = .325$
75–100	11	$\frac{11}{40} = .275$
100–125	3	$\frac{3}{40} = .075$
Total	40	1.000

A histogram of the frequency distribution in Table 5 is shown in Figure 8. The
height of the rectangle drawn on the class interval 0–25 is equal to the relative
frequency 5/40 = .125 and so on.

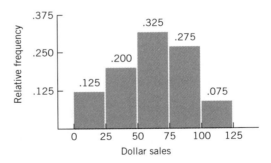

Figure 8. Histogram of the bookstore sales data
of Tables 4 and 5. Sample size = 40.

Visually, we note that the tallest block, or most frequent class interval, is
$50–$75. Also, the proportion .275 + .075 = .35 of the sales are $75 or more.

Histograms Based on Unequal Class Intervals

Frequency distributions and histograms with unequal class intervals are useful
for situations where most of the measurements are concentrated in one part
of the range while relatively few measurements are spread over the remainder.
For example, in plotting a histogram for birth weights, most of the weights

are expected to be in the range of 7 to 10 pounds. Given a very large number of birth weights, we could use class intervals of width .2 pound from 7 to 10 pounds and of width .5 pound outside of this interval. Tabulations of income, age, and other characteristics in official reports are often made with unequal class intervals.

When a frequency distribution has unequal class intervals as in Table 6, we must take area to equal relative frequency. If, instead, we stay with height of the rectangle equal to relative frequency, the resulting histogram gives a distorted view of the data. Consider the simple frequency distribution in Table 6.

TABLE 6 Frequency Table with Unequal Class Intervals

Class Interval	Frequency	Relative Frequency	Height
[0, 5)	30	.30	.30/5 = .060
[5, 10)	35	.35	.35/5 = .070
[10, 20)	35	.35	.35/10 = .035
Total	100	1.00	

The last two class intervals each contain 35 observations and so they have the same relative frequency. However, in the first case the 35 observations are spread over an interval of length $10 - 5 = 5$, whereas the 35 observations in the last interval are less dense because they are spread over an interval of length 10. If this difference in width is ignored and the relative frequency is used as the height, we obtain the incorrect and misleading histogram in Figure 9(a). One rectangle is twice as large as the other. For this reason, we plot histograms using area rather than height to represent relative frequency. The heights should be .06, .07, and .035, respectively. The correct histogram is shown in Figure 9(b).

To illustrate the calculation of height, consider the rectangle drawn on the

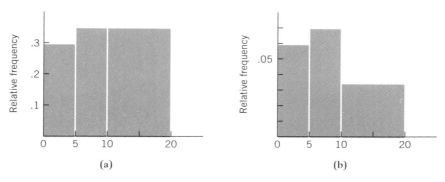

Figure 9. (a) Incorrect histogram and (b) Correct histogram for the data of Table 6

class interval 0–5. Its area is .06 × 5 = .30, which is the relative frequency of this class. Actually, we determine the height .06 as

$$\text{Height} = \frac{\text{Relative frequency}}{\text{Width of interval}} = \frac{.30}{5} = .06$$

The units on the vertical axis can be viewed as relative frequencies per unit of the horizontal scale. For instance, .06 is the relative frequency per unit for the interval 0–5.

For histograms with unequal class intervals, make the **area of a rectangle equal to the relative frequency.** This provides a fair and accurate representation of the distribution of data.

When the area of a rectangle equals the relative frequency, the histogram is called a **density histogram.** The total area of the density histogram equals the sum of the relative frequencies, which is 1.

The total area of a density histogram is 1.

Remark: When all class intervals have equal widths, the heights of the rectangles are proportional to the relative frequencies that the areas represent. The formal calculation of height, as area divided by the width, is then redundant. Instead, one can make the heights of the rectangles equal to the relative frequencies. The resulting histogram has the same shape as the density histogram.

Figure 10 shows one ingenious way of displaying two histograms for comparison. In spite of their complicated shapes, their back-to-back plot as a "tree" allows for easy visual comparison of the male and female age distributions. The female distribution has a larger right-hand tail; a larger proportion live beyond 80.

Stem-and-Leaf Display

A stem-and-leaf display provides a more efficient variant of the histogram for displaying data, especially when the observations are two-digit numbers. This plot is obtained by sorting the observations into rows according to their leading digit. To make a stem-and-leaf display of the exam scores in Table 7,

1. List the digits 0 through 9 in a column and draw a vertical line. These correspond to the leading digit.

2. For each observation, record its second digit to the right of this vertical line in the row where the first digit appears.

3. Finally, arrange the second digits in each row so they are in increasing order.

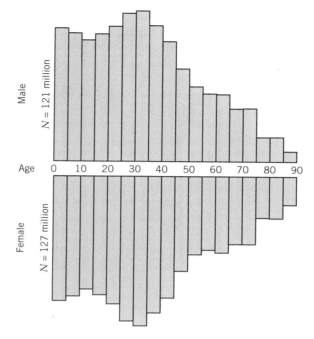

Figure 10. Population tree (histograms) of the male and
female age distributions in the United States in 1990
(*Source:*   U.S. Bureau of the Census.)

TABLE 7   Examination Scores of 50 Students

75	98	42	75	84	87	65	59	63
86	78	37	99	66	90	79	80	89
68	57	95	55	79	88	76	60	77
49	92	83	71	78	53	81	77	58
93	85	70	62	80	74	69	90	62
84	64	73	48	72				

The stem-and-leaf display for the data of Table 7 is shown in Table 8.

TABLE 8   Stem-and-Leaf Display for the
Examination Scores

3	7
4	289
5	35789
6	022345689
7	01234556778899
8	00134456789
9	0023589

In the stem-and-leaf display, the column of first digits to the left of the vertical line is viewed as the stem, and the second digits as the leaves. Viewed sidewise, it looks like a histogram with a cell width equal to 10. However, it is more informative than a histogram because the actual data points are retained. In fact, every observation can be recovered exactly from the stem-and-leaf display.

A stem-and-leaf display retains all the information in the leading digits of the data. If one stem is 3.5|0 2 3 7 8, the corresponding data are 3.50, 3.52, 3.53, 3.57, and 3.58. Leaves may also be two-digit at times. The stem 0.4|07 13 82 90 presents the data 0.407, 0.413, 0.482, and 0.490.

Further variants of the stem-and-leaf display are described in Exercises 3.24 and 3.25. This versatile display is one of the most applicable techniques of exploratory data analysis.

When the sample size is small or moderate, no information is lost with the stem-and-leaf diagram because you can see every data point. The major disadvantage is that, when the sample size is large, diagrams with hundreds of numbers in a row cannot be constructed in a legible manner.

Exercises

3.1 Recorded here are the blood types of 40 persons who have volunteered to donate blood at a plasma center.

O	O	A	B	A	O	A	A	A	O
B	O	B	O	O	A	O	O	A	A
A	A	AB	A	B	A	A	O	O	A
O	O	A	A	A	O	A	O	O	AB

(a) Summarize the data in a frequency table. Include calculations of the relative frequencies.

(b) Code O as 1, A as 2, B as 3, AB as 4. Are these data nominal or ordinal?

3.2 In a study of the job hazards in the roofing industry in California, records of the disabling injuries were classified according to the accident types. Of the total number of 1132 injuries, 329 were due to falls, 256 from burns, 219 from overexertion, 202 from being struck, 40 from foreign substance in eye, and 86 from other miscellaneous reasons. (*Source:* U.S. Dept. of HEW Publication NIOSH 75-176).

Present these data in a frequency table and also give the relative frequencies.

3.3 The following table shows how workers in one department get to work.

Mode of Transportation	Frequency
Drive alone	25
Car pool	3
Ride bus	7
Other	5

(a) Calculate the relative frequency of each mode of transportation.

(b) Construct a pie chart.

3.4 Of the $207 million raised by a major university's fund drive, $117 million came from individuals and bequests, $24 million from industry and business, and $66 million from foundations and associations. Present this information in the form of a pie chart.

3.5 Measurements of the lengths of fish (in inches), given to one decimal, are to be grouped into the classes of less than 5, 5–7, 8–10, 10–12, over 12, where the left endpoint is included but not the right.

Explain where difficulties might arise.

3.6 The number of burglaries reported during each week in a county are to be grouped into the classes of 0–1, 2–3, 4–5, 5–10, 11 or more.

Explain where difficulties might arise.

3.7 The times for persons to complete their transactions at a money machine (to the nearest tenth of a second) are to be grouped into the following classes: 0–30, 30–60, 60–90, 90–120, more than 120, where the left endpoint is included but not the right.

Explain where difficulties might arise.

3.8 The weights of the players on the university football team (to the nearest pound) are to be grouped into the following classes: 160–175, 175–190, 190–205, 205–220, 220–235, 235 or more, where the left endpoint is included but not the right.

Explain where difficulties might arise.

3.9 On flights from San Francisco to Chicago, the numbers of empty seats are to be grouped into the following classes: 0–4, 5–9, 10–14, 15–19, more than 19.

Is it possible to determine from this frequency distribution the exact number of flights on which there were

(a) Fewer than 10 empty seats?

(b) More than 14 empty seats?

(c) At least 5 empty seats?

(d) Exactly 9 empty seats?

(e) Between 5 and 15 empty seats inclusively?

3.10   A major West Coast power company surveyed 50 customers who were asked to respond to the statement, "People should rely mainly on themselves to solve problems caused by power outages" with one of the following responses:

1.   Definitely agree
2.   Somewhat agree
3.   Somewhat disagree
4.   Definitely disagree

The responses are as follows:

```
4  2  1  3  3  2  4  2  1  1  2  2  2  2  1  3  4
1  4  4  1  3  2  4  1  4  3  3  1  1  1  2  1  1
4  4  4  4  4  1  2  2  2  4  4  4  1  3  4  2
```

(a)   Construct a frequency table.
(b)   Are these data nominal or ordinal?

3.11   The numbers of peas in 60 pea pods, randomly taken from a day's pick, are

```
4  3  4  3  3  5  5  6  4  4  4  3
3  4  3  3  6  4  5  3  6  3  2  1
4  4  4  4  4  5  3  4  3  1  2  2
5  2  4  3  5  5  3  3  3  3  5  5
3  3  7  4  3  5  6  4  4  3  3  4
```

(a)   Construct a frequency distribution including the calculations of relative frequencies.
(b)   Display the data as a line diagram.

3.12   On seven occasions, the amounts of suspended solids (parts per million) detected in the effluent of a municipal wastewater treatment plant were 14, 12, 21, 28, 30, 65, 26. Display the data in a dot diagram.

3.13   Before the microwave ovens are sold, the manufacturer must check to ensure that the radiation coming through the door is below a specified safe limit. The amounts of radiation leakage (mw/cm^2), with the door closed, from 25 ovens are as follows (courtesy of John Cryer).

```
15   9  18  10   5
12   8   5   8  10
 7   2   1   5   3
 5  15  10  15   9
 8  18   1   2  11
```

Display the data in a dot diagram.

3.14   In August, the Jump River Electric Co. suffered 11 power outages that lasted the following number of hours.

   2.5,   2.0,   1.5,   3.0,   1.0,   1.5,   2.0,   1.5,   1.0,   10.0,   1.0

   Display the data in a dot diagram.

3.15   Loss of calcium is a serious problem for older women. At the beginning of a study on the influence of exercise on the rate of loss, the amounts of bone mineral content in the radius bone of the dominant hand of 7 elderly women were

   .90,   1.06,   1.06,   .79,   .85,   .91,   .96

   Display the data in a dot diagram.

3.16   From a department store's records, the total monthly finance charges in dollars were obtained from 240 customer accounts that included finance charges (left endpoints included).

Class Limits	No. of Customers
0–5	65
5–10	88
10–15	42
15–20	27
20–25	18

   Construct a histogram.

3.17   Among the major active volcanoes in the world, 53 are in North America. Their heights, in thousands of feet, are summarized in the following frequency table (left endpoints are included).

Class Limits	Height (1000 feet)
[0, 4.5)	19
[4.5, 9.0)	22
[9.0, 13.5)	9
[13.5, 18.0)	3
Total	53

   Construct a histogram. Comment on the shape of the histogram.

3.18   Weather Bureau data can help us understand what to expect in terms of total snowfall during the winter in Madison, Wisconsin. The ordered values, in inches, for 47 recent years are

25.62	29.01	38.71	31.97	31.69	29.67	35.07	22.49
31.51	32.86	21.10	40.34	38.79	31.04	21.43	26.19
23.62	37.60	26.44	34.02	30.71	29.67	30.72	27.24
31.00	35.54	36.08	34.57	21.11	32.55	36.48	28.15
34.39	32.11	31.61	31.68	33.77	38.97	31.86	33.42
24.59	23.43	36.55	39.09	32.13	43.34	33.54	

Construct a histogram using the class intervals [20, 25), [25, 30), etc., where the left endpoints are included. Comment on the shape of the histogram.

3.19   One of the major indicators of air pollution in large cities and industrial belts is the concentration of ozone in the atmosphere. From massive data collected by Los Angeles County authorities, 78 measurements of ozone concentration (in parts per hundred million) in the downtown Los Angeles area during two summers are recorded here (courtesy of G. Tiao). Each measurement is an average of hourly readings taken every fourth day.

3.5	1.4	6.6	6.0	4.2	4.4	5.3	5.6
6.8	2.5	5.4	4.4	5.4	4.7	3.5	4.0
2.4	3.0	5.6	4.7	6.5	3.0	4.1	3.4
6.8	1.7	5.3	4.7	7.4	6.0	6.7	11.7
5.5	1.1	5.1	5.6	5.5	1.4	3.9	6.6
6.2	7.5	6.2	6.0	5.8	2.8	6.1	4.1
5.7	5.8	3.1	5.8	1.6	2.5	8.1	6.6
9.4	3.4	5.8	7.6	1.4	3.7	2.0	3.7
6.8	3.1	4.7	3.8	5.9	3.3	6.2	7.6
6.6	4.4	5.7	4.5	3.7	9.4		

(a)   Construct a frequency distribution using the class intervals 0–2, 2–4, and so on, with the endpoint convention that the left endpoint is included but not the right endpoint. Calculate the relative frequencies.

(b)   Make a histogram.

3.20   Graduate students in a counseling course were asked to choose one of their personal habits that needed improvement. To reduce the effect of this habit, they were asked to first gather data on the frequency of occurrence and the circumstances. One student collected the following frequency data on fingernail biting over a two-week period.

Frequency	Activity
58	Watching television
21	Reading newspaper
14	Talking on phone
7	Driving a car
3	Grocery shopping
12	Other

Make a Pareto diagram showing the relationship between nail biting and type of activity.

3.21 The frequency distribution of the number of lives lost in major tornadoes in the United States between 1900 and 1993 appears below. (*Source:* U.S. Environmental Data Service.)

No. of Deaths	Frequency
≤ 24	10
25–49	24
50–74	22
75–99	13
100–149	7
150–199	2
200–249	4
250 and over	2
Total	84

(a) Plot a density histogram. Use 0–25 for the first interval, 25–50 for the second, . . . , and 250–400 for the last.
(b) Comment on the shape of the distribution.

3.22 The following data represent the scores of 40 students on a college qualification test (courtesy of R. W. Johnson).

162	171	138	145	144	126	145	162	174	178
167	98	161	152	182	136	165	137	133	143
184	166	115	115	95	190	119	144	176	135
194	147	160	158	178	162	131	106	157	154

Make a stem-and-leaf display.

3.23   The following is a stem-and-leaf display with two-digit leaves.

$$
\begin{array}{c|llll}
1 & & & & \\
2 & 37 & 68 & 91 & \\
3 & 19 & 44 & 71 & 72 & 80 \\
4 & 05 & 26 & 43 & 91 \\
5 & 04 & 70 \\
6 & 21 \\
\end{array}
$$

List the corresponding measurements.

3.24   If there are too many leaves on some stems in a stem-and-leaf display, we might double the number of stems. The leaves 0–4 could hang on one stem and 5–9 on the repeated stem. For example, for the observations

193   198   200   202   203   203   205   205   206   207
207   208   212   213   214   217   219   220   222   226   237

we would get the **double-stem display**

$$
\begin{array}{c|l}
19 & 3 \\
19 & 8 \\
20 & 0233 \\
20 & 556778 \\
21 & 234 \\
21 & 79 \\
22 & 02 \\
22 & 6 \\
23 & \\
23 & 7 \\
\end{array}
$$

Add the observations 231 and 187 to the display.

3.25   If the double-stem display still has too few stems, we may wish to construct a stem-and-leaf display with separate stems to hold leaves 0 and 1, 2 and 3, 4 and 5, 6 and 7, and a stem to hold 8 and 9. The resulting stem-and-leaf display is called a **five-stem display.** The following is a five-digit stem-and-leaf display.

$$
\begin{array}{c|l}
1 & 89 \\
2 & 011 \\
2 & 2233 \\
2 & 44455 \\
2 & 667 \\
2 & 89 \\
3 & 0 \\
\end{array}
$$

List the corresponding measurements.

3.26 The following table lists values of the Consumer Price Index for 27 large cities both for 1992 and 1995.

	1992	1995		1992	1995
Anchorage	128	139	Los Angeles	147	149
Atlanta	139	149	Miami	135	147
Baltimore	140	150	Milwaukee	137	154
Boston	149	157	Minneapolis	135	145
Buffalo	138	146	New York	150	158
Chicago	141	148	Philadelphia	147	158
Cincinnati	134	143	Pittsburgh	136	143
Cleveland	137	140	Portland	140	150
Dallas	134	145	St. Louis	135	145
Denver	130	144	San Diego	147	147
Detroit	136	144	San Francisco	143	149
Honolulu	155	168	Seattle	139	149
Houston	129	139	Washington, D.C.	145	153
Kansas City	134	142			

Construct a five-stem display for 1992.

3.27 The Census Bureau compiled the percentage of voting-age population that actually voted in 1994, for each of the 50 states and the District of Columbia (left endpoints are included).

Class Limits Percent Voting	Number of States
[33, 38)	6
[38, 43)	10
[43, 48)	15
[48, 53)	9
[53, 64)	11
Total	51

Construct a density histogram. Comment on the shape of the histogram.

3.28 In 1940 there were 6.4 million farms in the United States, but by 1993 the number was reduced to 2.1 million farms. Does the following graphic accurately compare these data? If not, present a correct graphic.

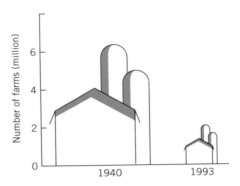

3.29   The graphic below shows how one aircraft manufacturer, Company C, advertised its dominance in the market. We give only the graphic for Company C and the leading competitor, Company L.

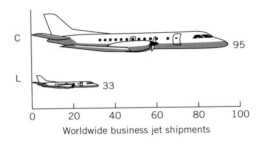

Does this graphic, for the number of business jet planes shipped during the previous year, accurately represent these data? If not, present a correct graphic.

4. NUMERICAL MEASURES OF CENTER

The graphical methods described in Section 3 summarize a data set pictorially so that we can visually assess the overall pattern. In this section, we will summarize a data set in terms of numerical measures. Such numerical measures are convenient because they condense the data, and are therefore especially useful for communicating features of the data either in writing or verbally.

We describe two types of numerical measures. The first type measures the center of the data—one single value to represent the whole data set. Two such measures are the sample mean and sample median. The second type of measure describes the amount of variation in the data. The measures of variation include the sample variance, sample standard deviation, sample range, and sample interquartile range.

To be able to describe the measures in general terms, we use the symbols

$$x_1, x_2, \ldots, x_i, \ldots, x_n$$

to represent a sample of n measurements. The subscript i on x_i denotes the position of the measurement; x_i is the ith observation in the list of n observations. That is, x_1 represents the value of the first measurement, x_2 represents the value of the second measurement, and so on.

Example 6   General notation for a data set
The birth weights, in pounds, of five babies born in a hospital one day are

$$9.2, \quad 6.4, \quad 10.5, \quad 8.1, \quad 7.8$$

Express this specific data set in the general notation and identify n.

SOLUTION AND DISCUSSION   In terms of symbols, we take

$$x_1 = 9.2, \quad x_2 = 6.4, \quad x_3 = 10.5, \quad x_4 = 8.1, \quad x_5 = 7.8$$

The last subscript, $n = 5$, is the size of the sample.

Always, the subscripts start with 1 for the first observation listed and run to n, the sample size.   ∎

Referring to Example 6, the sequence x_1, x_2, x_3, x_4, x_5 has no specific ordering with respect to the magnitude of the actual measurements. Certainly, the smallest baby need not be born first, the second smallest second, and so forth. Consequently, for any given sample, the values $x_1, x_2, \ldots, x_n$ are generally not in ascending or descending order.

To further simplify the writing of the sum of the measurements in a sample $x_1, x_2, \ldots, x_n$, we introduce the Greek letter Σ (sigma) as a shorthand for "the sum of." We write

$$\sum_{i=1}^{n} x_i = x_1 + x_2 + \cdots + x_n$$

which is read as "the sum of the measurements from position $i = 1$ to position $i = n$."

The following example illustrates the use of the summation notation.

Example 7   The summation notation
Example 6 referred to the birth weights of five babies:

$$x_1 = 9.2, \quad x_2 = 6.4, \quad x_3 = 10.5, \quad x_4 = 8.1, \quad x_5 = 7.8$$

Write the following sums using summation notation and give the numerical value.

(a) The total weight of all five babies
(b) The sum of the weights of the second, third, and fourth babies

SOLUTION AND DISCUSSION

(a)   The total weight of the five babies is

$$\sum_{i=1}^{5} x_i = x_1 + x_2 + x_3 + x_4 + x_5$$
$$= 9.2 + 6.4 + 10.5 + 8.1 + 7.8 = 42.0$$

(b)   The limits on the summation sign change for the sum of weights of the second, third, and fourth babies:

$$\sum_{i=2}^{4} x_i = x_2 + x_3 + x_4 = 6.4 + 10.5 + 8.1 = 25.0$$ ∎

Remark: When the number of terms being summed is understood from the context, we often simplify to $\sum x_i$, instead of $\sum_{i=1}^{n} x_i$. Some further operations with the Σ notation are discussed in Appendix A.

THE MEANING OF CENTER

Given a sample of n measurements $x_1, x_2, \ldots, x_n$, we would like to report a "center" value that is representative of the whole sample. How should "center" be defined?

Center as Middle Value in Position—Sample Median

One way to define the center of a sample is that it lies in the middle of the measurements in the sense of relative position. That is, center could be defined as a value such that at least 50% of the data are as small or smaller and at least 50% are as large or larger.

Five students were asked about the number of movies they watched last month. Their responses

$$4 \quad 9 \quad 1 \quad 5 \quad 11$$

have center 5 since the 3 observations 4, 1, and 5 are as small or smaller, and 3 observations are as larger or larger. This measure of center is called the *sample median*. It does not depend on the particular values of the largest and smallest observations in the sample.

Center as Balance of Deviations—Sample Mean

A second way of defining the center of a data set leads to the most common measure of center. It depends on the distances, or deviations, from each data point to this center.

Let c be a point somewhere in the middle of the data points displayed in a dot diagram. Figure 11 is a dot diagram for the number of movies watched data.

Figure 11. Possible center c defined in terms of the deviations

Then c naturally splits the data set into two groups of data points: (1) those points at or to the left of the point c and (2) those points at or to the right. We now call c the center if, for the left-hand group, the sum of distances of the points to c is equal to the sum of distances for the right-hand group.

The center defined in this way can be shown to be the average of all the data points in the data set. We call this notion of center the *sample mean* and denote it by $\bar{x}$. For the movies data, $\bar{x} = (1 + 4 + 5 + 9 + 11)/5 = 6$. For these data, we check that the sum of distances from the left-hand points, $5 + 2 + 1 = 8$, equals the sum of distances $3 + 5 = 8$ for the right-hand points. Unlike the median, the sample mean depends on the actual value of each observation in the data set.

SAMPLE MEAN—CALCULATION AND INTERPRETATION

The most commonly used measure of center for a sample is called the **sample mean**. It is denoted by $\bar{x}$ and read as x-bar. Simply put, the sample mean is the average of all of the measurements in the sample.

> The sample mean of a set of n measurements, $x_1, x_2, \ldots, x_n$ is the sum of these measurements divided by n. The sample mean is denoted by $\bar{x}$.
>
> $$\bar{x} = \frac{\sum_{i=1}^{n} x_i}{n} = \frac{\text{Sum of all the observations}}{\text{Sample size}} \quad \text{or} \quad \frac{\sum x_i}{n}$$

According to the concept of "average," the mean represents a center of a data set. If we picture the dot diagram of a data set as a thin weightless horizontal bar on which balls of equal size and weight are placed at the positions of the data points, then the mean $\bar{x}$ represents the point on which the bar will balance. The computation of the sample mean and its physical interpretation are illustrated in Example 8.

Example 8   Sample mean—Calculation and interpretation

The birth weights in pounds of five babies born in a hospital on a certain day are 9.2, 6.4, 10.5, 8.1, and 7.8. Obtain the sample mean and create a dot diagram.

The mean birth weight for these data is

$$\bar{x} = \frac{9.2 + 6.4 + 10.5 + 8.1 + 7.8}{5} = \frac{42.0}{5} = 8.4 \text{ pounds}$$

The dot diagram of the data appears in Figure 12, where the sample mean (marked by $\triangle$) is the balancing point or center of the picture.

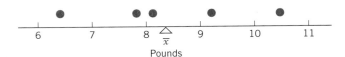

Figure 12. Dot diagram and the sample mean for the birth weight data

Median

> The **sample median** of a set of n measurements $x_1, \ldots, x_n$ is the middle value when the measurements are arranged from smallest to largest.

Roughly speaking, the median is the value that divides the data into two equal halves. In other words, 50% of the data lie below the median and 50% above it. If n is an odd number, there is a unique middle value and it is the median. If n is an even number, there are two middle values and the median is defined as their average. For instance, the data 3, 5, 7, 8 have two middle values, 5 and 7, so the median is $(5 + 7)/2 = 6$.

Example 9   Calculation of sample median
Find the median of the birth weight data given in Example 8.

The measurements, ordered from smallest to largest, are

$$6.4, \quad 7.8, \quad \boxed{8.1}, \quad 9.2, \quad 10.5$$

The middle value is 8.1, and the median is therefore 8.1 pounds.

Example 10   Choosing between the mean and median
Calculate the median of the survival times given in Example 4. Also calculate the mean and compare.

To find the median, first we order the data. The ordered values are

$$3, \quad 15, \quad 46, \quad 64, \quad 126, \quad 623$$

There are two middle values, so

$$\text{Median} = \frac{46 + 64}{2} = 55 \text{ days}$$

The sample mean is

$$\bar{x} = \frac{3 + 15 + 46 + 64 + 126 + 623}{6} = \frac{877}{6} = 146.2 \text{ days}$$

Note that one large survival time greatly inflates the mean. Only 1 out of the 6 patients survived longer than $\bar{x} = 146.2$ days. Here the median of 55 days appears to be a better indicator of the center than the mean. ∎

Example 10 demonstrates that the median is not affected by a few very small or very large observations, whereas the presence of such extremes will have a considerable effect on the mean. For histograms with one long tail, or dot diagrams with a few stragglers all on one side, the median is likely to be a more sensible measure of center than the mean. That is why government reports on income distribution quote the median income, rather than the mean, as a summary. A relatively small number of very highly paid persons can have a great effect on the mean salary.

Dividing the Data in Fourths—Quartiles

If the number of observations is quite large (greater than, say, 25 or 30), it is sometimes useful to extend the notion of the median and divide the *ordered* data set into quarters. The median cuts the data so that half of the values are to one side and half to the other side. A description should also locate the center half of the data. To this end, we introduce the **quartiles** which, as the name suggests, divide the ordered data set into fourths. Starting with the smallest observation, the **first quartile** is one-quarter of the way down the list. The **second quartile,** or median, is in the middle of the list, and the **third quartile** is three-quarters of the way to the end or largest observation. The first quartile is larger than just 25% of the observations, the second quartile, the median, is larger than just 50% of the observations, and the third quartile is larger than just 75% of the observations.

We calculate the quartiles according to the following procedure.

Determining the Quartiles Q_1, Q_2, Q_3

1. Sort the observations from smallest to largest and locate the median. The sample median is also the second quartile Q_2.
2. The first quartile, Q_1, is the median of the reduced data consisting of those ordered observations that lie below the position of the overall median in the sorted list determined in step 1.
3. The third quartile, Q_3, is the median of the reduced data consisting of those ordered observations whose position is above that of the overall median in the sorted list.

When the sample size is even, this rule simply says to use the first half of the sorted list to determine Q_1, and the second half to determine Q_3. When the sample size is odd, the single middle value, the total sample median, is removed when doing steps 2 and 3. There are numerous slight variations of this definition of a quartile but they all result in nearly the same value for the quartiles when the sample size is large.

Example 11
Calculating quartiles for traffic noise
The data from 50 measurements of the traffic noise level at an intersection are already ordered from smallest to largest in Table 9. Locate the quartiles.

TABLE 9 Measurements of Traffic Noise Level in Decibels

52.0	55.9	56.7	59.4	60.2	61.0	62.1	63.8	65.7	67.9
54.4	55.9	56.8	59.4	60.3	61.4	62.6	64.0	66.2	68.2
54.5	56.2	57.2	59.5	60.5	61.7	62.7	64.6	66.8	68.9
55.7	56.4	57.6	59.8	60.6	61.8	63.1	64.8	67.0	69.4
55.8	56.4	58.9	60.0	60.8	62.0	63.6	64.9	67.1	77.1

Courtesy of J. Bollinger.

SOLUTION AND DISCUSSION
We first determine the median. Since 50 is an even number, the median is the average of the two middle values, those in the 25th and 26th positions in the sorted list.

$$\text{Median} = \frac{60.8 + 61.0}{2} = 60.9$$

Next, since 50 is even, we find the median of the first half or 25 smallest observations. The middle value is at the 13th position and equals 57.2.

$$\text{First quartile} = Q_1 = 57.2$$

The middle value among the larger half of the observations is 64.6 so

$$\text{Third quartile} = Q_3 = 64.6$$

Only one-fourth of the time does the noise level exceed $Q_3 = 64.6$ decibels. ■

Other Percentiles (*Optional*) Rather than break a sample into just quarters, other divisions can also be made. The **sample 100pth percentile** divides the sample so that proportion p lies at this value or below and proportion $1 - p$ lies at or above.

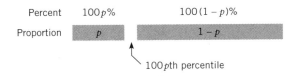

Taking $p = .5$, the sample $100(.5) = 50$th percentile has half of the observations that are the same or smaller and half that are the same or larger. The sample 10th percentile effects a 10%–90% division of the data set.

The following rule will simplify the calculation of the sample $100p$th percentile.

Calculating the Sample 100*p*th Percentile

1. Make a sorted list of the observations arranged from smallest to largest.
2. Calculate the product

$$\text{Number of observations} \times p$$

3. Round this product up to the next integer and then locate the corresponding observation in the ordered list. If this product is an integer, then calculate the average of the observation at this position and the observation in the next position.

Example 12   Calculating the 10th percentile of traffic noise

Refer to the traffic noise data in Example 11. Calculate the sample 10th percentile.

SOLUTION AND DISCUSSION   For the 10th percentile, we determine $50 \times .10 = 5$. The product is an integer so we take the average of the 5th and 6th observations, $(55.8 + 55.9)/2 = 55.85$, as the 10th percentile. Only 10% of the 50 measurements of noise level were quieter than 55.85 decibels.   ∎

Exercises

4.1   Calculate the mean and median for each of the following data sets.
   (a)   6, 10, 7, 14, 8
   (b)   2, 1, 4, 2, 1

4.2   Calculate the mean and median for each of the following data sets.
   (a)   4, 7, 3, 6, 5
   (b)   24, 28, 36, 30, 24, 29
   (c)   −2, 1, −1, 0, 3, −2, 1

4.3   Eight participants in a bike race had the following finishing times in minutes.

$$28, \quad 22, \quad 26, \quad 33, \quad 21, \quad 23, \quad 37, \quad 24$$

Find the mean and median for the finishing times.

4.4   With reference to the power outages in Exercise 3.14,

(a)   Find the sample mean.

(b)   Does the sample mean or the median give a better indication of the time of a "typical" power outage?

4.5   The monthly incomes in dollars for seven staff members of an insurance office are

950,   775,   925,   2500,   1150,   850,   975

(a)   Calculate the mean and median salary.

(b)   Which of the two is preferable as a measure of center, and why?

4.6   Records show that in Las Vegas, Nevada, the normal daily maximum temperature (°F) for each month starting in January is

56,   62,   68,   77,   87,   99,   105,   102,   95,   82,   66,   57

Verify that the mean of these figures is 79.67. Comment on the claim that the daily maximum temperature in Las Vegas averages a pleasant 79.67.

4.7   One year, the scores for the eight initial games of the National Basketball Association playoffs were

111–106,   104–92,   116–105,   119–113,
109–102,   111–97,   113–96,   101–89

Find the sample mean of the point spread, which is the number of points scored by the winner minus the number of points scored by the loser.

4.8   With reference to the radiation leakage data given in Exercise 3.13,

(a)   Calculate the sample mean.

(b)   Which gives a better indication of the amount of radiation leakage, the sample mean or the median?

4.9   The following table gives the number of days in which selected metropolitan areas failed to meet ozone standards in 1991 and 1988.

	1991	1988		1991	1988
Atlanta	4.0	13.2	Los Angeles	91.0	148.0
Atlantic City	2.1	6.0	Miami	0	3.0
Baltimore	2.1	19.3	Milwaukee	10.2	14.2
Boston	7.6	11.4	New York	13.3	19.4
Buffalo	0	8.3	Philadelphia	10.3	18.2
Chicago	11.8	16.2	Pittsburgh	0	14.8
Cincinnati	3.0	14.1	Portland	1.4	2.4
Cleveland	2.0	13.3	Sacramento	15.7	15.5
Dallas	4.2	4.3	St.Louis	1.0	7.3
Detroit	3.0	4.1	San Diego	7.1	7.1
Houston	16.6	8.3	San Francisco	1.0	4.1
			Washington, D.C.	6.1	13.8

(a)   Find the sample mean for 1991.

(b)   Comment on the effect of large observations.

4.10   With reference to Exercise 3.12, find the sample mean.

4.11   Old Faithful, the most famous geyser in Yellowstone Park, had the following durations (measured in seconds), in six consecutive eruptions:

$$240, \quad 248, \quad 113, \quad 268, \quad 117, \quad 253$$

(a)   Find the sample median.

(b)   Find the sample mean.

4.12   With reference to Exercise 4.7, find the sample median for the winning team's score.

4.13   Loss of calcium is a serious problem for older women. To investigate the amount of loss, a researcher measured the initial amount of bone mineral content in the radius bone of the dominant hand of elderly women and then the amount remaining after one year. The differences, representing the loss of bone mineral content, are given in the following table (courtesy of E. Smith).

8	7	13	3	6
4	8	6	3	4
0	1	11	7	1
8	6	12	13	10
9	11	3	2	9
7	1	16	3	2
10	15	2	5	8
17	8	2	5	5

(a)   Find the sample mean.

(b)   Does the sample mean or the median give a better indication of the amount of mineral loss?

4.14   Physical education researchers, interested in the development of the over-arm throw, measured the horizontal velocity of a thrown ball at the time of release. The results for first-grade children (in feet/second) (courtesy of L. Halverson and M. Roberton) are

Males

54.2, 39.6, 52.3, 48.4, 35.9, 30.4, 25.2, 45.4, 48.9, 48.9
45.8, 44.0, 52.5, 48.3, 59.9, 51.7, 38.6, 39.1, 49.9, 38.3

Females

30.3, 43.0, 25.7, 26.7, 27.3, 31.9, 53.7, 32.9, 19.4
23.7, 23.3, 23.3, 37.8, 39.5, 33.5, 30.4, 28.5

(a)   Find the sample median for males.

(b)   Find the sample median for females.

(c)   Find the sample median for the combined set of males and females.

4.15   On opening day one season, 10 major league baseball games were played and they lasted the following numbers of minutes.

167,   211,   187,   176,   170,   158,   198,   218,   145,   232

Find the sample median.

4.16   If you were to use the data on the length of major league baseball games in Exercise 4.15 to estimate the total amount of videotape needed to film another 10 major league baseball games, which is the more meaningful description, the sample mean or the sample median? Explain.

4.17   The following measurements of the diameters (in feet) of Indian mounds in southern Wisconsin were gathered by examining reports in the *Wisconsin Archeologist* (courtesy of J. Williams).

22,   24,   24,   30,   22,   20,   28,   30,   24,   34,   36,   15,   37

(a)   Plot a dot diagram.

(b)   Calculate the mean and median and then mark these on the dot diagram.

(c)   Calculate the quartiles.

4.18   With reference to Exercise 4.9, calculate the quartiles for 1991.

4.19   Refer to the data of college qualification test scores given in Exercise 3.22.

(a)   Find the median.

(b)   Find Q_1 and Q_3.

4.20   A large mail-order firm employs numerous persons to take phone orders. Computers on which orders are entered also automatically collect data on phone activity. One variable useful for planning staffing levels is the number of calls per shift handled by each employee. From the data collected on 25 workers, calls per shift were (courtesy of Lands' End)

118	118	57	92	127	109	96	68	73
69	106	91	93	94	102	105	100	104
80	50	96	82	72	108	73		

Calculate the sample mean.

4.21   With reference to Exercise 4.20, calculate the quartiles.

4.22   To investigate the economic impact of doing business with the state, a sample was taken of 15 small firms in the service sector that are vendors to the state. The data, on the percent of total sales due to sales to the state, are

$$27.0, \quad 12.0, \quad 15.0, \quad 1.0, \quad 0.1, \quad 1.0, \quad 5.0, \quad 0.1$$
$$8.0, \quad 5.0, \quad 1.0, \quad 1.0, \quad 3.0, \quad 3.0, \quad 7.0$$

(a) Find the sample median, first quartile, and third quartile.

(b) (*Optional*) Find the sample 90th percentile.

4.23 With reference to Exercise 4.13,

(a) Find the sample median, first quartile, and third quartile.

(b) (*Optional*) Find the sample 90th percentile.

4.24 *Some properties of the mean and median.*

1. If a fixed number c is added to all measurements in a data set, then the mean of the new measurements is

$$c + (\text{the original mean}).$$

2. If all measurements in a data set are multiplied by a fixed number d, then the mean of the new measurements is

$$d \times (\text{the original mean}).$$

(a) Verify these properties for the data set

$$5, \quad 9, \quad 9, \quad 8, \quad 10, \quad 7$$

taking $c = 4$ in property (1) and $d = 2$ in (2).

(b) The same properties also hold for the median. Verify these for the data set and the numbers c and d given in part (a).

4.25 On a day, the noon temperature measurements (in °F) reported by five weather stations in a state were

$$76, \quad 82, \quad 78, \quad 78, \quad 75$$

(a) Find the mean and median temperature in °F.

(b) The Celsius (°C) scale is related to the Fahrenheit (°F) scale by $C = \frac{5}{9}(F - 32)$. What are the mean and median temperatures in °C? (Answer without converting each temperature measurement to °C. Use the properties stated in Exercise 4.24.)

4.26 Given here are the mean and median salaries of machinists employed by two competing companies, A and B.

	Company	
	A	B
Mean salary	$35,000	$33,500
Median salary	$32,000	$34,000

Assume that the salaries are set in accordance with job competence and the overall quality of workers is about the same in the two companies.

(a) Which company offers a better prospect to a machinist having superior ability? Explain your answer.

(b) Where can a medium-quality machinist expect to earn more? Explain your answer.

4.27   Refer to the alligator data in Table D.7 of the Data Bank. Using the data on testosterone x_4 for male alligators,

(a) Make separate dot plots for the Lake Apopka and Lake Woodruff alligators.

(b) Calculate the sample means for each group.

(c) Do the concentrations of testosterone appear to differ between the two groups? What does this suggest the contamination has done to male alligators in the Lake Apopka habitat?

4.28   Refer to the alligator data in Table D.7 of the Data Bank. Using the data on testosterone x_4 from Lake Apopka,

(a) Make separate dot plots for the male and female alligators.

(b) Calculate the sample means for each group.

(c) Do the concentrations of testosterone appear to differ between the two groups? We would expect differences. What does your graph suggest the contamination has done to alligators in the Lake Apopka habitat?

5. NUMERICAL MEASURES OF VARIATION

The sample mean and median are two numerical measures for representing the center of a data set. But numerical measures are also required for a second important characteristic of a data set, the amount of variation about its center. For example, suppose a random sample of Brand A batteries is tested in a portable cassette player and their lifetimes measured. The sample mean, a measure of the center of the data, is found to be 3 hours. If the measured lifetimes are more or less the same, then the lifetime of a typical Brand A battery is about 3 hours. On the other hand, if the lifetimes vary over a wide range, then "the average life of a Brand A battery is 3 hours" does not contain enough information about the lifetime of a typical Brand A battery. We also need a numerical description of the variation.

Two data sets may exhibit similar positions of center, but may be remarkably different with respect to variability. For example, the dots in Figure 13(b) are more scattered than those in Figure 13(a). Any numerical description of variation should be larger for the second data set.

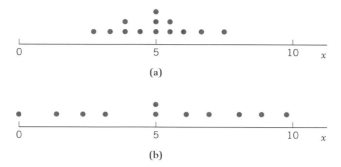

Figure 13. Dot diagrams with similar center values but different amounts of variation

One measure of variation, called the **sample range,** is simple:

Sample range = Largest observation − Smallest observation

Example 13   Sample range of bicycles returned
The weekly number of stolen bicycles recovered by campus police was recorded for five weeks.

$$3, \quad 7, \quad 5, \quad 8, \quad 7$$

Calculate the sample range of the number of bicycles returned.

SOLUTION AND DISCUSSION   We order the data set 3, 5, 7, 7, 8 and find the sample range is $8 - 3 = 5$. Notice that the sample range depends on only the two most extreme values, the largest and the smallest. None of the other observations are involved.   ■

The most commonly used measure of variation is called the *sample variance*. In contrast to the sample range, the calculation of the sample variance involves all the measurements in the sample.

Taking the sample mean, $\bar{x}$, as the measure of the center of the data, we define the deviation of a data point from the sample mean as

Deviation = Observation − Sample mean = $x - \bar{x}$

If all of the deviations are small, then all of the observations are closely clustered around $\bar{x}$ and the amount of variation in the data is small. On the other hand, if many of the deviations are large, any reasonable measure of variation should give a large value to reflect this situation. The sample variance is such a measure of variation and its value depends on all of the deviations.

For instance, the bicycle data set 3, 5, 7, 7, 8 has mean $\bar{x} = (3 + 5 + 7 + 7 + 8)/5 = 30/5 = 6$, so the deviations are calculated by subtracting 6 from each observation. See Table 10.

TABLE 10   Calculation
of Deviations

Observation x	Deviation $x - \bar{x}$
3	-3
5	-1
7	1
7	1
8	2

One might feel that the average of the deviations would provide a numerical measure of spread. However, some deviations are positive and some negative, and the total of the positive deviations exactly cancels the total of the negative ones. In fact, this was our motivation for choosing the sample mean as a measure of center in the discussion on page 52.

In the foregoing example, we see that the positive deviations add to 4 and the negative ones add to -4, so the total deviation is 0. For any data set, the total of the deviations is 0 (for a formal proof of this fact, see Appendix A).

$$\sum_{i=1}^{n} (\text{Deviations}) = \sum_{i=1}^{n} (x_i - \bar{x}) = 0$$

To obtain a measure of spread, we must eliminate the signs of the deviations before averaging. One way of removing the interference of signs is to square the numbers. A measure of spread called the **sample variance** is constructed by adding the squared deviations and dividing the total by the number of observations minus one.

Sample variance of n observations:

$$s^2 = \frac{\text{Sum of squared deviations}}{n - 1}$$

$$= \frac{\sum_{i=1}^{n} (x_i - \bar{x})^2}{n - 1}$$

Example 14 Calculation of sample variance

Calculate the sample variance of the bicycle data: 3, 5, 7, 7, 8.

SOLUTION AND DISCUSSION For this data set, $n = 5$. To find the sample variance, we first calculate the mean, then the deviations and the squared deviations. See Table 11.

TABLE 11 Calculation of Variance

Observation x	Deviation $x - \bar{x}$	(Deviation)^2 $(x - \bar{x})^2$
3	-3	9
5	-1	1
7	1	1
7	1	1
8	2	4

Total	30	0	16
	Σx	$\Sigma(x - \bar{x})$	$\Sigma(x - \bar{x})^2$

$$\bar{x} = \frac{30}{5} = 6$$

$$\text{Sample variance } s^2 = \frac{16}{5 - 1} = 4$$

The variance $s^2 = 4$ quantifies the variation in a squared deviation from the sample mean. ■

Remark: Although the sample variance is conceptualized as the **average squared deviation,** notice that the divisor is $n - 1$ rather than n. The divisor, $n - 1$, is called the degrees of freedom[2] associated with s^2.

Because the variance involves a sum of squares, its unit is the square of the unit in which the measurements are expressed. For Example 14 the variance is expressed in bicycles squared. If the data pertain to measurements of weight in pounds, the variance is expressed in (pounds)^2. To obtain a measure of variability in the same unit as the data, we take the positive square root of the variance, called the standard deviation. The standard deviation rather than the variance serves as a basic measure of variability.

[2]The deviations add to 0 so a specification of any $n - 1$ deviations allows us to recover the one that is left out. For instance, the first four deviations in Example 14 add to -2, so to make the total 0, the last one must be $+2$, as it really is. In the definition of s^2, the divisor $n - 1$ represents the number of deviations that can be viewed as free quantities.

Sample Standard Deviation

$$s = \sqrt{\text{variance}} = \sqrt{\frac{\sum_{i=1}^{n} (x_i - \bar{x})^2}{n - 1}}$$

Example 15   Calculation of standard deviation
Calculate the standard deviation for the bicycle data of Example 14.

SOLUTION AND DISCUSSION   We already calculated the variance $s^2 = 4$, so the standard deviation is $s = \sqrt{4} = 2$ bicycles. ∎

To show that a larger spread of the data does indeed result in a larger numerical value of the standard deviation, we consider another data set in Example 16.

Example 16   Comparing variation in two data sets
Calculate the standard deviation for the data: 1, 4, 5, 9, 11. Plot the dot diagram of this data set and also the data set of Example 14.

SOLUTION AND DISCUSSION   The standard deviation is calculated in Table 12. The dot diagrams, given in Figure 14, show that the data points of Example 14 have less spread than those of this example. This visual comparison is confirmed by a smaller value of s for the first data set.

TABLE 12   Calculation of s

x	$(x - \bar{x})$	$(x - \bar{x})^2$
1	-5	25
4	-2	4
5	-1	1
9	3	9
11	5	25
Total 30	0	64

$\bar{x} = 6$

$$s^2 = \frac{64}{4} = 16$$
$$s = \sqrt{16} = 4$$

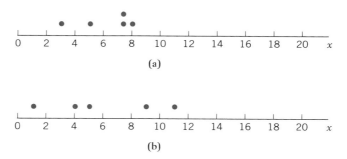

Figure 14. Dot diagrams of two data sets   ■

Computer Calculations

It is preferable to calculate the sample mean and standard deviation using a standard statistical package. Data can be entered and checked, and then the calculation proceeds blunder free. A portion of the output from the MINITAB package (see Exercise 7.32 on page 93)

	N	Mean	Median	StDev
C1	5	6.00	5.00	4.00

confirms our calculations in Table 12.

An alternative formula for the sample variance is

$$s^2 = \frac{1}{n-1}\left[\sum_{i=1}^{n} x_i^2 - \frac{\left(\sum_{i=1}^{n} x_i\right)^2}{n}\right]$$

This formula does not require the calculation of the individual deviations. In hand calculation, the use of this alternative formula often reduces the arithmetic work, especially when $\bar{x}$ turns out to be a number with many decimal places. The equivalence of the two formulas is shown in Appendix A1.2.

Example 17   Alternative calculation of variance
In a psychological experiment a stimulating signal of fixed intensity was used on six experimental subjects. Their reaction times, recorded in seconds, were 4, 2, 3, 3, 6, 3. Calculate the standard deviation for the data by using the alternative formula.

SOLUTION AND DISCUSSION   These calculations can be conveniently carried out in tabular form:

	x	x^2
	4	16
	2	4
	3	9
	3	9
	6	36
	3	9
Total	21	83
	$= \Sigma x$	$= \Sigma x^2$

$$s^2 = \frac{1}{n-1}\left[\Sigma x^2 - \frac{(\Sigma x)^2}{n}\right] = \frac{83 - (21)^2/6}{5} = \frac{83 - 73.5}{5} = \frac{9.5}{5} = 1.9$$

$$s = \sqrt{1.9} = 1.38 \text{ seconds}$$

The reader may do the calculations with the first formula and verify that the same result is obtained.   ■

STATISTICAL REASONING

Catching the Spies[3]

Greg was hired to work in a computer lab on the west coast of the United States. One day, he noticed that the accounting system was off by five cents. This trivial matter was ignored by others, but he wondered what caused the discrepancy. He discovered that someone had dialed into the computer system and altered entries in the system. Greg monitored many remote terminals to try to determine information about the caller. After finally tracking down the intruder's sign-on, he gathered data about the hacker's sign-on times. At this point, Greg created a graphical display, a dot diagram, to understand the variability in the sign-on times. The most common sign-on time was around noon, and the hacker stayed connected to the system for only about five minutes at a time.

Greg wondered why the hacker often signed on at noon rather than, say, late at night. Greg found a clue in the response delay between the sending and receiving of signals. He began taking samples of response delays between his workplace and various sites in the United States. For each site, he took several observations and computed some statistical summaries (such as a measure of the center

---

[3]This account is very similar to the facts in an actual case documented in a PBS *NOVA* program entitled "The KGB, the Computer, and Me" a few years ago.

Telegraph Colour Library/FPG International

of the observations) of the response delays. The hacker's response delays were much longer than all Greg's data from sites within the United States indicated. In other words, the data suggested that the hacker might be located outside the country.

The hacker used Greg's workplace computer to gain access to various government agencies' sensitive files. The FBI was called in to help track down the hacker. Progress was slow because calls needed to be tracked through complex telephone exchanges. Finally, through continued effort, the investigators discovered that the phone calls originated in Germany. Still, many of the German phone systems operated manually so any exhaustive search to trace the caller might take two hours. How could the investigators make sure the hacker remained on the line for two hours?

Greg's friend suggested that a fake file of sensitive government data be placed on Greg's computer system in the hope of keeping the hacker signed on long enough to trace the call. Indeed, the hacker viewed the data long enough for the German police to trace the call and arrest him and his accomplices. It turned out that these hackers did their computer spying after dinner, in the evening!

> **Statistical methods are *tools*. Tools are useful only if they are used in appropriate situations.**
>
> **We examine possible factors that can *explain* (1) the pattern of observations or (2) the variation of outcomes.**

In Example 16, we saw that one data set with a visibly greater amount of variation yields a larger numerical value of s. The issue there surrounds a comparison between different data sets. In the context of a single data set, can we relate the numerical value of s to the physical closeness of the data points to the center $\bar{x}$? To this end, we view one standard deviation as a benchmark distance from the mean $\bar{x}$. For bell-shaped histograms, an empirical rule relates the standard deviation to the proportion of the data that lie within particular intervals around $\bar{x}$.

Empirical Guidelines for Symmetric Bell-Shaped Histograms

Approximately	68%	of the data lie within $\bar{x} \pm s$
	95%	of the data lie within $\bar{x} \pm 2s$
	99.7%	of the data lie within $\bar{x} \pm 3s$

Example 18   Empirical guidelines for bookstore data
Examine the 40 bookstore sales receipts in Table 4 in the context of the empirical guideline.

SOLUTION AND DISCUSSION   Using a computer (see, for instance, Exercise 7.32), we obtain

$$\bar{x} = \$61.35$$
$$s = \$28.658, \qquad 2s = 2(28.658) = \$57.316$$

Going two standard deviations on each side of $\bar{x}$ results in the interval

$$\$61.35 - 57.316 = \$4.034 \qquad \text{to} \qquad \$118.666 = \$61.35 + 57.316$$

By actual count, all the observations except $3.20 and $124.27 fall in this interval. We find that 38/40 = .95 or 95% of the observations lie within two standard deviations of $\bar{x}$. This example is too good! Ordinarily, we would not find exactly 95% but something close to it. ■

OTHER MEASURES OF VARIATION

As mentioned earlier, the simplest measure of variation is the sample range.

Sample range = Largest observation − Smallest observation

The range gives the length of the interval that contains all the observations.

Example 19   Calculation of sample range
Calculate the range for the traffic noise data given in Example 11.

SOLUTION AND DISCUSSION   The data given in Table 9 contained

$$\text{Smallest observation} = 52.0$$
$$\text{Largest observation} = 77.1$$

Therefore, the length of the interval covered by all of the noise level observations is

$$\text{Sample range} = 77.1 - 52.0 = 25.1 \text{ decibels} \qquad \blacksquare$$

As a measure of spread, the range has two attractive features: It is extremely simple to compute and interpret. However, it suffers from the serious disadvantage that it is much too sensitive to the existence of a very large or very small observation in the data set. Also, it ignores the information present in the scatter of the intermediate points.

To circumvent the problem of using a measure that may be thrown far off the mark by one or two wild or unusual observations, a compromise is made by measuring the interval between the first and third quartiles.

Sample interquartile range = Third quartile − First quartile

The sample interquartile range represents the length of the interval covered by the center half of the observations. This measure of the amount of variation is not disturbed if a small fraction of the observations are very large or very small. The sample interquartile range is usually quoted in government reports on income and other distributions that have long tails in one direction, in preference to standard deviation as the measure of spread.

Example 20 Calculation of sample interquartile range
Calculate the sample interquartile range for the noise-level data given in Table 9.

SOLUTION AND DISCUSSION In Example 11, the quartiles were found to be $Q_1 = 57.2$ and $Q_3 = 64.6$. Therefore,

$$\text{Sample interquartile range} = Q_3 - Q_1$$
$$= 64.6 - 57.2$$
$$= 7.4 \text{ decibels}$$

The middle half of the noise level data is contained in an interval of length 7.4 decibels. $\qquad \blacksquare$

BOXPLOTS

A graphic display called a **boxplot** highlights the summary information in the quartiles. To create a boxplot, begin with the

Five-number summary: minimum, Q_1, Q_2, Q_3, maximum

The center half of the data, from the first to the third quartile, is represented by a rectangle (box) with the median indicated by a bar. A line extends from Q_3 to the maximum value and another from Q_1 to the minimum. Figure 15 gives the boxplot for the noise level data in Table 9. The long line to the right is a consequence of the single high noise level 77.1. The next largest is 67.1.

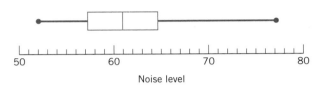

Figure 15. Boxplot of the noise level data in Table 9

Many computer programs create a **modified boxplot** where the lines extend to the smallest and largest observations only if these points are within 1.5 × (interquartile range) of the first and third quartiles, respectively.

Boxplots are particularly effective for displaying several samples alongside each other for the purpose of visual comparison.

Figure 16 displays the amount of reflected light in the near infrared band as recorded by satellite when flying over forest areas and urban areas, respectively. Because high readings tend to correspond to forest and low readings to urban areas, the readings have proven useful in classifying unknown areas.

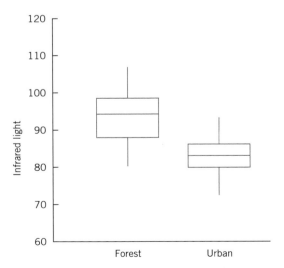

Figure 16. Boxplots of near infrared light reflected from forest and urban areas

The quartiles for forest areas are all larger than their urban area counterparts. Also, the interquartile range, the length of the box, is about 50% larger.

Exercises

5.1 On eight occasions, the times in minutes for a child to walk to a friend's house were

11   12   9   8   9   9   11   10

(a) Give three possible factors that may contribute to the variation in these times.

(b) Calculate the sample mean.

(c) Calculate the sample standard deviation.

5.2 Inquiries about available two-bedroom apartments in one area of a city produced the following monthly rents (in dollars) for seven properties.

720   595   915   1000   880   845   748

(a) Give three possible factors that may contribute to the variation in monthly rent.

(b) Calculate the sample mean.

(c) Calculate the sample standard deviation.

5.3 For the data set

9,   4,   5

(a) Calculate the deviations $(x - \bar{x})$ and check to see that they add up to 0.

(b) Calculate the sample variance and the standard deviation.

5.4 Repeat (a) and (b) of Exercise 5.3 for the data set

3,   8,   1

5.5 For the data set 10, 8, 16, 6,

(a) Calculate the deviations $(x - \bar{x})$ and check to see that they add up to 0.

(b) Calculate the variance and the standard deviation.

5.6 For the data of Exercise 5.3, calculate s^2 by using the alternative formula.

5.7 For the data of Exercise 5.5, calculate s^2 by using the alternative formula.

5.8 For each data set, calculate s^2.

(a) 2, 5, 4, 3, 3

(b) $-1, 2, 0, -2, 1, -1$

(c) 12, 11, 11, 12, 11, 11, 12

5.9   One year, the scores for the eight initial games of the National Basketball Association playoffs were

$$111-106, \quad 104-92, \quad 116-105, \quad 119-113,$$
$$109-102, \quad 111-97 \quad 113-96, \quad 101-89$$

Find the sample variance for the score of the winning team.

5.10   Find the standard deviation of the measurements of diameters given in Exercise 4.17.

5.11   In August, the Jump River Electric Co. suffered 11 power outages that lasted the following number of hours.

$$2.5, \quad 2, \quad 1.5, \quad 3, \quad 1, \quad 1.5, \quad 2, \quad 1.5, \quad 1, \quad 10, \quad 1$$

Calculate

(a)   The sample variance

(b)   The sample standard deviation

5.12   Loss of calcium is a serious problem for older women. At the beginning of a study on the influence of exercise on the rate of loss, the amounts of bone mineral content in the radius bone of the dominant hand of seven elderly women were

$$.90, \quad 1.06, \quad 1.06, \quad .79, \quad .85, \quad .91, \quad .96$$

Calculate

(a)   The sample variance

(b)   The sample standard deviation

5.13   With reference to the radiation leakage data given in Exercise 3.13, calculate

(a)   The sample variance

(b)   The sample standard deviation

5.14   With reference to the data on the length of 10 major league baseball games in Exercise 4.15,

(a)   Find the sample mean.

(b)   Find the sample variance.

(c)   Find the sample standard deviation.

5.15   With reference to Exercise 3.12,

(a)   Find the sample mean.

(b)   Find the sample standard deviation.

5.16   A sample of seven compact discs at the music store stated the following performance times (in minutes) for Beethoven's Ninth Symphony.

$$66.9, \quad 66.2, \quad 71.0, \quad 68.6, \quad 65.4, \quad 68.4, \quad 71.9$$

(a) Find the sample median.
(b) Find the sample mean.
(c) Find the sample standard deviation.

5.17 For the data set of Exercise 4.20, calculate the interquartile range.

5.18 For the data set of Exercise 3.22, calculate the interquartile range.

5.19 For the data set of Exercise 4.9, calculate the interquartile range for 1991.

5.20 Should you be surprised if the range is larger than twice the interquartile range?

5.21 Calculations with the test scores data of Exercise 3.22 give $\bar{x} = 150.125$ and $s = 24.677$.

(a) Find the proportion of the observations in the intervals $\bar{x} \pm 2s$ and $\bar{x} \pm 3s$.

(b) Compare your findings in part (a) with those suggested by the empirical guidelines for bell-shaped distributions.

5.22 Refer to the data on mineral loss in Exercise 4.13.

(a) Calculate $\bar{x}$ and s.

(b) Find the proportion of the observations that are in the intervals $\bar{x} \pm s$, $\bar{x} \pm 2s$, and $\bar{x} \pm 3s$.

(c) Compare the results of part (b) with the empirical guidelines.

5.23 Refer to the data of ozone measurements in Exercise 3.19.

(a) Calculate $\bar{x}$ and s.

(b) Find the proportion of the observations that are in the intervals $\bar{x} \pm s$, $\bar{x} \pm 2s$, and $\bar{x} \pm 3s$.

(c) Compare the results of part (b) with the empirical guidelines.

5.24 *Some properties of the standard deviation.*

1. If a fixed number c is added to all measurements in a data set, the deviations $(x - \bar{x})$ remain unchanged (see Exercise 4.24). Consequently, s^2 and s remain unchanged.

2. If all measurements in a data set are multiplied by a fixed number d, the deviations $(x - \bar{x})$ get multiplied by d. Consequently, s^2 gets multiplied by d^2, and s by $|d|$. (*Note:*   The standard deviation is never negative.)

Verify these properties for the data set

$$5, \quad 9, \quad 9, \quad 8, \quad 10, \quad 7$$

taking $c = 4$ in property (1) and $d = 2$ in (2).

5.25   Two cities provided the following information on public school teachers' salaries.

	Minimum	Q_1	Median	Q_3	Maximum
City A	18,400	24,000	28,300	30,400	36,300
City B	19,600	26,500	31,200	35,700	41,800

(a)   Construct a boxplot for the salaries in City A.

(b)   Construct a boxplot, on the same graph, for the salaries in City B.

(c)   Are there larger differences at the lower or the higher salary levels? Explain.

5.26   Make a boxplot of the bookstore sales data in Table 4.

5.27   Refer to the data on throwing speed in Exercise 4.14. Make separate boxplots to compare males and females.

5.28   Refer to Exercise 3.26 and the data on the consumer price index for various cities. Find the increase, for each city, by subtracting the 1992 value from the 1995 value.

(a)   Obtain the five-number summary: minimum, Q_1, Q_2, Q_3, and maximum. Which city had the largest increase? Were there any decreases?

(b)   Make a boxplot of the increases.

5.29   Refer to Exercise 3.26 and the data on the consumer price index for various cities. Find the increase, for each city, by subtracting the 1992 value from the 1995 value.

(a)   Find the sample mean and standard deviation of these differences.

(b)   What proportion of the increases lie within $\bar{x} \pm 2s$?

5.30   Refer to Exercise 4.9 and the data on the number of days that a metropolitan area failed to meet the ozone standard. Find the decrease, for each area, by subtracting the 1991 value from the 1988 value.

(a)   Obtain the five-number summary: minimum, Q_1, Q_2, Q_3, and maximum. Which metropolitan area had the largest decrease? Were there any increases?

(b)   Make a boxplot of the decreases.

5.31   Refer to Exercise 4.9 and the data on the number of days that a metropolitan area failed to meet the ozone standard. Find the decrease, for each area, by subtracting the 1991 value from the 1988 value.

(a) Find the sample mean and standard deviation of these differences.

(b) Comment on how a large observation can influence the value of the mean.

5.32 Presidents also take midterms! After two years of the President's term, members of Congress are up for election. The following table gives the number of net seats lost, by the party of the President, in the House of Representatives since the end of World War II.

Net House Seats Lost in Midterm Elections

1950	Truman (D)	55	1974	Nixon/Ford (R)	43
1954	Eisenhower (R)	16	1978	Carter (D)	11
1962	Kennedy (D)	4	1982	Reagan (R)	26
1966	Johnson (D)	47	1986	Reagan (R)	5
1970	Nixon (R)	12	1990	Bush (R)	8
			1994	Clinton (D)	52

For the data on the number of House seats lost,

(a) Calculate the sample mean.

(b) Calculate the standard deviation.

(c) Make a dot diagram.

(d) What is one striking feature of this data set that could be useful in predicting future midterm election results? (*Hint:* Would you have expected some elections to result in net gains?)

5.33 With reference to Exercise 5.32,

(a) Calculate the median number of lost House seats.

(b) Find the maximum and minimum losses and identify these with a President.

(c) Determine the range for the number of House seats lost.

6. CHECKING THE STABILITY OF THE OBSERVATIONS OVER TIME

The calculations for the sample mean and sample variance treat all the observations alike. The presumption is that there are no apparent trends, no steady increase or steady decrease, in the data over time and there are no unusual observations. Another way of saying this is that the process producing the observations is in **statistical control**. The concept of statistical control allows for variability in the observations, but requires that the pattern of variability not change over time. Variability should not increase or decrease with time nor should the center of the pattern change.

To check on the stability of the observations over time, observations should be plotted versus time, or at least the order in which they were taken. The resulting plot is called a **time plot,** or sometimes a **time series plot.**

Example 21   Time plot of police overtime
The Madison Police Department charts several important variables, one of which is the number of overtime hours due to extraordinary events. These events would include murders, major robberies, and so forth. Although any one event is not very predictable, there is some constancy when data are grouped into six-month periods.

The values of overtime hours for extraordinary events for the last eight years, beginning with 2200, 875, . . . , through 1223, are

2200	875	957	1758	868	398	1603	523
2034	1136	5326	1658	1945	344	807	1223

Is the extraordinary event overtime hours process in control? Construct a time plot and comment.

SOLUTION AND DISCUSSION   The time plot is shown in Figure 17. There does not appear to be any trend, but there is one large value of 5326 hours. This occurred when protests took place on campus in response to the bombing of a foreign capital. These events required city police to serve 1773 extraordinary overtime hours in just one 2-week period and 683 in the next period. That is, there was really one exceptional event, or special cause, that could be identified with the unusual point.

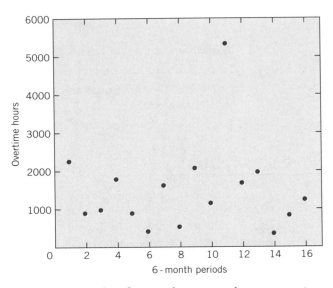

Figure 17. Time plot of extraordinary event hours versus time order

Example 22   Time plot of exchange rate
The exchange rate between the United States and Japan can be stated as the number of yen that can be purchased with $1. Although this rate changes daily, we quote the official value for the year:

Year	1985	1986	1987	1988	1989	1990	1991	1992	1993	1994
Exchange rate	238.5	168.4	144.6	128.2	138.1	145.0	134.6	126.8	111.1	102.2

Is this exchange rate in statistical control? Make a time plot and comment.

SOLUTION AND DISCUSSION   The time plot is shown in Figure 18. There is a rather strong downhill trend over time so the exchange rate is definitely not in statistical control. A dollar has purchased fewer and fewer yen over the years. It is the downward trend that is the primary feature and a cause for serious concern with regard to trade deficits.

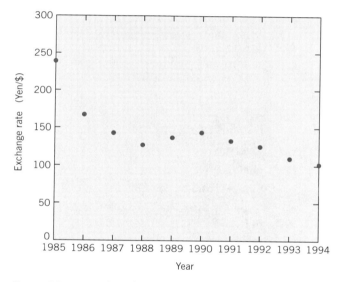

Figure 18. Time plot of the exchange rate   ■

Exercises

6.1   Make a time plot of the phone call data in Exercise 4.20 by treating the number of calls as arising in the order given. Comment on the statistical control.

6.2   A city department has introduced a quality improvement program and has allowed employees to get credit for overtime hours when attending meetings of their quality groups. The total number of overtime meeting hours, for each of the 26 pay periods in one year by row, were

30	215	162	97	194	163	60	41	100
43	96	69	80	42	162	75	95	65
57	131	54	114	64	114	38	140	

Make a time plot of the overtime meeting hours data.

6.3 The exchange rate between the United States and Germany can be stated as the number of marks that can be purchased with $1. The official values for the year are

Year	1986	1987	1988	1989	1990	1991	1992	1993	1994
Exchange rate	2.17	1.80	1.76	1.88	1.62	1.66	1.56	1.65	1.62

Is this exchange rate in statistical control? Make a time plot and comment.

💻 USING A COMPUTER

💾 6.4 **Computer-constructed graphics.** Time series plots can be constructed using many statistical software packages. We illustrate the commands for the MINITAB software using the extraordinary event overtime data from Example 21. These data are assumed to be stored in the file C2EX21.DAT, as on the data disk.

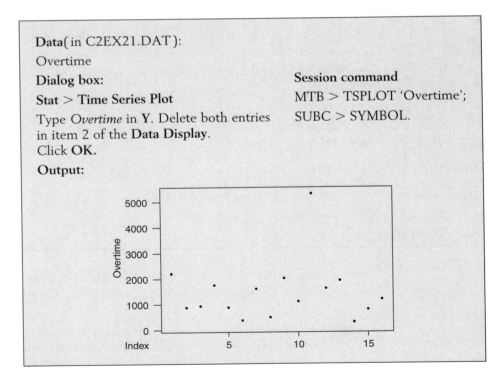

Data(in C2EX21.DAT):
Overtime
Dialog box: **Session command**
Stat > Time Series Plot   MTB > TSPLOT 'Overtime';
Type *Overtime* in **Y**. Delete both entries   SUBC > SYMBOL.
in item 2 of the **Data Display**.
Click **OK**.
Output:

Use the computer to construct the time plot in Exercise 6.2.

KEY IDEAS AND FORMULAS

Qualitative data refer to frequency counts in categories. These are summarized by calculating the

$$\text{Relative frequency} = \frac{\text{Frequency}}{\text{Total number of observations}}$$

for the individual categories.

Pareto diagrams display events according to their frequency to highlight the most important few that occur most of the time.

Data obtained as measurements on a numerical scale are either discrete or continuous.

A discrete data set is summarized by a frequency distribution that lists the distinct data points and the corresponding relative frequencies. Either a line diagram or a histogram can be used for a graphical display.

Continuous measurement data should be graphed as a dot diagram when the data set is small—say, fewer than 20 or 25 observations. Larger data sets are summarized by grouping the observations in class intervals, preferably of equal lengths. A list of the class intervals along with the corresponding relative frequencies provides a frequency distribution, which can be graphically displayed as a histogram.

A stem-and-leaf display is another effective means of display when the data set is not too large. It is more informative than a histogram because it retains the individual observations in each class interval instead of lumping them into a frequency count.

A summary of measurement data (discrete or continuous) should also include numerical measures of center and spread.

Two important measures of center are

$$\text{sample mean} \quad \bar{x} = \frac{\sum x}{n}$$

$$\text{sample median} = \text{middle value of the ordered data set}$$

The quartiles and, more generally, percentiles are other useful locators of the distribution of a data set. The second quartile is the same as the median.

The amount of variation or spread of a data set is measured by the sample standard deviation s. The sample variance s^2 is given by

$$s^2 = \frac{\sum (x - \bar{x})^2}{n - 1}$$

Also, $\quad s^2 = \frac{1}{n-1} \left[\sum x^2 - \frac{(\sum x)^2}{n} \right]$ (convenient for hand calculation)

$$\text{sample standard deviation} \quad s = +\sqrt{s^2}$$

The standard deviation indicates the amount of spread of the data points around the mean $\bar{x}$. If the histogram appears symmetric and bell-shaped, then the guidelines suggest that the interval

$$\bar{x} \pm s \quad \text{includes approximately 68\% of the data}$$
$$\bar{x} \pm 2s \quad \text{includes approximately 95\% of the data}$$
$$\bar{x} \pm 3s \quad \text{includes approximately 99.7\% of the data}$$

Two other measures of variation are

$$\text{sample range} = \text{largest observation} - \text{smallest observation}$$

and

$$\text{sample interquartile range} = \text{third quartile} - \text{first quartile}$$

The five-number summary—namely, the median, the first and third quartiles, the smallest observation, and the largest observation—serves as a useful indicator of the distribution of a data set. These quantities are displayed in a **boxplot**.

7. REVIEW EXERCISES

7.1 Recorded here are the numbers of civilians employed in the United States by major occupation groups for the years 1980 and 1992. (*Source: Statistical Abstract of the United States*, 1993.)

	Number of Workers in Millions	
	1980	1992
Goods producing	25.7	23.4
Service (private)	48.5	66.0
Government	16.2	18.6
Total	90.4	108.0

(a) For each year, calculate the relative frequencies of the occupation groups.

(b) Comment on changes in the occupation pattern between 1980 and 1992.

7.2 Table 13 gives data collected from the students attending an elementary statistics course at the University of Wisconsin. These data include sex, height, number of years in college, and the general area of intended major [Humanities (H); Social Science (S); Biological Science (B); Physical Science (P)].

(a) Summarize the data of "intended major" in a frequency table.

(b) Summarize the data of "year in college" in a frequency table and draw either a line diagram or a histogram.

TABLE 13   Class Data

Student No.	Sex	Height in Inches	Year in College	Intended Major	Student No.	Sex	Height in Inches	Year in College	Intended Major
1	F	67	3	S	26	M	67	1	B
2	M	72	3	P	27	M	68	3	P
3	M	70	4	S	28	M	72	4	B
4	M	70	1	B	29	F	68	3	P
5	F	61	4	P	30	F	66	2	B
6	F	66	3	B	31	F	65	2	B
7	M	71	3	H	32	M	64	4	B
8	M	67	4	B	33	M	72	1	H
9	M	65	3	S	34	M	67	4	B
10	F	67	3	B	35	M	73	3	S
11	M	74	3	H	36	F	71	4	B
12	M	68	3	S	37	M	71	3	B
13	M	74	2	P	38	M	69	2	S
14	F	64	4	P	39	F	69	4	P
15	M	69	3	S	40	M	74	4	S
16	M	64	3	B	41	M	73	3	B
17	M	72	4	P	42	M	68	3	B
18	M	71	3	B	43	F	66	2	S
19	F	67	2	S	44	M	73	2	P
20	M	70	4	S	45	M	73	2	S
21	M	66	4	S	46	M	67	4	S
22	F	67	2	B	47	F	62	3	S
23	M	68	4	S	48	M	68	2	B
24	M	71	3	H	49	M	71	3	S
25	M	75	1	S					

(c) Plot the dot diagrams of heights separately for the male and female students, and compare.

7.3 Refer to the data on power outages in Table D.1 in the Data Bank. Make a Pareto chart for the cause of the outage.

7.4 Based on a sample survey, the frequency distribution in Table 14 was obtained for the age of groom at first marriage in the state of Wisconsin. (*Source: Vital Statistics of the United States*, 1987, Vol. III, Marriage and Divorce.)

(a) Calculate the relative frequency of each class interval.

(b) Plot a density histogram. (*Hint:* Since the intervals have unequal widths, make the height of each rectangle equal to the relative frequency divided by the width of the interval.)

(c) What proportion of the grooms marry before age 25? After age 30?

TABLE 14

Age Interval[a]	Number of Grooms
< 18	61
18–20	1,124
20–25	11,768
25–30	9,796
30–35	3,300
35–45	1,244
45–55	123
55–65	58
65 and over	17
Total	27,491

[a] The intervals include the lower endpoints but not the upper. Take the first interval as 16–18 and the last as 65–75.

7.5 The dollar amounts claimed by businessmen for their lunches are to be grouped into the following classes: 0–5, 5–10, 10–15, 15–20, 20 or more, where the left endpoint is included but not the right. Is it possible to determine from this frequency distribution the exact number of lunches for which the amount claimed was

(a) Less than 15?

(b) 10 or more?

(c) 30 or more?

7.6 On five occasions, the amounts of time in seconds to order baseball tickets by phone were

$$340, \quad 129, \quad 247, \quad 388, \quad 95$$

Find the sample mean and standard deviation.

7.7 The stem-and-leaf display given here shows the final examination scores of students in a sociology course.

2	48
3	155
4	002
5	03368
6	0124479
7	22355689
8	004577
9	0025

(a) Find the median score.

(b) Find the quartiles Q_1 and Q_3.

(c) What proportion of the students scored below 70? 80 and over?

7.8 The following are the numbers of passengers on the minibus tour of Hollywood.

```
9  12  10  11  11   7  12   6  11   4  10  10  11   9  10
7  10   8   8   9   8   9  11   9   8   6  10   6   8  11
```

(a) Find the sample median.

(b) Find the sample mean.

(c) Find the sample variance.

7.9 The following table shows the age at inauguration of each U.S. President.

	Name	Age at Inauguration		Name	Age at Inauguration
1.	Washington	57	22.	Cleveland	47
2.	J. Adams	61	23.	B. Harrison	55
3.	Jefferson	57	24.	Cleveland	55
4.	Madison	57	25.	McKinley	54
5.	Monroe	58	26.	T. Roosevelt	42
6.	J. Q. Adams	57	27.	Taft	51
7.	Jackson	61	28.	Wilson	56
8.	Van Buren	54	29.	Harding	55
9.	W. H. Harrison	68	30.	Coolidge	51
10.	Tyler	51	31.	Hoover	54
11.	Polk	49	32.	F. D. Roosevelt	51
12.	Taylor	64	33.	Truman	60
13.	Fillmore	50	34.	Eisenhower	62
14.	Pierce	48	35.	Kennedy	43
15.	Buchanan	65	36.	L. Johnson	55
16.	Lincoln	52	37.	Nixon	56
17.	A. Johnson	56	38.	Ford	61
18.	Grant	46	39.	Carter	52
19.	Hayes	54	40.	Reagan	69
20.	Garfield	49	41.	Bush	64
21.	Arthur	50	42.	Clinton	46

(a) Make a stem-and-leaf display with a double stem.

(b) Find the median, Q_1, and Q_3.

7.10 Calculate the mean, variance, and standard deviation for each of the following data sets.

(a) 8, 9, 7, 9, 12

(b) 23, 28, 23, 26

(c) − 1.1, 1.6, .8, − .2, 2.9

7.11 (a) Calculate $\bar{x}$ and s for the data 9, 11, 7, 12, 11.

(b) Consider the data set 109, 111, 107, 112, 111, which is obtained by adding 100 to each number given in part (a). Use your results of part (a) and the properties stated in Exercises 4.24 and 5.24 to obtain $\bar{x}$ and s for this modified data set. Verify your results by direct calculations with this new data set.

(c) Consider the data set − 27, − 33, − 21, − 36, − 33, which is obtained by multiplying each number of part (a) by − 3. Repeat the problem given in part (b) for this new data set.

7.12 A reconnaissance study of radioactive materials was conducted in Alaska to call attention to anomalous concentrations of uranium in plutonic rocks. The amounts of uranium in 13 locations under the Darby Mountains are

$$7.92, \quad 10.29, \quad 19.89, \quad 17.73, \quad 10.36, \quad 13.50, \quad 8.81,$$
$$6.18, \quad 7.02, \quad 11.71, \quad 8.33, \quad 9.32, \quad 14.61$$

[*Source:* T. Miller and C. Bunker, U.S. Geological Survey, *Journal of Research* (1976), pp. 367–377.]

Find the:

(a) Mean

(b) Standard deviation

(c) Median and the quartiles

(d) Interquartile range

7.13 Refer to the class data in Exercise 7.2. Calculate the following.

(a) $\bar{x}$ and s for the heights of males

(b) $\bar{x}$ and s for the heights of females

(c) Median and the quartiles for the heights of males

(d) Median and the quartiles for the heights of females

7.14 In a genetic study, a regular food was placed in each of 20 vials and the number of flies of a particular genotype feeding on each vial recorded. The counts of flies were also recorded for another set of 20 vials that contained grape juice. The following data sets were obtained (courtesy of C. Denniston and J. Mitchell).

Number of Flies (Regular Food)

15	20	31	16	22	22	23	33	38	28
25	20	21	23	29	26	40	20	19	31

Number of Flies (Grape Juice)

6	19	0	2	11	12	13	12	5	16
2	7	13	20	18	19	19	9	9	9

(a)   Plot separate dot diagrams for the two data sets.

(b)   Make a visual comparison of the two distributions with respect to their centers and spreads.

(c)   Calculate $\bar{x}$ and s for each data set.

7.15   The data below were obtained from a detailed record of purchases of toothpaste over several years (courtesy of A. Banerjee). The usage times (in weeks) per ounce of toothpaste for a household taken from a consumer panel were:

.74   .45   .80   .95   .84   .82   .78   .82   .89   .75   .76   .81
.85   .75   .89   .76   .89   .99   .71   .77   .55   .85   .77   .87

(a)   Plot a dot diagram of the data.

(b)   Find the relative frequency of the usage times that do not exceed .80.

(c)   Calculate the mean and the standard deviation.

(d)   Calculate the median and the quartiles.

7.16   To study how first-grade students utilize their time when assigned to a math task, a researcher observes 24 students and records their times off-task out of 20 minutes (courtesy of T. Romberg).

Times Off-Task (minutes)

4	0	2	2	4	1
4	6	9	7	2	7
5	4	13	7	7	10
10	0	5	3	9	8

For this data set, find the

(a)   Mean and standard deviation

(b)   Median

(c)   Range

7.17   Blood cholesterol levels were recorded for 43 persons sampled in a medical study group and the following data obtained (courtesy of G. Metter).

239	212	249	227	218	310	281	330	226	233
223	161	195	233	249	284	245	174	154	256
196	299	210	301	199	258	205	195	227	244
355	234	195	179	357	282	265	286	286	176
195	163	297							

(a) Group the data into a frequency distribution.

(b) Plot the histogram and comment on the shape of the distribution.

7.18 For the blood cholesterol data in Exercise 7.17, calculate the

(a) $\bar{x}$ and s

(b) Median and other quartiles

(c) Range and interquartile range

7.19 Refer to Exercises 7.17 and 7.18.

(a) Determine the intervals $\bar{x} \pm s$, $\bar{x} \pm 2s$, and $\bar{x} \pm 3s$.

(b) Find the proportion of the measurements in Exercise 7.17 that lie in each of these intervals.

(c) Compare your findings with the empirical guidelines for bell-shaped distributions.

7.20 The following summary statistics were obtained from a data set.

$$\bar{x} = 80.5 \qquad \text{median} = 84.0$$
$$s = 10.5 \qquad Q_1 = 75.5$$
$$Q_3 = 96.0$$

Approximately what proportion of the observations are

(a) Below 96.0?

(b) Above 84.0?

(c) In the interval 59.5–101.5?

(d) In the interval 75.5–96.0?

(e) In the interval 49.0–112.0?

State which of your answers are based on the assumption of a bell-shaped distribution.

7.21 Acid precipitation is linked to the disappearance of sport fish and other organisms from lakes. Sources of air pollution, including automobile emissions and the burning of fossil fuels, add to the natural acidity of precipitation. The Wisconsin Department of Natural Resources initiated a precipitation monitoring program with the goal of developing appropriate air pollution controls to reduce the problem. The acidity of the first 50 rains monitored, measured on a pH scale from 1 (very acidic) to 7 (basic), are

3.58	3.80	4.01	4.01	4.05	4.05
4.12	4.18	4.20	4.21	4.27	4.28
4.30	4.32	4.33	4.35	4.35	4.41
4.42	4.45	4.45	4.50	4.50	4.50
4.50	4.51	4.52	4.52	4.52	4.57
4.58	4.60	4.61	4.61	4.62	4.62
4.65	4.70	4.70	4.70	4.70	4.72
4.78	4.78	4.80	5.07	5.20	5.26
5.41	5.48				

(a) Calculate the median and quartiles.

(b) (*Optional*) Find the 90th percentile.

(c) Determine the mean and standard deviation.

(d) Display the data in the form of a boxplot.

(e) Make a histogram.

7.22   Refer to Exercise 7.21.

(a) Determine the intervals $\bar{x} \pm s$, $\bar{x} \pm 2s$, and $\bar{x} \pm 3s$.

(b) What proportions of the measurements lie in those intervals?

(c) Compare your findings with the empirical guidelines for bell-shaped distributions.

7.23   One job hazard of firefighters is their exposure to the poisonous gas carbon monoxide (CO). In a study concerning the design specifications for respiratory breathing devices, Boston firefighters wore personal air samplers and measured maximum concentration of CO during actual fire situations. The 51 observations where the maximum level of CO reached or exceeded the presumably safe level of .05% (500 parts per million) are

.07	.07	.12	.95	.35	.10
.13	.06	.72	.13	.17	
.15	.27	.38	.09	.06	
.58	.31	.12	.86	.05	
.06	.20	.39	.12	.07	
.14	1.13	.10	.15	.20	
.05	.22	.10	.10	.19	
2.40	.57	.11	.40	.50	
.14	.12	.08	.29	.09	
2.70	.12	.11	.05	.22	

(*Source:* U.S. Dept. of HEW Report NIOSH 76-121.)

(a) Calculate the median and quartiles.

(b) Calculate the mean and standard deviation.

(c)   Group the data into a frequency distribution (unequal intervals are appropriate here).

(d)   Display the data in the form of boxplot.

7.24   The *z-scale* (or *standard scale*) measures the position of a data point relative to the mean and in units of the standard deviation. Specifically,

$$z\text{-value of a measurement} = \frac{\text{measurement} - \bar{x}}{s}$$

When two measurements originate from different sources, converting them to the z-scale helps to draw a sensible interpretation of their relative magnitudes. For instance, suppose a student scored 65 in a math course and 72 in a history course. These (raw) scores tell little about the student's performance. If the class averages and standard deviations were $\bar{x} = 60$, $s = 20$ in math and $\bar{x} = 78$, $s = 10$ in history, this student's

$$z\text{-score in math} = \frac{65 - 60}{20} = .25$$

$$z\text{-score in history} = \frac{72 - 78}{10} = -.60$$

Thus, the student was .25 standard deviation above the average in math and .6 standard deviation below the average in history.

(a)   If $\bar{x} = 490$ and $s = 120$, find the z-scores of 350 and 620.

(b)   For a z-score of 2.4, what is the raw score if $\bar{x} = 210$ and $s = 50$?

7.25   The winning times of the men's 400-meter freestyle swimming in the Olympics (1908–1992) appear in the following table.

Winning Times in Minutes and Seconds

Year	Time	Year	Time
1908	5:35.8	1960	4:18.3
1912	5:24.4	1964	4:12.2
1920	5:26.8	1968	4:09.0
1924	5:04.2	1972	4:00.27
1928	5:01.6	1976	3:51.93
1932	4:48.4	1980	3:51.31
1936	4:44.5	1984	3:51.23
1948	4:41.0	1988	3:46.95
1952	4:30.7	1992	3:45.00
1956	4:27.3		

(a) Draw a dot diagram and label the points according to time order.

(b) Explain why it is not reasonable to group the data into a frequency distribution.

7.26 The *mode* of a collection of observations is defined as the observed value with the largest relative frequency. The mode is sometimes used as a center value. There can be more than one mode in a data set. Find the mode for the data given in Exercise 3.11.

7.27 The speeds at which drivers were ticketed for speeding in a 30-mile-per-hour zone are given in the table (the left endpoint is included but not the right).

Speed	Frequency
37–39	15
39–41	10
41–43	9
43–45	5
45–47	5
47–51	3
51–57	3

(a) Construct a density histogram, where the area of a rectangle equals the relative frequency.

(b) What proportion of the tickets were issued for speeds of less than 45 miles per hour?

(c) Give two possible reasons why no speeding tickets were issued in the 32–35 miles per hour interval.

7.28 Mr. A recorded the times (in minutes) required to drive his car from the garage at home to his parking lot at work. The recorded values, from 50 occasions, are given below:

Time in minutes	Frequency
30–32	2
32–34	8
34–36	12
36–38	16
38–40	8
40–44	3
44–48	1

(a) Construct a density histogram (with the area of a rectangle above a class interval equal to the relative frequency of the interval).

(b) What are three factors that might affect Mr. A's times recorded above?

7.29 A movie critic classifies movies into one of the categories

action–adventure   comedy   mystery   romance

These categories are coded 1, 2, 3, and 4, respectively. Five current movies are classified as follows:

$$3 \quad 1 \quad 4 \quad 3 \quad 1$$

(a) Calculate the sample mean of the five current movies.

(b) Could you interpret the mean as being between comedy and mystery? Explain why it is not sensible to calculate a sample mean for these, or any other, nominal data.

7.30 The following are the tips (in dollars) a waiter received in 8 nights in a Chinese restaurant:

20.30,   15.60,   18.20,   14.70,   21.10,   16.50,   12.90,   41.40

(a) Calculate the sample mean and the sample median.

(b) Calculate the sample standard deviation.

(c) If one of the numbers above is misrecorded, which number is likely to be the wrong number? Which measure in (a) is affected the most? Explain.

7.31 The Dow Jones Industrial Average provides an indication of overall stock market level. The changes from year end to year end are given below for 1969–1970 through 1993–1994, by row.

Yearly Change in Dow Jones Industrial Average

38.56	51.28	129.82	−169.16	−234.62
236.17	152.24	−173.48	−26.16	33.73
25.25	−88.99	171.54	212.40	−47.37
335.10	349.28	42.88	229.74	584.63
−119.54	535.17	132.28	453.98	80.35

(a) Make a time plot of the data on changes.

(b) What year-to-year changes, if any, are outliers?

💻 USING A COMPUTER

💾 7.32 *Computer-aided statistical calculations.*   Calculations of the descriptive statistics such as $\bar{x}$ and s are increasingly tedious with larger data sets. Modern computers have come a long way in alleviating the drudgery of hand calculations. MINITAB is one computing package that is easily accessible to students because its commands are in simple English.

   Annual snowfall in Madison, Wisconsin, is measured from July to June. The data on total inches were introduced in Exercise 3.18. The data disk file C2P3-18.DAT contains the snowfall beginning with the first year at the new airport station. We present the commands and selected output, which gives n, $\bar{x}$, and s. We also give the command for a boxplot.

Data(in C2P3-18.DAT):
Snow

Dialog box:	**Session command**
Stat > Basic Statistics > Descriptive Statistics	MTB > DESCRIBE 'Snow'
Type *Snow* in **Variables.** Click **OK.**	SUBC > GBOXPLOT.

Output:

Descriptive Statistics

Variable	N	Mean	StDev
Snowdata	47	31.563	5.360

Use MINITAB (or some other package program) to find $\bar{x}$ and s for

   (a)   The ozone data in Exercise 3.19
   (b)   The acid rain data in Exercise 7.21

💾 7.33 *Computer-aided statistical calculations.*   Refer to Exercise 7.32. MINITAB also calculates several other descriptive statistics, including the five-number summary: minimum, first quartile, second quartile, third quartile, and maximum. (MINITAB uses a slightly different scheme to determine the first and third quartiles, but the difference is not of practical importance with large samples.) A more complete output for the snow data is

Descriptive Statistics

Variable	N	Mean	Median	StDev
Snowdata	47	31.563	31.860	5.360

Variable	Min	Max	Q1	Q3
Snowdata	21.100	43.340	28.150	35.070

The sequence of choices

> **Stat > Basic Statistics > Descriptive Statistics**
> Type *Snow* in **Variables.** Click **Graphs.** Click **Graphical Summary.**
> Click **OK.** Click **OK.**

produces an even more complete summary:

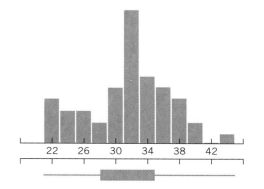

Descriptive Statistics

Variable:	Snowdata
Mean	31.5632
StDev	5.3599
Variance	28.7286
N	47
Minimum	21.1000
1st Quartile	28.1500
Median	31.8600
3rd Quartile	35.0700
Maximum	43.3400

Vancouver, British Columbia, has very cloudy winters but some months are very sunny. The numbers of hours of bright sunshine during September are given for 86 years.

294.9	181.2	142.2	185.9	173.2	79.4	141.3
190.5	153.8	236.2	202.9	138.2	172.1	163.4
226.4	217.8	205.2	221.7	124.2	199.2	228.2
192.0	118.5	235.3	123.7	155.3	213.2	181.4
164.6	161.6	193.6	180.1	119.2	175.1	201.0
167.0	154.0	144.9	193.2	184.5	209.6	222.6
201.2	215.1	169.0	116.3	148.2	147.6	227.1
164.4	128.5	210.7	152.8	170.2	166.6	169.1
178.6	153.4	215.1	177.0	130.9	192.0	149.7
164.8	192.4	260.3	249.7	202.2	150.5	112.7
171.5	155.5	179.9	168.4	199.2	187.8	171.9
186.3	232.5	215.3	265.0	218.4	237.9	187.8
251.7	191.0					

(a)   Use MINITAB (or some other package) to describe these data.

(b)   What is the average number of hours of bright sunshine per day?

(c)   Comment on the presence or absence of symmetry and identify any outliers.

7.34   Use MINITAB (or some other package program) to find $\bar{x}$ and s for the data set in Table 4.

7.35   Use MINITAB (or some other package program) to find $\bar{x}$ and s for the final times to run 1.5 miles in Table D.5 in the Data Bank.

7.36   The salmon fisheries support a primary industry in Alaska and their management is of high priority. Salmon are born in freshwater rivers and streams, but then swim out into the ocean for a few years before returning to spawn and die. To identify the origins of mature fish, researchers studied growth rings on their scales. The growth the first year in freshwater is measured by the width of the growth rings for that period of life. The growth ring for the first year in the ocean environment will give an indication of growth for that period. A set of these measurements is given in Table D.6 in the Data Bank.

(a)   Describe the freshwater growth for males by making a histogram and calculating the mean, standard deviation, and quartiles.

(b)   Describe the freshwater growth for females by making a histogram and calculating the mean, standard deviation, and quartiles.

(c)   Construct boxplots to compare the growth of males and females.

7.37   Refer to the alligator data in Table D.7 of the Data Bank. Using the data on x_5 for thirty-seven alligators,

(a)   Make a histogram.

(b)   Obtain the sample mean and standard deviation.

7.38   Refer to Exercise 7.37.

(a)   Obtain the quartiles.

(b)   (*Optional*)   Obtain the 90th percentile. How many of the alligators above the 90th percentile are female?

CHAPTER 3

Describing Bivariate Data

Chapter Objectives

After reading this chapter, you should be able to

▶ Create a two-way table of frequencies for categorical data.
▶ Calculate relative frequencies for two-way tables.
▶ Understand the potential dilemmas when combining tables.
▶ Construct and interpret a scatter diagram of bivariate observations.
▶ Calculate the sample correlation and interpret its value.
▶ Distinguish between correlation and causation.

1. INTRODUCTION

Investigations involving data are rarely limited to a single variable. To illustrate the concepts, we restrict our attention to the simplest case where two variables, or characteristics, are observed on each sampling unit. Some examples are

Gender and the type of occupation of college graduates

Smoking habit and lung capacity of adult males

Average daily carbohydrate intake and protein intake of college students

Age of an aircraft and the time required for repairs

To discover facts and extract information from these multivariable data, we need some new graphical and numerical tools. These new tools, together with those introduced in Chapter 2, will enable us to summarize data and communicate the essential numerical information contained in multivariable data.

We focus on finding and interpreting the relations between two variables. A person's political party and attitude toward a new health program is one example; the number of goals scored for and against a team is another. Observations on two variables, recorded on the same sampling unit, produce bivariate data.

Bivariate data arise when measurements on two variables are recorded for each sampling unit.

The two variables, or characteristics, may both be qualitative traits. For example, five persons classified according to hair and eye color

(brown, green)   (black, brown)   (blonde, blue)

(brown, blue)   (brown, green)

produce bivariate categorical data. Section 2 discusses the summarization of bivariate categorial data in terms of frequencies, or counts. In the example above, the combined category brown hair and green eyes has frequency 2, and so on.

Sections 4 and 5 are concerned with bivariate data where both components are quantitative or measurement data. A sample of values of (carbohydrate, protein) intake for college students is one example.

In this chapter, we concentrate on two types of data and displays for examining the relationships between the two variables:

1. Two-way tables of counts associated with two categorical variables
2. Two-dimensional scatter diagrams of pairs of measurement data

2. SUMMARIZATION OF BIVARIATE CATEGORICAL DATA

When two traits are observed for the individual sampling units and each trait is recorded in some qualitative categories, the resulting data can be summarized in the form of a two-way frequency table. The categories for one trait are marked along the left margin, those for the other along the upper margin, and the frequency counts are recorded in the cells. Data in this summary form are commonly called **cross-classified** or **cross-tabulated** data. In statistical terminology, these two-way tables are also called **contingency tables.**

Example 1 Relative frequency calculation for attitude data
A survey was conducted by sampling 400 persons who were questioned regarding union membership and attitude toward decreased national spending on social welfare programs. The cross-tabulated frequency counts are presented in Table 1.

TABLE 1 Cross-Tabulated Frequency Counts

	Support	Indifferent	Opposed	Total
Union	112	36	28	176
Nonunion	84	68	72	224
Total	196	104	100	400

The entries of this table are self-explanatory. For instance, of the 400 persons polled, there were 176 union members. Among these union members, 112 expressed support, 36 were indifferent, and 28 were opposed.
To gain further understanding of how the responses are distributed, calculate the relative frequencies of the cells.

SOLUTION AND DISCUSSION For this purpose, we divide each cell frequency by the sample size 400. The relative frequencies (for instance, $84/400 = .21$) are shown in Table 2.
Depending on the specific context of a cross-tabulation, one may also wish to examine the cell frequencies relative to a marginal total. In Example 1, you may wish to compare the attitude patterns of the union members with those of

TABLE 2 Relative Frequencies for the Data of Table 1

	Support	Indifferent	Opposed	Total
Union	.28	.09	.07	.44
Nonunion	.21	.17	.18	.56
Total	.49	.26	.25	1.00

the nonmembers. This is accomplished by calculating the relative frequencies separately for the two groups (for instance, 84/224 = .375), as Table 3 shows.

TABLE 3   Relative Frequencies by Group

	Support	Indifferent	Opposed	Total
Union	.636	.205	.159	1.00
Nonunion	.375	.304	.321	1.00

From the calculations in Table 3, it appears that the attitude patterns are different between the two groups—support seems to be stronger among union members than nonmembers.

Now the pertinent question is: Can these observed differences be explained by chance or are there real differences of attitude between the populations of members and nonmembers? We will pursue this aspect of statistical inference in Chapter 11.   ∎

SIMPSON'S PARADOX

Quite surprising and misleading conclusions can occur when data from different sources are combined into a single table. We illustrate this reversal of implications with graduate school admissions data.

Example 2   Care is needed when combining tables
We consider graduate school admissions at a large midwestern university but, to simplify, we use only two departments as the whole school. We are interested in comparing admission rates by gender and obtain the data for the school given in Table 4.

TABLE 4   School Admission Rates

	Admit	Not Admit	Total Applicants
Male	233	324	557
Female	88	194	282
Total	321	518	839

Does there appear to be a gender bias?

SOLUTION AND DISCUSSION   It is clear from these admission statistics that the proportion of males admitted, 233/557 = .418, is greater than the proportion of females admitted, 88/282 = .312.

Does this imply some type of discrimination? Not necessarily. By checking the admission records, we were able to further categorize the cases according to department in Table 5. Table 4 is the aggregate of these two sets of data.

TABLE 5 Admission Rates by Department

Mechanical Engineering				History			
	Admit	Not Admit	Total		Admit	Not Admit	Total
Male	151	35	186	Male	82	289	371
Female	16	2	18	Female	72	192	264
Total	167	37	204	Total	154	481	635

One of the two departments, mechanical engineering, has mostly male applicants. Even so, the proportion of males admitted, 151/186 = .812, is smaller than the proportion of females admitted, 16/18 = .889. The same is true for the history department, where the proportion of males admitted, 82/371 = .221, is again smaller than the proportion of females admitted, 72/264 = .273. When the data are studied department by department, the reverse conclusion holds; females have a higher admission rate in both cases!

To obtain the correct interpretation, these data need to be presented as the full three-way table of gender–admission action–department as given above. If department is ignored and the data are aggregated across this variable, "department" can act as an unrecorded or lurking variable. In this example, it has reversed the direction of possible gender bias and led to the erroneous conclusion that males have a higher admission rate than females.   ■

The reversal of the comparison, as for the admissions in Example 2, when data are combined from several groups, is called Simpson's paradox.

When data from several sources are aggregated into a single table, there is always the danger that unreported variables may cause a reversal of the findings. In practical applications, there is not always agreement on how much effort to expend following up on unreported variables. When two medical treatments are being compared, the results often need to be adjusted for the age, gender, and sometimes current health of the subjects and other variables.

Exercises

2.1 Nausea from air sickness affects some travelers. A drug company, wanting to establish the effectiveness of its motion sickness pill, randomly gives either its pill or a look-alike sugar pill (placebo) to 200 passengers.

	Degree of Nausea				Total
	None	Slight	Moderate	Severe	
Pill	43	36	18	3	100
Placebo	19	33	36	12	100
Total					

(a) Complete the marginal totals.

(b) Calculate the relative frequencies separately for each row.

(c) Comment on any apparent differences in response between the pill and the placebo.

2.2   Records of drivers with a major medical condition (diabetes, heart condition, or epilepsy) and also of a group of drivers with no known health conditions were retrieved from a motor vehicle department. Drivers in each group were classified according to their driving record in the last year.

Medical Condition	Traffic Violations		Total
	None	One or More	
Diabetes	119	41	160
Heart condition	121	39	160
Epilepsy	72	78	150
None (control)	157	43	200

Compare each medical condition with the control group by calculating the appropriate relative frequencies.

2.3   The aging of commercial aircraft can make them more vulnerable to "skin-cracking" rivets. A major manufacturer collected data on its three most popular models in active use to determine the magnitude of the problem.

Number of Aircraft Still in Service

Model	≤ 20 Years	> 20 Years
B7	90	123
B27	1214	435
B37	1042	9

Compare the aging of the three types of planes by calculating the relative frequencies.

2.4   A survey was conducted to study the attitudes of the faculty, academic staff, and students in regard to a proposed measure for reducing the heating and air-conditioning expenses on campus.

	Favor	Indifferent	Opposed	Total
Faculty	36	42	122	200
Academic staff	44	77	129	250
Students	106	178	116	400

Compare the attitude patterns of the three groups by computing the relative frequencies.

2.5   Groundwater from 18 wells was classified as low or high in alkalinity and low or high in dissolved iron. There were 9 wells with high alkalinity, 6 that were high in iron, and 5 that were high in both.
   (a)   Based on these data, complete the following two-way frequency table.
   (b)   Calculate the relative frequencies of the cells.
   (c)   Calculate the relative frequencies separately for each row.

	Iron	
Alkalinity	Low	High
Low		
High		

2.6   Interviews with 185 persons engaged in a stressful occupation revealed that 76 were alcoholics, 81 were mentally depressed, and 54 were both.
   (a)   Based on these records, complete the following two-way frequency table.
   (b)   Calculate the relative frequencies.

	Alcoholic	Not Alcoholic	Total
Depressed			
Not depressed			
Total			

2.7 Cross tabulate the "Class data" of Exercise 7.2 in Chapter 2 according to gender (M, F) and the general areas of intended major (H, S, B, P). Calculate the relative frequencies.

2.8 A psychologist interested in obese children gathered data on a group of children and their parents.

	Child	
Parent	Obese	Not obese
At least one obese	12	24
Neither obese	8	36

(a) Calculate the marginal totals.
(b) Convert the frequencies to relative frequencies.
(c) Calculate the relative frequencies separately for each row.

2.9 Of the 1992 world's largest diversified companies, in terms of sales, 21 are from the United States or Japan. These companies can be further categorized as those having sales of $50 million or more and those having less than $50 million in sales.

	Sales Less Than $50 Million	Sales $50 Million or More
Japan	7	9
United States	4	1

(a) Calculate the marginal totals.
(b) Convert the frequencies to relative frequencies.
(c) Comment on the pattern.

2.10   A large research hospital and a community hospital are located in your area. The surgery records for the two hospitals are summarized in the table. The outcome is "survived" if the patient lived at least six weeks.

	Died	Survived	Total
Research Hospital	90	2110	2200
Community Hospital	23	677	700

(a)   Calculate the proportion of patients who survive surgery at each of the hospitals.

(b)   Which hospital do these data suggest you should choose for surgery?

2.11   Refer to Exercise 2.10. Not all surgery cases, even of the same type, are equally serious. Large research hospitals tend to get the most serious surgery cases, whereas community hospitals tend to get more of the routine cases. Suppose that patients can be classified as being in either "Good" or "Poor" condition and the outcomes of surgery are

Survival Rates by Condition

Good Condition	Died	Survived	Total
Research Hospital	15	685	700
Community Hospital	16	584	600
Total	31	1269	1300

Poor Condition	Died	Survived	Total
Research Hospital	75	1425	1500
Community Hospital	7	93	100
Total	82	1518	1600

(a)   Calculate the proportions who survive for each hospital and each condition.

(b)   From these data, which hospital would you choose if you were in good condition? If you were in bad condition?

(c)   Compare your answer with that to Exercise 2.10. Explain this reversal as an example of Simpson's paradox and identify the lurking variable in Exercise 2.10.

3. A DESIGNED EXPERIMENT FOR MAKING A COMPARISON

We regularly encounter claims that, as a group, smokers have worse health records than nonsmokers with respect to one disease or another, or that a new medical treatment is better than the former one. Recalling the discussion in Chapter 1, Section 7, properly designed experiments can often provide data that are so conclusive that a comparison is clear-cut. An example of a comparative study will illustrate the design issue and how to conduct an experiment.

During the early development of a medicated skin patch to help smokers break the habit, a test was conducted with 112 volunteers. Because of the so-called **placebo effect**, of people tending to respond positively just because attention is paid to them, half the volunteers were given an unmedicated skin patch. The persons receiving the unmedicated patch are called the **control (or placebo) group**. The data will consist of a count of the number of persons who are abstinent at the end of the study.

> *PURPOSE:* To determine the effectiveness of a medicated nicotine patch for smoking cessation based on the end-of-therapy numbers of abstinent persons in the medicated and unmedicated groups

What is involved in comparing two approaches or methods for doing something? First the subjects, or experimental units, must be assigned to the two groups in such a manner that neither method is favored. One approach is to list the subjects' names on a piece of paper, cut the paper into strips, each with one name on it, and then draw one at a time until half the names are drawn. Ideally, we like to have equal sample size groups, so if there is an odd number of subjects, draw just over one-half. These subjects are assigned to the first approach. The other subjects are assigned to the second approach. This step, called **random assignment**, helps guarantee a valid comparison. Any subject likely to respond positively has the same chance of supporting the first approach as supporting the second approach. When subjects cannot be randomly assigned, we will never know whether an observed difference in the numbers of abstinent smokers is due to the approaches or to some other variables associated with the particular persons assigned to the two groups.

In the quit-smoking experiment, subjects were randomly assigned to the medicated and unmedicated (placebo) groups. As with many medical trials, this was a **double blind trial.** That is, the medical staff in contact with the patients was also kept unaware of which patients were getting the treated patch and which were not. At the end of the study, the number of persons in each group who were abstinent and who were smoking were recorded.

The data[1] collected from this experiment are summarized in Table 6.

The proportion abstinent is 21/57 = .368 for the medicated skin patch group and only 11/55 = .200 for the control. The medicated patch seems to work. Later in Chapter 11, we verify that the difference .368 − .200 = .168 is greater than can be explained by chance variation.

In any application where the subjects might learn from the subjects before them, it would be a poor idea to perform all the trials for Treatment 1 and then

---

[1] "Two Studies of the Clinical Effectiveness of the Nicotine Patch with Different Counseling Treatments," M. Fiore, S. Kenford, D. Jorenby, D. Wetter, S. Smith, and T. Baker. *Chest*, Vol. 105 (1994), pp. 524–533.

TABLE 6   Quitting Smoking

	Abstinent	Smoking	
Medicated patch	21	36	57
Unmedicated patch	11	44	55
	32	80	112

all those for Treatment 2. Learning or other uncontrolled variables must not be given the opportunity to systematically affect the experiment. We could write the name Treatment 1 on 57 and Treatment 2 on 55 individual slips of paper. The 112 slips of paper should be mixed and placed in a pile and then drawn one at a time to determine the sequence in which the trials are conducted.

Exercises

3.1   With reference to the quit-smoking experiment, suppose two new subjects are available. Explain how you would assign one subject to receive the placebo and one to receive the medicated patch.

3.2   With reference to the quit-smoking experiment,

(a)   Suppose the placebo trials were ignored and you were told only that 21 of 57 were abstinent after using the medicated patches. Would this now appear to be stronger evidence in favor of the patches?

(b)   Explain why the placebo trials provide a more valid reference for results of the medicated patch trials.

4. SCATTER DIAGRAM OF BIVARIATE MEASUREMENT DATA

We now turn to a description of data sets concerning two measurement variables, so each produces data on a numerical scale. For ease of reference, we will label one variable x and the other y. Thus, two numerical observations (x, y) are recorded for each sampling unit. These observations are *paired* in the sense that an (x, y) pair arises from the same sampling unit. An x observation from one pair and an x or y from another are unrelated. One example is the number of doctors (x) and number of registered nurses (y) at area hospitals. A sample of size 3 consists of three pairs of values, for instance

$$(52, 126) \qquad (37, 82) \qquad (45, 73)$$

For n sampling units, we can write the measurement pairs as

$$(x_1, y_1), (x_2, y_2), \ldots, (x_n, y_n)$$

The set of x measurements alone, if we disregard the y measurements, constitutes a data set for one variable. The methods of Chapter 2 can be employed for descriptive purposes including graphical presentation of the pattern of distribution of the measurements, calculation of the mean, standard deviation, and other quantities. Likewise, the y measurements can be studied disregarding the x measurements. However, a major purpose of collecting bivariate data is to answer such questions as:

Are the variables related?

What form of relationship is indicated by the data?

Can we quantify the strength of their relationship?

Studying either the x measurements by themselves or the y measurements by themselves would not help answer these questions.

An important first step in studying the relationship between two variables is to graph the data. To this end, the variable x is marked along the horizontal axis and y on the vertical axis on a graph paper. The pairs (x, y) of observations are then plotted as dots on the graph. The resulting diagram is called a **scatter diagram.** By looking at the scatter diagram, we can form a visual impression about the relation between the variables. For instance, we can observe whether the points band around a line or a curve, or whether they form a patternless cluster.

Example 3   The scatter diagram—A visual display of relationship

Recorded in Table 7 are the data of

TABLE 7   Data of Undergraduate GPA x
and GMAT Score y

x	y	x	y	x	y
3.63	447	2.36	399	2.80	444
3.59	588	2.36	482	3.13	416
3.30	563	2.66	420	3.01	471
3.40	553	2.68	414	2.79	490
3.50	572	2.48	533	2.89	431
3.78	591	2.46	509	2.91	446
3.44	692	2.63	504	2.75	546
3.48	528	2.44	336	2.73	467
3.47	552	2.13	408	3.12	463
3.35	520	2.41	469	3.08	440
3.39	543	2.55	538	3.03	419
				3.00	509

$$x = \text{Undergraduate GPA}$$

and

$$y = \text{Score on the Graduate Management Aptitude Test (GMAT)}$$

for applicants seeking admission to a Masters of Business Administration program. Construct a scatter diagram.

SOLUTION AND DISCUSSION   The scatter diagram is plotted in Figure 1. The southwest-to-northeast pattern of the points indicates a positive relation between x and y. That is, the applicants with a high GPA tend to have a high GMAT score. Evidently, the relation is far from a perfect mathematical relation.

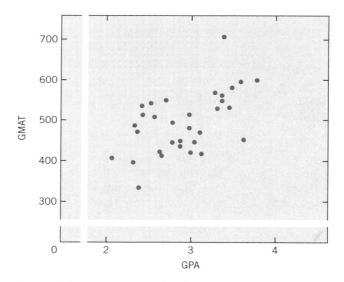

Figure 1. Scatter diagram of applicants' scores

Example 4   Multiple scatter diagram
Concern was raised by environmentalists that spills of contaminants were affecting wildlife in and around an adjacent lake. Estrogenic contaminants in the environment can have grave consequences on the ability of living things to reproduce. Researchers examined the reproductive development of young male alligators hatched from eggs taken from around (1) Lake Apopka, the lake that was contaminated, and (2) Lake Woodruff, which acted as a control. The contaminants were thought to influence sex steroid concentrations. The concentrations of two steroids, estradiol and testosterone, were determined by radioimmunoassay.

Lake Apopka

Estradiol	38	23	53	37	30
Testosterone	22	24	8	6	7

Lake Woodruff

Estradiol	29	64	19	36	27	16	15	72	85
Testosterone	47	20	60	75	12	54	33	53	100

(a) Make a scatter diagram of the two concentrations for the Lake Apopka alligators.

(b) Create a multiple scatter diagram by adding to the same plot the pairs of concentrations for the Lake Woodruff male alligators. Use a different symbol for the two lakes.

(c) Comment on any major differences between the two groups of male alligators.

SOLUTION AND DISCUSSION

(a) Figure 2(a) gives the scatter diagram for the Lake Apopka alligators.

(b) Figure 2(b) is the multiple scatter diagram with the points for Lake Woodruff marked as *B*.

(c) The most prominent feature of the data is that the male alligators from the contaminated lake have, generally, much lower levels of testosterone than those from the nearly pollution-free control lake. (The *A*'s are at the bottom third of the scatter diagram.) Low testosterone level in males has grave consequences regarding reproduction.

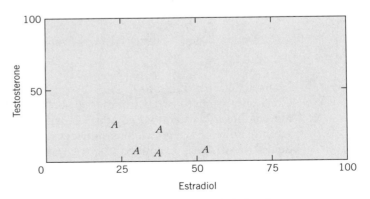

(a) Scatter diagram for Lake Apopka

Figure 2. Scatter diagrams. *A* = Lake Apopka. *B* = Lake Woodruff.

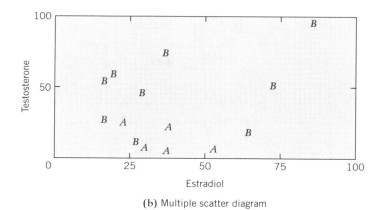

(b) Multiple scatter diagram

Figure 2. Continued

5. THE CORRELATION COEFFICIENT—A MEASURE OF LINEAR RELATION

The scatter diagram provides a visual impression of the nature of relation between the x and y values in a bivariate data set. In a great many cases, the points appear to band around a straight line. Our visual impression of the closeness of the scatter to a linear relation can be quantified by calculating a numerical measure called the correlation coefficient.

The correlation coefficient, denoted by r, is a measure of strength of the linear relation between the x and y variables. In Figure 1, there is a southwest-to-northeast pattern for r to quantify. Before introducing its formula, we outline some important features of the correlation coefficient and discuss the manner in which it serves to measure the strength of a linear relation.

1. The value of r is always between -1 and $+1$.

2. The magnitude of r indicates the strength of a linear relation, whereas its sign indicates the direction. More specifically,

    $r > 0$ if the pattern of (x, y) values is a band that runs from lower left to upper right.

    $r < 0$ if the pattern of (x, y) values is a band that runs from upper left to lower right.

    $r = +1$ if all (x, y) values lie exactly on a straight line with a positive slope (perfect positive linear relation).

    $r = -1$ if all (x, y) values lie exactly on a straight line with a negative slope (perfect negative linear relation).

    A high numerical value of r, that is, a value close to $+1$ or -1, represents a strong linear relation.

3. A value of r close to zero means that the linear association is very weak.

The correlation coefficient is close to zero when there is no visible pattern of relation; that is, the y values do not change in any direction as the x values change. A value of r near zero could also happen because the points band around a curve that is far from linear. After all, r measures linear association, and a markedly bent curve is far from linear.

Figure 3 shows the correspondence between the appearance of a scatter diagram and the value of r. Observe that (e) and (f) correspond to situations where $r = 0$. The zero correlation in (e) is due to the absence of any relation between x and y, whereas in (f) it is due to a relation following a curve that is far from linear.

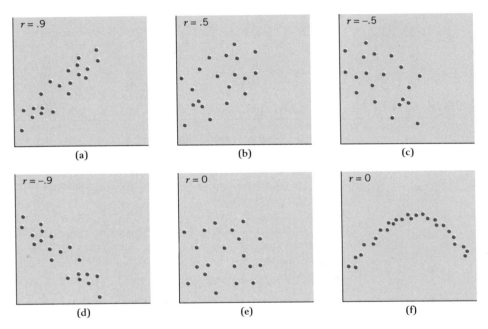

Figure 3. Correspondence between the values of r and the amount of scatter

CALCULATION OF r

The sample correlation coefficient is calculated from pairs of observations on two characteristics. When n pairs are available, we write

$$(x_1, y_1), \quad (x_2, y_2), \quad \ldots, \quad (x_n, y_n)$$

The correlation coefficient is best interpreted in terms of the standardized observations

$$\frac{\text{Observation} - \text{Sample mean}}{\text{Sample standard deviation}} = \frac{x_i - \bar{x}}{s_x}$$

where the subscript x on s distinguishes the sample standard deviation of the x observations, $s_x = \sqrt{\sum_{i=1}^{n} (x_i - \bar{x})^2 / (n-1)}$, from the sample standard deviation s_y of the y observations.

Since the difference $x_i - \bar{x}$ has the units of x and the sample standard deviation s_x also has the same units, the standardized observation is free of the units of measurement. The sample correlation coefficient r is just the sum of the products of the standardized x observation times the standardized y observation from each pair, divided by $n - 1$.

Sample Correlation Coefficient

$$r = \frac{1}{n-1} \sum_{i=1}^{n} \left(\frac{x_i - \bar{x}}{s_x} \right) \left(\frac{y_i - \bar{y}}{s_y} \right)$$

When most of the pairs of observations (x_i, y_i) are such that either both components are simultaneously above their sample means or both simultaneously below their sample means, the products of standardized values will tend to be positive and r will be positive. If one component of the pair tends to be large when the other is small, and vice versa, the correlation coefficient r will be negative.

We also present one alternative computational formula for r. The value of r may be calculated from n pairs of observations (x, y) according to the following formula, obtained by canceling the common term $n - 1$.

Alternative Formula for the Correlation Coefficient

$$r = \frac{S_{xy}}{\sqrt{S_{xx}}\,\sqrt{S_{yy}}}$$

where

$$S_{xy} = \Sigma (x - \bar{x})(y - \bar{y})$$
$$S_{xx} = \Sigma (x - \bar{x})^2, \qquad S_{yy} = \Sigma (y - \bar{y})^2$$

The quantities S_{xx} and S_{yy} are the sums of squared deviations of the x observations and the y observations, respectively. S_{xy} is the sum of cross-products of the x deviations with the y deviations. This formula will be examined in more detail in Chapter 12.

Example 5   Calculation of the sample correlation

Calculate r for the $n = 4$ pairs of observations.

$$(2, 5), \quad (1, 3), \quad (5, 6), \quad (0, 2)$$

SOLUTION AND DISCUSSION   We first determine the mean $\bar{x}$ and deviations $x - \bar{x}$, and then $\bar{y}$ and the deviations $y - \bar{y}$. See Table 8.

TABLE 8   Calculation of r

x	y	$x - \bar{x}$	$y - \bar{y}$	$(x - \bar{x})^2$	$(y - \bar{y})^2$	$(x - \bar{x})(y - \bar{y})$
2	5	0	1	0	1	0
1	3	-1	-1	1	1	1
5	6	3	2	9	4	6
0	2	-2	-2	4	4	4
Total 8	16	0	0	14	10	11
$\bar{x} = 2$	$\bar{y} = 4$			S_{xx}	S_{yy}	S_{xy}

Consequently,

$$r = \frac{S_{xy}}{\sqrt{S_{xx}}\sqrt{S_{yy}}} = \frac{11}{\sqrt{14}\sqrt{10}} = .930$$

The value .930 is large and it implies a strong association where both x and y tend to be small or both tend to be large.

It is sometimes convenient, when using hand-held calculators, to evaluate r using the alternative formulas for S_{xx}, S_{yy}, and S_{xy}.

$$S_{xx} = \Sigma\, x^2 - \frac{(\Sigma\, x)^2}{n}, \qquad S_{yy} = \Sigma\, y^2 - \frac{(\Sigma\, y)^2}{n}$$

$$S_{xy} = \Sigma\, xy - \frac{(\Sigma\, x)(\Sigma\, y)}{n}$$

This calculation is illustrated in Table 9.

TABLE 9   Alternative Calculation of r

x	y	x^2	y^2	xy
2	5	4	25	10
1	3	1	9	3
5	6	25	36	30
0	2	0	4	0
Total 8	16	30	74	43
$\Sigma\, x$	$\Sigma\, y$	$\Sigma\, x^2$	$\Sigma\, y^2$	$\Sigma\, xy$

$$r = \frac{43 - \dfrac{8 \times 16}{4}}{\sqrt{30 - \dfrac{8^2}{4}}\ \sqrt{74 - \dfrac{(16)^2}{4}}} = .930$$

■

We remind the reader that r measures the closeness of the pattern of scatter to a line. Figure 2(f) presents a strong relationship between x and y, but one that is not linear. The small value of r for these data does not properly reflect the strength of the relation. Clearly, r is not an appropriate summary of a curved pattern. Another situation where the sample correlation coefficient r is not appropriate occurs when the scatter diagram breaks into two clusters. Faced with separate clusters as depicted in Figure 4, it is best to try and determine the underlying cause. It may be that a part of the sample has come from one population and a part from another.

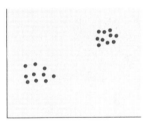

Figure 4. r is not appropriate—samples from two populations

CORRELATION AND CAUSATION

Data analysts often jump to unjustified conclusions by mistaking an observed correlation for a cause-and-effect relationship. A high sample correlation coefficient does not necessarily signify a causal relation between two variables. A classic example concerns an observed high positive correlation between the number of storks sighted and the number of births in a European city. It is hoped no one would use this evidence to conclude that storks bring babies, or worse yet, that killing storks would control population growth.

The observation that two variables tend to simultaneously vary in a certain direction does not imply the presence of a direct relationship between them. If we record the monthly number of homicides x and the monthly number of religious meetings y for several cities of widely varying sizes, the data will probably indicate a high positive correlation. It is the fluctuation of a third variable (namely, the city population) that causes x and y to vary in the same direction, despite the fact that x and y may be unrelated or even negatively related. The third variable, which in this example is actually causing the observed correlation between crime and religious meetings, is referred to picturesquely as a **lurking variable**. The false correlation that it produces is called **spurious correlation**. It is more a matter of common sense than of statistical reasoning to determine whether an observed correlation can be practically interpreted or is spurious.

> An observed correlation between two variables may be **spurious**. That is, it may be caused by the influence of a third variable.

When using the correlation coefficient as a measure of relationship, we must be careful to avoid the possibility that a lurking variable is affecting any of the variables under consideration.

Example 6   Correlation due to lurking variables
A scatter diagram of the total federal debt versus the number of golfers in the United States is shown in Figure 5(a). This diagram shows a strong correlation, with r greater than .99. Would cutting the number of golfers by taxing them or making the sport illegal reduce the federal debt? Comment.

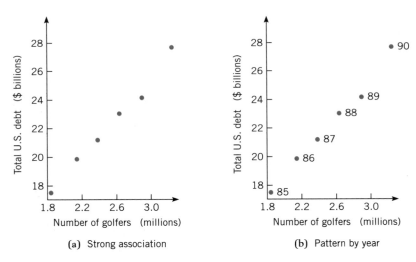

Figure 5. A scatter diagram of U.S. debt and number of golfers

SOLUTION AND DISCUSSION   We observe a strong association, but common sense suggests that there is no cause-and-effect relation between the federal debt and the number of golfers. In Figure 5(b) we have repeated the scatter diagram but have labeled each point according to the year. For example, 86 stands for 1986 and so on. The years march up the points from the lower left to the upper right of the diagram. Many things change over the course of a year, and the year is just a stand-in, or proxy, for all of them.

Adding the year notation to the points makes it clear that other variables are leading to the observed association. Such a graph can help discredit claims of causal relations. However, if the years had been scrambled along the curve represented by the points, then a causal relation might be indicated. ■

STATISTICAL REASONING

Car Safety Statistics and Lurking Variables

The Insurance Institute for Highway Safety regularly announces the safest and unsafest cars based on the number of fatalities for every 10,000 cars. One recent report named two Volvo models as the safest, followed by a Mercedes and then five luxury cars including Buick Riviera, Lexus S400, and Lincoln Town Car.

David Lissy/FPG International

The unsafest car list was headed by the Chevrolet Corvette, which was followed by the Pontiac LeMans and Ford Mustang. Next came the two-door Ford Fiesta and Ford Escort along with the Geo Metro. All of these cars are relatively small.

It must be acknowledged that there is some truth to the statement by the Institute that large cars are generally safer. However, there is more to the story — a point emphasized by talk show hosts on their national radio program.[2]

To make their point, they raised questions such as "How often does a teenager cruise in the family Volvo or other luxury car?" In short, the driver or aggressive driver behavior is a major lurking variable here. Besides a high negative correlation between driver age and aggressive driving behavior, the more mature drivers tend to own the cars on the safest list while younger drivers own those on the unsafe list. Even among drivers of the same age, there is a difference in typical driving behavior between Corvette owners and Town Car owners.

How much of the safety effect is due to driver and how much is due to car? To answer this question, we would need to perform a controlled experiment where a very large number of drivers of different ages are randomly assigned to cars. Some young drivers would likely be given Mercedes to drive while some middle-aged drivers would get Escorts and so on. Being unable to perform this experiment, we cannot eliminate driver as a lurking variable that is primarily responsible for the observed pattern of car fatalities.

Exercises

5.1   Would you expect a positive, negative, or nearly zero correlation between

(a)   Ability of a sales manager and amount of sales

(b)   Number of years of education and age of women at birth of first child

(c)   Number of years of education and age at first full-time job

5.2   If the value of r is small, can we conclude that there is not a strong relationship between the two variables?

5.3   There is a small negative correlation between divorce rate (the number of divorces in 1000 marriages) and the death rate (number of deaths per thousand persons) in each state. Can it be concluded that

(a)   States with a high divorce rate tend to have high death rates?

(b)   States with a high divorce rate tend to have low death rates?

5.4   Refer to Exercise 5.3 and the negative correlation between the divorce rate and death rate in a state.

(a)   Can it be concluded that if divorce rates can be increased, the death rates will drop?

[2]This interpretation was mentioned by Tom and Ray Magliozzi on their public radio program *Car Talk*.

(b) Explain how the age distribution in each state could be a lurking variable.

5.5 For the data set

x	0	1	6	3	5
y	4	3	0	2	1

(a) Construct a scatter diagram.
(b) Guess the sign and value of the correlation coefficient.
(c) Calculate the correlation coefficient.

5.6 Refer to the alligator data in Table D.7 of the Data Bank. Using the data on x_3 and x_4 for male and female alligators from Lake Apopka,

(a) Make a scatter diagram of the pairs of concentrations for the male alligators. Calculate the sample correlation coefficient.

(b) Create a multiple scatter diagram by adding, on the same plot, the pairs of concentrations for the female alligators. Use a different symbol for females. Calculate the sample correlation coefficient for this latter group.

(c) Comment on any major differences between the male and female alligators.

5.7 (a) Construct scatter diagrams of the data sets,

(i)

x	-1	3	1	5	2
y	2	4	0	6	3

(ii)

x	-1	3	1	5	2
y	6	0	3	2	4

(b) Calculate r for the data set (i).
(c) Guess the value of r for the data set (ii) and then calculate r. (*Note:* The x and y values are the same for both sets, but they are paired differently in the two cases.)

5.8 Match the following values of r with the correct diagrams in Figure 6 on page 120.
(a) $r = -.3$ (b) $r = .1$ (c) $r = .9$

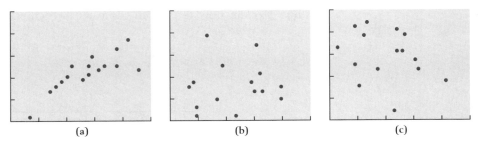

Figure 6. Scatter diagrams

5.9   Is the correlation in Figure 7 about (a) .1, (b) .5, (c) .9, or (d) − .7?

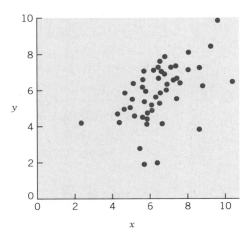

Figure 7. Scatter diagram

5.10   Calculations from a data set of $n = 48$ pairs of (x, y) values have provided the following results.

$$\Sigma (x - \bar{x})^2 = 260.2, \quad \Sigma (y - \bar{y})^2 = 403.7, \quad \Sigma (x - \bar{x})(y - \bar{y}) = 298.8$$

Obtain the correlation coefficient.

5.11   For a data set of (x, y) pairs, one finds that

$$n = 26, \qquad \Sigma x = 1287, \qquad \Sigma y = 1207$$
$$\Sigma x^2 = 66,831, \qquad \Sigma y^2 = 59,059, \qquad \Sigma xy = 62,262$$

Calculate the correlation coefficient.

5.12   Heating and combustion analyses were performed to study the composition of moon rocks collected by Apollo 14 and 15 crews. Recorded here are the determinations of hydrogen (H) and carbon (C) in parts per million (ppm) for 11 specimens. [*Source:* U.S. Geological Survey, *Journal of Research* **2** (1974).]

Hydrogen (ppm)	120	82	90	8	38	20	2.8	66	2.0	20	85
Carbon (ppm)	105	110	99	22	50	50	7.3	74	7.7	45	51

Calculate r.

5.13 Over the years, a traffic officer noticed that cars with fuzzy dice hanging on the rear-view mirror always seemed to be speeding. Perhaps tongue in cheek, he suggested that outlawing the sale of fuzzy dice would reduce the number of cars exceeding the speed limit. Comment on lurking variables.

5.14 Recorded here are the scores of 16 students at the midterm and final examinations of an intermediate statistics course.

Midterm	81	75	71	61	96	56	85	18
Final	80	82	83	57	100	30	68	56

Midterm	70	77	71	91	88	79	77	68
Final	40	87	65	86	82	57	75	47

(a) Plot the scatter diagram and identify any unusual points.
(b) Calculate r.

5.15 In each of the following instances, would you expect a positive, negative, or zero correlation?
(a) Number of salespersons and total dollar sales for real estate firms
(b) Total payroll and percent of wins of National League baseball teams
(c) The amount spent on a week of TV advertising and sales of a cola
(d) Age of adults and their ability to maintain a strenuous exercise program

5.16 In an experiment to study the relation between the dose of a stimulant and the time a stimulated subject takes to respond to an auditory signal, the following data were recorded.

Dose (milligrams)	1	3	4	7	9	12	13	14
Reaction time (seconds)	3.5	2.4	2.1	1.3	1.2	2.2	2.6	4.2

(a)   Calculate the correlation coefficient.

(b)   Plot the data and comment on the usefulness of r as a measure of relation.

5.17   The following table gives the federal deficit (billions of dollars) and the number of golfers (millions) in the United States.

Year	Deficit (billion $)	Golfers (million)
1985	17.5	1.81
1986	19.9	2.12
1987	21.2	2.35
1988	23.0	2.60
1989	24.2	2.87
1990	27.8	3.21

Source: Statistical Abstract of the United States 1992.

Determine the correlation between number of golfers (in millions) and the federal deficit (in billions of dollars).

6. DESIGNING COMPARATIVE EXPERIMENTS—STUDENT PROJECTS

The quit-smoking example illustrated an experiment where the experimental units, subjects who wished to quit smoking, were divided into two groups. Members of one group received the treatment with a medicated patch and the members of the other group received the unmedicated patch. Let us review the key features of the design in another context. Suppose we wish to compare two different car speaker systems and the eight cars pictured in Figure 8 are available.

The cars are already numbered 1 to 8 so we could choose cars to receive speaker system 1 by a blindfolded selection of four cards from among eight cards

Figure 8. Eight cars available for experiment

that are numbered 1 through 8. Those cars corresponding to the numbers drawn would have speaker system 1 installed and the other cars would have the second speaker system installed. This design is called the **two sample design** and it is similar to the smoking patch study. Each experimental unit, here a car, receives one of the two treatments (speaker systems). The assignment is done in an impartial blindfolded manner.

There is, however, a second experimental design option called a **matched pair design.** The size of the car may influence which of the two speaker systems is better. An inspection of the available cars shows that numbers 1 and 2 are compacts, numbers 3 and 4 are mid-sized, numbers 5 and 6 are subcompacts, and numbers 7 and 8 are large-size cars. Having identified a factor that will likely affect speaker performance, we could remove this factor by pairing the two cars of the same size to obtain four matched pairs. One of each pair of cars would get speaker system 1 and the other car would get system 2. The choice should be made in an impartial manner by a separate flip of a coin for each pair. If heads, then the odd-numbered car in the pair would get system 1. Otherwise the even-numbered car would get system 1. The speaker systems would be judged on a pair by pair basis. We could either record which method is better for the pair or even assign a number for the difference in quality.

When you design your own comparative experiment, you need to decide whether to use a matched pair comparison or a two sample design. The following two student design project descriptions illustrate the design issues that will arise in your experiment.

Student Design Project 1.   Paul is an avid golfer and he wanted to know whether his new and expensive brand P putter was better than his low-cost generic G model. He decided to make the same putts with each of the two putters. That is, Paul selected a matched pair design. Because it was winter, the golf courses were snow covered so he putted inside by trying several locations at different directions and distances from the hole.

At each location, Paul decided somewhat arbitrarily to take three shots with one putter and then three shots with the other. The choice of which putter to use first was determined by a coin flip at each position. A putt is sunk if the ball goes in the hole.

TABLE 10   Numbers of Putts Sunk with Brand P and Brand G Putters

	Location					
	1	2	3	4	5	6
Brand P	2	3	2	2	3	2
Brand G	3	1	1	2	1	1
Difference	−1	2	1	0	2	1

Paul recorded the pairs of values for the number of putts sunk

$$(P, G) = (\text{No. sunk with brand P, No. sunk with brand G})$$

at each location. The difference $= P - G$ was also calculated as shown in Table 10 for the first six locations.

A histogram of the differences, from the complete experiment with a total of 30 locations, is shown in Figure 9.

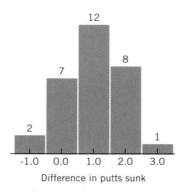

Figure 9. Histogram of differences $P - G$ of number of putts sunk

There are positive and negative differences and they are centered a little above zero. The conclusion that putter P is better is not clear-cut. In later chapters, we will learn how to quantify the amount of difference and the strength of evidence when the data exhibit variation.   ■

Student Design Project 2.   After reading about a nationally publicized study on how listening to music can aid learning, Beth decided to conduct her own experiment with twenty students. She made up a list of ten objects for them to memorize while listening to music. Then, after fifteen minutes, they were asked to write down all of the objects they could remember. Beth wanted to compare the memorization results when listening to classical music, represented by a Mozart compact disc, and rock music, a Rolling Stones compact disc.

The list of objects could not really be used twice with the same person so she selected a two sample design. Beth wrote 1 on ten slips of paper and 2 on another ten. Each of the twenty persons (subjects) drew a slip and those with 1 listened to Mozart and those with 2 heard the Rolling Stones while memorizing the list. This accomplished the randomization of treatments—the type of music—to the subjects.

Beth recorded the number of objects correctly recalled by each subject. These are shown in the combined dot diagram in Figure 10.

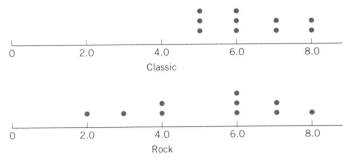

Figure 10. Number of objects recalled

Randomly assigning treatments to subjects or experimental units with cards or slips of paper is a key step. It helps to prevent uncontrolled variables from exerting a systematic effect on the response. For instance, if the better memorizers preferred rock music and were allowed to select the music in student project 2, then rock music would likely fare better in the experiment even if music had no effect.

Any comparative experiment conducted for a student project should include the following steps.

Designing a Comparative Experiment

1. Give a statement of purpose.
2. Select a response to measure.
3. Decide between (i) a matched pair design and (ii) a two sample design.
4. Identify the experimental units (subjects) and treatments. Randomize the assignment of treatments to the units.

Exercises

6.1 In the golf experiment, comment on why it is a good idea
   (a) to have a warm-up period before starting.
   (b) to conduct the experiment over a period of several days.

6.2 In the music and memorizing experiment,
   (a) comment on why there should be a fixed length of time for subjects to memorize the list.

(b) comment on why there should be a fixed waiting period before having subjects write down the objects they remember.

(c) would you expect a different general outcome for different choices of classical and rock music? Why or why not?

KEY IDEAS

Cross-classified data can be described by calculating the relative frequencies.

The **correlation coefficient** r measures how closely the scatter approximates a straight-line pattern.

A positive value of correlation indicates a tendency of large values of x to occur with large values of y, and also for small values of both to occur together.

A negative value of correlation indicates a tendency of large values of x to occur with small values of y and vice versa.

A high correlation does not necessarily imply a causal relation.

Designed experiments are useful for comparing two treatments. There are **matched pair** and **two sample** designs available.

A **control** or **placebo** group provides a baseline reference.

KEY FORMULAS

For pairs of measurements (x, y), the **correlation coefficient** is defined in terms of the standardized values $(x - \bar{x})/s_x$ and $(y - \bar{y})/s_y$, where s_x is the standard deviation of the x values and s_y is the standard deviation of the y values.

$$r = \frac{1}{n-1} \sum_{i=1}^{n} \left(\frac{x - \bar{x}}{s_x} \right) \left(\frac{y - \bar{y}}{s_y} \right)$$

An alternative calculation formula is

$$\text{sample correlation} \quad r = \frac{S_{xy}}{\sqrt{S_{xx}} \sqrt{S_{yy}}}$$

where $S_{xx} = \Sigma (x - \bar{x})^2$, $S_{yy} = \Sigma (y - \bar{y})^2$, and $S_{xy} = \Sigma (x - \bar{x})(y - \bar{y})$.

7. REVIEW EXERCISES

7.1 Applicants for welfare are allowed an appeals process when they believe they have been unfairly treated. At the hearing, the applicant may choose self-representation or representation by an attorney. The appeal may result in an increase, decrease, or no change in benefit recommendation. Court records of 320 appeals cases provided the following data.

Type of Representation	Amount of Aid			Total
	Increased	Unchanged	Decreased	
Self	59	108	17	
Attorney	70	63	3	
Total				

Calculate the relative frequencies for each row and compare the patterns of the appeals decisions between the two types of representation.

7.2 Table 11 gives the numbers of civilian employed persons by major occupation group and gender for the years 1983 and 1993. (*Source:* U.S. Bureau of Labor Statistics, *Employment and Earnings.*)

TABLE 11 Numbers of Civilian Employed Persons in the United States (in Thousands)

Occupation	1983		1993	
	Male	Female	Male	Female
Managerial/Professional	13,933	9,659	16,839	15,441
Technical/Sales	11,078	20,187	13,311	23,503
Service	5,530	8,326	6,688	9,833
Precision production	11,328	1,000	12,185	1,141
Operators/Laborers	11,809	4,282	12,862	4,176
Farming/Forest/Fishing	3,108	592	2,814	512
Total	56,786	44,046	64,699	54,606

(a) Consider females only for each year separately. Calculate the percentages of females in the various occupation groups and compare between 1983 and 1993.

(b) For each occupation group, calculate the percentage of employees who are female and compare between 1983 and 1993.

7.3 To study the effect of soil condition on the growth of a new hybrid plant, saplings were planted in three types of soil and their subsequent growth classified in three categories.

	Soil Type		
Growth	Clay	Sand	Loam
Poor	16	8	14
Average	31	16	21
Good	18	36	25
Total	65	60	60

Calculate the appropriate relative frequencies and compare the quality of growth for different soil types.

7.4   A high-risk group of 1083 male volunteers were included in a major clinical trial for testing a new vaccine for type B hepatitis. The vaccine was given to 549 persons randomly selected from the group, and the others were injected with a neutral substance (placebo). Eleven of the vaccinated people and 70 of the nonvaccinated ones later got the disease. (*Source: Newsweek*, Oct. 13, 1980.)

(a)   Present these data in the following two-way frequency table.

(b)   Compare the rates of incidence of hepatitis between the two subgroups.

	Hepatitis	No Hepatitis	Total
Vaccinated			
Not vaccinated			
Total			

7.5   Given the following (x, y) values

x	1	3	6	5	2	7
y	5	4	4	2	7	2

(a)   Plot the scatter diagram.

(b)   Calculate the correlation coefficient.

7.6   Calculate r for the data using both formulas.

x	9	3	5	0	8
y	4	2	3	1	3

7.7   Calculating from a data set of 20 pairs of (x, y) values, one obtains

$$\Sigma x = 156, \qquad \Sigma y = 1178$$
$$\Sigma x^2 = 1262, \qquad \Sigma y^2 = 69{,}390, \qquad \Sigma xy = 9203$$

Find the correlation coefficient.

7.8   As part of a study of the psychobiological correlates of success in athletes, the following measurements (courtesy of W. Morgan) are obtained from members of the U.S. Olympic wrestling team.

Anger x	6	7	5	21	13	5	13	14
Vigor y	28	23	29	22	20	19	28	19

(a)   Plot the scatter diagram.
(b)   Calculate r.

7.9   The tar yield of cigarettes is often assayed by the following method: A motorized smoking machine takes a two-second puff once every minute until a fixed butt length remains. The total tar yield is determined by laboratory analysis of the pool of smoke taken by the machine. Of course, the process is repeated on several cigarettes of a brand to determine the average tar yield. Given here are the data of average tar yield and the average number of puffs for six brands of filter cigarettes.

Average tar (milligrams)	12.2	14.3	15.7	12.6	13.5	14.0
Average no. of puffs	8.5	9.9	10.7	9.0	9.3	9.5

(a)   Plot the scatter diagram.
(b)   Calculate r.

Remark: Fewer puffs taken by the smoking machine means a faster burn time. The amount of tar inhaled by a human smoker depends largely on how often the smoker puffs.

7.10   A director of student counseling is interested in the relationship between the numerical score x and the social science score y on college qualification tests. Data on 36 students (courtesy of R. W. Johnson) are recorded.

x	41	39	53	67	61	67	46	50	55	72	63	59
y	29	19	30	27	28	27	22	29	24	33	25	20

(continued)

x	53	62	65	48	32	64	59	54	52	64	51	62
y	28	22	27	22	27	28	30	29	21	36	20	29

x	56	38	52	40	65	61	64	64	53	51	58	65
y	34	21	25	24	32	29	27	26	24	25	34	28

(a)   Plot the scatter diagram.

(b)   Calculate r. (Use of a computer is recommended; see Exercise 7.13.)

7.11   Would you expect a positive, negative, or nearly zero correlation for each of the following? Give reasons for your answers.

(a)   The weight of an automobile and the average number of miles it gets per gallon of gasoline

(b)   Intelligence scores of husbands and wives

(c)   The age of an aircraft and the proportion of time it is available for flying (part of the time it is grounded for maintenance or repair)

(d)   Stock prices for IBM and General Motors

(e)   The temperature at a baseball game and beer sales

7.12   Examine each of the following situations and state whether you would expect to find a high correlation between the variables. Give reasons why an observed correlation cannot be interpreted as a direct relationship between the variables and indicate possible lurking variables.

(a)   Correlation between the data on the incidence rate x of cancer and per capita consumption y of beer, collected from different states

(b)   Correlation between the police budget x and the number of crimes y recorded during the last 10 years in a city like Houston, Texas

(c)   Correlation between the gross national product x and the number of divorces y in the country recorded during the last 10 years

(d)   Correlation between the concentration x of air pollutants and the number of riders y on public transportation facilities when the data are collected from several cities that vary greatly in size

(e)   Correlation between the wholesale price index x and the average speed y of winning cars in the Indianapolis 500 during the last 10 years

🖥 USING A COMPUTER

💾 7.13   *Computer-aided statistical calculations.*   The MINITAB commands for scatter diagram, and correlation coefficient are illustrated along with the computer output.

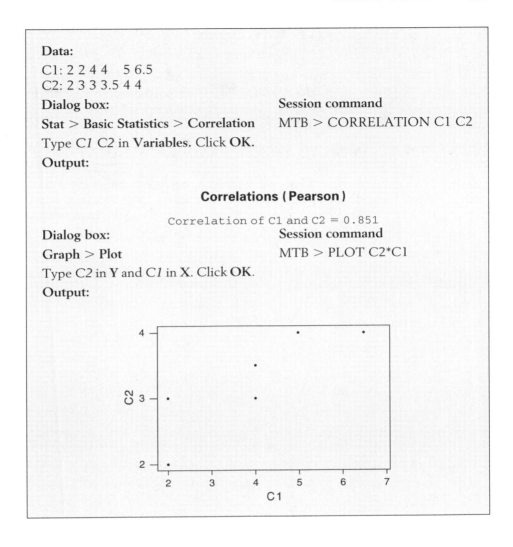

Use MINITAB (or another package program) to obtain the scatter diagram and the correlation coefficient for:

(a) The GPA and GMAT scores data of Table 7 in Example 3

(b) The hydrogen x and carbon y data in Exercise 5.12

USING A COMPUTER

7.14 Describe the association between marine and freshwater growth of salmon given in Table D.6 in the Data Bank. Use MINITAB or some other computer package to obtain the

(a)   Scatter diagram

(b)   Correlation coefficient

■ 7.15   Use MINITAB or some other computer package to obtain the scatter diagram and the correlation coefficient for the final and initial number of sit-ups given in Table D.5 in the Data Bank.

ACTIVITIES TO IMPROVE UNDERSTANDING: STUDENT PROJECTS

Refer to Section 6. Propose a project for comparing two ways of doing something. This could concern one of your regular recreational activities or a task you must perform on a regular basis. Advertisers regularly suggest comparisons of products.

(a)   Design the project following the four steps set out in Section 6.

(b)   Conduct the experiment and construct a graphical display to summarize the data. In later chapters, we learn more about interpreting your data.

CHAPTER 4

Probability—The Basis for Inference

Chapter Objectives

After reading this chapter, you should be able to:

▶ Specify the sample space associated with an experiment and determine whether it is discrete or continuous.
▶ Identify events as collections of elementary outcomes.
▶ Discuss probability as a numerical measure of uncertainty and how it can be assessed using the equally likely uniform model or long-run relative frequency.
▶ Define the complement of an event, the union of two events, and the intersection of two events.
▶ Calculate probabilities using the law of the complement, the addition law, and the multiplication law.
▶ Discuss and compute conditional probabilities.
▶ Determine whether two events are independent and be able to assign probabilities according to the product rule for independence.

1. INTRODUCTION AND OVERVIEW OF PROBABILITY

In Chapter 1, we distinguished between *sample* and the *statistical population* from which it was selected. While the sample can be described by the methods presented in Chapter 2, the primary objective of sampling is to generalize from the sample to make conclusions about the population. It is the ideas of probability that provide the basis for these generalizations or inferences.

In everyday conversations, we intuitively speak quantitatively when we use expressions such as "most likely," "fifty–fifty chance," and "unlikely" to indicate the chance an event will occur. Probability, as a subject, provides approaches to quantifying uncertainty regarding the occurrence of specific events. Any event is assigned a probability on a scale from 0 to 1. A very low value indicates that the event is very unlikely to occur, whereas a high value near 1 means that it is very likely to occur.

We will use several examples to link many ideas from probability and give an overview of the material discussed in this chapter.

Example 1   Random choice and chocolate chip cookies
Chris takes two cookies from a cookie jar that contains 18 plain and 2 chocolate chip cookies. Suppose Chris takes the two chocolate chip cookies and then claims to have chosen the two cookies at random. Do you believe this claim?

SOLUTION AND DISCUSSION   The probability of obtaining two chocolate chip cookies, when the selection is random, appears to be very small but we cannot yet give a numerical value. Granted that the probability is small, we are inclined not to believe Chris's claim.

The question of how to obtain the numerical value of the probability is addressed in the next two examples.   ■

In Example 1, suppose we use a red ball to represent a chocolate chip cookie and a white ball to represent a plain cookie. The probability of selecting two chocolate chip cookies, when they are chosen at random, is the same as the solution to the **urn model** problem in Example 2.

Example 2   Recasting the cookie problem as an urn model problem
Consider an urn consisting of 2 red (R) balls and 18 white (W) balls. Two balls will be selected in a blindfolded, or random, manner. The event of interest, A, is that the two selected balls are red. Find the probability that the event A will occur.

SOLUTION AND DISCUSSION   Let $P(A)$ be the probability of the event A. What are the possible ways of obtaining the numerical value of $P(A)$? We will outline three general approaches that are not only useful in obtaining the value of $P(A)$, but that also apply to many other problems.

Three Solutions to the Urn Model Problem

SOLUTION 1   **Long-Run Relative Frequency Approach to Specifying a Probability**   One can repeat the whole experiment, of choosing two balls at random from the urn with 2 R's and 18 W's, many times. Then the relative frequency of the event that two R's were selected can be used as an approximation of $P(A)$. For example, if the experiment is repeated 1000 times and the event occurs 5 times, then $P(A)$ should be reasonably close to the relative frequency $\frac{5}{1000} = .005$. The probability here is interpreted as a long-run relative frequency.

SOLUTION 2   **Counting the Outcomes, a Sample Space Approach to Specifying a Probability**   In many situations, including the urn model problem here, there are certain symmetries in the problem that we can use to derive the value of $P(A)$ objectively. For example, we could label the red balls as R_1 and R_2 and the white balls as W_1, W_2, . . . , W_{18}. Now, each ball has a different label. We can then list all the possible ways (outcomes) of obtaining two balls out of these twenty balls. For example, the outcome $W_7 R_2$ means that the first ball selected is W_7 and the second ball is R_2. The set of all possible outcomes is called the **sample space**. Since the two balls are chosen at random, each outcome has the same probability (that is, each outcome is equally likely to occur). Hence, $P(A)$ can be obtained as the proportion

$$P(A) = \frac{\text{Number of outcomes in } A}{\text{Total number of possible outcomes}}$$

We will show how the counting is done in a simpler urn model in Example 7 in the next section. For the moment, observe that the event A here has two outcomes, $R_1 R_2$ and $R_2 R_1$. Using a counting method to be described later, we can show that the number of possible outcomes is $20 \times 19 = 380$. Hence, $P(A) = 2/380 = .00526$. As anticipated, the event A is unlikely to occur.

Note that the formula for $P(A)$ given above applies only to the equally likely outcome case. For instance, suppose that all the red balls are labeled R and all the white balls are labeled W. Obviously, the probability of drawing a W is $\frac{18}{20}$, not $\frac{1}{2}$, because there are 18 W's in this urn with 20 balls. Even though the sample space in this case contains only two possible outcomes, red ball and white ball, the outcomes are not equally likely. The formula for $P(A)$, for the equally likely case, is not applicable under this second specification of the sample space.

SOLUTION 3   **A Probability Law Approach for Specifying a Probability**   Here we use another objective approach to obtain the value of $P(A)$. In doing so, we explain how a certain probability law is developed for an urn model problem.

The event A can be described as "the 1st ball is R and the 2nd ball is R" when two balls are drawn one at a time and without replacement. For A to occur, the event "the 1st ball is R" must occur. Intuitively, the probability of this occurrence is

$$P[\text{the 1st ball is R}] = \frac{2}{20}$$

because a ball is chosen at random from twenty balls, of which two are R's. We know that $\frac{2}{20}$ is too high for $P(A)$ because even if the first ball is R, the second ball is not necessarily R. Thus, $P(A)$ can be only a certain fraction of the probability $\frac{2}{20}$. We reason that given that the first ball is R, only one R and 18 W's are left in the urn. That is, the probability that the second ball is R given that one red ball has already been taken out, is $\frac{1}{19}$. That is, only $\frac{1}{19}$ of the probability $\frac{2}{20}$ (of obtaining a red ball first) describes the occurrence of the event $A = [$ 1st ball is R and 2nd ball is R $]$. In other words,

$P[$ the 1st ball is R and 2nd ball is R $]$

$$= P[\text{the 1st ball is R}]P[\text{2nd ball is R given that first ball is R}]$$

$$= \frac{2}{20} \times \frac{1}{19} = .00526$$

This answer agrees with Solution 2 above, which used a counting method.

The first equality between probabilities, in the last expression above, is a special case of the **multiplication law for probabilities.** Later, we will discover that the probability of an event can often be derived in an objective manner similar to this reasoning.

The third solution, based on the laws of probability, will usually be the preferable method for obtaining the probability of an event.   ■

Next, we consider an example in a different context, but which has the same probability problem as the one discussed in Example 2.

Example 3   Defective answering machines

Two digital answering machines are selected at random from 20 that are available for inspection. If both of the answering machines selected are found to be defective, can we believe the manager's claim that these are the only defectives among the 20 available?

SOLUTION AND DISCUSSION   On the surface, this problem appears to be quite different from Example 2. However, suppose we call a defective machine "red" and a nondefective machine "white." Then the probability of obtaining two defective machines, in a random selection of two machines out of the 20 machines when only two are defective, is exactly the urn model problem discussed in Example 2. The probability is still .00526.

If the store manager's claim is correct, we are unlikely to observe two defective machines in our random selection. We are faced with two possible conclusions:

1.   The manager's claim is correct and we just happened to observe an unlikely event; or

2.   The manager's claim is not correct. There are more defective machines among those not inspected.

Most people would not believe the first conclusion because of the extremely low probability of observing two defective machines.   ■

Examples 1–3 demonstrate the importance of being able to obtain probabilities from a generic model, such as the urn model. The same solution, or a similar argument, can readily be applied in a wide variety of practical situations having a similar structure provided the correspondence with red and white balls can be identified.

Not all problems can be cast as sampling without replacement from an urn. The next example describes another important type of problem.

Example 4   Using probability to judge a hypnotist's evidence
Suppose it has been observed that, half of the time, a certain type of muscular pain goes away by itself. A hypnotist claims that her method is effective in relieving the pain. To provide experimental evidence, she hypnotizes 15 patients and 12 report that the pain has stopped. Does this demonstrate that hypnotism is effective in stopping the pain?

SOLUTION AND DISCUSSION   Let us scrutinize the claim and evidence from a statistical point of view. If indeed hypnotism had nothing to offer, there would be a 50–50 chance that a patient would be cured. Observing 12 cures among 15 patients amounts to getting 12 heads in 15 tosses of a fair coin. Later, we will see that the probability of at least 12 heads is .018, indicating that the event is very unlikely to happen.

If we tentatively assume the model (or hypothesis) that her method of hypnotism is ineffective, then 12 or more cures is very unlikely. Rather than agree that an unlikely event has occurred, we conclude that the experimental evidence strongly supports the hypnotist's claim.   ■

The probability of the event of interest, 12 or more patients cured, in Example 4 was translated to the probability of an event relating to a coin-tossing experiment. In fact, many probability problems have common structures that allow them to be recast into questions relating to coin tossing. Familiarity with these models enables us to understand how to compute probabilities in many practical problems. Moreover, the basic generic models, urn models or coin tossing, can sometimes provide an explanation of the chance mechanism that generates the data.

This introduction, including the four examples, shows why an understanding of basic probability is very useful in drawing meaningful inferences and providing solutions to many practical problems involving sampling. There is also some complexity to this subject and in the rest of the chapter we expand our discussion of the concepts and methods mentioned in our overview. This material is important for understanding the development of the many useful inferential statistical procedures that are the core subject of this book.

2. THE PROBABILITY OF AN EVENT

The probability of an event is a numerical value that measures the chances that the event will occur. The chances of the event occurring must be judged against all possible outcomes of an experiment, survey, or data-gathering procedure. This will require a careful listing of all possible outcomes.

SAMPLE SPACE AND OUTCOMES

We first define **experiment** in a broad sense as any data-gathering procedure. It may be checking historical records in the library, conducting a survey of students regarding satisfaction with housing, or performing a chemical experiment in the laboratory.

An **experiment** is the process of observing a phenomenon that has variation in its outcomes.

The experiment in Example 1 consisted of recording whether each of the two cookies, selected from the 20 in the jar, was chocolate chip or plain. Example 3 described the selection of two answering machines from 20 and recording their status, defective or nondefective. The experiment in Example 4 concerned the number of persons who received pain relief after being hypnotized.

For ease of reference, we define a number of terms for describing a wide range of situations and experiments where the outcome cannot be predicted with certainty. We have briefly mentioned many of these terms in the introduction.

The **sample space** associated with an experiment is the collection of all possible distinct outcomes of the experiment.

Each outcome is called an **elementary outcome, a simple event,** or an **element of the sample space.**

An **event** is the set of elementary outcomes possessing a designated feature.

The elementary outcomes, which together comprise the sample space, are the ultimate breakdown of the potential results of the experiment. For general discussion, we write

S for the sample space

A, B, and so on, for events.

> An event A occurs when any one of the elementary outcomes in A occurs.

Example 5   Events for coin tossing
Toss a coin twice and record the outcome head (H) or tail (T) for each toss. Let A denote the event of getting exactly one head, and B the event of getting no heads at all. List the sample space and give the compositions of A and B.

SOLUTION AND   For two tosses of a coin, the elementary outcomes can be conveniently identified
DISCUSSION   by means of a **tree diagram**.

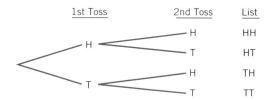

The sample space can then be listed as S = { HH, HT, TH, TT }. The order in which the elements of S are listed is inconsequential. It is the collection that matters.

Consider the event A of getting exactly one head. Scanning the above list, we see that only the elements HT and TH satisfy this requirement. Therefore, the event A has the composition

$$A = \{HT, TH\}$$

which is, of course, a subset of S. The event of getting no heads consists of a single elementary outcome, so B = { TT }. That is, B is a simple event, or elementary outcome, as well as an event. The term "event" is general and includes both cases. ∎

Example 6   Assigning probabilities to choices of pasta—Menu planning
A restaurant near campus serves three types of pasta (spaghetti, rigatoni, and fettucine) with one of two sauces (tomato or white). A customer will order a pasta dish.

(a)   What is the sample space of possible customer selections?

(b)   List the outcomes for the following events:

$$A = [\text{A spaghetti dish}]$$
$$B = [\text{A white-sauce dish}]$$

(c)   How should the owner go about determining the probability that a customer will order any given dish on the menu?

For the pasta, let S = spaghetti, R = rigatoni, and F = fettucine, and for the sauce let T = tomato and W = white. A dish with rigatoni and tomato sauce will be denoted by RT, and so on.

(a) The sample space of possible customer selections is

$$\{ST, RT, FT, SW, RW, FW\}$$

This sample space has six possible outcomes.

(b) $$A = \text{spaghetti dish} = \{ST, SW\}$$
$$B = \text{white-sauce dish} = \{SW, RW, FW\}$$

(c) If the customer always chooses a pasta dish at random, then each dish would have the same probability of being chosen, namely $\frac{1}{6}$. However, most customers do not choose their pasta dish at random. The restaurant owner should look at the history of the number of dishes of each type that were ordered. The relative frequency of any dish can serve as an approximation to the probability of selecting that particular dish. Most likely, the relative frequencies would not even be close to being uniform across the pasta dishes. In fact, one or two may be ordered by the vast majority of customers.

Probabilities should be assigned on the basis of an understanding of the context of a problem, not on mathematical convenience. Don't treat the outcomes as equally likely when they clearly are not. ■

When a potential outcome can be any point within an interval, the collection of all outcomes is called a **continuous sample space**. In contrast, the sample spaces in Examples 1–6 are all finite sample spaces. For an outcome that is a count that has no upper limit, we could take $\mathcal{S} = \{0, 1, \ldots\}$. Along with other sample spaces where the outcomes can be arranged in a sequence, each of these is called a **discrete sample space**. We will develop the ideas of probability in the context of finite sample spaces.

The most intuitive quantification of the idea of the probability of an event is to consider the number of times the event would occur if the experiment were repeated many times.

The **probability of an event** is a numerical value that represents the proportion of times the event is expected to occur when the experiment is repeated under identical conditions.

The probability of event A is denoted by $P(A)$.

Since a proportion must lie between 0 and 1, the probability of an event is a number between 0 and 1. To explore a few other important properties of probability, let us refer to the experiment in Example 5 of tossing a coin twice. The event A of getting exactly one head consists of the elementary outcomes HT and TH. Consequently, A occurs if either of these occurs. Because

$$\begin{bmatrix} \text{Proportion of times} \\ A \text{ occurs} \end{bmatrix} = \begin{bmatrix} \text{Proportion of times} \\ \text{HT occurs} \end{bmatrix} + \begin{bmatrix} \text{Proportion of times} \\ \text{TH occurs} \end{bmatrix}$$

the number that we assign as $P(A)$ must be the sum of the two numbers $P(\text{HT})$ and $P(\text{TH})$. Guided by this example, we state some general properties of probability.

The probability of an event is the sum of the probabilities assigned to all the elementary outcomes contained in the event.

Next, since the sample space S includes all conceivable outcomes, in every trial of the experiment some element of S must occur. Viewed as an event, S is certain to occur, and therefore its probability is 1.

The sum of the probabilities of all the elements of S must be 1.

To express the general case, we write e for a generic elementary outcome. In summary, any assignment of probability must obey certain rules.

Probability must satisfy these rules:

1. $0 \le P(A) \le 1$, for all events A
2. $P(A) = \sum_{\text{all } e \text{ in } A} P(e)$
3. $P(S) = \sum_{\text{all } e \text{ in } S} P(e) = 1$

We have deduced these basic properties of probability by reasoning from the definition that the probability of an event is the proportion of times the event is expected to occur in many repeated trials of the experiment.

To be of any practical importance, the numbers assigned as probabilities to the elementary outcomes must do more than just satisfy the three rules of probability. The probability assigned to any event of interest must be in agreement with the proportion of times the event is expected to occur. We discuss the implementation of this concept when (i) the outcomes are equally likely and (ii) when the experiment has already been repeated a large number of times. Example 2 illustrated both of these approaches.

ASSIGNING PROBABILITY—EQUALLY LIKELY OUTCOMES AND THE UNIFORM PROBABILITY MODEL

Often, the description of an experiment ensures that each elementary outcome is as likely to occur as any other. For example, consider the experiment of rolling a fair die and recording the top face. The sample space can be listed as

$$S = \{1, 2, 3, 4, 5, 6\}$$

where 1 stands for one dot, 2 for two dots, and so on.

Without actually rolling a die, we can deduce the probabilities. Because a fair die is a symmetric cube, each of its six faces is as likely to appear as any other. In other words, each face is expected to occur one-sixth of the time. The probability assignments should therefore be

$$P(1) = P(2) = P(3) = P(4) = P(5) = P(6) = \frac{1}{6}$$

and any other assignment would contradict the statement that the die is fair. We say that rolling a fair die conforms to a uniform probability model because the total probability 1 is evenly apportioned to all the elementary outcomes.

What is the probability of getting a number higher than 3? Letting A denote this event, A consists of three outcomes $\{4, 5, 6\}$ so

$$P(A) = P(4) + P(5) + P(6) = \frac{1}{6} + \frac{1}{6} + \frac{1}{6} = \frac{3}{6}$$

When the elementary outcomes are modeled as equally likely, we have a uniform probability model. If there are k elementary outcomes in S, each is assigned the probability of $1/k$.

An event A consisting of m elementary outcomes is then assigned

$$P(A) = \frac{m}{k} = \frac{\text{Number of elementary outcomes in } A}{\text{Number of elementary outcomes in } S}$$

Referring to the possible pasta dishes in Example 6, it was determined that there were six possible outcomes. Is it possible to determine the number of outcomes without listing the sample space? For each choice of pasta, there were two dishes according to the number of choices for the sauce. A tree diagram sets out the possible selections of a pasta dish:

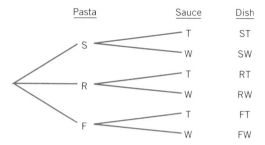

Since there are 3 choices for the pasta at the first branch, and 2 choices for the sauce at the second branch, there are $3 \times 2 = 6$ different dishes. The same reasoning based on a tree diagram produces a general counting rule.

$m \times n$ Counting Rule

When a first task has m possible outcomes and a second task has n possible outcomes, then there are $m \times n$ outcomes when performing both tasks.

Referring to Example 2, there are 20 ways to select the first ball from an urn containing 20 balls. After the first ball is removed, there are 19 left so there are 19 ways to select the second ball. Hence, there are $20 \times 19 = 380$ ways to select two balls, without replacement from the urn. We have used the counting rule to determine the number of choices without having to list a large number of possible outcomes!

Example 7   Using the counting rule for equally likely outcomes

Consider the problem of drawing two balls at random, without replacement, from an urn containing two red balls and three white balls.

(a)   List the sample space and the outcomes contained in each of the three events

$$A = [\text{Both balls are red}]$$
$$B = [\text{The first ball is red and the second ball is white}]$$
$$C = [\text{One of the balls is red}]$$

(b)   Find the probability of each of the events A, B, and C.

SOLUTION AND DISCUSSION   We first give the five balls different labels. Denote the two red balls as R_1 and R_2, and the three white balls as W_1, W_2, and W_3. Further, denote the outcome where the first ball is R_2 and the second ball is W_3 as $R_2 W_3$, and so on. The advantage of this labeling is that the outcomes are equally likely because the balls are drawn at random.

(a) The sample space is

$$S = \{R_1R_2, R_1W_1, R_1W_2, R_1W_3, R_2R_1, R_2W_1, R_2W_2, R_2W_3,$$
$$W_1R_1, W_1R_2, W_1W_2, W_1W_3, W_2R_1, W_2R_2, W_2W_1, W_2W_3,$$
$$W_3R_1, W_3R_2, W_3W_1, W_3W_2\}$$

The compositions of the events of interest are

$$A = \{R_1R_2, R_2R_1\}$$
$$B = \{R_1W_1, R_1W_2, R_1W_3, R_2W_1, R_2W_2, R_2W_3\}$$
$$C = \{R_1R_2, R_1W_1, R_1W_2, R_1W_3, R_2R_1, R_2W_1, R_2W_2, R_2W_3,$$
$$W_1R_1, W_1R_2, W_2R_1, W_2R_2, W_3R_1, W_3R_2\}$$

Note that we could have determined the number of outcomes in the sample space without having to make the complete list. According to the $m \times n$ counting rule, there are 5 ways of choosing the first ball from the urn containing five balls. After the first ball is removed there are 4 ways of choosing the second ball. The counting rule says there are $5 \times 4 = 20$ outcomes in S.

(b) Because the balls are drawn at random, each of the 20 possible outcomes is equally likely to occur and each has the same probability, $\frac{1}{20}$.

The event $A = \{R_1R_2, R_2R_1\}$ consists of two outcomes so, according to the probability law for the equally likely case,

$$\text{Probability of an event} = \frac{\text{Number of outcomes in the event}}{\text{Number of outcomes in the sample space}}$$

$$P(A) = \frac{2}{20} = .1$$

Similarly, the probabilities of the other two events are

$$P(B) = \frac{6}{20} = .3$$

and

$$P(C) = \frac{14}{20} = .7$$

Again, for the event $B =$ [First ball is red and second ball is white], there are 2 possible ways of choosing the first red ball and 3 possible ways of choosing the second white ball. By the counting rule, the number of outcomes in B is $2 \times 3 = 6$. ■

ASSIGNING PROBABILITY—LONG-RUN RELATIVE FREQUENCY

In many situations, it is not possible to construct a sample space where the elementary outcomes are equally likely. If one corner of a die is cut off, it would be unreasonable to assume that the faces remain equally likely.

When speaking of the probability (or risk) that a man will die in his thirties, one may choose to identify the occurrence of death at each decade or even each year of age as an elementary outcome. However, no sound reasoning can be provided in favor of a uniform probability model. In fact, from extensive mortality studies, demographers have found considerable disparity in the risk of death for different age groups.

When the assumption of equally likely elementary outcomes is not tenable, how do we assess the probability of an event? One recourse is to repeat the experiment many times and observe the proportion of times the event occurs. This was our approach in Solution 1 to Example 2 and for assigning probabilities to pasta dishes in Example 6.

Letting N denote the number of repetitions (or trials) of an experiment, we set

$$\text{Relative frequency of event } A \text{ in } N \text{ trials} = \frac{\text{No. of times } A \text{ occurs in } N \text{ trials}}{N}$$

Let A be the event of getting a 6 when rolling a die. We performed the experiment many times by repeatedly rolling the die. Our first ten rolls gave 5, 6, 2, 1, 3, 1, 2, 5, 1, and 2. For simplicity, we will calculate the relative frequency at every fifth roll. At the end of the first five rolls, 6 had appeared once, so the observed relative frequency of A would be $\frac{1}{5} = .20$. At the end of ten trials, the relative frequency of A is $\frac{1}{10} = .10$. Continuing until the die is rolled 100 times, we found that 6 occurred 16 times, so the relative frequency of A was $\frac{16}{100} = .16$.

Figure 1(a) on page 146 shows a plot of the relative frequency of obtaining a 6 versus the number of trials when the relative frequency is updated at the end of every five trials through 55 repetitions of the experiment. Consistent with our previous discussion, the first relative frequency in this plot is .20 and the second is .10. Notice there is considerable variability in the relative frequency when the number of trials, N, is small.

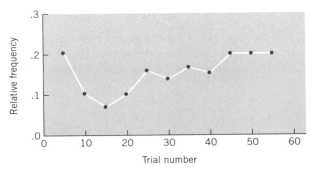

(a) Relative frequency updated after every 5 trials

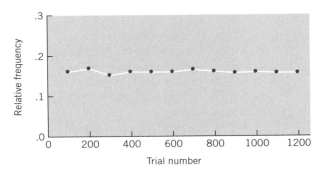

(b) Relative frequency updated after every 100 trials

Figure 1. Stabilization of relative frequency: Obtaining a 6 in a roll of a die

Figure 1(b) shows the relative frequency versus the number of trials, with the relative frequency updated at the end of every hundred rolls. The first relative frequency in this plot is .16 as discussed previously. In the second 100 rolls of the die, 6 came up 18 times, so the relative frequency after 200 trials is determined by collecting the first two sets together to give

$$\frac{16 + 18}{200} = \frac{34}{200} = .17$$

This is the second relative frequency plotted in Figure 1(b). From Figure 1(b), we see that fluctuations in the relative frequency decrease with increasing N. In this case, the relative frequency is relatively stable after about 200 repetitions of the experiment.

For any experiment, two persons separately performing the same experiment N times are not going to get exactly the same results, and therefore the relative frequency graphs will be different. However, the numerical value at which the relative frequency stabilizes in the long run will be the same. This concept, called *long-run stability of relative frequency*, is the key point illustrated in Figure 1.

> ## Probability as Long-Run Relative Frequency
>
> We define $P(A)$, the probability of an event A, as the value to which the relative frequency stabilizes with increasing number of trials.
>
> Although we will never know $P(A)$ exactly, it can be estimated accurately by repeating the experiment many times.

Example 8   Assigning probability to making an accident claim
An insurance company checked its records for a recent year and found that of 12,299 automobile insurance policies in effect, 2073 made a claim.[1] Among insured drivers under age 25, there were 1053 claims out of 5192 policies.

Assuming the claim history is valid for the current year, approximate the probability that

(a) an insured driver will make a claim.

(b) an insured driver under 25 years will make a claim.

SOLUTION AND DISCUSSION

(a) A large number of policies, 12,299, were in effect and each can be considered as a trial with the outcome being either "no claim" or "claim." Since 2073 insured drivers made a claim, we approximate the probability by the long-run relative frequency

$$P(\text{claim}) = \frac{2073}{12{,}299} = .169$$

(b) Restricting attention to drivers under 25 years of age, we use the same argument to approximate the probability of a claim as

$$\frac{1053}{5192} = .203$$

The higher than average claim rate for persons under 25 years old is typical experience for all insurance companies. They need to charge higher rates for groups with higher claim rates because, in the long run, relatively more claims will be made by drivers in the group. ∎

The property of the long-run stabilization of relative frequencies is based on the findings of experimenters in many fields who have undertaken the strain of studying the behavior of the relative frequencies under prolonged repetitions of their experiments. French gamblers, who provided much of the early impetus for the study of probability, performed experiments tossing dice and coins, drawing cards, and playing other games of chance thousands and thousands of times. They observed the stabilization property of relative frequency and applied this knowl-

[1] Data courtesy of J. Hickman.

edge to achieve an understanding of the uncertainty involved in these games. Demographers have compiled and studied volumes of mortality data to examine the relative frequency of the occurrence of such events as death in particular age groups. In each context, the relative frequencies were found to stabilize at specific numerical values as the number of cases studied increased. Life and accident insurance companies actually depend on the stability property of relative frequencies.

Exercises

2.1 Match the proposed probability of A with the appropriate verbal description. (More than one description may apply.)

Probability		Verbal Description
(a)	.95	(i) No chance of happening
(b)	.03	(ii) Very likely to happen
(c)	4.0	(iii) As much chance of occurring as not
(d)	− .2	(iv) Very little chance of happening
(e)	.4	(v) May occur but by no means certain
(f)	.5	(vi) An incorrect assignment
(g)	0	

2.2 For each numerical value assigned to the probability of an event, identify the verbal statements that are appropriate.

(a) $\sqrt{2}$ (b) $\dfrac{1}{\sqrt{2}}$ (c) $\dfrac{1}{186}$ (d) 1.1 (e) $\dfrac{1}{2}$ (f) $\dfrac{43}{45}$

Verbal statements: (i) cannot be a probability, (ii) the event is very unlikely to happen, (iii) 50–50 chance of happening, (iv) sure to happen, (v) more likely to happen than not

2.3 Identify the statement that best describes each $P(A)$.
    (a) $P(A) = .03$ (i) $P(A)$ is incorrect.
    (b) $P(A) = .30$ (ii) A rarely occurs.
    (c) $P(A) = 3.0$ (iii) A occurs moderately often.

2.4 Construct a sample space for each of the following experiments.
    (a) Someone claims to be able to taste the difference among bottled, tap, and canned draft beer of the same brand. A glass of each is

poured and given to the subject in an unknown order. The subject is asked to identify the contents of each glass. The number of correct identifications will be recorded.

(b) Record the number of traffic fatalities in a state next year.

(c) Observe the length of time a new video recorder will continue to work satisfactorily without service.

Which of these sample spaces are discrete and which are continuous?

2.5 Identify these events in Exercise 2.4.

(a) Not more than one correct identification

(b) Fewer accidents than last year

(*Note:* If you don't know last year's value, use 345.)

(c) Longer than the 90-day warranty but less than 425.4 days

2.6 When bidding on two projects, the president and vice president of a construction company make the following probability assessments for winning the contracts.

President	Vice President
$P(\text{win none}) = .2$	$P(\text{win none}) = .1$
$P(\text{win only one}) = .5$	$P(\text{win Project 1}) = .2$
$P(\text{win both}) = .3$	$P(\text{win Project 2}) = .4$
	$P(\text{win both}) = .3$

For both cases, examine whether the probability assignment is permissible.

2.7 Bob, John, and Linda are the finalists in the spelling contest of a local school district. The winner and the first runner-up will be sent to a statewide competition.

(a) List the sample space concerning the outcomes of the local contest.

(b) Give the composition of each of the following events.

$$A = [\text{Linda wins the local contest}]$$
$$B = [\text{Bob does not go to the state contest}]$$

2.8 Consider the following experiment: A coin will be tossed twice. If both tosses show heads, the experiment will stop. If one head is obtained in the two tosses, the coin will be tossed one more time, and in the case of both tails in the two tosses, the coin will be tossed two more times.

(a) Make a tree diagram and list the sample space.

(b) Give the composition of the following events.

$$A = [\text{Two heads}]$$
$$B = [\text{Two tails}]$$

2.9   There are four elementary outcomes in a sample space. The first has probability .1; the second, .5; and the third, .1. What is the probability of the fourth elementary outcome?

2.10   *Probability and odds.*   The probability of an event is often expressed in terms of odds. Specifically, when we say that the odds are k to m that an event will occur, we mean that the probability of the event is $k/(k + m)$. For instance, "the odds are 4 to 1 that candidate Jones will win" means that $P(\text{Jones will win}) = \frac{4}{5} = .8$. Express the following statements in terms of probability.

(a)   The odds are 2 to 1 that there will be fair weather tomorrow.

(b)   The odds are 5 to 2 that the city council will delay the funding of a new sports arena.

2.11   Consider the experiment of tossing a coin three times.

(a)   List the sample space by drawing a tree diagram.

(b)   Assign probabilities to the elementary outcomes.

(c)   Find the probability of getting exactly one head.

2.12   A letter is chosen at random from the word "COLLEGE." What is the probability that it is a vowel?

2.13   A stack contains eight tickets numbered 1, 1, 1, 2, 2, 3, 3, 3. One ticket will be drawn at random and its number will be noted.

(a)   List the sample space and assign probabilities to the elementary outcomes.

(b)   What is the probability of drawing an odd-numbered ticket?

2.14   Suppose you are eating at a pizza parlor with two friends. You have agreed to the following rule to decide who will pay the bill. Each person will toss a coin. The person who gets a result that is different from the other two will pay the bill. If all three tosses yield the same result, the bill will be shared by all. Find the probability that

(a)   Only you will have to pay.

(b)   All three will share.

2.15   A white and a colored die are tossed. The possible outcomes are shown in the illustration at the top of page 151.

(a)   Identify the events $A = [\text{Sum} = 6]$, $B = [\text{Sum} = 7]$, $C = [\text{Sum is even}]$, $D = [\text{Same number on each die}]$.

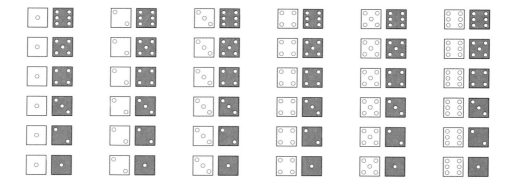

(b) If both dice are "fair," assign probability to each elementary outcome.

(c) Obtain $P(A)$, $P(B)$, $P(C)$, $P(D)$.

2.16 A roulette wheel has 34 slots, 2 of which are green, 16 are red, and 16 are black. A successful bet on black or red doubles the money, whereas one on green fetches 30 times as much. If you play the game once by betting $2 on the black, what is the probability that

(a) You will lose your $2?

(b) You will win $2?

2.17 Based on data from the Center for Health Statistics, the 1988 birth rates in the 50 states are grouped in the following frequency table.

Birth rate (per thousand)	12–13	13–14	14–15	15–16	16–17	17–18	18–19	19 and over	Total
No. of states	1	5	13	17	4	5	3	2	50

Endpoint convention: Lower point is included, upper is not.

If one state is selected at random, what is the probability that the birth rate there is

(a) Under 16?

(b) Under 18 but not under 15?

(c) 17 or over?

2.18 Fifteen persons reporting to a Red Cross center one day are typed for blood, and the following counts are found.

Blood group	O	A	B	AB	Total
No. of persons	3	5	6	1	15

If one person is randomly selected, what is the probability that this person's blood group is

(a) AB?

(b) Either A or B?

(c) Not O?

2.19 Of three agricultural plots, one has high acidity in soil, one low, and one moderate. One plot will be randomly selected for planting beans, and then another plot will be randomly selected for planting peppers.

(a) Make a tree diagram to list all possible assignments.

(b) What is the probability that peppers get assigned to a plot of low or moderate acidity?

(c) What is the probability that one of the crops is assigned to high and the other to low acidic soil?

2.20 Children joining a kindergarten class will be checked one after another to see whether they have been inoculated for polio (I) or not (N). Suppose that the checking is to be continued until one noninoculated child is found or four children have been checked, whichever occurs first. List the sample space for this experiment.

2.21 (a) Consider the simplistic model that human births are evenly distributed over the 12 calendar months. If a person is randomly selected, say, from a phone directory, what is the probability that his or her birthday would be in November or December?

(b) The following record shows a classification of 41,208 births in Wisconsin (courtesy of Professor Jerome Klotz). Calculate the relative frequency of births for each month and comment on the plausibility of the uniform probability model.

Jan.	3,478	July	3,476
Feb.	3,333	Aug.	3,495
March	3,771	Sept.	3,490
April	3,542	Oct.	3,331
May	3,479	Nov.	3,188
June	3,304	Dec.	3,321
		Total	41,208

2.22 Explain why the long-run relative frequency interpretation of probability does not apply to the following situations.

(a) The proportion of days the Dow Jones average of industrial stock prices exceeds 7900

(b) The proportion of income-tax returns containing improper deductions if the data are collected only in the slack season (say, January)

(c)   The proportion of cars that do not meet emission standards, if the data are collected from service stations where the mechanics have been asked to check the emission while attending to other requested services

2.23   Three jars contain different chemicals R, S, and T. Their identifying labels were accidentally dropped during transportation. Suppose that a careless technician decides to put the labels on these jars at random without inspecting their contents.

(a)   List the sample space.

(b)   State the compositions of the events:

$$A = [\text{There is exactly one match}]$$
$$B = [\text{All jars receive wrong labels}]$$

2.24   Refer to Exercise 2.23.

(a)   Assign probabilities to the elementary outcomes.

(b)   Find $P(A)$ and $P(B)$.

3. EVENT RELATIONS AND TWO LAWS OF PROBABILITY

Later, when making probability calculations to support generalizations from the actual sample to the complete population, we will need to calculate probabilities of combined events, such as whether the count of no-shows for a flight is either large or low.

Recall that the probability of an event A is the sum of the probabilities of all the elementary outcomes that are in A. It often turns out, however, that the event of interest has a complex structure that requires tedious enumeration of its elementary outcomes. On the other hand, this event may be related to other events that can be handled more easily. The purpose of this section is to first introduce the three most basic event relations: **complement, union,** and **intersection.** These event relations will then motivate some laws of probability.

The event operations are conveniently described in graphical terms. We first represent the sample space as a collection of points in a diagram, each identified with a specific elementary outcome. The geometric pattern of the plotted points is irrelevant. What is important is that each point be clearly tagged to indicate which elementary outcome it represents, and to watch that no elementary outcome is missed or duplicated in the diagram. To represent an event A, identify the points that correspond to the elementary outcomes in A, enclose them in a boundary line, and attach the tag A. This representation, called a **Venn diagram,** is illustrated in Example 9.

Example 9   Making a Venn diagram

An urn contains 10 balls labeled 1, 2, . . . , 10. Suppose one ball is drawn at random from the urn. Make a Venn diagram for the experiment and indicate the following events:

A = [The number on the ball is an even number]

B = [The number on the ball is among the first five numbers]

C = [The number on the ball is one of the last three numbers]

SOLUTION   The sample space is

$$S = \{1, 2, 3, 4, 5, 6, 7, 8, 9, 10\}$$

and the events are

$$A = \{2, 4, 6, 8, 10\} \qquad B = \{1, 2, 3, 4, 5\} \qquad C = \{8, 9, 10\}$$

These are indicated in Figure 2.

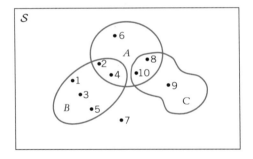

Figure 2. Venn diagram for Example 9

We now proceed to define the three basic event operations and introduce the corresponding symbols.

The event **not A,** or the complement of an event A, denoted by $\overline{A}$, is the set of all elementary outcomes that are not in A. The occurrence of $\overline{A}$ means that *A does not occur.*

The event **A or B,** the union of two events A and B, denoted by A or B, is the set of all elementary outcomes that are in A, B, or both. The occurrence of A or B means that *either A or B or both occur.*

The event **A and B,** the intersection of two events A and B, denoted by $A\,B$, is the set of all elementary outcomes that are in A and B. The occurrence of $A\,B$ means that *both A and B occur.*

Note that A or B is a larger set containing A as well as B, whereas $A\,B$ is the common part of the sets A and B. Also it is evident from the definitions that A or B and B or A represent the same event, while $A\,B$ and $B\,A$ are both expressions for *A and B*. The operations of union and intersection can be extended to more than two events. For instance, *A or B or C* stands for the set of all outcomes that are in *at least one* of *A*, *B*, and *C*, whereas *ABC* represents the *simultaneous occurrence* of all three events.

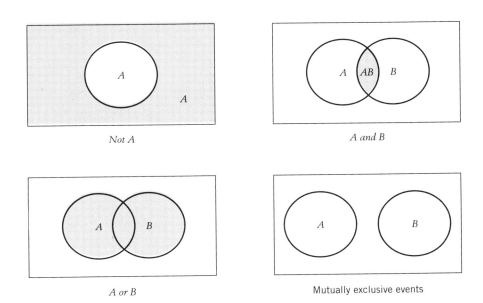

Not A

A and B

A or B

Mutually exclusive events

Two events *A* and *B* are called mutually exclusive if their intersection *AB* is empty. Because mutually exclusive events have no elementary outcomes in common, they cannot occur simultaneously.

Let us now examine how probabilities behave as the operations of complementation, union, and intersection are applied to events. It would be worthwhile for the reader to review the properties of probability listed in Section 2. In particular, recall that $P(A)$ is the sum of probabilities of the elementary outcomes that are in *A*, and $P(S) = 1$.

First, let us examine how $P(\overline{A})$ is related to $P(A)$. The sum $P(A) + P(\overline{A})$ is the sum of the probabilities of all elementary outcomes that are in *A* plus the sum of the probabilities of elementary outcomes not in *A*. Together, these two sets comprise S and we must have $P(S) = 1$. Consequently, $P(A) + P(\overline{A}) = 1$, and we arrive at the following law.

Law of the Complement

$$P(A) = 1 - P(\overline{A})$$

This law or formula is useful in calculating $P(A)$ when $\overline{A}$ is of a simpler form than A so that $P(\overline{A})$ is easier to calculate.

Recall that A *or* B is composed of points (or elementary outcomes) that are in A, in B, or in both A and B. Consequently, $P(A$ *or* $B)$ is the sum of the probabilities assigned to these elementary outcomes, each probability taken *just once*. Now, the sum $P(A) + P(B)$ includes contributions from all these points, but it double counts those in the region AB (see the figure of A *or* B). To adjust for this double counting, we must therefore subtract $P(AB)$ from the sum $P(A) + P(B)$. This results in the following law.

Addition Law

$$P(A \text{ or } B) = P(A) + P(B) - P(AB)$$

If the events A and B are mutually exclusive, their intersection AB is empty, so $P(AB) = 0$, and we obtain

---

Special Addition Law for Mutually Exclusive Events

$$P(A \text{ or } B) = P(A) + P(B)$$

The addition law expresses the probability of a larger event A *or* B in terms of the probabilities of the smaller events A, B, and AB. Some applications of these two laws are given in the following examples.

Example 10   Probability and guessing

A child is presented with three word-association problems. With each problem, two answers are suggested—one is correct and the other wrong. If the child has no understanding of the words whatsoever and answers the problems by guessing, what is the probability of getting at least one correct answer?

SOLUTION AND DISCUSSION   Let us denote a correct answer by C and a wrong answer by W. The elementary outcomes are conveniently enumerated by the tree diagram on page 157.

There are 8 elementary outcomes in the sample space and, because they are equally likely, each has the probability $\frac{1}{8}$. Let A denote the event of getting at least one correct answer. Scanning our list, we see that A contains 7 elementary outcomes, all except WWW. Our direct calculation yields $P(A) = \frac{7}{8}$.

Now let us see how this probability calculation could be considerably simplified. First, making a complete list of the sample space is not necessary. Since the elementary outcomes are equally likely, we need only determine that there are a total of 8 elements in $\mathcal{S}$. How can we obtain this count without making a

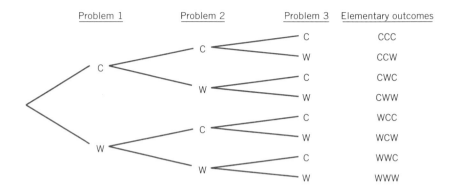

Problem 1	Problem 2	Problem 3	Elementary outcomes

list? Note that an outcome is represented by three letters. There are 2 choices for each letter—namely, C or W. We then have $2 \times 2 \times 2 = 8$ ways of filling the three slots. The tree diagram and the $m \times n$ counting rule explain this multiplication rule of counting. Evidently, the event A contains many elementary outcomes. On the other hand, $\overline{A}$ is the event of getting all answers wrong. It consists of the single elementary outcome WWW, so $P(\overline{A}) = \frac{1}{8}$. According to the law of the complement,

$$P(A) = 1 - P(\overline{A})$$
$$= 1 - \frac{1}{8} = \frac{7}{8}$$ ∎

Example 11   Determining probabilities from Venn diagrams

The accompanying Venn diagram shows three events A, B, and C and also the probabilities of the various intersections. (For instance, $P(AB) = .07$, $P(A\overline{B}) = .13$.) Determine:

(a) $P(A)$
(b) $P(B\overline{C})$
(c) $P(A \text{ or } B)$

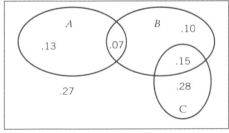

SOLUTION AND DISCUSSION   To calculate a probability, first identify the set in the Venn diagram. Then add the probabilities of those intersections that together comprise the stated event. We obtain

(a) $P(A) = .13 + .07 = .20$
(b) $P(B\overline{C}) = .10 + .07 = .17$
(c) $P(A \text{ or } B) = .13 + .07 + .10 + .15 = .45$

This example shows how the probability of an event can be computed without knowing all the outcomes involved. ∎

Example 12   Expressing events in set notation
Refer to Example 11. Express the following events in set notation and find their probabilities.

(a)   Both B and C occur.
(b) C occurs and B does not.
(c)   Exactly one of the three events occurs.

SOLUTION   The stated events and their probabilities are

(a) BC; $P(BC) = .15$
(b) $\bar{B}C$; $P(\bar{B}C) = .28$
(c) $(A\bar{B}\bar{C})$ or $(\bar{A}B\bar{C})$ or $(\bar{A}\bar{B}C)$
        The probability $= .13 + .10 + .28 = .51$ ∎

Example 13   Combining events
Refer to Example 9, where one ball is drawn at random from an urn that contains 10 balls labeled 1, 2, . . . , 10. With

A = [The number on the ball is an even number]
B = [The number on the ball is among the first five numbers]
C = [The number on the ball is one of the last three numbers]

determine the composition of the events

(a) $\bar{A}$ (b) $\bar{B}$ (c) AB (d) A or B (e) BC
(f) AC (g) B or C

SOLUTION AND DISCUSSION   From the Venn diagram in Figure 2,

$A = \{2, 4, 6, 8, 10\}$ $B = \{1, 2, 3, 4, 5\}$ $C = \{8, 9, 10\}$

(a) $\bar{A} = \{1, 3, 5, 7, 9\}$, the set of outcomes with odd numbers
(b) $\bar{B} = \{6, 7, 8, 9, 10\}$ (c) $AB = \{2, 4\}$
(d) A or $B = \{1, 2, 3, 4, 5, 6, 8, 10\}$
(e) BC is an empty set. That is, B and C are mutually exclusive.
(f) $AC = \{8, 10\}$ (g) B or $C = \{1, 2, 3, 4, 5, 8, 9, 10\}$ ∎

Example 14   The probability of combined events
Find the probabilities of the events listed in Example 13. Verify the laws of probability where they are applicable.

SOLUTION AND DISCUSSION   A ball is drawn at random from an urn having 10 balls so each ball has the same chance of being selected. The equally likely model applies and we can assign probabilities according to the rule

$$\text{Probability of an event} = \frac{\text{Number of outcomes in the event}}{\text{Number of outcomes in the sample space}}$$

We first assign probabilities to the events A, B, and C by counting the number of outcomes in each.

$$P(A) = \frac{5}{10} = .5 \qquad P(B) = \frac{5}{10} = .5 \qquad P(C) = \frac{3}{10} = .3$$

Now, referring to Example 13, we count the number of outcomes in each specified event.

(a) $P(\overline{A}) = \frac{5}{10} = .5$ Since $P(A) \stackrel{.}{=} .5$, we have
$$.5 = 1 - P(A) = P(\overline{A}).$$

(b) $P(\overline{B}) = \frac{5}{10} = .5$ Similarly, $P(\overline{B}) = 1 - P(B)$.

(c) $P(AB) = \frac{2}{10} = .2$ Observe that $P(A)P(B) = (.5)(.5) = .25$.
In general, $P(A)P(B) \neq P(AB)$.

(d) $P(A \text{ or } B) = \frac{8}{10} = .8$ Now $P(A) + P(B) - P(AB)$
$$= .5 + .5 - .2 = .8.$$
Thus $P(A \text{ or } B) = P(A) + P(B) - P(AB)$.

(e) $P(BC) = 0$ There is no chance that both
B and C occur at the same time.

(f) $P(AC) = \frac{2}{10} = .2$

(g) $P(B \text{ or } C) = \frac{8}{10} = .8$ You should verify that
$$P(B \text{ or } C) = P(B) + P(C) - P(BC). \quad \blacksquare$$

Example 15 More probabilities for combined events
Refer to Example 13, where one ball is drawn at random from an urn that contains 10 balls labeled 1, 2, . . . , 10, and

$A = $ [The number on the ball is an even number]
$B = $ [The number on the ball is among the first five numbers]
$C = $ [The number on the ball is one of the last three numbers]

(a) Determine the composition of the events

$$A \text{ or } B \text{ or } C \qquad \overline{A \text{ or } B \text{ or } C} \qquad \overline{A}\,\overline{B} \qquad \overline{A \text{ or } B}$$

(b) Find the probabilities of these events. Verify the laws of probability where they are applicable.

SOLUTION AND (a)
DISCUSSION

$$A \text{ or } B \text{ or } C = \{1, 2, 3, 4, 5, 6, 8, 9, 10\}$$
$$\overline{A \text{ or } B \text{ or } C} = \{7\}$$
$$\overline{A}\,\overline{B} = \{7, 9\} \qquad \overline{A \text{ or } B} = \{7, 9\}$$

Observe that $\overline{A \text{ or } B} = \overline{A}\,\overline{B}$. In other words, not at least one of A or B is equivalent to A does not occur *and* B does not occur. You should check that $\overline{A \text{ or } B \text{ or } C} = \overline{A}\,\overline{B}\,\overline{C}$ in this example.

(b)

$P(A \text{ or } B \text{ or } C) = \frac{9}{10} = .9$

$P(\overline{A \text{ or } B \text{ or } C}) = \frac{1}{10} = .1$

$P(\overline{A}\overline{B}) = \frac{2}{10} = .2$

$P(\overline{A \text{ or } B}) = \frac{2}{10} = .2$ From part (d) of Example 14, $P(A \text{ or } B) = .8$. Then $P(\overline{A \text{ or } B})$ is the same as $1 - P(A \text{ or } B) = 1 - .8 = .2$, which is the same answer as $P(\overline{A}\overline{B})$.   ■

Exercises

3.1 Sun visors come in red (r), orange (o), yellow (y), green (g), blue (b), purple (p), and white (w). For the next visor sold, consider the two events $A = \{o, g, b, w\}$ and $B = \{r, o, b\}$.

(a) Specify the sample space.

(b) Draw a Venn diagram and show the events A and B.

(c) Determine the composition of the following events: (i) AB (ii) $\overline{B}$ (iii) $A\overline{B}$ (iv) $A \text{ or } B$.

3.2 Eight pizza delivery persons d1, d2, . . . , d8 are working in the campus area. Which one will deliver the next pizza order?

(a) Specify the sample space.

(b) Consider the three events $A = \{d1, d2, d5, d6, d7\}$, $B = \{d2, d3, d6, d7\}$, and $C = \{d6, d8\}$. Draw the Venn diagram and show these events.

(c) The probabilities that each delivery person will be the one to deliver the next pizza order are

$$P(d1) = .16 \qquad P(d2) = P(d3) = P(d4) = .08$$
$$P(d5) = P(d6) = P(d7) = P(d8) = .15$$

Give the composition and determine the probability of (i) $\overline{B}$ (ii) BC (iii) $A \text{ or } C$ (iv) $\overline{A} \text{ or } C$.

3.3 Refer to Exercise 3.2. Corresponding to each verbal description given here, write the event in set notation, give its composition, and find its probability.

(a) C does not occur.

(b) Both A and B occur.

(c) A occurs and B does not occur.

(d) Neither A nor C occurs.

3.4   Suppose you have had interviews for summer jobs at a grocery store, a department store, and a lumber yard. Let G, D, and L denote the events of your getting an offer from the grocery store, the department store, and the lumber yard, respectively. Express the following events in set notation.

(a) You get offers from the grocery store and the lumber yard.

(b) You get offers from the grocery store and the lumber yard, but fail to get an offer from the department store.

(c) You do not get offers from the department store and the lumber yard.

3.5   Four applicants will be interviewed for a position with an oil company. They have the following characteristics.

1.   Chemistry major, male, GPA 3.5

2.   Geology major, female, GPA 3.8

3.   Chemical engineering major, female, GPA 3.7

4.   Petroleum engineering major, male, GPA 3.2

One of the candidates will be hired.

(a)   Draw a Venn diagram and exhibit these events:

$$A = [\text{An engineering major is hired}]$$
$$B = [\text{The GPA of the selected candidate is higher than 3.6}]$$
$$C = [\text{A male candidate is hired}]$$

(b)   Give the composition of the events A or B and AB.

3.6   For the experiment of Exercise 3.5, give a verbal description of each of the following events and also state the composition of the event.

(a) $\overline{C}$

(b) $C\overline{A}$

(c) A or $\overline{C}$

3.7   Consumer complaints that cannot be handled by one of its 9 regional centers are referred to the national office of the company. Denote the nine regional centers by $c1$, $c2$, . . . , $c9$. Some regional centers submit more unresolved complaints than the others. The probabilities that a complaint came from each center are

$$P(c1) = P(c2) = .08 \qquad P(c3) = P(c4) = P(c5) = .1$$
$$P(c6) = P(c7) = .2 \qquad P(c8) = P(c9) = .07$$

Let $A = \{c1, c5, c8\}$ and $B = \{c2, c5, c8, c9\}$.

(a)   Calculate $P(A)$, $P(B)$, and $P(AB)$.

(b)   Using the addition law of probability, calculate $P(A$ or $B)$.

(c) List the composition of the event A or B and calculate $P(A \text{ or } B)$ by adding the probabilities of the elementary outcomes.

(d) Calculate $P(\overline{B})$ from $P(B)$ and also by listing the composition of $\overline{B}$.

3.8 Refer to Exercise 3.1. Suppose the probabilities for the color of the next visor sold are

$$P(r) = P(o) = P(y) = .1 \qquad P(g) = P(b) = .15$$
$$P(p) = .05 \qquad P(w) = .35$$

(a) Find $P(A)$, $P(B)$, and $P(AB)$.

(b) Employing the laws of probability and the results of part (a), calculate $P(\overline{A})$ and $P(A \text{ or } B)$.

(c) Verify your answers to part (b) by adding the probabilities of the elementary outcomes in each of $\overline{A}$ and A or B.

3.9 For two events A and B, the following probabilities are specified.

$$P(A) = .52 \qquad P(B) = .36 \qquad P(AB) = .20$$

(a) Enter these probabilities in the following table.

	B	$\overline{B}$
A		
$\overline{A}$		

(b) Determine the probabilities of $A\overline{B}$, $\overline{A}B$, and $\overline{A}\overline{B}$ and fill in the table.

3.10 Refer to Exercise 3.9. Express the following events in set notation and find their probabilities.

(a) B occurs and A does not occur.

(b) Neither A nor B occurs.

(c) Either A occurs or B does not occur.

3.11 The following table shows the probabilities concerning two events A and B.

	B	$\overline{B}$
A	.25	.12
$\overline{A}$		
	.40	

(a) Determine the missing entries.
(b) What is the probability that A occurs and B does not occur?
(c) Find the probability that either A or B occurs.
(d) Find the probability that one of these events occurs and the other does not.

3.12 If $P(A) = .6$ and $P(B) = .5$, can A and B be mutually exclusive? Why or why not?

3.13 From the probabilities shown in this Venn diagram, determine the probabilities of the following events.

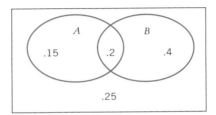

(a) A does not occur.
(b) A occurs and B does not occur.
(c) Exactly one of the events A and B occurs.

3.14 In a class of 64 seniors and graduate students, 38 are men and 15 are graduate students of whom 8 are women. If a student is randomly selected from this class, what is the probability that the selected student is (a) a senior? (b) a male graduate student?

3.15 Of 15 nursing homes in a city, 6 are in violation of sanitary standards, 8 are in violation of security standards, and 5 are in violation of both. If a nursing home is chosen at random, what is the probability that it is in compliance with both security and sanitary standards?

3.16 Given the probability that A occurs is $\frac{1}{3}$, the probability that B does not occur is $\frac{1}{4}$, and the probability that either A or B occurs is $\frac{8}{9}$, find
(a) The probability that A does not occur
(b) The probability that both A and B occur
(c) The probability that A occurs and B does not occur

3.17 The medical records of the male diabetic patients reporting to a clinic during one year provide the following percentages.

Age of Patient	Light Case		Serious Case	
	Diabetes in Parents		Diabetes in Parents	
	Yes	No	Yes	No
Below 40	15	10	8	2
Above 40	15	20	20	10

Suppose a patient is chosen at random from this group, and the events A, B, and C are defined as follows.

$$A = [\text{He has a serious case}]$$
$$B = [\text{He is below 40}]$$
$$C = [\text{His parents are diabetic}]$$

(a) Find the probabilities $P(A)$, $P(B)$, $P(BC)$, $P(ABC)$.

(b) Describe the following events verbally and find their probabilities: (i) $\overline{AB}$, (ii) $\overline{A}$ or $\overline{C}$, (iii) $\overline{A}B\overline{C}$.

3.18 The following frequency table shows the classification of 58 landfills in a state according to their concentration of the three hazardous chemicals arsenic, barium, and mercury.

Arsenic	Barium			
	High		Low	
	Mercury		Mercury	
	High	Low	High	Low
High	1	3	5	9
Low	4	8	10	18

If a landfill is selected at random, find the probability that it has a

(a) High concentration of barium

(b) High concentration of mercury and low concentrations of both arsenic and barium

(c) High concentration of any two of the chemicals and low concentration of the third

(d) High concentration of any one of the chemicals and low concentrations of the other two

3.19   A mail-order firm offers a mystery gift box to customers who place a purchase order of $20 or more. Each box contains one of the following five assortments.

1.   Keychain and utility knife
2.   Name tag and flashlight
3.   Letter opener and flashlight
4.   Utility knife and letter opener
5.   Memo pad and letter opener

(a)   If a customer places two separate orders of $20 and receives two gift boxes at random, list the sample space and assign probabilities to the simple events.

(b)   State the compositions of the following events and find their probabilities.

$$A = [\text{The customer gets a flashlight}]$$
$$B = [\text{The customer gets a letter opener}]$$
$$AB = [\text{The customer gets a flashlight and a letter opener}]$$

3.20   Refer to Exercise 3.19. Let C denote the event that the customer gets either a flashlight or a letter opener or both.

(a)   Relate C to the events A and B, and calculate $P(C)$ by employing an appropriate law of probability.

(b)   State the composition of C and calculate its probability by adding the probabilities of the simple events.

4. CONDITIONAL PROBABILITY

In the urn problem of Example 7, in Section 2, where 2 balls are drawn from 2 red and 3 white balls, consider the conditional probability that the second ball drawn is red given that the first ball is red. Since the first ball drawn is red, there are 1 red and 3 white balls in the urn at the time the second ball is drawn. The probability that the second ball is red given that the first ball is red is then $\frac{1}{4}$.

The probability of an event A must often be modified after information is obtained as to whether or not a related event B has taken place. Information about some aspect of the experimental results may therefore necessitate a revision of the probability of an event concerning some other aspect of the results. The revised probability of A when it is known that B has occurred is called the conditional probability of *A* given *B.*

How do we find the conditional probability that A occurs given that B has occurred? The following example will lead us to a formula for the conditional probability of A given B, which is denoted by $P(A|B)$.

Example 16   Conditional probability of hypertension given overweight
A group of executives is classified according to the status of body weight and incidence of hypertension. The proportions in the various categories appear in Table 1.

TABLE 1   Body Weight and Hypertension

	Overweight	Normal Weight	Underweight	Total
Hypertensive	.10	.08	.02	.20
Not hypertensive	.15	.45	.20	.80
Total	.25	.53	.22	1.00

(a)   What is the probability that a person selected at random from this group will have hypertension?

(b)   A person, selected at random from this group, is found to be overweight. What is the probability that this person is also hypertensive?

SOLUTION AND DISCUSSION   Let A denote the event that a person is hypertensive, and let B denote the event that a person is overweight.

(a)   Because 20% of the group is hypertensive and the individual is selected at random from this group, we conclude that $P(A) = .2$. This is the unconditional probability of A.

(b)   When we are given the information that the selected person is overweight, the categories in the second and third columns of Table 1 are not relevant to this person. The first column shows that among the subgroup of overweight persons, the proportion having hypertension is .10/.25. Therefore, given the information that the person is in this subgroup, the probability that he or she is hypertensive is

$$P(A|B) = \frac{.10}{.25} = .4$$

Noting that $P(AB) = .10$ and $P(B) = .25$, we have derived $P(A|B)$ by taking the ratio $P(AB)/P(B)$. In other words, $P(A|B)$ is the proportion of the population having the characteristic A among all those having the characteristic B. ∎

The **conditional probability** of A given B is denoted by $P(A|B)$ and defined by the formula

$$P(A|B) = \frac{P(AB)}{P(B)}$$

Equivalently, this formula can be written

$$P(AB) = P(B)P(A|B)$$

This latter version is called the **multiplication law of probability.**

Similarly, the conditional probability of B given A can be expressed

$$P(B|A) = \frac{P(AB)}{P(A)}$$

which gives the relation $P(AB) = P(A)P(B|A)$. Thus, the multiplication law of probability states that the conditional probability of an event multiplied by the probability of the conditioning event gives the probability of the intersection.

The multiplication law can be used in either of two ways, depending on convenience. When it is easy to compute $P(B)$ and $P(AB)$ directly, these values can be used to compute $P(A|B)$, as in Example 16. On the other hand, if it is easy to calculate $P(B)$ and $P(A|B)$ directly, these values can be used to compute $P(AB)$.

Example 17   Conditional probability of survival
The box "How Long Will a Baby Live?" shows the probabilities of death within 10-year age groups.

(a)   What is the probability that a newborn child will survive beyond age 90?

(b)   What is the probability that a person who has just turned 80 will survive beyond age 90?

SOLUTION AND DISCUSSION   (a)   Let A denote the event "Survive beyond 90." Adding the probabilities of death in the age groups 90–100 and beyond, we find

$$P(A) = .147 + .013 = .160$$

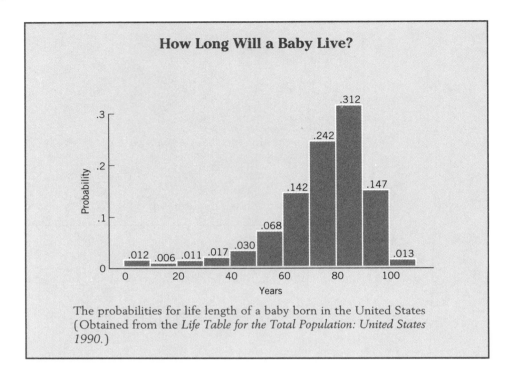

How Long Will a Baby Live?

The probabilities for life length of a baby born in the United States (Obtained from the *Life Table for the Total Population: United States 1990.*)

(b)   Letting B denote the event "Survive beyond 80," we see that the required probability is the conditional probability $P(A|B)$. Because $AB = A$, $P(A) = .160$ and

$$P(B) = .312 + .147 + .013 = .472$$

we obtain

$$P(A|B) = \frac{P(AB)}{P(B)} = \frac{.160}{.472} = .339$$

A person is much more likely to live more than 90 years if it is given that they survive beyond 80 years.   ∎

Example 18   Probability calculations when sampling accounts
A list of important customers contains 25 names. Among them, 20 persons have their accounts in good standing while 5 are delinquent. Two persons will be selected at random from this list and the status of their accounts checked. Calculate the probability that

(a)   Both accounts are delinquent.
(b)   One account is delinquent and the other is in good standing.

SOLUTION AND DISCUSSION   We will use the symbols D for delinquent and G for good standing, and attach subscripts to identify the order of the selection. For instance, G_1D_2 will represent the event that the first account checked is in good standing and the second delinquent.

(a)   Here the problem is to calculate $P(D_1D_2)$. Evidently, D_1D_2 is the intersection of the two events D_1 and D_2. Using the multiplication law, we write

$$P(D_1D_2) = P(D_1)P(D_2|D_1)$$

To calculate $P(D_1)$, we need only consider selecting one account at random from 20 good and 5 delinquent accounts. Clearly, $P(D_1) = \frac{5}{25}$. The next step is to calculate $P(D_2|D_1)$. Given that D_1 has occurred, there will remain 20 good and 4 delinquent accounts at the time the second selection is made. Therefore, the conditional probability of D_2 given D_1 is $P(D_2|D_1) = \frac{4}{24}$. Multiplying these two probabilities, we get

$$P(\text{both delinquent}) = P(D_1D_2) = \frac{5}{25} \times \frac{4}{24} = \frac{1}{30} = .033$$

(b)   The event [exactly one delinquent] is the union of the two mutually exclusive events G_1D_2 and D_1G_2. The probability of each of these can be calculated by the multiplication law as in part (a). Specifically,

$$P(G_1D_2) = P(G_1)P(D_2|G_1) = \frac{20}{25} \times \frac{5}{24} = \frac{1}{6}$$

$$P(D_1G_2) = P(D_1)P(G_2|D_1) = \frac{5}{25} \times \frac{20}{24} = \frac{1}{6}$$

By the special addition law for mutually exclusive events, the required probability is $P(G_1D_2) + P(D_1G_2) = \frac{2}{6} = .333$. Note that this example is yet another example of the urn problems and solutions discussed in Sections 1 and 2.   ∎

Remark: In solving the problems of Example 18, we have avoided listing the sample space corresponding to the selection of two accounts from a list of 25. A judicious use of the multiplication law has made it possible to focus attention on one draw at a time, thus simplifying the probability calculations.

STATISTICAL REASONING

A High Conditional Probability Can Improve Prediction

A now reformed but formerly less than reputable member of society was being interviewed on a nationwide TV program. This former professional thief was

Barry Marcus/FPG International

arrested for breaking and entering many homes in a west coast city. Until he was caught, he made a considerable amount of money from his activities and considered himself to be quite successful. When asked by the interviewer, he revealed his method of operation. It was based on his intuitive understanding of conditional probabilities.

On the air, the former robber described how he would hang around downtown and look for cars where the driver had left the keys in the car. When he found one, he would write down the license number and use this information to obtain the driver's home address.

The thief had noticed that people who left their keys in the car often did not lock the door between the garage and their house. His estimate was that this garage door was unlocked in 80% of the cases where owners left car keys in their car. This is a much higher percentage than in the general population. The thief could choose an appropriate time and most of the houses he targeted were easy to enter.

Don't leave keys in your car and lock your doors. Another message is that you will be a more difficult target for crime if you avoid being too predictable.

5. INDEPENDENCE

The idea of independent events plays a key role when modeling variation by probability. It is also a key component in the development of inferential statistics in later chapters.

The coin-tossing model conforms to our intuitive notion of independence because a coin does not remember how it fell last time.

Example 19   Probabilities of events concerning coin tossing
Consider the experiment of tossing a fair coin twice.

(a)   Find the composition of the events

$$A = [\text{Head on the first toss}]$$
$$B = [\text{Tail on the second toss}]$$

(b)   Find the probabilities of A, B, and AB.

(c)   Verify that $P(AB) = P(A)P(B)$ in this example.

SOLUTION AND   (a)   Here the sample space is
DISCUSSION

$$S = \{TT, TH, HT, HH\}$$
$$A = \{HT, HH\} \qquad B = \{TT, HT\}$$

(b)   The event $AB = \{HT\}$ contains the outcome common to A and B.
Since the coin is fair, the four outcomes in S are equally likely.
Hence

$$P(A) = \frac{2}{4} = .5 \qquad P(B) = \frac{2}{4} = .5 \qquad P(AB) = \frac{1}{4} = .25$$

(c)   In this case, $P(A)P(B) = (.5)(.5) = .25$, which is equal to $P(AB)$.
Is it only a coincidence that $P(A)P(B) = P(AB)$? In Example 14,
part (c), we saw that equality need not hold.   ■

What is so special about the coin-tossing example that the equality
$P(A)P(B) = P(AB)$ holds? Examine the events A and B closely. We see that
A depends only on the first toss. Since B depends only on the second toss, its
probability will not change whether A occurs or does not occur.
Looked at in terms of conditional probability,

$$P(B|A) = \frac{P([\text{Head on the first toss}] \; and \; [\text{Tail on the second toss}])}{P[\text{Head on the first toss}]}$$

$$= \frac{P[\text{Head on the first toss}] \; P[\text{Tail on the second toss}]}{P[\text{Head on the first toss}]}$$

$$= P[\text{Tail on the second toss}] = P(B)$$

The probability of B is unchanged, given that A has occurred.
More generally, whenever $P(AB) = P(A)P(B)$ holds for any two events A
and B, the conditional probability of A, given that B has occurred, is unchanged.

$$P(A|B) = \frac{P(AB)}{P(B)} = \frac{P(A)P(B)}{P(B)} = P(A)$$

Also, $P(B|A) = P(B)$ as above so the occurrence of either event does not influ-
ence or change the probability of the other. Consequently, we say that two events
A and B are **independent.**

> Two events A and B are **independent** if $P(AB) = P(A)P(B)$.

Example 20   Demonstrating dependence: Body weight and hypertension
Referring to Example 16, are the two events $A = $ [hypertensive] and $B = $ [overweight] independent?

SOLUTION AND DISCUSSION   In that example, $P(A) = .2$, $P(B) = .25$, and we determined that $P(AB) = .10$ so

$$P(AB) = .10 \neq .05 = (.2)(.25) = P(A)P(B)$$

The events do not satisfy the condition for independence. We say they are *dependent*.

Looked at in terms of conditional probabilities,

$$P(A|B) = \frac{P(AB)}{P(B)} = \frac{.10}{.25} = .4 > P(A)$$

We might interpret this as positive dependence, since knowing B has occurred (the person is overweight) has increased the probability of A (the person is hypertensive.)   ■

Example 21   Exploiting independence: The probability of solving a business problem
A business is experiencing a problem. To overcome the difficulties, two employees are assigned to work independently to solve the problem. The probability that the first employee will solve the problem is .9 and the probability the second will solve it is .8.

It is reasonable to assume that the success or failure of the first employee to solve the problem is independent of the success or failure of the second employee. Under this assumption, determine the probability that at least one of the two employees will solve the problem.

SOLUTION AND DISCUSSION   Define the two events

$$A = \text{[First employee solves the problem]}$$
$$B = \text{[Second employee solves the problem]}$$

The event of interest is then *A or B*. Clearly, A and B can occur at the same time so the two events are not mutually exclusive. By the addition law,

$$P(A \text{ or } B) = P(A) + P(B) - P(AB)$$

We are given that $P(A) = .9$ and $P(B) = .8$. Since A and B are independent, we can evaluate $P(AB)$ as the product of $P(A)$ and $P(B)$:

$$P(AB) = P(A)P(B) = (.9)(.8) = .72$$

Consequently,

$$P(A \text{ or } B) = P(A) + P(B) - P(AB) = .9 + .8 - .72 = .98$$

The probability is much larger than if just the first employee were assigned to solve the problem.

A second solution to the problem proceeds by considering the complementary event. Let $C = A \text{ or } B$. Then $\bar{C}$ is the event that $A \text{ or } B$ does not occur. It has to be the case that [First employee does not solve the problem] and [Second employee does not solve the problem]. That is,

$$\bar{C} = \bar{A}\bar{B} \quad (\text{see Example 15})$$

Because the employees are working independently, $\bar{A}$ and $\bar{B}$ are independent events. By the definition of independence,

$$P(\bar{A}\bar{B}) = P(\bar{A})P(\bar{B})$$

Also, by the probability law of the complement

$$P(\bar{A}) = 1 - P(A) = 1 - .9 = .1$$
$$P(\bar{B}) = 1 - P(B) = 1 - .8 = .2$$

so $P(\bar{C}) = P(\bar{A}\bar{B}) = P(\bar{A})P(\bar{B}) = (.1)(.2) = .02$. Finally, by the probability law of the complement,

$$P(A \text{ or } B) = P(C) = 1 - P(\bar{C}) = 1 - .02 = .98$$

which matches the previous solution.

This example shows how the laws of probability can be used in different ways to solve a problem. Which solution do you prefer? What if three persons were assigned to work independently instead of two and we are interested in finding the probability that the problem is solved by at least one of the employees? You would find the second method of solution is much easier in this latter case. ∎

Caution: Do not confuse the terms "mutually exclusive events" and "independent events." We say A and B are mutually exclusive when their intersection AB is empty so $P(AB) = 0$. That is, mutually exclusive events cannot happen at the same time. On the other hand, if A and B are independent, $P(AB) = P(A)P(B)$. Independent events can happen at the same time. Both these properties cannot hold simultaneously as long as A and B have nonzero probabilities.

We introduced the condition of independence in the context of checking a given assignment of probability to see if $P(A|B) = P(A)$. A second use of this condition is in the assignment of probability when the experiment consists of two physically unrelated parts. When events A and B refer to unrelated parts of an experiment, AB is assigned the probability $P(AB) = P(A)P(B)$. Example 22 illustrates the assignment of probability by this product formula. Example 23 gives another application.

STATISTICAL REASONING

Airplane Accidents and Independence

One type of passenger airplane has four engines but it can still fly with only one engine. Suppose each engine has probability .001 of failing during flight. If the failures of engines are independent, then the probability that all four engines will fail is

$$(.001)(.001)(.001)(.001) = .000000000001$$

which is extremely small. Now, if an accident happens because all four engines fail, what would be a possible conclusion?

Chas Schneider/FPG International

We should suspect the assumption of independence. For instance, bad fuel or a lightning strike could cause more than one engine to fail. A common cause of failure invalidates the independence of failures and the assignment of probability by taking the product of the individual probabilities.

One accident did occur when all engines on an aircraft failed. An investigation of the accident revealed that the same mechanic had serviced all of the engines and made the same mistake when replacing parts in all of them. This incident prompted the airline to require different mechanics for each engine of an aircraft. This improved practice helps to ensure independence of operation for the engines.

Example 22   Probabilities when sampling with replacement
In the context of Example 18, suppose that a box contains 25 cards identifying the accounts. One card is drawn at random. It is returned to the box and then another card is drawn at random. What is the probability that both draws produce delinquent accounts?

SOLUTION AND   As before, we will use the letter D for delinquent and G for good standing. Be-
DISCUSSION   cause the first card is returned to the box, the contents of the box remain unchanged. Hence, with each draw, $P(D) = \frac{5}{25}$, and the results of the two draws are independent. Instead of working with conditional probability as we did in Example 18, we can use the property of independence to calculate

$$P(D_1 D_2) = P(D_1)P(D_2) = \frac{5}{25} \times \frac{5}{25} = .04$$

Remark 1.   Evidently, this method of probability calculation extends to any number of draws if after each draw, the selected card is returned to the box. For instance, the probability that the first draw produces a D and the next two draws produce G's is

$$P(D_1 G_2 G_3) = \frac{5}{25} \times \frac{20}{25} \times \frac{20}{25} = .128$$

Remark 2.   Sampling with replacement is seldom used in practice, but it serves as a conceptual frame for simple probability calculations when a problem concerns sampling from a large population. For example, consider drawing 3 cards from a box containing 2500 cards, of which 2000 are G's and 500 are D's. Whether or not a selected card is returned to the box before the next draw makes little difference in the probabilities. The model of independence serves as a reasonable approximation.   ∎

The connection between dependent trials and the size of the population merits further emphasis.

Example 23   Dependence and sampling without replacement
If the outcome of a single trial of any experiment is restricted to just two possible outcomes, it can be modeled as drawing a single ball from an urn containing only red (R) and white (W) balls. In the previous example, these two possible outcomes were good and defective. Consider taking a sample of size 3, without replacement, from each of the two populations:

1.   Small population where the urn contains 7 W and 3 R.
2.   Large population where the urn contains 7000 W and 3000 R.

Compare with a sample of size 3 generated from a spinner having a probability .7 of white, where R or W is determined by a separate spin for each trial.

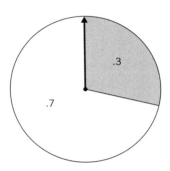

(a)   Calculate the probability assigned to each possible sample.
(b)   Let D = [at least one W]. Calculate the probability of D.

SOLUTION AND DISCUSSION

(a)   We will write RWR for the outcome where the first and third draws are R and the second is W. Applying the general multiplication rule $P(ABC) = P(A)P(B|A)P(C|AB)$, when sampling the small population, we get

$$P(RWR) = P(R)P(W|R)P(R|RW) = \frac{3}{10} \times \frac{7}{9} \times \frac{2}{8} = \frac{42}{720}$$

For the larger population,

$$P(RWR) = \frac{3000}{10,000} \times \frac{7000}{9999} \times \frac{2999}{9998} \approx (.3) \times (.7) \times (.3) = (.3)^2(.7)$$

When the population size is large, the assumption of independence produces a very good approximation.

Under the spinner model, the probability of R is .3 for the first trial and this probability is the same for all trials. A spinner is a classic representation of a device with no memory, so that the outcome of the current trial is independent of the outcomes of all the previous trials. According to the product rule for independence, we assign

$$P(RWR) = (.3) \times (.7) \times (.3)$$

Notice that the spinner model is equivalent to sampling with replacement from either of the two finite populations.

The results for all eight possible samples are shown in Table 2.

TABLE 2 A Comparison of Finite Populations and the Spinner Model

Draw 3 Balls without replacement	Small Population	Large Population	Spinner
	$\begin{matrix}3R\\7W\end{matrix}$	$\begin{matrix}30\%R\\70\%W\end{matrix}$ e.g., 3000R 7000W	30% R W
	$P(ABC) =$ $P(A)P(B|A)P(C|AB)$	$P(ABC) \approx$ $P(A)P(B)P(C)$	$P(ABC) =$ $P(A)P(B)P(C)$
Outcome	Not independent	Approximately independent	Independent
RRR	$\dfrac{3}{10} \times \dfrac{2}{9} \times \dfrac{1}{8} = \dfrac{6}{720}$	$\dfrac{3000}{10,000} \times \dfrac{2999}{9999} \times \dfrac{2998}{9998} \approx (.3)(.3)(.3)$	$(.3)(.3)(.3)$
RRW	$\dfrac{3}{10} \times \dfrac{2}{9} \times \dfrac{7}{8} = \dfrac{42}{720}$	$\approx (.3)(.3)(.7)$	$(.3)^2(.7)$
RWR	$\dfrac{3}{10} \times \dfrac{7}{9} \times \dfrac{2}{8} = \dfrac{42}{720}$	$\approx (.3)(.7)(.3)$	$(.3)^2(.7)$
WRR	$\dfrac{7}{10} \times \dfrac{3}{9} \times \dfrac{2}{8} = \dfrac{42}{720}$	$\approx (.7)(.3)(.3)$	$(.3)^2(.7)$
RWW	$\dfrac{3}{10} \times \dfrac{7}{9} \times \dfrac{6}{8} = \dfrac{126}{720}$	$\approx (.3)(.7)(.7)$	$(.3)(.7)^2$
WRW	$\dfrac{7}{10} \times \dfrac{3}{9} \times \dfrac{6}{8} = \dfrac{126}{720}$	$\approx (.7)(.3)(.7)$	$(.3)(.7)^2$
WWR	$\dfrac{7}{10} \times \dfrac{6}{9} \times \dfrac{3}{8} = \dfrac{126}{720}$	$\approx (.7)(.7)(.3)$	$(.3)(.7)^2$
WWW	$\dfrac{7}{10} \times \dfrac{6}{9} \times \dfrac{5}{8} = \dfrac{210}{720}$	$\approx (.7)(.7)(.7)$	$(.7)^3$
$P(D) = 1 - P(\bar{D})$	$= 1 - \dfrac{6}{720}$	$\approx 1 - (.3)^3$	$= 1 - (.3)^3$

If $A = [$1st is $R]$ $B = [$2nd is $R]$ $C = [$3rd is $R]$ $D = [$at least one $W]$

then $ABC = \{RRR\} = \bar{D}$ $D = \{RRW, RWR, WRR, RWW, WRW, WWR, WWW\}$

(b) The event D is complicated, whereas $\bar{D} = \{RRR\}$ consists of a single outcome. By the law of the complement,

$$P(D) = 1 - P(\bar{D}) = 1 - \frac{3}{10} \times \frac{2}{9} \times \frac{1}{8} = 1 - \frac{6}{720}$$

In the second case, $P(D)$ is approximately $1 - (.3) \times (.3) \times (.3)$ and this answer is exact for the spinner model. ∎

**Chart I:
Computing the Probability of an Event A**

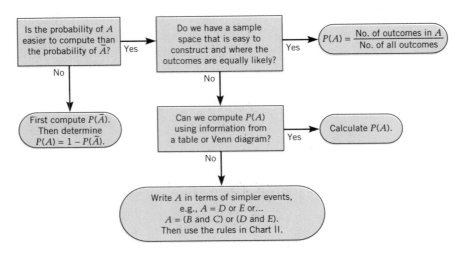

**Chart II:
Computing Probabilities for Two Events A, B**

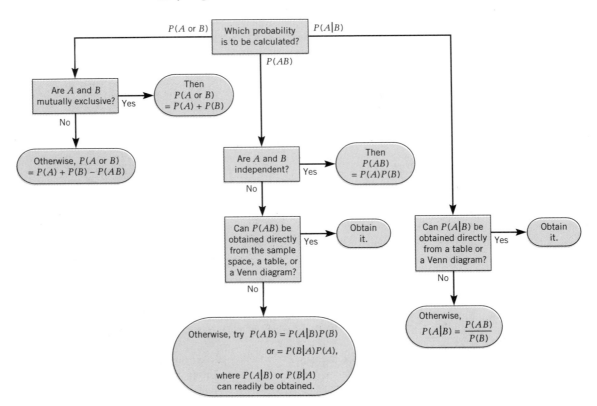

Table 2 summarizes sampling from a small finite population, a large but finite population, and the spinner model. Dependence does matter when sampling without replacement from a small population.

Exercises

5.1   Suppose that $P(A) = .68$, $P(B) = .55$, and $P(AB) = .32$. Find
  (a)   The conditional probability that B occurs given that A occurs.
  (b)   The conditional probability that B does not occur given that A occurs.
  (c)   The conditional probability that B occurs given that A does not occur.

5.2   Suppose that $P(A) = .4$, $P(B) = .6$, and the probability that A or B occurs is .7. Find
  (a)   The conditional probability that A occurs given that B occurs.
  (b)   The conditional probability that B occurs given that A does not occur.

5.3   The following data relate to the proportions in a population of drivers.

$$A = [\text{Defensive driver training last year}]$$
$$B = [\text{Accident in current year}]$$

The probabilities are given in the accompanying Venn diagram. Find $P(B|A)$. Are A and B independent?

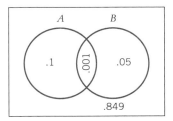

5.4   Suppose $P(A) = .45$, $P(B) = .32$, and $P(\bar{A}B) = .20$.
  (a)   Determine all the probabilities needed to fill in the accompanying table.

	B	$\bar{B}$	
A			.45
$\bar{A}$	.20		
	.32		

  (b)   Find the conditional probability of A given that B does not occur.

5.5   For two events A and B, the following probabilities are given.

$$P(A) = .5 \qquad P(B) = .25 \qquad P(A|B) = .8$$

Use the appropriate laws of probability to calculate
(a) $P(\overline{A})$
(b) $P(AB)$
(c) $P(A \text{ or } B)$

5.6   Records of student patients at a dentist's office concerning fear of visiting the dentist suggest the following proportions.

	School		
	Elementary	Middle	High
Fear	.12	.08	.05
Do not fear	.28	.25	.22

For a student selected at random, consider the events

$$A = [\text{fear}] \qquad M = [\text{middle school}]$$

(a)   Find the probabilities

$$P(A) \qquad P(AM) \qquad P(M) \qquad P(A \text{ or } M)$$

(b)   Are A and M independent?

5.7   An urn contains two green balls and three red balls. Suppose two balls will be drawn at random one after another and *without replacement* (i.e., the first ball is *not* returned to the urn before the second one is drawn).
(a)   Find the probabilities of the events

$$A = [\text{A green ball appears in the first draw}]$$
$$B = [\text{A green ball appears in the second draw}]$$

(b)   Are the two events independent? Why or why not?

5.8   Refer to Exercise 5.7. Now suppose two balls will be drawn *with replacement* (i.e., the first ball drawn will be returned to the urn before the second draw). Repeat parts (a) and (b).

5.9   In a county, men constitute 58% of the labor force. The rates of unemployment are 6.2% and 4.3% among males and females, respectively.
(a)   In the context of selecting a worker at random from the county labor force, state the probabilities represented by the foregoing percentages. (Use symbols such as M for male, E for employed.)

(b) What is the overall rate of unemployment in the county?

(c) If a worker selected at random is found to be unemployed, what is the probability that the worker is a woman?

5.10   If $P(A) = .5$, $P(B) = .5$, and $P(A \text{ or } B) = .8$,

(a) Are A and B independent? Why or why not?

(b) Can A and B be mutually exclusive? Why or why not?

5.11   Suppose $P(A) = .50$ and $P(B) = .22$.

(a) Determine $P(A \text{ or } B)$ if A and B are independent.

(b) Determine $P(A \text{ or } B)$ if A and B are mutually exclusive.

(c) Find $P(A|\overline{B})$ if A and B are mutually exclusive.

5.12   In a region, 15% of the adult population are smokers, 0.86% are smokers with emphysema, and 0.24% are nonsmokers with emphysema.

(a) What is the probability that a person, selected at random, has emphysema?

(b) Given that the selected person is a smoker, what is the probability that this person has emphysema?

(c) Given that the selected person is not a smoker, what is the probability that this person has emphysema?

5.13   Refer to Exercise 3.17.

(a) Suppose a patient will be chosen at random from the group of patients who are below age 40. What is the probability that this patient will have a serious case of the disease? Explain how this can be interpreted as a conditional probability.

(b) Calculate the following conditional probabilities and interpret them in light of your answer to part (a): $P(\overline{A}|B)$, $P(C|A)$.

5.14   Refer to Exercise 3.15.

(a) If a nursing home selected at random is found to comply with security standards, what is the probability that it violates sanitary standards?

(b) If a nursing home selected at random is found to violate at least one of the two standards, what is the probability that it complies with security standards?

5.15   There are 10 batteries in a drawer. Suppose three batteries are chosen at random and are found to be defective. Do you believe that the other seven batteries in the drawer are good ones? Justify your answer by computing the probability that all three chosen batteries are defective, assuming only three defective batteries are in the drawer.

5.16   Suppose you choose two apples at random from a bag of twelve apples and discover that the two apples are rotten inside. Do you believe that these are the only two apples that are rotten inside (i.e., the rest of the apples are good)? Use a probability argument to support your answer.

5.17   Of 20 rats in a cage, 12 are males and 9 are infected with a virus that causes hemorrhagic fever. Of the 12 male rats, 7 are infected with the virus. One rat is randomly selected from the cage.

(a)   If the selected rat is found to be infected, what is the probability that it is a female?

(b)   If the selected rat is found to be a male, what is the probability that it is infected?

(c)   Are the events "the selected rat is infected" and "the selected rat is male" independent? Why or why not?

5.18   The following probabilities are given for two events A and B.

$$P(A) = \frac{1}{2} \qquad P(B) = \frac{1}{4} \qquad P(A|B) = \frac{3}{4}$$

(a)   Using the appropriate laws of probability, calculate $P(\overline{A})$, $P(AB)$, and $P(A \text{ or } B)$.

(b)   Draw a Venn diagram to determine $P(\overline{A}B)$.

5.19   Of three events, A, B, and C, suppose events A and B are independent and events B and C are mutually exclusive. Their probabilities are $P(A) = .7$, $P(B) = .1$, and $P(C) = .3$. Express the following events in set notation and calculate their probabilities.

(a)   Both B and C occur.

(b)   At least one of A and B occurs.

(c) B does not occur.

(d)   All three events occur.

5.20   Approximately 40% of the Wisconsin population has type O blood. If 4 persons are selected at random to be donors, find $P[\text{at least one type O}]$.

5.21   Refer to the Statistical Reasoning example on page 174 concerning airplane engines. Suppose the probability that a certain engine will fail during flight is now .002. What is the probability that all four engines will fail?

5.22   An accountant screens large batches of bills according to the following sampling inspection plan. She inspects 4 bills chosen at random from each batch and passes the batch if, among the 4, none is irregular. Find the probability that a batch will be passed if, in fact

(a)   5% of its bills are irregular

(b)   20% of its bills are irregular

5.23   An electronic scanner is successful in detecting flaws in a material in 70% of the cases. Three material specimens containing flaws will be tested with the scanner. Assume that the tests are independent.

(a) List the sample space and assign probabilities to the simple events.

(b) Find the probability that the scanner is successful in at least two of the three cases.

5.24 Refer to Exercise 3.18. Given that a landfill, selected at random, is found to have a high concentration of mercury, what is the probability that its concentration is

(a) High in barium?

(b) Low in both arsenic and barium?

(c) High in either arsenic or barium?

5.25 Of the patients reporting to a clinic with the symptoms of sore throat and fever, 25% have strep throat, 50% have an allergy, and 10% have both.

(a) What is the probability that a patient selected at random has either strep throat, an allergy, or both?

(b) Are the events "strep throat" and "allergy" independent?

5.26 Suppose that a letter sent from City A has a probability .8 of reaching City B in three days.

(a) Three letters are sent from City A to City B on three different occasions. What is the probability that exactly two letters reach City B in three days?

(b) If the three letters are mailed together at the same time and location, how does the conclusion in part (a) change? Explain.

6. ANOTHER COUNTING TOOL: THE RULE OF COMBINATIONS

In our earlier examples of probability calculations, we have used the phrase "randomly selected" to mean that all possible selections are equally likely. It usually is not difficult to enumerate all the elementary outcomes when both the population size and sample size are small numbers. With larger numbers, making a list of all the possible choices becomes a tedious job.

The simple $m \times n$ counting rule, described in Section 2, often helps us count the number of outcomes in an event without having to list the sample space. Here, we introduce another useful counting rule.

We begin with an example where the population size and the sample size are both small numbers so all possible samples can be conveniently listed.

Example 24 Combinations—Selections where order is not important
There are five qualified applicants for two editorial positions on a college newspaper. Two of these applicants are men and three are women.

(a) List all possible combinations of 2 out of 5 persons.

(b) If the positions are filled by randomly selecting two of the five applicants, what is the probability that neither of the men is selected?

SOLUTION AND DISCUSSION   Suppose the three women applicants are identified as a, b, and c, and the two men as d and e. Two members are selected at random from the

$$\text{Population: } \{a, \underbrace{\quad b, \quad c,}_{\text{women}} \underbrace{d, \quad e}_{\text{men}}\}$$

(a)   The possible samples, or combinations, may be listed as

$$\{a, b\}, \quad \{b, c\}, \quad \{c, d\}, \quad \{d, e\},$$
$$\{a, c\}, \quad \{b, d\}, \quad \{c, e\},$$
$$\{a, d\}, \quad \{b, e\},$$
$$\{a, e\}$$

As the list shows, our sample space has 10 elementary outcomes. That is, there are 10 possible combinations when choosing 2 of 5 persons.

(b)   The notion of random selection entails that these are all equally likely, so each is assigned the probability $\frac{1}{10}$. Let A represent the event that two women are selected. Scanning our list, we see that A consists of the three elementary outcomes

$$\{a, b\}, \quad \{a, c\}, \quad \{b, c\}$$

Consequently,

$$P(A) = \frac{\text{Number of elements in } A}{\text{Number of elements in } S} = \frac{3}{10} = .3 \qquad \blacksquare$$

Note that our probability calculation in Example 24 only requires knowledge of the two counts: the number of elements in S and the number of elements in A. Can we arrive at these counts without formally listing the sample space? An important counting rule comes to our aid.

The Rule of Combinations

Notation: The number of possible choices of r objects from a group of N distinct objects is denoted by $\binom{N}{r}$, which is read as "N choose r."

Formula:

$$\binom{N}{r} = \frac{N \times (N - 1) \times \cdots \times (N - r + 1)}{r \times (r - 1) \times \cdots \times 2 \times 1}$$

More specifically, the numerator of the formula $\binom{N}{r}$ is the product of r consecutive integers starting with N and proceeding downward. The denominator is also the product of r consecutive integers, but starting with r and proceeding down to 1. $\binom{5}{2} = \dfrac{5 \times 4}{2 \times 1} = 10$, as verified in Example 24.

As another example, $\binom{5}{3} = \dfrac{5 \times 4 \times 3}{3 \times 2 \times 1} = 10.$

Although not immediately apparent, there is a certain symmetry in the counts $\binom{N}{r}$. The process of selecting r objects is the same as choosing $(N - r)$ objects to leave behind. Because every choice of r objects corresponds to a choice of $(N - r)$ objects,

$$\binom{N}{r} = \binom{N}{N - r}$$

$$\begin{array}{ccc} & N & \\ r \swarrow & & \searrow N - r \\ \text{choose} & & \text{leave} \end{array}$$

This relation often simplifies calculations. Since $\binom{N}{N} = 1$, we take $\binom{N}{0} = 1.$

Example 25   Evaluating the number of combinations
Calculate the values of $\binom{6}{2}$, $\binom{15}{4}$, and $\binom{15}{11}$.

SOLUTION

$$\binom{6}{2} = \frac{6 \times 5}{2 \times 1} = 15, \qquad \binom{15}{4} = \frac{15 \times 14 \times 13 \times 12}{4 \times 3 \times 2 \times 1} = 1365$$

Using the relation $\binom{N}{r} = \binom{N}{N - r}$, we have

$$\binom{15}{11} = \binom{15}{4} = 1365 \qquad\blacksquare$$

Example 26   Probabilities of being selected under random selection
The dean's office has received 12 nominations from which to designate 4 student representatives to serve on a campus curriculum committee. Among the nominees, 5 are liberal arts majors and 7 are science majors.

(a) How many ways can 4 students be selected from the group of 12?

(b) How many selections are possible that would include 1 science major and 3 liberal arts majors?

(c) If the selection process were random, what is the probability that 1 science major and 3 liberal arts majors would be selected?

SOLUTION AND DISCUSSION

(a) According to the counting rule $\binom{N}{r}$, the number of ways 4 students can be selected out of 12 is

$$\binom{12}{4} = \frac{12 \times 11 \times 10 \times 9}{4 \times 3 \times 2 \times 1} = 495$$

(b) One science major can be chosen from 7 in $\binom{7}{1} = 7$ ways. Also, 3 liberal arts majors can be chosen from 5 in

$$\binom{5}{3} = \frac{5 \times 4 \times 3}{3 \times 2 \times 1} = 10 \text{ ways}$$

Each of the 7 choices of a science major can accompany each of the 10 choices of 3 liberal arts majors. Reasoning from the $m \times n$ counting rule, we conclude that the number of possible samples with the stated composition is

$$\binom{7}{1} \times \binom{5}{3} = 7 \times 10 = 70$$

(c) Random sampling requires that the 495 possible samples are all equally likely. Of these, 70 are favorable to the event $A =$ [1 science and 3 liberal arts majors]. Consequently,

$$P(A) = \frac{70}{495} = .141$$

Similar calculations have been used to study bias in the selection of juries. ∎

Example 27 Coin tossing and the rule of combinations
A fair coin will be tossed 5 times. Find the probability that exactly 2 heads will appear.

SOLUTION AND DISCUSSION

We write HHTTT for the outcome where the first toss is a head (H), the second toss is a head, and the last three tosses are tails (T). The event of having 2 heads in 5 tosses consists of the outcomes HHTTT, HTHTT, and so on. How many outcomes are in this list? Can this number be determined without making a complete listing of all the outcomes?

One view of the problem of listing an outcome in the event is that of putting 2 H's in the five places that describe an outcome. The other 3 positions receive a T. From this point of view, there are $\binom{5}{2} = 10$ outcomes with exactly 2 heads.

Further, to count the number of outcomes in the sample space, we note that there are 2 choices for each of the five positions. By a simple extension of the $m \times n$ counting rule, there are $2 \times 2 \times 2 \times 2 \times 2 = 2^5 = 32$ outcomes. These are all equally likely since the coin is fair. Therefore

$$P[2 \text{ heads in 5 tosses}] = \frac{10}{32} = .3125$$

Alternatively, we can obtain this probability using the fact that the results of different tosses are independent. The outcome HHTTT is assigned probability

$$P(HHTTT) = P(H)P(H)P(T)P(T)P(T) = (.5)(.5)(.5)(.5)(.5) = (.5)^5$$

Similarly, any sequence with 2 H's and 3 T's will have the same probability. Since

$$P[2 \text{ heads in 5 tosses}] = P(HHTTT) + P(HTHTT) + \cdots + P(TTTHH)$$

and there are $\binom{5}{2} = 10$ terms in the sum,

$$P[2 \text{ heads in 5 tosses}] = 10(.5)^5 = .3125$$

This latter approach pertains to the binomial distribution, which will be introduced in the next chapter. ■

The notion of a random sample from a finite population is crucial to statistical inference. To generalize from a sample to the population, it is imperative that the sampling process be impartial. This criterion is evidently met if we allow the selection process to be such that all possible samples are given equal opportunity to be selected. This is precisely the idea behind the term *random sampling*, and a formal definition can be phrased as follows.

A sample of size n selected from a population of N distinct objects is said to be a **random sample** if each collection of size n has the same probability $1 / \binom{N}{n}$ of being selected.

Note that this is a conceptual rather than an operational definition of a random sample. In Chapter 1, Section 4, we showed how to choose a random sample using a table of random digits.

Exercises

6.1 Evaluate:

(a) $\binom{8}{2}$ (b) $\binom{10}{4}$ (c) $\binom{20}{3}$

(d) $\binom{20}{17}$ (e) $\binom{30}{4}$ (f) $\binom{30}{26}$

6.2   List all the samples from $\{a, b, c, d, e\}$ when (a) 3 out of 5 are selected, (b) Compare with Example 24. Count the number of samples in each case.

6.3   Of 9 available candidates for membership in a university committee, 5 are men and 4 women. The committee is to consist of 4 persons.

(a)   How many different selections of the committee are possible?

(b)   How many selections are possible if the committee must have 2 men and 2 women?

6.4   If a coin is tossed 12 times, the outcome can be recorded as a 12-character sequence of H's and T's according to the results of the successive tosses. In how many ways can there be 4 H's and 8 T's? (Put differently, in how many ways can one choose 4 positions out of 12 to put the letter H?)

6.5   Out of 12 people applying for an assembly job, 3 cannot do the work. Suppose two persons will be hired.

(a)   How many distinct pairs are possible?

(b)   In how many of the pairs will 0 or 1 people not be able to do the work?

(c)   If two persons are chosen in a random manner, what is the probability that neither will be able to do the job?

6.6   After a preliminary screening, the list of qualified jurors consists of 10 males and 7 females. The 5 jurors the judge selects from this list are all males. Did the selection process seem to discriminate against females? Answer this by computing the probability of having no female members in the jury if the selection is random.

6.7   A batch of 18 items contains 4 defectives. If three items are sampled at random, find the probability of the event

(a) $A = $ [None of the defectives appear]

(b) $B = $ [Exactly two defectives appear]

6.8   ***Ordered sampling versus unordered sampling.***   Refer to Exercise 6.7. Suppose the sampling of three items is done by randomly choosing one item after another and without replacement. The event A can then be described as $G_1 G_2 G_3$, where G denotes "good" and the subscripts refer to the order of the draws. Use the method of Example 18 to calculate $P(A)$ and $P(B)$. Verify that you get the same results as in Exercise 6.7.

This illustrates the following fact: To arrive at a random sample, we may randomly draw one object at a time without replacement and then disregard the order of the draws.

6.9   A college senior is selected, at random, from each state. Next, one senior is selected at random from the group of 50. Does this procedure produce a senior selected at random from those in the United States?

6.10   An instructor will choose 3 problems from a set of 6 containing 3 hard and 3 easy problems. If the selection is made at random, what is the probability that only the hard problems are chosen?

6.11   Are the following methods of selection likely to produce a random sample of 5 students from your school? Explain.

(a)   Pick 5 students throwing flying discs on the mall.

(b)   Pick 5 students who are studying in the library on Friday night.

(c)   Select 5 students sitting near you in your statistics course.

USING A COMPUTER

6.12   *Using the computer to generate a random sample.* MINITAB will select random integers between specified limits that include the endpoints. (The additional subcommand REPLACE allows the sampling to be done with replacement.)

   We consider drawing a random sample of 10 integers between 1 and 85, both inclusive. The sample is drawn with replacement.

Dialog box: **Session command**

Calc > Random Data > Integer   MTB > SET C1
Type *10* in **Generate,** *1* in **Minimum**   1:85
and *85* in **Maximum.** Click **OK.**   END
                                                   SAMPLE 10 FROM C1 SET C2;
                                                   REPLACE.
                                                   PRINT C2

Output:

                17 22 4 13 50 53 61 6 17 70

Select 15 random integers between 1 and 129:

(a)   With replacement

(b)   Without replacement (Drop "REPLACE." subcommand.)

KEY IDEAS

The probability model of an experiment is described by

1.   The sample space, a list or statement of all possible distinct outcomes.

2.   Assignment of probabilities to all the elementary outcomes. $P(e) \geq 0$ and $\Sigma P(e) = 1$, where the sum extends over all e in S.

The probability of an event A is the sum of the probabilities of all the elementary outcomes that are in A.

A uniform probability model holds when all the elementary outcomes in S are equiprobable. With a **uniform probability model,**

$$P(A) = \frac{\text{Number of outcomes in } A}{\text{Number of outcomes in } S}$$

$P(A)$, viewed as the long-run relative frequency of A, can be approximately determined by repeating the experiment a large number of times.

The three basic laws of probability are

Law of the Complement	$P(A) = 1 - P(\overline{A})$
Addition Law	$P(A \text{ or } B) = P(A) + P(B) - P(AB)$
Multiplication Law	$P(AB) = P(B)P(A|B)$

These are useful in probability calculations when events are formed with the operations of complement, union, and intersection.

The concept of conditional probability is useful to determine how the probability of an event A must be revised when another event B has occurred.

Conditional probability of A given B

$$P(A|B) = \frac{P(AB)}{P(B)}$$

This forms the basis of the multiplication law of probability and the notion of

independent events $P(AB) = P(A)P(B)$

The notion of **random sampling** is formalized by requiring that all possible samples are equally likely to be selected. The rule of combinations facilitates the calculation of probabilities in the context of random sampling from N distinct units.

7. REVIEW EXERCISES

7.1   Describe the sample space for each of the following experiments.

(a)   The record of your football team after its first game next season.

(b)   The number of students out of 20 who will pass beginning swimming and graduate to the intermediate class.

(c)   In an unemployment survey, 1000 persons will be asked to answer "yes" or "no" to the question "Are you employed?" Only the number answering "no" will be recorded.

(d)   A geophysicist wants to determine the natural gas reserve in a particular area. The volume will be given in cubic feet.

7.2   For the experiments in Exercise 7.1, which sample spaces are discrete and which are continuous?

7.3 Identify these events in the corresponding parts of Exercise 7.1.
   (a) Don't lose
   (b) At least half the students pass.
   (c) Less than or equal to 5.5% unemployment
   (d) Between 1 and 2 million cubic feet

7.4 Examine each of these probability assignments and state what makes it improper.
   (a) Concerning tomorrow's weather,

$$P(\text{rain}) = .2, \qquad P(\text{cloudy but no rain}) = .4, \qquad P(\text{sunny}) = .6$$

   (b) Concerning your passing the statistics course,

$$P(\text{pass}) = 2 \qquad P(\text{fail}) = .2$$

   (c) Concerning your grades in statistics and economics courses,

$$P(\text{A in statistics}) = .5 \qquad P(\text{A in economics}) = .8$$
$$P(\text{A's in both statistics and economics}) = .6$$

7.5 A driver is stopped for erratic driving, and the alcohol content of his blood is checked. Specify the sample space and the event $A = [\text{Level exceeds legal limit}]$ if the legal limit is .10.

7.6 The Wimbledon men's tennis championship ends when one player wins three sets.
   (a) How many elementary outcomes end in three sets? in four?
   (b) If the players are evenly matched, what is the probability that the tennis match ends in four sets?

7.7 There are four tickets numbered 1, 2, 3, and 4. Suppose a two-digit number will be formed by first drawing one ticket at random and then drawing a second ticket at random from the remaining three. (For instance, if the first ticket drawn shows 3 and the second shows 1, the number recorded is 31.) List the sample space and determine the following probabilities.
   (a) An even number
   (b) A number larger than 20
   (c) A number between 22 and 30

7.8 To compare two varieties of wheat, say, *a* and *b*, a field trial will be conducted on four square plots located in two rows and two columns (see page 192). Each variety will be planted on two of these plots.
   (a) List all possible assignments for variety *a*.
   (b) If the assignments are made completely at random, find the probability that the plots receiving variety *a* are
      (i) In the same column
      (ii) In different rows and different columns

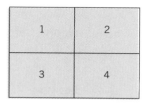

Plot arrangement

7.9   Refer to Exercise 7.8. Instead of a completely random choice, suppose a plot is chosen at random from each row and assigned to variety *a*. Find the probability that the plots receiving *a* are in the same column.

7.10   The Chevalier de Méré, a French nobleman of the seventeenth century, reasoned that in a single throw of a fair die, $P(1) = \frac{1}{6}$, so in two throws, $P(1 \text{ appears at least once}) = \frac{1}{6} + \frac{1}{6} = \frac{1}{3}$. What is wrong with the above reasoning? Use the sample space of Exercise 2.15 to obtain the correct answer.

7.11   A letter is chosen at random from the word "STATISTICIAN."
       (a)   What is the probability that it is a vowel?
       (b)   What is the probability that it is a T?

7.12   Does the uniform model apply to the following observations? Explain.
       (a)   Day of week on which maximum pollution reading, for nitrous oxides, occurs downtown in a large city
       (b)   Day of week on which monthly maximum temperature occurs
       (c)   Week of year for peak retail sales of new cars

7.13   A three-digit number is formed by arranging the digits 1, 5, and 6 in a random order.
       (a)   List the sample space.
       (b)   Find the probability of getting a number larger than 400.
       (c)   What is the probability that an even number is obtained?

7.14   A late shopper for Valentine's Day flowers calls by phone to have a flower wrapped. The store has only 4 roses, of which 3 will open by the next day, and 6 tulips, of which 2 will open by the next day.
       (a)   Construct a Venn diagram and show the events $A = $ [rose], and $B = $ [will open the next day].
       (b)   If the store selects one flower at random, find the probability that it will not open by the next day.

7.15   In checking the conditions of a used car, let A denote the event that the car has a faulty transmission, B the event that it has faulty brakes, and C the event that it has a faulty exhaust system. Describe in words what the following events represent.
       (a) $A \text{ or } B$ (b) ABC (c) $\overline{A}\,\overline{B}\,\overline{C}$ (d) $\overline{A} \text{ or } \overline{B}$

7.16 Express the following statements in the notations of the event operations.

(a) *A* occurs and *B* does not

(b) Neither *A* nor *B* occurs

(c) Exactly one of the events *A* and *B* occurs

7.17 Suppose each of the numbers .13, .47, and .68 represents the probability of one of the events *A*, *AB*, and *A or B*. Connect the probabilities to the appropriate events.

7.18 From the probabilities exhibited in this Venn diagram, find $P(\overline{A})$, $P(AB)$, $P(B\ or\ C)$, and $P(BC)$.

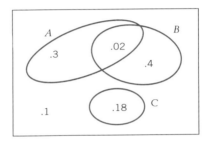

7.19 Using event relations, express the following events in terms of the three events *A*, *B*, and C.

(a) All three events occur.

(b) At least one of the three events occurs.

(c) *A* and *B* occur and C does not.

(d) Only *B* occurs.

7.20 Concerning three events *A*, *B*, and C, the probabilities of the various intersections are given in the accompanying table. [For instance, $P(AB\overline{C}) = .10$.]

	B		$\overline{B}$	
	C	$\overline{C}$	C	$\overline{C}$
A	.05	.10	.12	.17
$\overline{A}$	.13	.15	.18	.10

(a) Draw a Venn diagram, identify the intersections, and mark the probabilities.

(b) Determine the probabilities

$$P(AB) \qquad P(A\overline{C}) \qquad P(C)$$

(c) Fill in the accompanying probability table concerning the events A and B.

	B	$\bar{B}$
A		
$\bar{A}$		

7.21 Referring to Exercise 7.20, calculate the probabilities of the following events.

(a) Both B and C occur.

(b) Either B or C occurs.

(c) B occurs and C does not occur.

(d) Only one of the three events A, B, and C occurs.

7.22 Concerning three events A, B, and C, the following probabilities are specified.

$$P(A) = .51 \qquad P(AB) = .17 \qquad P(ABC) = .12$$
$$P(B) = .45 \qquad P(BC) = .20$$
$$P(C) = .50 \qquad P(AC) = .33$$

Draw a Venn diagram and determine the probabilities of all the intersections that appear in the diagram. Also, make a probability table like the one given in Exercise 7.20.

7.23 Referring to Exercise 7.22, find the probability that

(a) B occurs and C does not occur.

(b) At least one of the events A and B occurs.

(c) Exactly two of the events A, B, and C occur.

7.24 Suppose a fair die has its even-numbered faces painted red, and the odd-numbered faces are white. Consider the experiment of rolling the die once and the events.

$$A = [\text{2 or 3 shows up}]$$
$$B = [\text{A red face shows up}]$$

Find the following probabilities:

(a) $P(A)$ (b) $P(B)$ (c) $P(AB)$

(d) $P(A|B)$ (e) $P(A \text{ or } B)$

7.25 Given $P(AB) = .4$ and $P(B) = .8$ find $P(A|B)$. If further, $P(A) = .5$, are A and B independent?

7.26 Suppose three events A, B, and C are such that B and C are mutually exclusive and

$$P(A) = .6 \qquad P(B) = .3 \qquad P(C) = .25$$

$$P(A|B) = \frac{2}{3} \qquad P(\bar{A}C) = .1$$

(a) Show the events in a Venn diagram.

(b) Determine the probabilities of all the intersections and mark them in the Venn diagram.

(c) Find the probability that only one of the three events occurs.

7.27 Refer to Exercise 7.26. For each pair of events given as follows, determine whether or not the events are independent.

(a) A, C

(b) $A\bar{B}$, C

7.28 Concerning the events A and B, the following probabilities are given.

$$P(B) = \frac{1}{3} \qquad P(A|B) = \frac{2}{3} \qquad P(A|\bar{B}) = \frac{3}{7}$$

Determine (a) $P(A\bar{B})$, (b) $P(A)$, and (c) $P(\bar{B}|A)$.

7.29 Refer to the probability table given in Exercise 7.20 concerning three events, A, B, and C.

(a) Find the conditional probability of A given that B does not occur.

(b) Find the conditional probability of B given that both A and C occur.

(c) Determine whether the events A and C are independent.

7.30 Mr. Hope, a character apprehended by Sherlock Holmes, was driven by revenge to commit two murders. He presented two seemingly identical pills, one containing a deadly poison, to an adversary who selected one while Mr. Hope took the other. The entire procedure was then to be repeated with the second victim. Mr. Hope felt that Providence would protect him, but what is the probability of the success of his endeavor?

7.31 In an optical sensory experiment, a subject shows a fast response (F), a delayed response (D), or no response at all (N). The experiment will be performed on two subjects.

(a) Using a tree diagram, list the sample space.

(b) Suppose, for each subject, $P(F) = .4$, $P(D) = .3$, $P(N) = .3$, and the responses of different subjects are independent. Assign probabilities to the elementary outcomes.

(i) Find the probability that at least one of the subjects shows a fast response.

(ii) Find the probability that both of the subjects respond.

7.32 A file cabinet has eight students' folders arranged alphabetically according to their last names. Three folders are to be selected at random.

(a) How many different selections are possible?

(b) Find the probability that the selected folders are all adjacent. (*Hint:* Enumerate the selection of adjacent folders.)

7.33 An Internal Revenue Service agent receives a batch of 15 tax returns that were flagged by computer for possible tax evasions. Suppose, unknown to the agent, 6 of these returns have illegal deductions and the other 9 are in good standing. If the agent randomly selects 4 of these returns for audit, what is the probability that

(a) None of the returns that contain illegal deductions are selected?

(b) At least 2 have illegal deductions?

7.34 Referring to Exercise 7.33, now suppose that the agent will randomly select 4 returns to be audited one after another. What is the probability that the first 2 returns audited are in good standing and the last 2 have illegal deductions?

7.35 Chris commutes to work on two different routes A and B. If Chris comes home by route A, then he will be home before 6 P.M. with probability .9. If Chris chooses route B, then he will be home before 6 P.M. with probability .7. In the past, the proportion of times that Chris chose route A is .6.

(a) What proportion of times is Chris home from his office before 6 P.M.?

(b) If Chris is home after 6 P.M. today, what is the probability that he took route B?

7.36 *Birthdays.* It is somewhat surprising to learn the probability that two persons in a class share the same birthday. As an approximation, assume that the 365 days are equally likely birthdays.

(a) What is the probability that, among 3 persons, at least two have the same birthday? (*Hint:* The reasoning associated with a tree diagram shows that there are $365 \times 365 \times 365$ possible birthday outcomes. Of these, $365 \times 364 \times 363$ correspond to no common birthday.)

(b) Generalize the above reasoning to N persons. Show that

$$P[\text{No common birthday}] = \frac{365 \times 364 \times \cdots \times (365 - N + 1)}{(365)^N}$$

(Some numerical values are

N	5	9	18	22	23
$P[\text{No common birthday}]$	.973	.905	.653	.524	.493

We see that with $N = 23$ persons, the probability is greater than $\frac{1}{2}$ that at least two share a common birthday.)

CHAPTER 5

Random Variables and Probability Distributions

Chapter Objectives

After reading this chapter, you should be able to

▶ Understand a random variable as a model for variation.
▶ Describe a probability distribution as a model for discrete random variables.
▶ Compute the mean, variance, and standard deviation.
▶ Identify experiments whose outcomes are of only two types and which satisfy the conditions of Bernoulli trials.
▶ Understand the binomial distribution as a model for some counts.
▶ Use the table of binomial probabilities.

1. INTRODUCTION

The probability model for an experiment has two basic components:

1. The sample space consisting of all elementary outcomes
2. An assignment of probability to each event of interest

In Chapter 4, we encountered some experiments where the outcomes are categorical. The sample space with four outcomes {TT, TH, HT, HH} results when two coins are tossed. Regarding a person's attitude toward a proposed new state law, the outcome could be the response: oppose, neutral, or support. Even in these cases, the feature of interest for the outcome is usually numerical. It could be the number of heads (H) for the coin flips. If 100 persons are asked about the proposed law, the numerical feature is the summary consisting of the frequencies of the oppose, neutral, and support categories.

Of course, the outcomes of many experiments are naturally numerical. The total number of credits earned by a student, monthly apartment rent, and median starting salary for recent graduates are examples. Although the elementary outcomes may or may not be numerical, it will be a numerical aspect of the outcome or *numerical summary* of the experiment that will provide a basis for any statistical analysis.

In this chapter, we show how to model the variation in the outcome of an experiment that produces discrete data. This development will rest on the *probability distribution*, which describes how probability is spread over the possible numerical values associated with the outcomes. For example, using the probability distribution, we may examine which intervals of possible values are most likely to occur and which intervals are unlikely. Having done so, we can observe the numerical outcome with the intent of obtaining support for or against a possible probability distribution. This support will depend on whether a likely or unlikely value occurs. In later chapters, we will develop this method for making inferences.

Example 1   **Introducing random variables and probability distributions**
One house will be selected at random in a large community having 10,000 homes. Because the selection is random, each house has the same probability of being selected.

(a) One numerical characteristic of interest may be the number of cars owned by residents. Give some other numerical characteristics of a house that may be of interest.

(b) Suppose that the proportions of homes with 0, 1, 2, and 3 cars are .2, .4, .3, and .1, respectively. Express the probability distribution of the number of cars owned by residents of the house selected.

SOLUTION AND DISCUSSION   (a) Houses have several characteristics that can be expressed by numbers.
(i) The number of bedrooms in the house selected

(ii) The number of persons under 18 years old living in the house selected

(iii) The market value of the house selected

(iv) The water consumption, in a day, for the house selected

Each of these numerical characteristics of the house selected is a *random variable* in the sense that its value is associated with a randomly selected house. The selected house varies from one sample selection to another so the numerical value is not known in advance.

(b) For the randomly selected house, denote the number of cars by X. Because the selection is random, each house has the same probability of being selected. The proportion of homes with 0 cars, .2, is then the probability $P[X = 0]$. Continuing, we obtain the summary presented in Table 1, which is called a *probability distribution*.

TABLE 1 Probability Distribution of Number of Cars

Value of X	Probability
0	.2
1	.4
2	.3
3	.1

A probability distribution describes how probability is distributed over the possible values of the numerical characteristic. The probability distribution in Example 1 is the same as the relative frequency table of the population whose individual measurement is the number of cars owned by residents of a house in the community. When the population consists of units that physically exist, this will always be the case.

A different probability distribution would result if the characteristic of interest was the number of bedrooms in the house selected. This probability distribution would describe, for example, the probability of selecting a house with 4 bedrooms.

Example 1 introduced the ideas of a **random variable** and its **probability distribution**. These key statistical objects are explained in this chapter.

2. RANDOM VARIABLES AND THEIR PROBABILITY DISTRIBUTIONS

Based on our reasoning in Example 1, we call the correspondence between each outcome of an experiment and the numerical value of a characteristic, a random variable if it is numerical valued. If not, we can often assign numbers in a meaningful way. Probability is then assigned to each possible value.

RANDOM VARIABLE

The idea of a random variable formalizes a numerical valued outcome or numerical summary of an experiment.

A random variable X associates a numerical value with each outcome of an experiment.

We use letters such as X and Y for a random quantity. Once a random variable is defined and an experiment conducted, the random variable takes the value associated with the outcome. The word "random" reminds us that beforehand, the precise value that will occur is unknown because the outcome is uncertain.

Example 2 Number of heads as a random variable
Consider X to be the number of heads obtained in three tosses of a coin. List the numerical values of X and the corresponding elementary outcomes.

SOLUTION AND DISCUSSION First, X is a variable since the number of heads in three tosses of a coin can have any of the values 0, 1, 2, or 3. Second, this variable is random in the sense that the value that would occur in a given instance cannot be predicted with certainty. We can, though, make a list of the elementary outcomes and the associated values of X.

Outcome	Value of X
HHH	3
HHT	2
HTH	2
HTT	1
THH	2
THT	1
TTH	1
TTT	0

Note that, for each elementary outcome there is only one value of X. However, several elementary outcomes may yield the same value. Scanning our list, we now identify the events (i.e., the collections of the elementary outcomes) that correspond to the distinct values of X.

Numerical Value of X as an Event		Composition of the Event
$[X = 0]$	$=$	{TTT}
$[X = 1]$	$=$	{HTT, THT, TTH}
$[X = 2]$	$=$	{HHT, HTH, THH}
$[X = 3]$	$=$	{HHH}

Guided by this example, we observe the following general facts.

The events corresponding to the distinct values of X are mutually exclusive.

The union of these events is the entire sample space.

Typically, the possible values of a random variable X can be determined directly from the description of the random variable without listing the sample space. However, to assign probabilities to these values, treated as events, it is sometimes helpful to refer to the sample space.

Example 3   At an intersection, an observer will count the number X of cars passing by until a new Mercedes is spotted. The possible values of X are then 1, 2, 3, . . . , where the list never terminates.

A random variable is said to be a **discrete random variable** if its possible values proceed in steps with either a finite or an infinite number of steps. Essentially, in our discussions, the discrete random variables will be counts. Referring to Example 1, the number of bedrooms, number of cars, and number of persons under 18 years old are discrete, as are the random variables in Examples 2 and 3.

If a random variable represents a measurement on a continuous scale, so that all values in an interval are possible, it is called a **continuous random variable.** In Example 1, the market value of the house and the daily water consumption are continuous random variables.

The definitions of discrete and continuous random variables are analogous to the definitions of discrete and continuous data in Chapter 2. When a discrete random variable is observed, its numerical value becomes part of a discrete data set.

The next step is to assign probabilities to the values of a discrete random variable. The treatment of continuous variables is different and is postponed until Chapter 6.

PROBABILITY DISTRIBUTION OF A DISCRETE RANDOM VARIABLE

Given a random variable X, we now employ the ideas of probability to determine the chances of observing various values. These probabilities are described by a probability distribution.

> The probability distribution, or simply the distribution, of a discrete random variable X is a list of the distinct numerical values of X along with their associated probabilities.
>
> Often, a formula can be used in place of a detailed list.

A probability distribution can arise (i) as an approximation based on long-run frequency calculated from a large empirical study, (ii) from sampling an actual physical population, or (iii) from a theoretical argument often based on equally likely outcomes. We illustrate all three cases.

Example 4   A probability distribution based on an empirical study
Let X denote the number of magazines to which a college senior subscribes. From a survey of 400 college seniors, suppose the frequency distribution of Table 2 was observed. Approximate the probability distribution of X.

TABLE 2   Frequency Distribution of the Number X of Magazine Subscriptions

Magazine Subscriptions (x)	Frequency	Relative Frequency[a]
0	61	.15
1	153	.38
2	106	.27
3	56	.14
4	24	.06
Total	400	1.00

[a]Rounded to second decimal.

SOLUTION AND DISCUSSION   Viewing the relative frequencies as empirical estimates of the probabilities, we have essentially obtained an approximate determination of the probability distribution of X. The true probability distribution would emerge if a vast number (ideally, the entire population) of seniors were surveyed.   ∎

Example 5   A probability distribution based on a physical population
Refer to Example 1. One house will be selected at random in a large community having 10,000 homes. The proportions of homes with 0, 1, 2, and 3 cars are .2, .4, .3, and .1, respectively. Obtain the probability distribution of the number of cars owned by residents of the house selected.

SOLUTION AND DISCUSSION   Denote the number of cars by X. Because the selection is random, each house has the same probability of being selected. The proportion of homes with 0 cars, .2, is then the probability $P[X = 0] = f(0)$. Continuing, we obtain the summary presented in Table 3.

TABLE 3   Probability Distribution
of Number of Cars

Value of X x	Probability $f(x)$
0	.2
1	.4
2	.3
3	.1

Since the population consists of units that physically exist, the probability distribution corresponds to the relative frequency table for the number of cars owned by residents. ∎

Example 6   A probability distribution determined by theoretical argument
If X represents the number of heads obtained in three tosses of a fair coin, find the probability distribution of X.

SOLUTION AND DISCUSSION   In Example 2, we have already listed the eight elementary outcomes and the associated values of X. The distinct values of X are 0, 1, 2, and 3. We now calculate their probabilities.

The model of a fair coin entails that the eight elementary outcomes are equally likely, so each is assigned the probability $\frac{1}{8}$. The event $[X = 0]$ has the single outcome TTT, so its probability is $\frac{1}{8}$. Similarly, the probabilities of $[X = 1]$, $[X = 2]$, and $[X = 3]$ are found to be $\frac{3}{8}, \frac{3}{8}$, and $\frac{1}{8}$, respectively. Collecting these results, we obtain the probability distribution of X displayed in Table 4.

TABLE 4   The Probability Distribution of X,
the Number of Heads in Three
Tosses of a Coin

Value of X	Probability
0	$\frac{1}{8}$
1	$\frac{3}{8}$
2	$\frac{3}{8}$
3	$\frac{1}{8}$
Total	1

∎

For general discussion, we will use the notation x_1, x_2, and so on to designate the distinct values of a random variable X. The probability that a particular value x_i occurs will be denoted by $f(x_i)$. As in Example 6, if X can take k possible values x_1, . . . , x_k with the corresponding probabilities $f(x_1)$, . . . , $f(x_k)$, the probability distribution of X can be displayed in the format of Table 5.

TABLE 5   Form of a Discrete
Probability Distribution

Value x	Probability $f(x)$
x_1	$f(x_1)$
x_2	$f(x_2)$
.	.
.	.
.	.
x_k	$f(x_k)$
Total	1

Since the quantities $f(x_i)$ represent probabilities, they must all be numbers between 0 and 1. Furthermore, when summed over all possible values of X, these probabilities must add up to 1.

The **probability distribution** of a discrete random variable X is described as the function

$$f(x_i) = P[X = x_i]$$

which gives the probability for each value and satisfies:

1. $f(x_i) \geq 0$, for each value x_i of X

2. $\sum_{i=1}^{k} f(x_i) = 1$

A probability distribution describes the manner in which the total probability 1 gets apportioned to the individual values of the random variable.

A graphical presentation of a probability distribution helps reveal any pattern in the distribution of probabilities. Is there symmetry about some value or a long tail to one side? Is the distribution peaked with a few values having high probabilities, or is it uniform?

We consider a display similar in form to a relative frequency histogram discussed in Chapter 2. It will also facilitate the building of the concept of a continuous distribution. To draw a **probability histogram,** we first mark the values of X

on the horizontal axis. With each value x_i as center, a vertical rectangle is drawn whose area equals the probability $f(x_i)$. The probability histogram for the distribution of Example 6 is shown in Figure 1.

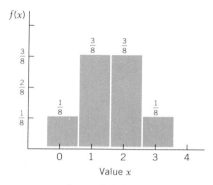

Figure 1. The probability histogram of X, the number of heads in three tosses of a coin

Example 7   Probability distribution for number of infested trees

Suppose 30% of the trees in a forest are infested with a parasite. Four trees are randomly sampled. Let X denote the number of the trees sampled that have the parasite. Obtain the probability distribution of X and plot the probability histogram.

SOLUTION AND DISCUSSION   Because each tree may be either infested (I) or not infested (N), the number of elementary outcomes concerning a sample of 4 trees is $2 \times 2 \times 2 \times 2 = 16$. These can be conveniently enumerated in the scheme of Example 5, Chapter 4 (called, coincidentally, a tree diagram). However, we list them here according to the count X.

$X = 0$	$X = 1$	$X = 2$	$X = 3$	$X = 4$
NNNN	NNNI	NNII	NIII	IIII
	NNIN	NINI	INII	
	NINN	NIIN	IINI	
	INNN	INNI	IIIN	
		ININ		
		IINN		

Our object here is to calculate the probability of each value of X. To this end, we first reflect on the assignment of probabilities to the elementary outcomes.

For a single tree selected at random, we obviously have $P(I) = .3$ and $P(N) = .7$ because 30% of the population of trees is infested. Moreover, as the population is vast while the sample size is very small, the observations on 4 trees can, for all practical purposes, be treated as independent. That is, knowledge that the first tree in the sample is infested does not change the probability that the second is infested and so on.

Invoking independence and the multiplication law of probability, we calculate $P(NNNN) = .7 \times .7 \times .7 \times .7 = .2401$, so $P[X = 0] = .2401$. The event $[X = 1]$ has 4 elementary outcomes, each containing three N's and one I. Since $P(NNNI) = (.7)^3 \times (.3) = .1029$ and the same result holds for each of these 4 elementary outcomes, we get $P[X = 1] = 4 \times .1029 = .4116$. In the same manner,

$$P[X = 2] = 6 \times (.7)^2 \times (.3)^2 = .2646$$
$$P[X = 3] = 4 \times (.7) \times (.3)^3 = .0756$$
$$P[X = 4] = (.3)^4 = .0081$$

Collecting these results, we obtain the probability distribution of X presented in Table 6 and the probability histogram plotted in Figure 2.

TABLE 6   The Probability Distribution of X in Example 7

x	$f(x)$
0	.2401
1	.4116
2	.2646
3	.0756
4	.0081
Total	1.0000

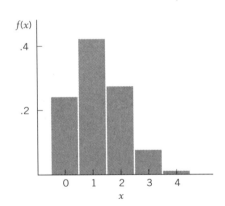

Figure 2. Probability histogram for observed number of trees infested

The reader should bear in mind an important distinction between a relative frequency distribution and the probability distribution. The former is a sample-based entity and is therefore susceptible to variation on different occasions of sampling. By contrast, the probability distribution is a stable entity that refers to the entire population. It is a model for describing the variation in the population and, consequently, it has a key role in the development of inference procedures.

The probability distribution of X can be used to calculate the probabilities of

TABLE 7 A Probability
Distribution

Value x	Probability $f(x)$
0	.02
1	.23
2	.40
3	.25
4	.10

events defined in terms of X. To illustrate this, consider the probability distribution of Table 7. What is the probability that X is equal to or larger than 2?
The event $[X \geq 2]$ is composed of $[X = 2]$, $[X = 3]$, and $[X = 4]$. Thus,

$$P[X \geq 2] = f(2) + f(3) + f(4)$$
$$= .40 + .25 + .10 = .75$$

Similarly, we also calculate

$$P[X \leq 2] = f(0) + f(1) + f(2)$$
$$= .02 + .23 + .40 = .65$$

Exercises

2.1   Identify each of the following as a discrete or continuous random variable.
   (a)   Number of cars serviced at a garage during one day
   (b)   Amount of precipitation produced by a storm system
   (c)   Number of hunting accidents in a state during the deer hunting
         season
   (d)   Time a mechanic takes to replace a defective muffler
   (e)   Number of correct answers that a student gives on a quiz containing
         20 problems
   (f)   Number of cars ticketed for illegal parking on campus today

2.2   Identify the variable as a discrete or continuous random variable.
   (a)   The loss of weight following a diet program
   (b)   The magnitude of an earthquake as measured on the open-ended
         Richter scale
   (c)   The seating capacity of an airplane
   (d)   The number of cars sold at a dealership on one day
   (e)   The percentage of fruit juice in a drink mix

2.3   Two of the integers {2, 4, 6, 7, 8} are chosen at random. Let X denote the difference of the smaller from the larger number.

   (a)   List all choices and the corresponding values of X.

   (b)   List the distinct values of X and determine their probabilities.

2.4   Three contestants A, B, and C are rated by two judges. Each judge assigns the ratings 1 for best, 2 for intermediate, and 3 for worst. Let X denote the total score for contestant A (the sum of the ratings received from the two judges).

   (a)   List all pairs of ratings that contestant A can receive.

   (b)   List the distinct values of X.

2.5   Each week a grocery shopper buys either canned (C) or bottled (B) soft drinks. The type of soft drink purchased in 3 consecutive weeks is to be recorded.

   (a)   List the sample space.

   (b)   If a different type of soft drink is purchased than in the previous week, we say that there is a switch. Let X denote the number of switches. Determine the value of X for each elementary outcome. (*Example:*   For BBB, X = 0; for BCB, X = 2.)

2.6   A child psychologist, interested in how friends are selected, studies groups of three children. For one group, Ann, Barb, and Carol, each is asked which of the other two she likes best.

   (a)   Make a list of the outcomes. (Use A, B, and C to denote the three children.)

   (b)   Let X be the number of times Carol is chosen. List the values of X.

2.7   Listed as follows are the elementary outcomes of an experiment, their probabilities, and the value of a random variable X at each outcome.

Elementary Outcome	Probability	Value of X
e_1	.08	2
e_2	.29	0
e_3	.15	2
e_4	.08	0
e_5	.16	4
e_6	.11	0
e_7	.13	0

Obtain the probability distribution of X.

2.8   Two of the integers {1, 2, 6, 7, 9} are chosen at random. Let X denote the sum of the two integers.

(a)   List all choices and the corresponding values of X.

(b)   List the distinct values of X.

(c)   Obtain the probability distribution of X.

2.9   Let the random variable X represent the sum of the points in two tosses of a die.

(a)   List the possible values of X.

(b)   For each value of X, list the corresponding elementary outcomes.

(c)   Obtain the probability distribution of X.

2.10   Determine whether the following are legitimate probability distributions.

(a)		(b)		(c)		(d)	
x	$f(x)$	x	$f(x)$	x	$f(x)$	x	$f(x)$
2	.2	1	.4	−2	.25	0	.3
8	.6	3	.5	0	.50	1	−.1
13	.1	9	.3	2	.25	2	.8
15	.1	10	.2	4	0		

2.11   The probability distribution of x is given by the function

$$f(x) = \frac{1}{15} \binom{4}{x} \quad \text{for } x = 1, 2, 3, 4$$

Find (a) $P[X = 2]$, (b) $P[X \text{ is odd}]$.

2.12   Refer to Exercise 2.5. Suppose that for each purchase, $P(B) = \frac{1}{2}$ and the decisions in different weeks are independent. Assign probabilities to the elementary outcomes and obtain the distribution of X.

2.13   Refer to Exercise 2.6. Assuming each choice is equally likely, determine the probability distribution of X.

2.14   From the six marbles numbered as shown,

①   ①   ①   ①   ②   ②

two marbles will be drawn at random without replacement. Let X denote the sum of the numbers on the selected marbles. List the possible values of X and determine the probability distribution.

2.15   Suppose, for a loaded die, the probabilities of the faces

1,   2,   3,   4,   5,   6

are in the ratios $1:3:5:2:1:1$. Let X denote the number appearing on a single roll of the die.

(a) Determine the probability distribution of X.

(b) What is the probability of getting an even number?

2.16 A surprise quiz contains three multiple-choice questions: Question 1 has three suggested answers, Question 2 has four, and Question 3 has two. A completely unprepared student decides to choose the answers at random. Let X denote the number of questions the student answers correctly.

(a) List the possible values of X.

(b) Find the probability distribution of X.

(c) Find $P[\text{at least 1 correct}] = P[X \geq 1]$.

(d) Plot the probability histogram.

2.17 A probability distribution is partially given in the following table with the additional information that the even values of X are equally likely. Determine the missing entries in the table.

x	$f(x)$
1	.2
2	
3	.2
4	
5	.3
6	

2.18 Consider the following setting of a random selection: A box contains 100 cards, of which 25 are numbered 1, 28 are numbered 2, 30 are numbered 3, 17 are numbered 4. One card will be drawn from the box and its number X observed. Give the probability distribution of X.

2.19 Two probability distributions are shown in the following tables. For each case, describe a specific setting of random selection (like the one given in Exercise 2.18) that yields the given probability distribution.

(a)		(b)	
x	$f(x)$	x	$f(x)$
2	.36	-2	3/11
4	.42	0	4/11
6	.22	4	2/11
		5	2/11

2.20   In a study of the life length of a species of mice, 120 newborn mice are observed. The numbers staying alive past the first, second, third, and fourth years are 106, 72, 25, and 0, respectively. Let X denote the life length (in discrete units of whole years) of this species of mice. Using these data, make an empirical determination of the probability distribution of X.

2.21   Use the probability distribution given here to calculate
(a) $P[X \leq 3]$
(b) $P[X \geq 2]$
(c) $P[1 \leq X \leq 3]$

x	$f(x)$
0	.12
1	.25
2	.43
3	.12
4	.08

2.22   Of seven candidates seeking three positions at a counseling center, four have degrees in social science and three do not. If three candidates are selected at random, find the probability distribution of X, the number having social science degrees among the selected persons.

2.23   Based on recent records, the manager of a car painting center has determined the following probability distribution for the number of customers per day.

x	$f(x)$
0	.05
1	.20
2	.30
3	.25
4	.15
5	.05

(a)   If the center has the capacity to serve two customers per day, what is the probability that one or more customers will be turned away on a given day?

(b)   What is the probability that the center's capacity will not be fully utilized on a day?

(c)   By how much must the capacity be increased so the probability of turning a customer away is no more than .10?

2.24   Suppose X denotes the number of telephone receivers in a single-family residential dwelling. From an examination of the phone subscription records of 381 residences in a city, the following frequency distribution is obtained.

No. of Receivers (x)	No. of Residences (Frequency)
0	2
1	82
2	161
3	89
4	47

(a) Based on these data, obtain an approximate determination of the probability distribution of X.

(b) Why is this regarded as an approximation?

(c) Plot the probability histogram.

2.25   Consider the following experiment. Suppose 100 students are each given the same paragraph of an essay and are asked to count the number of times the letter "e" appears in that paragraph. Each student counts the letters by himself or herself without knowing the other students' counts. Usually, the students do not all come up with the same e-counts even though there should be only one answer, the true e-count! Why? If we write down the distribution of the e-count, what is the likely shape of the distribution? Is it likely to be symmetric with respect to the true e-count, or skewed to the left, or skewed to the right? Explain.

3. EXPECTATION (MEAN) AND STANDARD DEVIATION

We will now introduce a numerical measure for the center of a probability distribution and another for its spread. In Chapter 2, we discussed the concepts of mean, as a measure of the center of a data set, and standard deviation, as a measure of spread. Because probability distributions are theoretical models in which the probabilities can be viewed as long-run relative frequencies, the sample measures of center and spread have their population counterparts.

We consider a die-tossing experiment to motivate the meaning and formula for the population mean. Specifically, we show the connection to the long-run relative frequency interpretation of probability. We first refer to the calculation of the mean of a data set. Suppose a die is tossed 20 times and the following data obtained.

$$4, \quad 3, \quad 4, \quad 2, \quad 5, \quad 1, \quad 6, \quad 6, \quad 5, \quad 2$$
$$2, \quad 6, \quad 5, \quad 4, \quad 6, \quad 2, \quad 1, \quad 6, \quad 2, \quad 4$$

The mean of these observations, called the sample mean, is calculated as

$$\bar{x} = \frac{\text{Sum of the observations}}{\text{Sample size}} = \frac{76}{20} = 3.8$$

Alternatively, we can first count the frequency of each point and use the relative frequencies to calculate the mean as

$$\bar{x} = 1\left(\frac{2}{20}\right) + 2\left(\frac{5}{20}\right) + 3\left(\frac{1}{20}\right) + 4\left(\frac{4}{20}\right) + 5\left(\frac{3}{20}\right) + 6\left(\frac{5}{20}\right) = 3.8$$

This second calculation illustrates the formula

$$\text{Sample mean } \bar{x} = \Sigma\,(\text{value} \times \text{relative frequency})$$

Rather than stopping with 20 tosses, if we imagine a very large number of tosses of a die, the relative frequencies will approach the probabilities, each of which is $\frac{1}{6}$ for a fair die. The mean of the (infinite) collection of tosses of a fair die should then be calculated as

$$1\left(\frac{1}{6}\right) + 2\left(\frac{1}{6}\right) + \cdots + 6\left(\frac{1}{6}\right) = \Sigma\,(\text{value} \times \text{probability}) = 3.5$$

Motivated by this example and the stability of long-run relative frequency, it is then natural to define the mean of a random variable X or its probability distribution as

$$\Sigma\,(\text{value} \times \text{probability}) \quad \text{or} \quad \Sigma\,x_i\,f(x_i)$$

where x_i's denote the distinct values of X. The mean of a probability distribution is also called the population mean for the variable X and is denoted by the Greek letter μ.

The mean of a random variable X is also called its expected value and, alternatively, denoted by $E(X)$. That is, the mean μ and expected value $E(X)$ are the same quantity and will be used interchangeably.

The Mean or the Expected Value of X

$$\mu = E(X) = \sum\,(\text{value} \times \text{probability}) = \sum\,x_i\,f(x_i)$$

Here the sum extends over all the distinct values x_i of X.

Example 8   Calculation of population mean

With X denoting the number of heads in three tosses of a fair coin, calculate the mean of X.

SOLUTION AND DISCUSSION   The probability distribution of X was recorded in Table 4. From the calculations exhibited in Table 8 (page 214), we find that the population mean is 1.5.

TABLE 8   Mean of the
          Distribution of
          Table 4

x	$f(x)$	$xf(x)$
0	$\dfrac{1}{8}$	0
1	$\dfrac{3}{8}$	$\dfrac{3}{8}$
2	$\dfrac{3}{8}$	$\dfrac{6}{8}$
3	$\dfrac{1}{8}$	$\dfrac{3}{8}$
Total	1	$\dfrac{12}{8} = 1.5 = \mu$

The distribution of probability is symmetric about 1.5, the midpoint between 1 and 2. As it should, the mean equals this value.   ■

The mean of a probability distribution has a physical interpretation. If a metal sheet is cut in the shape of the probability histogram, then μ represents the point on the base at which the sheet will balance. For instance, the mean $\mu = 1.5$ calculated in Example 8 is exactly at the center of mass for the distribution depicted in Figure 1 (page 205). Because the amount of probability corresponds to the amount of mass in a bar, we interpret the balance point μ as the center of the probability distribution.

Like many concepts of probability, the idea of the mean or expectation originated from studies of gambling. When X refers to the financial gain in a game of chance, such as playing poker or participating in a state lottery, the name "expected gain" is more appealing than "mean gain." In the realm of statistics, both the names "mean" and "expected value" are widely used.

Example 9   Expected value—Setting a premium
A trip insurance policy pays \$1000 to the customer in case of a loss due to theft or damage on a five-day trip. If the risk of such a loss is assessed to be 1 in 200, what is a fair premium for this policy?

SOLUTION AND   The probability that the company will be liable to pay \$1000 to a customer is
DISCUSSION $\frac{1}{200}$ = .005. Therefore, the probability distribution of X, the payment per customer, is as follows.

Payment x	Probability $f(x)$
$0	.995
$1000	.005

We calculate

$$E(X) = 0 \times .995 + 1000 \times .005$$
$$= \$5.00$$

The company's expected cost per customer is $5.00 and, therefore, a premium equal to this amount is viewed as the fair premium. If this premium is charged and no other costs are involved, then the company will neither make a profit nor lose money in the long run. In practice, the premium is set at a higher price because it must include administrative costs and intended profit.   ■

The concept of expected value also leads to a numerical measure for the spread of a probability distribution—namely, the standard deviation. When we define the standard deviation of a probability distribution, the reasoning parallels that for the standard deviation discussed in Chapter 2.

Because the mean μ is the center of the distribution of X, we express variation of X in terms of the deviation $(X - \mu)$. We define the variance of X as the expected value of the squared deviation $(X - \mu)^2$. To calculate this expected value, we note that

$(X - \mu)^2$ Takes Value	With Probability
$(x_1 - \mu)^2$	$f(x_1)$
$(x_2 - \mu)^2$	$f(x_2)$
.	.
.	.
.	.
$(x_k - \mu)^2$	$f(x_k)$

The expected value of $(X - \mu)^2$ is obtained by multiplying each value $(x_i - \mu)^2$ by the probability $f(x_i)$ and then summing these products. This motivates the definition:

$$\text{Variance of } X = \Sigma \, (\text{deviation})^2 \times (\text{probability})$$
$$= \Sigma \, (x_i - \mu)^2 f(x_i)$$

The variance of X is abbreviated as Var(X) and is also denoted by σ^2. The standard deviation of X is the positive square root of the variance and is denoted by sd(X) or σ (a Greek lowercase sigma).

Variance and Standard Deviation of X

$$\sigma^2 = \text{Var}(X) = \sum (x_i - \mu)^2 f(x_i)$$

$$\sigma = \text{sd}(X) = +\sqrt{\text{Var}(X)}$$

The variance of X is also called the **population variance** and σ denotes the **population standard deviation**.

Example 10   Calculation of population variance
Calculate the variance and the standard deviation of the distribution of X that appears in the left two columns of Table 9.

SOLUTION AND DISCUSSION   We calculate the mean μ, the deviations $(x - \mu)$, $(x - \mu)^2$, and finally $(x - \mu)^2 f(x)$. The details are shown in Table 9.

TABLE 9   Calculation of Variance and Standard Deviation

x	$f(x)$	$xf(x)$	$(x - \mu)$	$(x - \mu)^2$	$(x - \mu)^2 f(x)$
0	.1	0	-2	4	.4
1	.2	.2	-1	1	.2
2	.4	.8	0	0	0
3	.2	.6	1	1	.2
4	.1	.4	2	4	.4
Total	1.0	$2.0 = \mu$			$1.2 = \sigma^2$

$$\text{Var}(X) = \sigma^2 = 1.2$$
$$\text{sd}(X) = \sigma = \sqrt{1.2} = 1.095$$

The population variance and standard deviation are never negative.   ∎

An alternative formula for σ^2 often simplifies the numerical work.

Alternative Formula for Variance of X

$$\sigma^2 = \sum_i x_i^2 f(x_i) - \mu^2$$

Example 11   Alternative calculation of variance
Use the alternative formula to calculate σ^2 for the probability distribution in Example 10.

SOLUTION AND   See Table 10.
DISCUSSION

TABLE 10   Calculation of Variance by the
Alternative Formula

x	$f(x)$	$xf(x)$	$x^2f(x)$
0	.1	.0	0
1	.2	.2	.2
2	.4	.8	1.6
3	.2	.6	1.8
4	.1	.4	1.6
Total	1.0	$2.0 = \mu$	$5.2 = \Sigma x^2 f(x)$

$$\sigma^2 = 5.2 - (2.0)^2$$
$$= 1.2$$
$$\sigma = \sqrt{1.2} = 1.095$$

These values are exactly the same as those obtained in Example 10 using the definition of variance in terms of expected squared deviation.   ■

The standard deviation σ, rather than σ^2, is the appropriate measure of spread. Its unit is the same as that of X. For instance, if X refers to income in dollars, σ will have the unit (dollar), whereas σ^2 has the rather artificial unit (dollar)^2.

Exercises

3.1   Given the following probability distribution

x	$f(x)$
0	.4
1	.3
2	.2
3	.1

(a) Construct the probability histogram.

(b) Find $E(X)$, σ^2, and σ.

3.2 Find the mean and standard deviation of the following distribution.

x	$f(x)$
0	.3
1	.5
2	.1
3	.1

3.3 In bidding for a remodeling project, a carpenter determines that he will have a net profit of $5000 if he gets the contract and a net loss of $56 if his bid fails. If the probability of his getting the contract is $\frac{1}{4}$, calculate his expected return.

3.4 A book club announces a sweepstakes to attract new subscribers. The prizes and the corresponding chances are listed here (typically, the prizes are listed in bold print in an advertisement flyer, whereas the chances are entered in fine print or not mentioned at all).

Prize	Chance
$50,000	1 in one million
$ 5,000	1 in 250,000
$ 100	1 in 5,000
$ 20	1 in 500

Suppose you have just mailed in a sweepstakes ticket and X stands for your winnings.

(a) List the probability distribution of X. (*Caution:* What is not listed is the chance of winning nothing, but you can figure that out from the given information.)

(b) Calculate your expected winnings.

3.5 Calculate the mean and standard deviation for the probability distribution of Example 7.

3.6 An insurance policy pays $800 for the loss due to theft of a canoe. If the probability of a theft is assessed to be .05, find the expected payment. If the insurance company charges $50 for the policy, what is the expected profit per policy?

3.7 A construction company submits bids for two projects. Listed here are the profit and the probability of winning each project. Assume that the outcomes of the two bids are independent.

	Profit	Chance of Winning Bid
Project A	$ 75,000	.50
Project B	$120,000	.65

(a) List the possible outcomes (win/not win) for the two projects and find their probabilities.

(b) Let X denote the company's total profit out of the two contracts. Determine the probability distribution of X.

(c) If it costs the company $2000 for preparatory surveys and paperwork for the two bids, what is the expected net profit?

3.8 Refer to Exercise 3.7, but suppose that the projects are scheduled consecutively with A in the first year and B in the second year. The company's chance of winning project A is still .50. Instead of the assumption of independence, now assume that if the company wins project A, its chance of winning B becomes .80 due to a boost of its image, whereas its chance drops to .40 in case it fails to win A. Under this premise, do parts (a)–(c).

3.9 Upon examination of the claims records of 280 policyholders over a period of five years, an insurance company makes an empirical determination of the probability distribution of X = number of claims in five years.

(a) Calculate the expected value of X.

(b) Calculate the standard deviation of X.

x	$f(x)$
0	.307
1	.286
2	.204
3	.114
4	.064
5	.018
6	.007

3.10 The probability distribution of a random variable X is given by the function

$$f(x) = \frac{60}{77} \left(\frac{1}{x} \right) \quad \text{for } x = 2, 3, 4, 5$$

Calculate the mean and standard deviation of this distribution.

3.11   The probability distribution of a random variable X is given by the function

$$f(x) = \frac{1}{84}\binom{5}{x}\binom{4}{3-x} \quad \text{for } x = 0, 1, 2, 3$$

(a)   Calculate the numerical probabilities and list the distribution.

(b)   Calculate the mean and standard deviation of X.

3.12   Given here are the probability distributions of two random variables X and Y.

(a)   From the X distribution, determine the distribution of the random variable $8 - 2X$ and verify that it coincides with the Y distribution. (Hence, identify $Y = 8 - 2X$.)

x	$f(x)$	y	$f(y)$
0	.1	2	.2
1	.3	4	.4
2	.4	6	.3
3	.2	8	.1

(b)   Calculate the mean and standard deviation of X (call these μ_X and σ_X, respectively).

(c)   From the Y distribution, calculate the mean and standard deviation of Y (call these μ_Y and σ_Y, respectively).

(d)   If $Y = a + bX$, then according to theory, we must have the relations $\mu_Y = a + b\mu_X$ and $\sigma_Y = |b|\sigma_X$. Verify these relations from your results in (b) and (c).

3.13   A salesman of home computers will contact four customers during a week. Each contact can result in either a sale, with probability .2, or no sale, with probability .8. Assume that customer contacts are independent.

(a)   List the elementary outcomes and assign probabilities.

(b)   If X denotes the number of computers sold during the week, obtain the probability distribution of X.

(c)   Calculate the expected value of X.

3.14   Refer to Exercise 3.13. Suppose these computers are priced at $2000, and let Y denote the salesman's total sales (in dollars) during a week.

(a)   Give the probability distribution of Y.

(b)   Calculate $E(Y)$ and see that it is the same as $2000 \times E(X)$.

3.15 *Definition:*   The **median** of a distribution is the value m_0 of the random variable such that $P[X \le m_0] \ge .5$ and $P[X \ge m_0] \ge .5$. In other words, the probability at or below m_0 is at least .5, and the probability at or above m_0 is at least .5. Find the median of the distribution given in Exercise 3.1.

3.16 Given the two probability distributions

x	$f(x)$	y	$f(y)$
1	.2	0	.1
2	.6	1	.2
3	.2	2	.4
		3	.2
		4	.1

(a) Construct probability histograms. Which distribution has a larger spread?

(b) Verify that both distributions have the same mean.

(c) Compare the two standard deviations.

3.17 Suppose the number of days, X, that it takes for the post office to deliver a letter from City A to City B has the following distribution:

x	$f(x)$
2	.5
3	.3
4	.1
5	.05
6	.05

(a) Find the mean and the standard deviation of X.

(b) Find the probability that a letter takes at most 3 days to be delivered from City A to City B.

(c) If two letters are to be mailed from City A in different months, what is the probability that both letters will arrive in City B in less than 4 days after they were mailed?

4. SUCCESSES AND FAILURES—BERNOULLI TRIALS

Often, an experiment can have only two possible outcomes. Example 7 concerned individual trees that were either infested or not but proportion .30 of the population was infested. Also, only two outcomes are possible for a single trial in the scenario of Example 2. In these circumstances, a simple probability model can be developed for the chance variation in the outcomes. Moreover, the population proportion need not be known as in the previous examples. Instead, the probability distribution will involve this unknown population proportion as a parameter.

Sampling situations where the elements of a population belong to one of only two categories abound in virtually all walks of life. We count the number of elements in a sample that belong to the category of interest. A few examples are

Inspect a specified number of items coming off a production line and count the number of defectives.

Survey a sample of voters and observe how many favor a reduction of public spending on welfare.

Analyze the blood specimens of a number of rodents and count how many carry a particular viral infection.

Examine the case histories of a number of births and count how many involved delivery by Caesarean section.

Selecting a single element of the population is envisioned as a trial of the (sampling) experiment, so that each trial can result in one of two possible outcomes. Our ultimate goal is to develop a probability model for the number of outcomes in one category when repeated trials are performed.

An organization of the key terminologies concerning the successive repetitions of an experiment is now in order. We call each repetition by the simpler name—a trial. Furthermore, the two possible outcomes of a trial are now assigned the technical names **success** (S) and **failure** (F) just to emphasize the point that they are the only two possible results. These names bear no connotation of success or failure in real life. Customarily, the outcome of primary interest in a study is labeled success (even if it is a disastrous event). In a study of the rate of unemployment, the status of being unemployed may be attributed the statistical name success!

Further conditions on the repeated trials are necessary to arrive at our intended probability distribution. Repeated trials that obey these conditions are called **Bernoulli trials** after the Swiss mathematician Jacob Bernoulli.

Bernoulli Trials

1. Each trial yields one of two outcomes, technically called success (S) and failure (F).

2. For each trial, the probability of success $P(S)$ is the same and is denoted by $p = P(S)$. The probability of failure is then $P(F) = 1 - p$ for each trial and is denoted by q, so that $p + q = 1$.

3. Trials are independent. The probability of success in a trial does not change given any information about the outcomes of other trials.

Perhaps the simplest example of Bernoulli trials is the prototype model of tossing a coin, where the occurrences *head* and *tail* can be labeled S and F, respectively. For a fair coin, we have $p = q = \frac{1}{2}$.

Example 12   Sampling from a population with two categories of elements
Consider a lot (population) of items in which each item can be classified as either defective or nondefective. Suppose that a lot consists of 15 items, of which 5 are defective and 10 are nondefective.
Do the conditions for Bernoulli trials apply when sampling (1) with replacement and (2) without replacement?

SOLUTION AND DISCUSSION

1. *Sampling with replacement.* An item is drawn at random (i.e., in a manner that all items in the lot are equally likely to be selected). The quality of the item is recorded and it is returned to the lot before the next drawing. The conditions for Bernoulli trials are satisfied. If the occurrence of a defective is labeled S, we have $P(S) = \frac{5}{15}$.

2. *Sampling without replacement.* In situation (1), suppose that 3 items are drawn one at a time but without replacement. Then the condition concerning the independence of trials is violated. For the first drawing, $P(S) = \frac{5}{15}$. If the first draw produces S, the lot then consists of 14 items, 4 of which are defective. Given this information about the result of the first draw, the conditional probability of obtaining an S on the second draw is then $\frac{4}{14} \neq \frac{5}{15}$, which establishes the lack of independence.

This violation of the condition of independence loses its thrust when the population is vast and only a small fraction of it is sampled. Consider sampling 3 items without replacement from a lot of 1500 items, 500 of which are defective. With S_1 denoting the occurrence of an S in the first draw and S_2 that in the second, we have

$$P(S_1) = \frac{500}{1500} = \frac{5}{15}$$

and

$$P(S_2|S_1) = \frac{499}{1499}$$

For most practical purposes, the latter fraction can be approximated by $\frac{5}{15}$. Strictly speaking, there has been a violation of the independence of trials, but it is to such a negligible extent that the model of Bernoulli trials can be assumed as a good approximation.   ■

Example 12 reinforces the important points also illustrated, by the urn and spinner models, in Example 23 of Chapter 4.

If elements are sampled from a dichotomous population at random and with replacement, the conditions for Bernoulli trials are satisfied.

When the sampling is made without replacement, the condition of the independence of trials is violated. However, if the population is large and only a small fraction of it (less than 10%, as a rule of thumb) is sampled, the effect of this violation is negligible and the model of the Bernoulli trials can be taken as a good approximation.

Example 13 further illustrates the kinds of approximations that are sometimes employed when using the model of the Bernoulli trials.

Example 13

Testing a new antibiotic—Bernoulli trials?
Suppose that a newly developed antibiotic is to be tried on 10 patients who have a certain disease and the possible outcomes in each case are cure (S) or no cure (F).
Comment on the applicability of the Bernoulli trial model.

SOLUTION AND DISCUSSION

Each patient has a distinct physical condition and genetic constitution that cannot be perfectly matched by any other patient. Therefore, strictly speaking, it may not be possible to regard the trials made on 10 different patients as 10 repetitions of an experiment under identical conditions, as the definition of Bernoulli trials demands. If patient information is unavailable, the Bernoulli model is reasonable.

In general, we must remember that the conditions of a probability model are abstractions that help to realistically simplify the complex mechanism governing the outcomes of an experiment. Identification with Bernoulli trials in such situations is to be viewed as an approximation of the real world, and its merit rests on how successfully the model explains chance variations in the outcomes.   ■

Exercises

4.1 Is the model of Bernoulli trials plausible in each of the following situations? Discuss in what manner (if any) a serious violation of the assumptions can occur.

(a) Beetles of a common strain are sprayed with a given concentration of an insecticide and the occurrence of death or survival is recorded in each case.

(b) A word association test is given to 10 first grade children and the amount of time each child takes to complete the test is recorded.

(c) Items coming off an assembly line are inspected and classified as defective or nondefective.

(d) Going house by house down the block and recording whether the newspaper was delivered on time

4.2 In each case, examine whether or not repetitions of the stated experiment conform to the model of Bernoulli trials. Where the model is appropriate, determine the numerical value of p or indicate how it can be determined.

(a) Roll a fair die and observe the number that shows up.

(b) Roll a fair die and observe whether or not the number 5 shows up.

(c) Roll two fair dice and observe the total of the points that show up.

(d) Roll two fair dice and observe whether or not a total of 7 points is obtained.

(e) Roll a loaded die and observe whether or not the number 5 shows up.

4.3 A jar contains 25 candies, of which 7 are brown, 10 are yellow, and 8 are of other colors. Consider 4 successive draws of 1 candy at random from the jar and suppose the appearance of a yellow candy is the event of interest. For each of the following situations, state whether or not the model of Bernoulli trials is reasonable, and if so, determine the numerical value of p.

(a) After each draw, the selected candy is returned to the jar.

(b) After each draw, the selected candy is not returned to the jar.

(c) After each draw, the selected candy is returned to the jar and one new candy of the same color is added in the jar.

4.4 Refer to Exercise 4.3 and suppose instead that the mix consists of 2500 candies, of which 700 are brown, 1000 are yellow, and 800 are of other colors. Repeat parts (a)–(c) of Exercise 4.3 in this setting.

4.5 From four agricultural plots, two will be selected at random for a pesticide treatment. The other two plots will serve as controls. For each plot, denote by S the event that it is treated with the pesticide. Consider the assignment of treatment or control to a single plot as a trial.

(a) Is $P(S)$ the same for all trials? If so, what is the numerical value of $P(S)$?

(b) Are the trials independent? Why or why not?

4.6   Refer to Exercise 4.5. Now suppose for each plot a fair coin will be tossed. If a head shows up, the plot will be treated; otherwise, it will be a control. With this manner of treatment allocation, answer questions (a) and (b).

4.7   A market researcher intends to study the consumer preference between regular and decaffeinated coffee. Examine the plausibility of the model of Bernoulli trials in the following situations.

(a) One hundred consumers are randomly selected and each is asked to report the types of coffee (regular or decaffeinated) purchased in the five most recent occasions. If we consider each purchase as a trial, this inquiry deals with 500 trials.

(b) Five hundred consumers are randomly selected and each is asked about the most recent purchase of coffee. Here again the inquiry deals with 500 trials.

4.8   A backpacking party carries three emergency signal flares, each of which will light with a probability of .99. Assuming that the flares operate independently, find

(a) The probability that at least one flare lights

(b) The probability that exactly two flares light

4.9   Consider Bernoulli trials with success probability $p = \frac{1}{4}$.

(a) Find the probability that four trials result in all failures.

(b) Given that the first four trials result in all failures, what is the conditional probability that the next four trials are all successes?

(c) Find the probability that the first success occurs in the fourth trial.

4.10 If in three Bernoulli trials $P[$ all three are successes $] = .027$, what is the probability that all three are failures?

4.11 Consider four Bernoulli trials with success probability $p = .7$ in each trial. Find the probability that

(a) All four trials result in successes.

(b) All are failures.

(c) There is at least one success.

4.12 A new driver, who did not take driver's education, has probability .8 of passing the driver's license exam. If tries are independent, find the probability that the driver (a) will not pass in two attempts, (b) will not pass in three attempts.

4.13   An animal either dies (D) or survives (S) in the course of a surgical experiment. The experiment is to be performed first with two animals. If both survive, no further trials are to be made. If exactly one animal survives, one more animal is to undergo the experiment. Finally, if both animals die, two additional animals are to be tried.

(a)   List the sample space.

(b)   Assume that the trials are independent and the probability of survival in each trial is $\frac{1}{3}$. Assign probabilities to the elementary outcomes.

(c)   Let X denote the number of survivors. Obtain the probability distribution of X by referring to part (b).

4.14   The accompanying table shows the percentages of residents in a large community when classified according to gender and presence of a particular allergy.

| | Allergy | |
	Present	Absent
Male	16	36
Female	9	39

Suppose that the selection of a person is considered a trial and the presence of the allergy is considered a success. For each case, identify the numerical value of p and find the required probability.

(a)   Four persons are selected at random. What is the probability that none has the allergy?

(b)   Four males are selected at random. What is the probability that none has the allergy?

(c)   Two males and two females are selected at random. What is the probability that none has the allergy?

5. THE BINOMIAL DISTRIBUTION

This section deals with a basic distribution that models chance variation in repetitions of an experiment that has only two possible outcomes. The random variable X of interest is the frequency count of one of the categories. Previously, its distribution was calculated under the assumption that the population proportion is known. For instance, the probability distribution of Table 6, from Example 7, resulted from the specification that 30% of the population of trees are infested with the parasite. In a practical situation, however, the population proportion is usually an unknown quantity. When this is so, the probability distribution of X

cannot be numerically determined. However, we will see that it is possible to construct a model for the probability distribution of X that contains the unknown population proportion as a **parameter**. The probability model serves as the major vehicle of drawing inferences about the population from observations of the random variable X.

A **probability model** is an assumed form of the probability distribution that describes the chance behavior for a random variable X.

Probabilities are expressed in terms of relevant population quantities, called the **parameters**.

Consider a **fixed number** n of Bernoulli trials with the success probability p in each trial. The number of successes obtained in n trials is a random variable that we denote by X. The probability distribution of this random variable X is called a binomial distribution.

The binomial distribution depends on the two quantities n and p. For instance, the distribution appearing in Table 4 is precisely the binomial distribution with $n = 3$ and $p = .5$, whereas that in Table 6 is the binomial distribution with $n = 4$ and $p = .3$.

The Binomial Distribution

Denote

$$n = \text{a fixed number of Bernoulli trials}$$
$$p = \text{the probability of success in each trial}$$
$$X = \text{the (random) number of successes in } n \text{ trials}$$

The random variable X is called a **binomial random variable.** Its distribution is called a **binomial distribution.**

A review of the developments in Example 7 will help motivate a formula for the general binomial distribution.

Example 14   Example 7 revisited—The binomial distribution
The random variable X represents the number of infested trees among a random sample of $n = 4$ trees from the forest. Instead of the numerical value .3, we now denote the population proportion of infested trees by the symbol p. Furthermore, we relabel the outcome "infested" as a success (S) and "not infested" as a failure (F). The elementary outcomes of sampling 4 trees, the associated probabilities, and the value of X are listed as follows.

FFFF	SFFF	SSFF	SSSF	SSSS
	FSFF	SFSF	SSFS	
	FFSF	SFFS	SFSS	
	FFFS	FSSF	FSSS	
		FSFS		
		FFSS		

Value of X	0	1	2	3	4
Probability of each outcome	q^4	pq^3	p^2q^2	p^3q	p^4
Number of outcomes	1 $= \binom{4}{0}$	4 $= \binom{4}{1}$	6 $= \binom{4}{2}$	4 $= \binom{4}{3}$	1 $= \binom{4}{4}$

Because the population of trees is vast, the trials can be treated as independent. Also, for an individual trial, $P(S) = p$ and $P(F) = q = 1 - p$. The event $[X = 0]$ has one outcome FFFF, whose probability is

$$P[X = 0] = P(\text{FFFF}) = q \times q \times q \times q = q^4$$

To arrive at an expression for $P[X = 1]$, we consider the outcomes listed in the second column. The probability of SFFF is

$$P(\text{SFFF}) = p \times q \times q \times q = pq^3$$

and the same result holds for every outcome in this column. There are 4 outcomes so we obtain $P[X = 1] = 4pq^3$. The factor 4 is the number of outcomes with one S and three F's. Even without making a complete list of the outcomes, we can obtain this count. Every outcome has 4 places and the 1 place where S occurs can be selected from the total of 4 in $\binom{4}{1} = 4$ ways, while the remaining three places must be filled with an F. Continuing in the same line of reasoning, we see that the value $X = 2$ occurs with $\binom{4}{2} = 6$ outcomes, each of which has a probability of p^2q^2. Therefore, $P[X = 2] = \binom{4}{2}p^2q^2$. After we work out the remaining terms, the binomial distribution with $n = 4$ trials can be presented as in Table 11.

TABLE 11   Binomial Distribution with $n = 4$ Trials

Value x	0	1	2	3	4
Probability $f(x)$	$\binom{4}{0}p^0q^4$	$\binom{4}{1}p^1q^3$	$\binom{4}{2}p^2q^2$	$\binom{4}{3}p^3q^1$	$\binom{4}{4}p^4q^0$

It would be instructive for the reader to verify that the numerical probabilities appearing in Table 6 are obtained by substituting $p = .3$ and $q = .7$ in the entries of Table 11.   ∎

Extending the reasoning of Example 14 to the case of a general number n of Bernoulli trials, we observe that there are $\binom{n}{x}$ outcomes that have exactly x successes and $(n - x)$ failures. The probability of every such outcome is $p^x q^{n-x}$. Therefore,

$$f(x) = P[X = x] = \binom{n}{x} p^x q^{n-x} \quad \text{for } x = 0, 1, \ldots , n$$

is the formula for the binomial probability distribution with n trials.

The binomial distribution with n trials and success probability p is described by the function

$$f(x) = P[X = x] = \binom{n}{x} p^x (1 - p)^{n-x}$$

for the possible values $x = 0, 1, \ldots , n$.

Example 15   The binomial distribution and genetics
According to the Mendelian theory of inherited characteristics, a cross fertiliza-tion of related species of red- and white-flowered plants produces a generation whose offspring contain 25% red-flowered plants. Suppose that a horticulturist wishes to cross 5 pairs of the cross-fertilized species. Of the resulting 5 offspring, what is the probability that

(a)   There will be no red-flowered plants?

(b)   There will be 4 or more red-flowered plants?

SOLUTION AND DISCUSSION   Because the trials are conducted on different parent plants, it is natural to assume that they are independent. Let the random variable X denote the number of red-flowered plants among the 5 offspring. If we identify the occurrence of a red as a success S, the Mendelian theory specifies that $P(S) = p = \frac{1}{4}$, and hence X has a binomial distribution with $n = 5$ and $p = .25$. The required probabilities are therefore

(a) $P[X = 0] = f(0) = (.75)^5 = .237$

(b) $P[X \geq 4] = f(4) + f(5) = \binom{5}{4}(.25)^4(.75)^1 + \binom{5}{5}(.25)^5(.75)^0$

$$= .015 + .001 = .016 \qquad ∎$$

To illustrate the manner in which the values of p influence the shape of the binomial distribution, the probability histograms for three binomial distributions with $n = 6$ and $p = .5, .3,$ and $.7,$ respectively, are presented in Figure 3. When $p = .5$, the binomial distribution is symmetric with the highest probability occurring at the center [see Figure 3(a)].

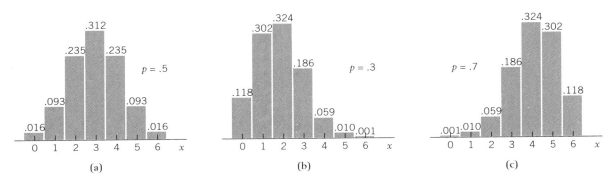

Figure 3. Binomial distributions for $n = 6$

For values of p smaller than .5, more probability is shifted toward the smaller values of x and the distribution has a longer tail to the right. Figure 3(b), where the binomial histogram for $p = .3$ is plotted, illustrates this tendency. On the other hand, Figure 3(c) with $p = .7$ illustrates the opposite tendency. The value of p is higher than .5, more probability mass is shifted toward higher values of x, and the distribution has a longer tail to the left. Considering the histograms in Figures 3(b) and 3(c), we note that the value of p in one histogram is the same as the value of q in the other. The probabilities in one histogram are exactly the same as those in the other, but their order is reversed. This illustrates a general property of the binomial distribution. When p and q are interchanged, the distribution of probabilities is reversed.

HOW TO USE THE BINOMIAL TABLE (APPENDIX B, TABLE 2)

Although the binomial distribution is easily evaluated on a computer and some hand calculators, we provide a short table in Appendix B, Table 2. It covers selected sample sizes n ranging from 1 to 25 and several values of p. For a given pair (n, p), the table entry corresponding to each c represents the cumulative probability $P[X \leq c] = \sum_{x=0}^{c} f(x)$, as is explained in the following scheme.

The Binomial Distribution	
Value x	Probability $f(x)$
0	$f(0)$
1	$f(1)$
2	$f(2)$
.	.
.	.
.	.
n	$f(n)$
Total	1

Appendix B, Table 2 provides

c	Table Entry $\displaystyle\sum_{x=0}^{c} f(x) = P[X \le c]$
0	$f(0)$
1	$f(0) + f(1)$
2	$f(0) + f(1) + f(2)$
.	.
.	.
.	.
n	1.000

The probability of an individual value x can be obtained from this table by a subtraction of two consecutive entries. For example,

$$P[X = 2] = f(2) = \left(\begin{array}{c} \text{table entry at} \\ c = 2 \end{array} \right) - \left(\begin{array}{c} \text{table entry at} \\ c = 1 \end{array} \right)$$

Example 16   Binomial distribution for number cured

Suppose it is known that a new treatment is successful in curing a muscular pain in 50% of the cases. If it is tried on 15 patients, find the probability that

(a) At most 6 will be cured.

(b) The number cured will be no fewer than 6 and no more than 10.

(c) Twelve or more will be cured.

SOLUTION AND DISCUSSION   Designating the cure of a patient by S and assuming that the results for individual patients are independent, we note that the binomial distribution with $n = 15$ and $p = .5$ is appropriate for $X =$ number of patients who are cured. To compute the required probabilities, we consult the binomial table for $n = 15$ and $p = .5$.

(a) $P[X \le 6] = .304$, which is directly obtained by reading from the row $c = 6$.

(b) We are to calculate:

$$P[6 \le X \le 10] = f(6) + f(7) + f(8) + f(9) + f(10)$$

$$= \sum_{x=6}^{10} f(x)$$

The table entry corresponding to $c = 10$ gives

$$P[X \le 10] = \sum_{x=0}^{10} f(x) = .941$$

and the entry corresponding to $c = 5$ yields

$$P[X \le 5] = \sum_{x=0}^{5} f(x) = .151$$

Because their difference represents the sum $\sum_{x=6}^{10} f(x)$, we obtain

$$\begin{aligned} P[6 \le X \le 10] &= P(X \le 10] - P[X \le 5] \\ &= .941 - .151 \\ &= .790 \end{aligned}$$

(c)   To find $P[X \ge 12]$ we use the law of the complement:

$$\begin{aligned} P[X \ge 12] &= 1 - P[X \le 11] \\ &= 1 - .982 \\ &= .018 \end{aligned}$$

Note that $[X < 12]$ is the same event as $[X \le 11]$.

(*An aside:*   Refer to our "muscular pain" example in Section 1 of Chapter 4. The mystery surrounding the numerical probability .018 is now resolved.)   ■

STATISTICAL REASONING

A Smelly Fur Coat

In a TV episode of *People's Court*, where actual cases are shown live, a plaintiff explained to Judge W that she had purchased an expensive fur coat from the defendant's store. Later, she discovered that the fur coat had an unpleasant odor and tried to return it for a refund. However, the defendant refused to take the coat back because she could not smell anything unusual.

After listening to the rest of the story as told by both the defendant and the plaintiff, Judge W called a recess. When he returned to the bench, he told the court that during the recess he had asked five persons to smell the coat and that none of them smelled anything unusual. At this point, Judge W ruled in favor of the defendant and told the plaintiff to keep the coat.

What are some of the questions that one might ask concerning the basis of the judge's decision?

Some Possible Questions

1.   Did Judge W ask the five people all at the same time as a single group? If so, then the responses from the five individuals are not likely to be independent. After one person heard another person's answer, his or her own answer is likely to be biased by it. Imagine what the last person would answer if the first four persons said that they did not smell anything unusual.

Brian Hennessey/FPG International

2. If the five persons had listened to the court proceedings before the recess, they may have already formed some opinions that would bias their answer concerning the smelly coat.

3. If the judge asked the five persons one by one in a private room, then the number of answers of "no unusual smell" would correspond to the number of heads when tossing a coin five times. The trials would be independent.

4. As a yardstick, we could calculate some probabilities assuming the responses are independent. For instance, suppose that the probability of detecting an unusual smell is .5. The probability of five "no unusual smell" answers corresponds to the probability of five heads in five tosses of a fair coin, $(.5)^5$ or .03125. Is this probability small enough to justify Judge W's decision? In other words, is it a rare event that all five people would respond "no unusual smell"? You might want to calculate this probability if the chance that an individual person can detect a smell is just .2.

When decisions are based on the outcome of an experiment that cannot be determined beforehand, we need to ensure that the experiment was conducted in an impartial and objective manner.

If several trials are to be performed, steps should be taken to make them independent to produce the maximum amount of information.

THE MEAN AND STANDARD DEVIATION OF THE BINOMIAL DISTRIBUTION

Although we already have a general formula that gives the binomial probabilities for any n and p, in later chapters we will need to know the mean and the standard deviation of the binomial distribution. The expression np for the mean is apparent from the following intuitive reasoning: If a fair coin is tossed 100 times, the expected number of heads is $100 \times \frac{1}{2} = 50$. Likewise, if the probability of an event is p, then in n trials the event is expected to happen np times. The formula for the standard deviation requires some mathematical derivation, which we omit.

The binomial distribution with n trials and success probability p has

$$\text{Mean} = np$$
$$\text{Variance} = npq \quad (\text{Recall: } q = 1 - p)$$
$$\text{sd} = \sqrt{npq}$$

Example 17   Calculation of binomial mean and variance
For the binomial distribution with $n = 3$ and $p = .5$, calculate the mean and the standard deviation.

SOLUTION AND DISCUSSION   Employing the formulas, we obtain:

$$\text{Mean} = np = 3 \times .5 = 1.5$$
$$\text{sd} = \sqrt{npq} = \sqrt{3 \times .5 \times .5} = \sqrt{.75} = .866$$

The mean agrees with the results of Example 8. The reader may wish to check the standard deviation by numerical calculations using the definition of σ.   ∎

Exercises

5.1   For each situation (a)–(d), state whether or not a binomial distribution holds for the random variable X. Also, identify the numerical values of n and p when a binomial distribution holds.

(a)   A fair die is rolled 9 times, and X denotes the number of times 5 shows up.

(b)   A fair coin is tossed until a head appears, and X denotes the number of tosses.

(c) In a jar, there are ten marbles, of which three are numbered 1, three are numbered 2, two are numbered 3, and two are numbered 4. Three marbles are drawn at random, one after another and with replacement, and X denotes the count of the selected marbles that are numbered either 1 or 2.

(d) The same experiment as described in part (c), but now X denotes the sum of the numbers on the selected marbles.

5.2 Construct a tree diagram for three Bernoulli trials. Attach probabilities in terms of p and q to each outcome and then table the binomial distribution for $n = 3$.

5.3 In each case, find the probability of x successes in n Bernoulli trials with success probability p for each trial.

(a) $x = 2$, $n = 4$, $p = \frac{1}{3}$

(b) $x = 3$, $n = 6$, $p = .25$

(c) $x = 2$, $n = 6$, $p = .75$

5.4 (a) Plot the probability histograms for the binomial distributions for $n = 5$ and $p = .2, .5,$ and $.8$.

(b) Locate the means.

(c) Find $P[X \geq 4]$ for each of the three cases.

5.5 If X has the binomial distribution with $n = 5$ and $p = .35$, calculate the probabilities $f(x) = P[X = x]$ for $x = 0, 1, \ldots, 5$ and find

(a) $P[X \leq 2]$

(b) $P[X \geq 2]$

(c) $P[X = 2 \text{ or } X = 4]$

5.6 Refer to Exercise 5.5. What is the most probable value of X (called the mode of a distribution)?

5.7 For the binomial distribution with $n = 4$ and $p = .25$, find the probability of

(a) Three or more successes

(b) At most three successes

(c) Two or more failures

5.8 Refer to the Statistical Reasoning example on page 233 concerning the smelly fur coat. Suppose the smell is faint enough so that only 30% of the population can detect it. What is the probability that, of four randomly selected persons, none could detect the unusual smell?

5.9   Suppose 15% of the trees in a forest have severe leaf damage from air pollution. If 5 trees are selected at random, find the probability that

(a)   Three of the selected trees have severe leaf damage.

(b)   No more than two have severe leaf damage.

5.10   Rh-positive blood appears in 85% of the white population in the United States. If 7 people are sampled at random from that population, find the probability that

(a)   At least 5 of them have Rh-positive blood.

(b)   At most 3 of them have Rh-negative blood—that is, an absence of Rh-positive.

5.11   Using the binomial table, find the probability of

(a)   Four successes in 12 trials when $p = .3$

(b)   Eight failures in 12 trials when $p = .7$

(c)   Eight successes in 12 trials when $p = .3$

Explain why you get identical answers in parts (b) and (c).

5.12   Using the binomial table, find the probability of

(a)   Five or fewer successes in 9 trials when $p = .7$

(b)   No more than 11 and no less than 6 successes in 16 trials when $p = .6$

(c)   Exactly 30% successes in 20 trials when $p = .3$

5.13   If $p = .7$, find the probability that

(a)   More than 5 trials are needed to obtain 3 successes. (*Hint:*   In other words, the event is "at most 2 successes in 5 trials.")

(b)   More than 9 trials are needed to obtain 7 successes.

5.14   A survey report states that 70% of adult women visit their doctors for a physical examination at least once in two years. If 19 adult women are randomly selected, find the probability that

(a)   Fewer than 14 of them have had a physical examination in the past two years.

(b)   At least 17 of them have had a physical examination in the past two years.

5.15   Calculate the mean and standard deviation of the binomial distribution with

(a) $n = 16, \quad p = .5$

(b) $n = 25, \quad p = .1$

(c) $n = 25, \quad p = .9$

5.16   (a)   For the binomial distribution with $n = 3$ and $p = .6$, list the probability distribution $(x, f(x))$ in a table.

(b)   From this table, calculate the mean and standard deviation by using the methods of Section 3.

(c)   Check your results with the formulas: mean $= np$, sd $= \sqrt{npq}$.

5.17   Suppose that 20% of the college seniors support an increase in federal funding for care of the elderly. If 20 college seniors are randomly selected, what is the probability that at most 3 of them support increased funding for care of the elderly?

5.18   Referring to Exercise 5.17, find

(a)   The expected number of college seniors, in a random sample of 20, supporting the increased funding.

(b)   The probability that the number of sampled college seniors supporting the increased funding equals the expected number.

5.19   Suppose that, for a particular type of cancer, chemotherapy provides a 5-year survival rate of 80% if the disease could be detected at an early stage. Among 18 patients diagnosed to have this form of cancer at an early stage who are just starting the chemotherapy, find the probability that

(a)   Fourteen will survive beyond 5 years.

(b)   Six will die within 5 years.

(c)   The number of patients surviving beyond 5 years will be between 9 and 13 (both inclusive).

5.20   Referring to Exercise 5.19, find the expectation and standard deviation of the number of 5-year survivors.

5.21   According to a report of the American Medical Association, 7.85% of practicing physicians in 1987 were in the specialty area of family practice. Assuming that the same rate prevails, find the mean and standard deviation of the number of physicians specializing in family practice out of a current random selection of 545 medical graduates.

5.22   According to the Mendelian theory of inheritance of genes, offspring of a dihybrid cross of peas could be any of the four types: round-yellow (RY), wrinkled-yellow (WY), round-green (RG), and wrinkled-green (WG), and their probabilities are in the ratios $9:3:3:1$.

(a)   If X denotes the number of RY offspring from 130 such crosses, find the mean and standard deviation of X.

(b)   If Y denotes the number of WG offspring from 85 such crosses, find the mean and standard deviation of Y.

5.23   The following table (see Exercise 4.14) shows the percentages of residents in a large community when classified according to gender and presence of a particular allergy. For each part below, find the mean and standard deviation of the specified random variable.

| | Allergy | |
	Present	Absent
Male	16	36
Female	9	39

(a) X stands for the number of persons having the allergy in a random sample of 40 persons.

(b) Y stands for the number of males having the allergy in a random sample of 40 males.

(c) Z stands for the number of females not having the allergy in a random sample of 40 females.

5.24 For a binomial distribution,

(a) If $n = 80$ and $p = .35$, find the mean and standard deviation.

(b) If mean $= 54$ and standard deviation $= 6$, find the numerical values of n and p.

5.25 Many computer packages produce binomial probabilities. The single MINITAB command

```
CDF;
BINOMIAL WITH N = 5, P = 0.25.
```

produces the cumulative probabilities

Cumulative Distribution Function

```
Binomial with n = 5 and p = 0.250000

       x     P( X <= x)
    0.00        0.2373
    1.00        0.6328
    2.00        0.8965
    3.00        0.9844
    4.00        0.9990
    5.00        1.0000
```

Alternatively, you can choose
Calc > Probability distributions > Binomial
Click **Cumulative probability** and fill in dialog box.

The command

```
PDF;
BINOMIAL WITH N = 5, P = 0.25.
```

produces the individual probabilities

Probability Density Function

Binomial with n = 5 and p = 0.250000

x	P (X = x)
0.00	0.2373
1.00	0.3955
2.00	0.2637
3.00	0.0879
4.00	0.0146
5.00	0.0010

Using the computer, calculate

(a) $P[X \leq 8]$ and $P[X = 8]$ when $p = .64$ and $n = 12$

(b) $P[10 \leq X \leq 15]$ when $p = .42$ and $n = 30$

6. THE BINOMIAL DISTRIBUTION IN CONTEXT

Requests for credit cards must be processed to determine whether the applicant meets certain financial standards. In many instances, as when the applicant already has a long-term good credit record, only a short review is required. Usually this consists of a credit score assigned on the basis of the answers to questions on the application and then a computerized check of credit records. Many other cases require a full review with manual checks of information to determine the creditworthiness of the applicant.

Each week, a large financial institution selects a sample of 20 incoming applications and counts the number requiring full review. From data collected over several weeks, it is observed that about 40% of the applications require full review. If we take this long-run relative frequency as the probability, what is an unusually large number of full reviews and what is an unusually small number?

Let X be the number in the sample that require a full review. From the binomial table, with $n = 20$ and $p = .4$, we get

$$P[X \leq 3] = .016$$
$$P[X \geq 13] = 1 - P[X \leq 12] = 1 - .979 = .021$$

Taken together, the probability of X being 3 or less or 13 or more is .037, so those values should occur less than four times in 100 samples. That is, they could be considered unusual. In Exercise 6.1, you will be asked to show that either including 4 or including 12 will lead to a combined probability greater than .05.

That is, the large and small values should then occur more than 1 in 20 times. For many people, this would be too frequent to be considered rare or unusual.

For the count X, we expect $np = 20 \times .4 = 8$ applications in the sample to require a full review. The standard deviation of this count is $\sqrt{20(.4)(1 - .4)} = 2.191$. Alternatively, when n is moderate or large, we could describe unusual as two or more standard deviations from the mean. A value at least two standard deviations, or $2(2.191) = 4.382$, above the mean of 8 must be 13 or more. A value 2 or more standard deviations below the mean must be 3 or less. These values correspond exactly to the values above that we called unusual. In other cases, the two standard deviations approach provides a reasonable and widely used approximation.

p CHARTS FOR TREND

A series of sample proportions should be visually inspected for trend. A graph called a p **chart** helps identify times when the population proportion has changed from its long-time stable value. Because many sample proportions will be graphed, it is customary to set control limits at 3 rather than 2 standard deviations.

When p_0 is the expected or long-run proportion, we obtain a lower control limit by dividing the lower limit for X, $np_0 - 3\sqrt{np_0(1 - p_0)}$, by the sample size. Doing the same with the upper bound, we obtain an upper control limit.

$$\text{Lower control limit} \qquad \text{Upper control limit}$$
$$p_0 - 3\sqrt{\frac{p_0(1 - p_0)}{n}} \qquad\qquad p_0 + 3\sqrt{\frac{p_0(1 - p_0)}{n}}$$

In the context of credit applications that require a full review, $p_0 = .4$ so the control limits are

$$p_0 - 3\sqrt{\frac{p_0(1 - p_0)}{n}} = .4 - 3\sqrt{\frac{.4(1 - .4)}{20}} = .4 - 3(.110) = .07$$
$$p_0 + 3\sqrt{\frac{p_0(1 - p_0)}{n}} = .4 + 3\sqrt{\frac{.4(1 - .4)}{20}} = .4 + 3(.110) = .73$$

The **centerline** is drawn as a solid line at the expected or long-run proportion $p_0 = .4$, and the two control limits, each at a distance of three standard deviations of the sample proportion from the centerline, are also drawn as horizontal lines as in Figure 4 (page 242). Sample proportions that fall outside of the control limits are considered unusual and should result in a search for a cause that may include a change in the mix of type of persons requesting credit cards.

The number of applications requiring full review, out of the twenty in the sample, were recorded for 19 weeks.

11   7   8   4   9   10   4   8   8   7   10   6   9   10   7   7   6   9   10

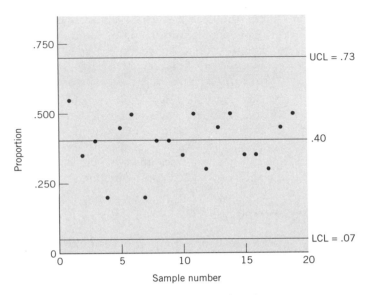

Figure 4. A p chart for the proportion of applications requiring a full review

After converting to sample proportions by dividing by 20, we can graph the points in a p chart as in Figure 4. All the points are in control, and the financial institution does not appear to have reached the point where the mix of applicants includes more marginal cases.

Exercises

6.1   Refer to the credit card application approval process where unusual values are defined.

  (a)   Show that if 4 is included as an unusual value, then the probability $P[X \leq 4 \text{ or } X \geq 13]$ is greater than .05.

  (b)   Show that if 12 is included as an unusual value, then the probability $P[X \leq 3 \text{ or } X \geq 12]$ is greater than .05.

6.2   Refer to the credit card application approval process.

  (a)   Make a p chart using the centerline and control limits calculated for $p_0 = .4$.

  (b)   Suppose the next five weeks bring 12, 10, 15, 11, and 16 applications requiring full review. Graph the corresponding proportions on your p chart.

  (c)   Identify any weeks where the chart signals "out of control."

6.3   Several fast-food restaurants advertise quarter-pound hamburgers. This could be interpreted as meaning half the hamburgers made have an uncooked weight of at least a quarter-pound and half have a weight that is less. An inspector checks 20 uncooked hamburgers at each restaurant.

    (a)   Make a p chart using the centerline and control limits calculated for $p_0 = .5$.

    (b)   Suppose that five restaurants have 8, 11, 7, 15, and 10 underweight hamburgers in samples of size 20. Graph the corresponding proportions of your p chart.

    (c)   Identify any restaurants where the chart signals "out of control."

6.4   Refer to Exercise 6.3.

    (a)   What are the unusual values for the number of underweight hamburgers in the sample if they correspond to proportions outside of the control limits of the p chart?

    (b)   Use the binomial table to find the probability of observing one of these unusual values.

6.5   Refer to Exercise 6.3. A syndicated newspaper story reported that inspectors found 22 of 24 hamburgers underweight at restaurant W and fined that restaurant. Draw new control limits on your chart, from Exercise 6.3, for one new sample of size 24. Plot the new proportion and determine whether this point is "out of control."

KEY IDEAS AND FORMULAS

The outcomes of an experiment are quantified by assigning each of them a numerical value related to a characteristic of interest. The rule for assigning the numerical value is called a random variable **X**.

For a discrete random variable, the probability distribution of X describes the manner in which probability is distributed over the possible values of X. Specifically, it is a list or formula giving the pairs x and $f(x) = P[X = x]$.

A probability distribution serves as a model for explaining variation in a population.

A probability distribution has mean

$$\mu = \Sigma \,(\text{value})(\text{probability}) = \Sigma \, xf(x)$$

which is interpreted as the population mean. This quantity is also called the expected value $E(X)$. Although X is a variable, $E(X)$ is a constant.

The population variance is

$$\sigma^2 = E(X - \mu)^2 = \Sigma(x - \mu)^2 f(x)$$

The standard deviation σ is its square root. The standard deviation is a measure of the spread or variation of the population.

Bernoulli trials are defined by the characteristics: (1) two possible outcomes, **success** (S) or **failure** (F) for each trial; (2) a constant probability of success; and (3) independence of trials.

Sampling from a finite population without replacement violates the requirement of independence. If the population is large and the sample size small, the trials can be treated as independent for all practical purposes.

The number of successes X in a fixed number of Bernoulli trials is called a **binomial random variable.** Its probability distribution, called the **binomial distribution,** is given by

$$f(x) = \binom{n}{x} p^x q^{n-x} \quad \text{for } x = 0, \ldots, n$$

where n = number of trials, p = the probability of success in each trial, and $q = 1 - p$.

The **binomial distribution** has

$$\text{Mean} = np$$
$$\text{Standard deviation} = \sqrt{npq}$$

A **p chart** displays sample proportions to reveal trends or changes in the population proportion over time.

7. REVIEW EXERCISES

7.1 Let X denote the difference (no. of heads − no. of tails) in three tosses of a coin.

(a) List the possible values of X.

(b) List the elementary outcomes associated with each value of X.

7.2 Suppose there are two boxes. Box 1 contains 20 articles, of which 6 are defective, and Box 2 contains 30 articles, of which 5 are defective. One article is randomly selected from each box, and the selections from the two boxes are independent. Let X denote the total number of defective articles obtained.

(a) List the possible values of X and identify the elementary outcomes associated with each value.

(b) Determine the probability distribution of X.

7.3 Refer to Exercise 7.2, but now suppose that the contents of the two boxes are pooled together into a single larger box. Then two articles are drawn at random and without replacement. Let W denote the number of defective articles in the sample. Obtain the probability distribution of W.

7.4 Refer to Exercise 7.2 and now suppose that the sampling is done in two stages: First, a box is selected at random and then, from the selected box, two articles are drawn at random and without replacement. Let Y denote the number of defective articles in the sample.

(a) List the elementary outcomes concerning the possible selections of the box and the possible compositions of the sample (use a tree diagram). Find their probabilities. (*Hint:* Use conditional probability and the multiplication rule.)

(b) Determine the probability distribution of Y.

7.5 In an assortment of 11 lightbulbs, there are 4 with broken filaments. A customer takes 3 bulbs from the assortment without inspecting the filaments. Find the probability distribution of the number X of defective bulbs that the customer may get.

7.6 For the following probability distribution

x	$f(x)$
0	.3
1	.4
2	.3

(a) Calculate μ.
(b) Calculate σ^2 and σ.
(c) Plot the probability histogram and locate μ.

7.7 For the following probability distribution

x	$f(x)$
2	.1
3	.3
4	.3
5	.2
6	.1

(a) Calculate $E(X)$.
(b) Calculate $sd(X)$.
(c) Draw the probability histogram and locate the mean.

7.8 Refer to Exercise 7.7.

(a) List the x values that lie in the interval $\mu - \sigma$ to $\mu + \sigma$ and calculate $P[\mu - \sigma \le X \le \mu + \sigma]$.

(b) List the x values that lie in the interval $\mu - 2\sigma$ to $\mu + 2\sigma$ and calculate $P[\mu - 2\sigma \le X \le \mu + 2\sigma]$.

7.9   A student buys a lottery ticket for $1. For every 1000 tickets sold, two bicycles are to be given away in a drawing.

(a)   What is the probability that the student will win a bicycle?

(b)   If each bicycle is worth $160, determine the student's expected gain.

7.10   In the finals of a tennis match, the winner will get $60,000 and the loser $15,000. Find the expected winnings of player B if (a) the two finalists are evenly matched and (b) player B has probability .9 of winning.

7.11   A lawyer feels that the probability is .3 she can win a wage discrimination suit. If she wins the case, she will make $15,000, but if she loses, she gets nothing.

(a)   What is the lawyer's expected gain?

(b)   If the lawyer has to spend $2500 in preparing the case, what is her expected net gain?

7.12   The number of overnight emergency calls X to the answering service of a heating and air conditioning firm have the probabilities .05, .1, .15, .35, and .20, and .15 for 0, 1, 2, 3, 4, and 5 calls, respectively.

(a)   Find the probability of fewer than 3 calls.

(b)   Determine $E(X)$ and sd(X).

7.13   A store conducts a lottery with 5000 cards. The prizes and the corresponding number of cards are as listed in the table. Suppose you have received one of the cards (presumably, selected at random) and let X denote your prize.

Prize	Number of Cards
$4000	1
$1000	3
$ 100	95
$ 5	425
$ 0	4476
Total	5000

(a)   Obtain the probability distribution of X.

(b)   Calculate the expected value of X.

(c)   If you have to pay $6 to get a card, find the probability that you will come out a loser.

7.14   A botany student is asked to match the popular names of three house plants with their obscure botanical names. Suppose the student never heard of these names and is trying to match by sheer guess. Let X denote the number of correct matches.

(a) Obtain the probability distribution of X.

(b) What is the expected number of matches?

7.15 A roulette wheel has 38 slots, of which 18 are red, 18 black, and 2 green. A gambler will play three times, each time betting $5 on red. The gambler gets $10 if red occurs, and loses the bet otherwise. Let X denote the net gain of the gambler in 3 plays (for instance, if he loses all three times, then $X = -15$).

(a) Obtain the probability distribution of X.

(b) Calculate the expected value of X.

(c) Will the expected net gain be different if the gambler alternates his bets between red and black? Why or why not?

7.16 Suppose that X can take the values 0, 1, 2, 3, and 4, and the probability distribution of X is, incompletely, specified by the function

$$f(x) = \frac{1}{4} \left(\frac{3}{4} \right)^x \quad \text{for } x = 0, 1, 2, 3$$

Find (a) $f(4)$, (b) $P[X \geq 2]$, (c) $E(X)$, and (d) $sd(X)$.

7.17 Let $\overline{X}$ = average number of dots resulting from two tosses of a fair die. For instance, if the faces 4 and 5 show, the corresponding value of $\overline{X}$ is $(4 + 5)/2 = 4.5$. Obtain the probability distribution of $\overline{X}$.

7.18 Refer to Exercise 7.17. On the same graph, plot the probability histograms of

$$X_1 = \text{No. of points on the first toss of the die}$$
$$\overline{X} = \text{Average number of points in two tosses of the die}$$

7.19 *The cumulative probabilities for a distribution.*   A probability distribution can also be described by a function that gives the accumulated probability at or below each value of X. Specifically,

$$F(c) = P[X \leq c] = \sum_{x \leq c} f(x)$$

Cumulative distribution function at c

$$= \text{Sum of probabilities of all values } x \leq c$$

For the probability distribution given here, we calculate

$$F(1) = P[X \leq 1] = f(1) \qquad = .07$$
$$F(2) = P[X \leq 2] = f(1) + f(2) = .19$$

x	$f(x)$	$F(x)$
1	.07	.07
2	.12	.19
3	.25	
4	.28	
5	.18	
6	.10	

(a)   Complete the $F(x)$ column in this table.

(b)   Now cover the $f(x)$ column with a strip of paper. From the $F(x)$ values, reconstruct the probability function $f(x)$.
[*Hint:* $f(x) = F(x) - F(x - 1)$.]

7.20   Let the random variable Y denote the proportion of times a head occurs in three tosses of a coin, that is, $Y = ($no. of heads in 3 tosses$)/3$.

(a)   Obtain the probability distribution of Y.

(b)   Draw the probability histogram.

(c)   Calculate $E(Y)$ and sd(Y).

7.21   Is the model of Bernoulli trials plausible in each of the following situations? Identify any serious violations of the conditions.

(a)   A dentist records whether each tooth in the lower jaw has a cavity or has none.

(b)   Persons applying for a driver's license will be recorded as writing left- or right-handed.

(c)   For each person taking a seat at a lunch counter, observe the time it takes to be served.

(d)   Each day of the first week in April is recorded as being either clear or cloudy.

(e)   Cars selected at random will or will not pass state safety inspection.

7.22   Give an example (different from those appearing in Exercise 7.21) of repeated trials with two possible outcomes where:

(a)   The model of Bernoulli trials is reasonable.

(b)   The condition of independence is violated.

(c)   The condition of equal $P(S)$ is violated.

7.23   If the probability of having a male child is .5, find the probability that the third child is the first son.

7.24   If the probability of getting caught copying someone else's exam is .2, find the probability of not getting caught in three attempts. Assume independence.

7.25   A stoplight on the way to class is red 60% of the time. What is the probability of hitting a red light
   (a)   2 days in a row?
   (b)   3 days in a row?
   (c)   2 out of 3 days?

7.26   A basketball team scores 40% of the times it gets the ball. Find the probability that the first basket occurs on its third possession. (Assume independence.)

7.27   If in three Bernoulli trials, the probability that the first two trials are both failures is 4/49, what is the probability that the first two are successes and the third is a failure?

7.28   The proportion of people having the blood type O in a large southern city is .4. For two randomly selected donors
   (a)   Find the probability of at least one type O.
   (b)   Find the expected number of type O.
   (c)   Repeat parts (a) and (b) if there are three donors.

7.29   A viral infection is spread by contact with an infected person. Let the probability that a healthy person gets the infection, in one contact, be $p = .4$.
   (a)   An infected person has contact with five healthy persons. Specify the distribution of X = no. of persons who contract the infection.
   (b)   Find $P[X \leq 3]$, $P[X = 0]$, and $E(X)$.

7.30   The probability that a voter will believe a rumor about a politician is .2. If 20 voters are told individually, find the probability that
   (a)   None of the 20 believes the rumor.
   (b)   Seven or more believe the rumor.
   (c)   Determine the mean and standard deviation of the number who believe the rumor.

7.31   A fair die will be rolled 4 times. Find the probability that 6 appears no more than twice.

7.32   A school newspaper claims that 80% of the students support its view on a campus issue. A random sample of 20 students is taken, and 12 students agree with the newspaper. Find $P[12 \text{ or less agree}]$, if 80% support the view, and comment on the plausibility of the claim.

7.33   For each situation, state whether a binomial distribution is reasonable for the random variable X. Justify your answer.
   (a)   A multiple-choice examination consists of 10 problems, each of which has 5 suggested answers. A student marks answers by pure

guesses (i.e., one answer is chosen at random out of the 5), and X denotes the number of marked answers that are wrong.

(b) A multiple-choice examination has two parts: Part 1 has 8 problems, each with 5 suggested answers, and Part 2 has 10 problems, each with 4 suggested answers. A student marks answers by pure guesses, and X denotes the total number of problems that the student correctly answers.

(c) Twenty married couples are interviewed about exercise, and X denotes the number of persons (out of the 40 people interviewed) who are joggers.

7.34 For the binomial distribution with $n = 14$ and $p = .4$, determine
(a) $P[4 \leq X \leq 9]$ (b) $P[4 < X \leq 9]$ (c) $P[4 < X < 9]$
(d) $E(X)$ (e) sd(X)

7.35 Using the binomial table,
(a) List the probability distribution for $n = 5$ and $p = .4$.
(b) Plot the probability histogram.
(c) Calculate $E(X)$ and Var(X) from the entries in the list from part (a).
(d) Calculate $E(X) = np$ and Var$(X) = npq$ and compare your answer with part (c).

7.36 Using the binomial table, find the probability of
(a) Three successes in 8 trials when $p = .4$
(b) Seven failures in 16 trials when $p = .6$
(c) Three or fewer successes in 9 trials when $p = .4$
(d) More than 12 successes in 16 trials when $p = .7$
(e) The number of successes between 8 and 13 (both inclusive) in 16 trials with $p = .6$

7.37 Using the binomial table, find the probability of
(a) Three or fewer successes for $p = .1, .2, .3, .4$, and .5 when $n = 12$
(b) Three or fewer successes for $p = .1, .2, .3, .4$, and .5 when $n = 18$

7.38 A sociologist believes that only half of the high school seniors capable of graduating from college go to college. Of 17 high school seniors who have the ability to graduate from college, find the probability that 10 or more will go to college if the sociologist is correct. Assume that the seniors will make their decisions independently. Also find the expected number.

7.39 Only 30% of the people in a large city think that its mass transit system is adequate. If 20 persons are selected at random, find the probability that 5 or fewer will think that the system is adequate. Find the probability that exactly 6 will think that the system is adequate.

7.40 Jones claims to have extrasensory perception (ESP). To test the claim, a psychologist shows Jones five cards that carry different pictures. Then Jones is blindfolded and the psychologist selects one card and asks Jones to identify the picture. This process is repeated 16 times. Suppose, in reality, that Jones has no ESP, but responds by sheer guesses. What is the probability that the identifications are

(a) Correct at most 3 times?

(b) Wrong at least 10 times?

7.41 Referring to Exercise 7.40, find the expected value and standard deviation of the number of correct identifications.

7.42 An inspector will sample bags of potato chips to see if they fall short of the weight, 14 ounces, printed on the bag. Samples of 20 bags will be selected and the number with weight less than 14 ounces will be recorded.

(a) Make a p chart using the centerline and control limits calculated for $p_0 = .5$.

(b) Suppose that samples from ten different days have

$$11 \quad 8 \quad 14 \quad 10 \quad 13 \quad 12 \quad 7 \quad 14 \quad 10 \quad 13$$

underweight bags. Graph the corresponding proportions on your p chart.

(c) Identify any days where the chart signals "out of control."

CHAPTER 6

Normal Distributions

Chapter Objectives

After reading this chapter, you should be able to

▶ Distinguish between probability models for continuous random variables and those for discrete random variables.

▶ Discuss the concept of a probability density function.

▶ Describe the standard normal distribution and calculate probabilities using the table of standard normal probabilities.

▶ Evaluate probabilities concerning normal random variables.

▶ Use the normal distribution to approximate binomial probabilities.

1. INTRODUCTION

The variation in outcomes that can be represented as values of a discrete random variable is relatively easy to model. Because the possible values of discrete random variables proceed in steps, usually integers representing a count, probabilities are assigned directly to these values. Probability statements, population means, and standard deviations can be obtained directly from the resulting probability distribution.

However, many interesting variables are essentially recorded on a continuous scale. They can take on any value in an interval. Weight, strength, life length, and temperature are examples of continuous variables.

Manipulating continuous random variables and probabilities is more difficult than it is for discrete random variables. Computer-generated tables of probabilities do simplify the mathematical details.

How do we assign probabilities to the values of a continuous variable? There are so many values in any interval—many more than can be counted with integers—that the probability of any particular one must be vanishingly small. Rather than deal with great quantities of extremely small numbers, we take an entirely different approach. It makes more sense to assign probabilities to intervals of values rather than to single values.

2. PROBABILITY MODEL FOR A CONTINUOUS RANDOM VARIABLE

Just as probability is conceived as the long-run relative frequency, the idea of a continuous probability distribution draws from the density histogram for a large number of measurements. The reader may wish to review Section 3 of Chapter 2, where grouping of data in class intervals and construction of a density histogram were discussed. We have remarked that with an increasing number of observations in a data set, histograms can be constructed with class intervals having smaller widths. We will now pursue this point to motivate the idea of a continuous probability distribution. To focus the discussion, let us consider that the weight X of a newborn baby is the continuous random variable of our interest. How do we conceptualize the probability distribution of X? Initially, suppose that the birth weights of 100 babies are recorded, the data grouped in class intervals of 1 pound, and the density histogram in Figure 1(a) is obtained. Recall that a density histogram has the following properties:

1. The total area under the histogram is 1.
2. For two points a and b such that each is a boundary point of some class, the relative frequency of measurements in the interval a to b is the area under the histogram above this interval.

For example, Figure 1(a) shows that the interval 7.5–9.5 pounds contains a proportion .28 + .25 = .53 of the 100 measurements.

Next, we suppose that the number of measurements is increased to 5000 and

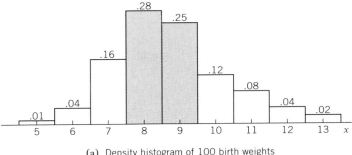

(a) Density histogram of 100 birth weights
with a class interval of 1 pound

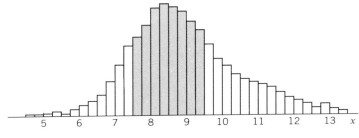

(b) Density histogram of 5000 birth weights
with a class interval of .25 pound

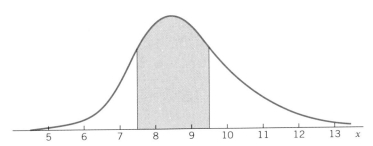

(c) Probability density curve for the continuous random variable X = birth weight

Figure 1. Probability density curve viewed as a limiting form of density histograms

they are grouped in class intervals of .25 pound. The resulting density histogram appears in Figure 1(b). This is a refinement of the histogram in part (a): It is constructed from a larger set of observations and exhibits relative frequencies for finer class intervals. (Narrowing the class interval without increasing the number of observations would obscure the overall shape of the distribution.) The refined histogram 1(b) again has the properties 1 and 2 stated above.

By proceeding in this manner, we can imagine even further refinements of density histograms with larger numbers of observations and smaller class intervals.

In pursuing this conceptual argument, we ignore the difficulty that the accuracy of the measuring device is limited. In the course of refining the histograms, the jumps between consecutive rectangles tend to dampen out, and the top of the density histogram approximates the shape of a smooth curve, as illustrated in Figure 1(c). Because probability is interpreted as long-run relative frequency, the curve obtained as the limiting form of the density histograms represents the manner in which the total probability 1 is distributed over the interval of possible values of the random variable X. This curve is called the probability density curve of the continuous random variable X. The mathematical function $f(x)$ whose graph produces this curve is called the probability density function of the continuous random variable X or the probability model for a continuous random variable X. In other words, the curve specified by $f(x)$ describes the statistical population from which data will be drawn. One usually proceeds by tentatively assuming a probability model suggested by data from the current or a similar source. However, any model that we tentatively assume must be checked for conformance with the data that are collected. We address this issue in Section 7.

The properties 1 and 2 that we stated earlier for a density histogram are shared by a probability density curve that is, after all, conceived as a limiting smoothed form of a histogram. Also, since a histogram can never protrude below the x-axis, we have the further fact that $f(x)$ is nonnegative for all x.

The probability density function $f(x)$ describes the distribution of probability for a continuous random variable. It has the properties

1.   The total area under the probability density curve is 1.
2. $P[a \leq X \leq b]$ = area under the probability density curve between a and b.
3. $f(x) \geq 0$ for all x.

Unlike the description of a discrete probability distribution, the probability density $f(x)$ for a continuous random variable does not represent the probability that the random variable will exactly equal the value x, or the event $[X = x]$. Instead, a probability density function relates the probability of an interval $[a, b]$ to the area under the curve in a strip over this interval. A single point x, being an interval with a width of 0, supports 0 area, so $P[X = x] = 0$.

With a continuous random variable, the probability that $X = x$ is always 0. It is only meaningful to speak about the probability that X lies in an interval.

The deduction that the probability at every single point is zero needs some clarification. In the birth-weight example, the statement $P[X = 8.5 \text{ pounds}] = 0$ probably seems shocking. Does this statement mean that no child can have a birth weight of 8.5 pounds? To resolve this paradox, we need to recognize that the accuracy of every measuring device is limited, so that here the number 8.5 is actually indistinguishable from all numbers in an interval surrounding it, say, $[8.495, 8.505]$. Thus, the question really concerns the probability of an interval surrounding 8.5, and the area under the curve is no longer 0.

When determining the probability of an interval a to b, we need not be concerned if either or both endpoints are included in the interval. Since the probabilities of $X = a$ and $X = b$ are both equal to 0,

$$P[a \leq X \leq b] = P[a < X \leq b] = P[a \leq X < b] = P[a < X < b]$$

In contrast, these probabilities may not be equal for a discrete distribution.

Fortunately, for important distributions, areas have been extensively tabulated. In most tables, the entire area to the left of each point is tabulated. To obtain the probabilities of other intervals, we must apply the following rules.

$P[a < X < b]$
= (Area to left of b) − (Area to left of a)

$P[b < X]$
= 1 − (Area to left of b)

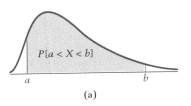

(a)

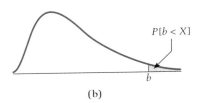

(b)

FEATURES OF A CONTINUOUS DISTRIBUTION

As is true for density histograms, the probability density curves of continuous random variables could possess a wide variety of shapes. A few of these are illustrated in Figure 2 (page 258). Many statisticians use the term **skewed** to describe a probability density curve with a long tail in one direction.

A continuous random variable X also has a mean, or expected value $E(X)$, as well as a variance and a standard deviation. Their interpretations are the same as in the case of discrete random variables, but their formal definitions involve integral calculus and are therefore not pursued here. However, it is instructive to see in Figure 3 (page 258) that the mean $\mu = E(X)$ marks the balance point of the probability mass. The median, another measure of center, is the value of X that divides the area under the curve into halves.

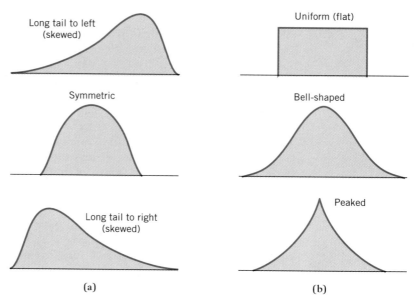

Figure 2. Different shapes of probability density curves: **(a)** Symmetry and deviations from symmetry; **(b)** Different peakedness

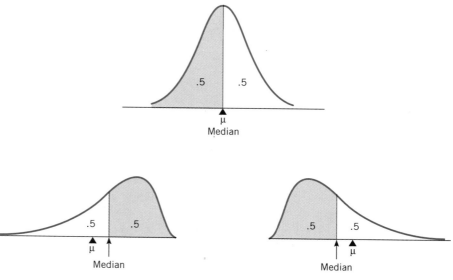

Figure 3. Mean as the balance point and median as the point of equal division of the probability mass

Exercises

2.1   When we go to buy a car we usually are given information such as:

        gas mileage in city driving: 27 miles per gallon

        gas mileage in highway driving: 45 miles per gallon

    (a)   How do the manufacturers arrive at these figures?

    (b)   Name at least three variables that may affect the gas mileage of a car.

2.2   Which of the functions sketched in (a)–(d) could be a probability density function for a continuous random variable? Why or why not?

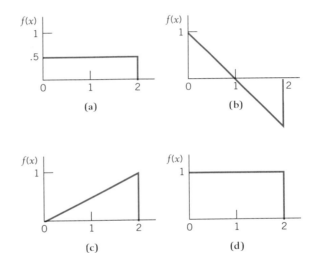

2.3   Determine the following probabilities from the curve $f(x)$ diagrammed in Exercise 2.2(a).

    (a) $P[0 < X < .5]$ (b) $P[.5 < X < 1]$

    (c) $P[1.5 < X < 2]$ (d) $P[X = 1]$

2.4   For the curve $f(x)$ graphed in Exercise 2.2(c), which of the two intervals $[0 < X < .5]$ or $[1.5 < X < 2]$ is assigned a higher probability?

2.5   Determine the median for the probability distribution depicted in Exercise 2.2(a).

2.6   If a student is more likely to be late than on time for the 1:20 P.M. history class,

    (a)   Determine whether the median of the student's arrival time distribution is earlier than, equal to, or later than 1:20 P.M.

    (b)   On the basis of the given information, can you determine whether the mean of the student's arrival time distribution is earlier than, equal to, or later than 1:20 P.M.? Comment.

2.7   Which of the distributions in Figure 3 are compatible with the following statements?

    (a)   The distribution of starting salaries for computer programmers has a mean of \$35,000, but half of the newly employed programmers make less than \$30,000 annually.

    (b)   In spite of recent large increases in salary, half of the professional football players still make less than the average salary.

3. THE NORMAL DISTRIBUTIONS—GENERAL FEATURES

The normal distribution, which may already be familiar to some readers as the curve with the bell shape, is sometimes associated with the names of Pierre Laplace and Carl Gauss, who figured prominently in its historical development. Gauss derived the normal distribution mathematically as the probability distribution of the error of measurements, which he called the "normal law of errors." Subsequently, astronomers, physicists, and, somewhat later, data collectors in a wide variety of fields found that their histograms exhibited the common feature of first rising gradually in height to a maximum and then decreasing in a symmetric manner. Although the normal curve is not unique in exhibiting this form, it has been found to provide a reasonable approximation in a great many situations. Unfortunately, at one time during the early stages of the development of statistics, it had many overzealous admirers. Apparently, they thought that all real-life data must conform to the bell-shaped normal curve, or otherwise, the process of data collection should be suspect. It is in this context that the distribution became known as the **normal distribution.** However, scrutiny of data has often revealed inadequacies of the normal distribution. In fact, the universality of the normal distribution is only a myth, and examples of quite nonnormal distributions abound in virtually every field of study. Still, the normal distribution plays a central role in statistics, and inference procedures derived from it have wide applicability and form the backbone of current methods of statistical analysis.

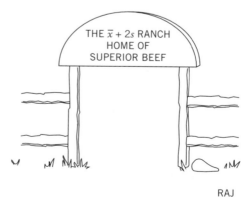

RAJ

Does the $\bar{x}$ ranch have average beef?

There is a normal distribution for each value of its mean μ and its standard deviation σ. A few details of the normal curve merit special attention. The curve is symmetric about its mean μ, which locates the peak of the bell (see Figure 4). The interval running one standard deviation in each direction from μ has a probability of .683, the interval from $\mu - 2\sigma$ to $\mu + 2\sigma$ has a probability of .954, and the interval from $\mu - 3\sigma$ to $\mu + 3\sigma$ has a probability of .997. It is these probabilities that give rise to the empirical guidelines stated in Chapter 2. The curve never reaches 0 for any value of x, but because the tail areas outside the interval $(\mu - 3\sigma, \mu + 3\sigma)$ are very small, we usually terminate the graph at these points.

A normal distribution has a bell-shaped density[1] as shown in Figure 4. It has

$$\text{mean} = \mu$$
$$\text{standard deviation} = \sigma$$

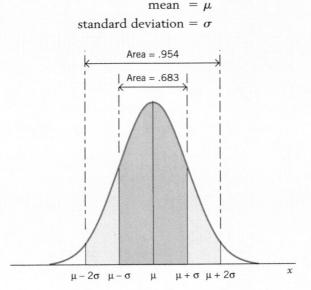

Figure 4. Normal distribution

The probability of the interval extending

one sd on each side of the mean: $P[\mu - \sigma < X < \mu + \sigma] = .683$
two sd on each side of the mean: $P[\mu - 2\sigma < X < \mu + 2\sigma] = .954$
three sd on each side of the mean: $P[\mu - 3\sigma < X < \mu + 3\sigma] = .997$

[1]The formula, which need not concern us, is

$$f(x) = \frac{1}{\sqrt{2\pi}\sigma} e^{-\frac{1}{2}\left(\frac{x-\mu}{\sigma}\right)^2} \quad \text{for } -\infty < x < \infty$$

where π is the area of a circle having unit radius, or approximately 3.1416, and e is approximately 2.7183.

Notation

The normal distribution with a mean of μ and a standard deviation of σ is denoted by $N(\mu, \sigma)$.

Interpreting the parameters, we can see in Figure 5 that a change of mean from μ_1 to a larger value μ_2 merely slides the bell-shaped curve along the axis until a new center is established at μ_2. There is no change in the shape of the curve.

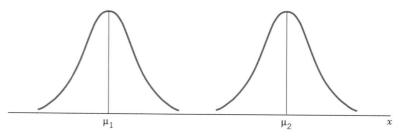

Figure 5. Two normal distributions with different means, but the same standard deviation

A different value for the standard deviation results in a different maximum height of the curve and changes the amount of the area in any fixed interval about μ (see Figure 6). The position of the center does not change if only σ is changed.

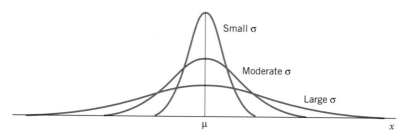

Figure 6. Decreasing σ increases the maximum height and the concentration of probability about μ.

4. THE STANDARD NORMAL DISTRIBUTION

As we will see in Section 5, probability calculations concerning a general normal distribution having mean μ and standard deviation σ can be transformed into a problem concerning a simple normal distribution. This particular normal distribution, which has mean 0 and standard deviation 1, is called the **standard normal distribution.** Consequently, we begin by examining probability calculations related to the standard normal distribution. The standard normal curve is illustrated in Figure 7.

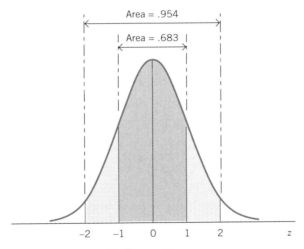

Figure 7. The standard normal curve

The **standard normal distribution** has a bell-shaped density with

$$\text{mean } \mu = 0$$
$$\text{standard deviation } \sigma = 1$$

The standard normal distribution is denoted by $N(0, 1)$.

USE OF THE STANDARD NORMAL TABLE (APPENDIX B, TABLE 3)

For any specified value z on the axis supporting the standard normal curve, the area under the curve to the left of z gives the probability $P[Z \leq z]$; see Figure 8.

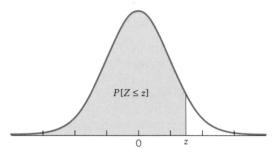

Figure 8. $P[Z \leq z]$ and area

The probability $P[Z \leq z]$ has been tabulated, in the standard normal table, for a large number of values of z. The values of z are located on the *top and sides* of the table and the probabilities are located *inside* the table.

Example 1   Standard normal probabilities for tail events
Find $P[Z \leq 1.37]$ and $P[Z > 1.37]$.

SOLUTION AND DISCUSSION   In the standard normal table we locate 1.3 on the side and .07 on the top margin (see Table 1 below): Then .9147 is located within the table. We see that the probability or area to the left of 1.37 is .9147. Consequently, $P[Z \leq 1.37] =$ .9147.

TABLE 1   How to Read from Appendix B, Table 3, for $P[Z \leq 1.37]$; $z = 1.37 = 1.3 + .07$

z	.00	$\cdots$	.07	$\cdots$
.0				
.				
.				
.				
1.3			.9147	
.				
.				
.				

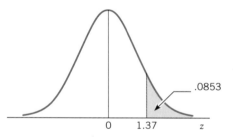

Figure 9. An upper-tail normal probability

Because $[Z > 1.37]$ is the complement of $[Z \leq 1.37]$,

$$P[Z > 1.37] = 1 - P[Z \leq 1.37] = 1 - .9147 = .0853$$

as we can see in Figure 9. An alternative method is to use symmetry to show that $P[Z > 1.37] = P[Z < -1.37]$, which can be obtained directly from the normal table. ∎

To obtain the probability of an interval $[a, b]$, notice that

$$P[a \leq Z \leq b] = [\text{Area to left of } b] - [\text{Area to left of } a]$$

The following properties can be observed from the symmetry of the standard normal curve about 0, as exhibited in Figure 10.

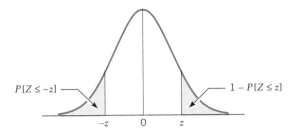

$P[Z \leq -z]$ $1 - P[Z \leq z]$

$-z$ 0 z

Figure 10. Equal normal tail probabilities

1. $P[Z \leq 0] = .5$
2. $P[Z \leq -z] = 1 - P[Z \leq z] = P[Z \geq z]$

The next series of examples illustrates the use of these properties and the standard normal table, Appendix B, Table 3.

Example 2   Standard normal probability of an interval
Calculate $P[-.15 < Z < 1.60]$.

SOLUTION AND DISCUSSION   From Appendix B, Table 3, we see that

$$P[Z \leq 1.60] = \text{Area to left of } 1.60 = .9452$$

and that

$$P[Z \leq -.15] = \text{Area to left of } -.15 = .4404$$

Therefore,

$$P[-.15 < Z < 1.60] = .9452 - .4404 = .5048$$

which is the shaded area in Figure 11.

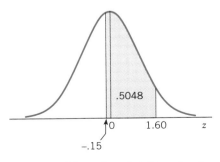

.5048

0   1.60 z

$-.15$

Figure 11. Normal probability of an interval

Example 3   Standard normal probability outside of an interval
Find $P[Z < -1.9 \text{ or } Z > 2.1]$.

SOLUTION AND   The two events $[Z < -1.9]$ and $[Z > 2.1]$ are mutually exclusive, so we add
DISCUSSION   their probabilities:

$$P[Z < -1.9 \text{ or } Z > 2.1] = P[Z < -1.9] + P[Z > 2.1]$$

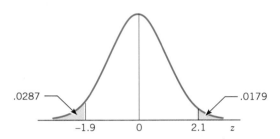

Figure 12. Normal probabilities for Example 3

As indicated in Figure 12, $P[Z > 2.1]$ is the area to the right of 2.1, which is $1 - [\text{Area to left of } 2.1] = 1 - .9821 = .0179$. The normal table gives $P[Z < -1.9] = .0287$ directly. Adding these two quantities, we get

$$P[Z < -1.9 \text{ or } Z > 2.1] = .0287 + .0179 = .0466 \qquad \blacksquare$$

Example 4   Determining an upper percentile of the standard normal
Locate the value of z that satisfies $P[Z > z] = .025$.

SOLUTION AND   If we use the property that the total area is 1, the area to the left of z must
DISCUSSION   be $1 - .0250 = .9750$. The marginal value with the tabular entry .9750 is
$z = 1.96$ (diagrammed in Figure 13).

z	.00	$\cdots$	.06	$\cdots$
.0				
.				
.				
.				
1.9	$\longleftarrow$	$\text{----------} .9750$		
.				
.				
.				

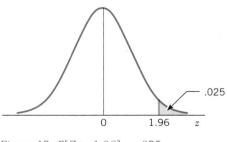

Figure 13. $P[Z > 1.96] = .025$

$\blacksquare$

Example 5 Find z for given equal tail probabilities
Obtain the value of z for which $P[-z \le Z \le z] = .90$.

SOLUTION AND DISCUSSION We observe from the symmetry of the curve that

$$P[Z < -z] = P[Z > z] = .05$$

From the normal table, we see that $z = 1.65$ gives $P[Z < -1.65] = .0495$ and $z = 1.64$ gives $P[Z < -1.64] = .0505$. Because .05 is halfway between these two probabilities, we interpolate between the two z-values to obtain $z = 1.645$ (see Figure 14).

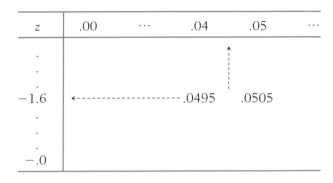

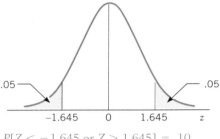

$P[Z < -1.645 \text{ or } Z > 1.645] = .10$
Figure 14.

The interpolated value 1.645, the z-value with probability .05 to its right, appears frequently in statistical calculations. Except for this particular interpolated value, we will always round z to two decimal digits before entering the standard normal table. Accurate computer calculations are preferred to interpolation (see Exercise 4.12).

Suggestion: The preceding examples illustrate the usefulness of a sketch to depict an area under the standard normal curve. A correct diagram shows how to combine the left-side areas given in the normal table.

Exercises

4.1 Find the area under the standard normal curve to the left of
(a) $z = 1.17$ (b) $z = .16$
(c) $z = -1.83$ (d) $z = -2.3$

4.2 Find the area under the standard normal curve to the left of
(a) $z = .83$ (b) $z = 1.03$
(c) $z = -1.03$ (d) $z = -1.35$

4.3 Find the area under the standard normal curve to the right of
     (a) $z = 1.17$ (b) $z = .60$ (c) $z = -1.13$

4.4 Find the area under the standard normal curve to the right of
     (a) $z = .83$ (b) $z = 2.83$ (c) $z = -1.23$

4.5 Find the area under the standard normal curve over the interval
     (a) $z = -.65$ to $z = .65$
     (b) $z = -1.04$ to $z = 1.04$
     (c) $z = .32$ to $z = 2.65$

4.6 Find the area under the standard normal curve over the interval
     (a) $z = -.44$ to $z = .44$
     (b) $z = -1.33$ to $z = 1.33$
     (c) $z = .40$ to $z = 2.03$

4.7 Identify the z-values in the following diagrams of the standard normal distribution.

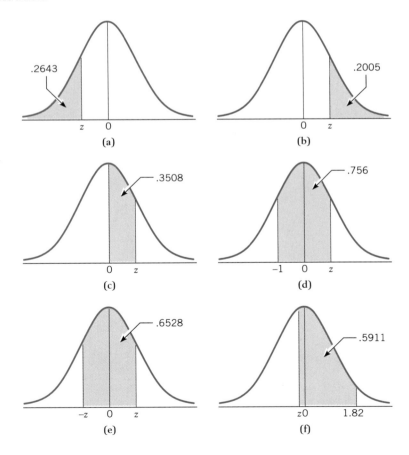

4.8   Identify the z-values in the following diagrams of the standard normal distribution.

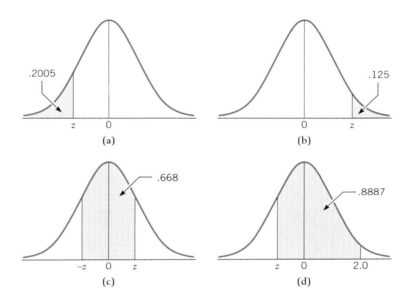

4.9   For a standard normal random variable Z, find
   (a) $P[Z < .42]$ (b) $P[Z < -.42]$
   (c) $P[Z > 1.69]$ (d) $P[Z > -1.69]$
   (e) $P[-1.2 < Z < 2.1]$ (f) $P[.05 < Z < .8]$
   (g) $P[-1.62 < Z < -.51]$ (h) $P[|Z| < 1.64]$

4.10 Find the z-value in each of the following cases.
   (a) $P[Z < z] = .1736$ (b) $P[Z > z] = .10$
   (c) $P[-z < Z < z] = .954$ (d) $P[-.6 < Z < z] = .50$

4.11 Find
   (a) $P[Z < .14]$
   (b) The 14th percentile of the standard normal distribution. That is, find the z that satisfies $P[Z \le z] = .14$.
   (c) $P[Z < .86]$
   (d) The 86th percentile of the standard normal distribution

💻 USING A COMPUTER

💾 4.12 ***Using MINITAB to Calculate Standard Normal Probabilities.*** We illustrate the calculation of $P[Z \le 1.273]$.

> **Dialog box:** **Session command**
> **Calc > Probability Distributions > Normal** MTB > CDF 1.273
> Click **Cumulative Distribution.**
> Type *1.273* in **Input constant.** Click **OK.**
> **Output:**
>
> **Cumulative Distribution Function**
> ```
> Normal with mean = 0 and standard deviation = 1.00000
>          x        P( X <= x )
>     1.2730          0.8985
> ```

Find (a) $P[Z \leq -1.5832]$ (b) $P[Z > .4726]$

5. PROBABILITY CALCULATIONS WITH NORMAL DISTRIBUTIONS

Fortunately, no new tables are required for probability calculations regarding the general normal distribution. Any normal distribution can be set in correspondence to the standard normal by the following relation.

> If X is distributed as $N(\mu, \sigma)$, then the standardized variable
>
> $$Z = \frac{X - \mu}{\sigma}$$
>
> has the standard normal distribution.

This property of the normal distribution allows us to cast a probability problem concerning X into one concerning Z. To find the probability that X lies in a given interval, convert the interval to the z-scale and then calculate the probability by using the standard normal table (Appendix B, Table 3).

Example 6   Converting a normal probability to a standard normal probability
Given that X has the normal distribution $N(60, 4)$, find $P[55 \leq X \leq 63]$.

SOLUTION AND DISCUSSION   Here, the standardized variable is $Z = \dfrac{X - 60}{4}$. The distribution of X is shown in Figure 15, where the z-scale and standard normal are displayed below the x-scale. In particular,

$x = 55$ gives $z = \dfrac{55 - 60}{4} = -1.25$

$x = 63$ gives $z = \dfrac{63 - 60}{4} = .75$

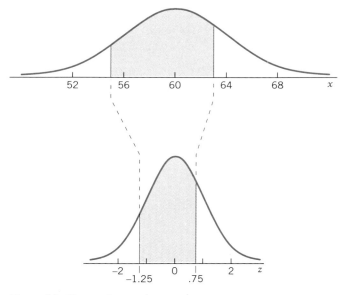

Figure 15. Converting to the z-scale

Therefore,

$$P[\,55 \le X \le 63\,] = P[\,-1.25 \le Z \le .75\,]$$

Using the normal table, we find $P[\,Z \le .75\,] = .7734$ and $P[\,Z \le -1.25\,] = .1056$, so the required probability is $.7734 - .1056 = .6678$. ∎

The working steps employed in Example 6 can be formalized into the rule,

---

If X is distributed as $N(\mu, \sigma)$, then

$$P[\,a \le X \le b\,] = P\left[\,\dfrac{a - \mu}{\sigma} \le Z \le \dfrac{b - \mu}{\sigma}\,\right]$$

where Z has the standard normal distribution.

Example 7 Calories in lunch salad

The number of calories in a salad on the lunch menu is normally distributed with mean = 200 and sd = 5. Find the probability that the salad you select will contain

(a) More than 208 calories.

(b) Between 190 and 200 calories.

SOLUTION AND DISCUSSION   Letting X denote the number of calories in the salad, we have the standardized variable

$$Z = \frac{X - 200}{5}$$

(a)   The z-value corresponding to $x = 208$ is

$$z = \frac{208 - 200}{5} = 1.6$$

Therefore,

$$\begin{aligned} P[X > 208] &= P[Z > 1.6] \\ &= 1 - P[Z \le 1.6] \\ &= 1 - .9452 = .0548 \end{aligned}$$

(b)   The z-values corresponding to $x = 190$ and $x = 200$ are

$$\frac{190 - 200}{5} = -2.0 \quad \text{and} \quad \frac{200 - 200}{5} = 0$$

respectively. We calculate

$$\begin{aligned} P[190 \le X \le 200] &= P[-2.0 \le Z \le 0] \\ &= .5 - .0228 = .4772 \end{aligned}$$

Calories vary in all foods. Labels and diet charts give averages. Should they also give distributions?   ∎

Example 8   Percentile of a normal distribution

The raw scores in a national aptitude test are normally distributed with mean = 506 and sd = 81.

(a)   What proportion of the candidates scored below 574?

(b)   Find the 14th percentile of the scores.

SOLUTION AND DISCUSSION   If we denote the raw score by X, the standardized score

$$Z = \frac{X - 506}{81}$$

is distributed as $N(0, 1)$.

(a)   The z-score corresponding to 574 is

$$z = \frac{574 - 506}{81} = .84$$

So

$$P[X < 574] = P[Z < .84] = .7995$$

Thus, 79.95% or about 80% of the candidates scored below 574. In other words, the score 574 nearly locates the 80th percentile.

(b)   We first find the 14th percentile in the z-scale and then convert it to the x-scale. From the standard normal table, we find

$$P[Z \leq -1.08] = .1401$$

The standardized score $z = -1.08$ corresponds to

$$x = 506 + 81(-1.08)$$
$$= 418.52$$

Therefore, the 14th percentile score is about 418.5.   ∎

It is often helpful to refer a random variable X to a standard scale even when it is not normal. One reason is that the standardized variable is free of units of measurement. If X is a real estate salesperson's commission in dollars, then the mean is also expressed in dollars and the difference

Random variable − Mean

is also expressed in dollars. Because the population standard deviation is also in dollars, their ratio is free of the unit dollars.

The **standardized variable**

$$Z = \frac{X - \mu}{\sigma} = \frac{\text{Variable} - \text{Mean}}{\text{Standard deviation}}$$

has mean 0 and sd 1. It does not depend on the unit of measurement.

Exercises

5.1   If X is normally distributed with $\mu = 80$ and $\sigma = 4$, find
   (a) $P[X < 75]$ (b) $P[X \leq 87]$
   (c) $P[X > 86]$ (d) $P[X > 71]$
   (e) $P[73 \leq X \leq 89]$ (f) $P[81 < X < 84]$

5.2   If X is normally distributed with a mean of 30 and a standard deviation of 5, find
   (a) $P[X < 28]$ (b) $P[X \leq 39]$
   (c) $P[X > 37]$ (d) $P[X > 21]$
   (e) $P[22 \leq X \leq 41]$ (f) $P[22 < X < 41]$

5.3   If X has a normal distribution with $\mu = 150$ and $\sigma = 5$, find b such that
      (a) $P[X < b] = .975$ (b) $P[X > b] = .025$
      (c) $P[X < b] = .305$

5.4   If X is normally distributed with a mean of 110 and a standard deviation of 3, find b such that
      (a) $P[X < b] = .7995$ (b) $P[X > b] = .001$
      (c) $P[X < b] = .063$

5.5   Scores on a certain nationwide college entrance examination follow a normal distribution with a mean of 500 and a standard deviation of 100. Find the probability that a student will score
      (a) Over 650   (b) Less than 250
      (c) Between 325 and 675

5.6   Refer to Exercise 5.5.
      (a) If a school only admits students who score over 670, what proportion of the student pool would be eligible for admission?
      (b) What limit would you set that makes 50% of the students eligible?
      (c) What should be the limit if only the top 15% are to be eligible?

5.7   According to the children's growth chart that doctors use as a reference, the heights of two-year-old boys are nearly normally distributed with a mean of 34.5 inches and a standard deviation of 1.3 inches. If a two-year-old boy is selected at random, what is the probability that he will be between 33.5 and 36.7 inches tall?

5.8   The time it takes a symphony orchestra to play Beethoven's *Ninth Symphony* has a normal distribution with a mean of 64.3 minutes and a standard deviation of 1.15 minutes. The next time it is played, what is the probability that it will take between 62.5 and 67.7 minutes?

5.9   The weights of apples served at a restaurant are normally distributed with a mean of 5 ounces and standard deviation of 1.1 ounces. What is the probability that the next person served will be given an apple that weighs less than 4 ounces?

5.10 The diameter of hail hitting the ground during a storm is normally distributed with a mean of .5 inch and a standard deviation of .1 inch. What is the probability that
      (a) A hailstone picked up at random will have a diameter greater than .75 inch?
      (b) Two hailstones picked up in a row will have diameters greater than .6 inch? (Assume independence of the two diameters.)
      (c) By the end of the storm, what proportion of the hailstones would have had diameters greater than .75 inch?

5.11 According to government reports (USDHEW 79-1659), the heights of adult male residents of the United States are approximately normally distributed with a mean of 69.0 inches and a standard deviation of 2.8 inches. If a clothing manufacturer wants to limit his market to the central 80% of the adult male population, what range of heights should be targeted?

5.12 The time for an emergency medical squad to arrive at the sports center at the edge of town is distributed as a normal variable with $\mu = 17$ minutes and $\sigma = 3$ minutes.

   (a) Determine the probability that the time to arrive is

      (i) More than 22 minutes

      (ii) Between 13 and 21 minutes

      (iii) Between 15.5 and 18.5 minutes

   (b) Which arrival period of duration 1 minute is assigned the highest probability by the normal distribution?

5.13 Find the standardized variable Z if X has

   (a) Mean 8 and standard deviation 2

   (b) Mean 350 and standard deviation 25

   (c) Mean 666 and variance 100

5.14 Males 18–24 years old have a mean height of 70 inches with a standard deviation of 2.8 inches. Females 18–24 years old have a mean height of 65 inches with a standard deviation of 2.4 inches. (Based on *Statistical Abstract of the United States 1990*, Table 201.)

   (a) Find the standardized variable for the heights of males.

   (b) Find the standardized variable for the heights of females.

   (c) For a 66-inch-tall person, find the value of the standardized variable for males.

   (d) For a 66-inch-tall person, find the value of the standardized variable for females. Compare your answer with part (c) and comment.

6. THE NORMAL APPROXIMATION TO THE BINOMIAL

When we encounter a similar situation repeatedly, and are interested only in whether an outcome called success occurs, then the total number of successes may follow a binomial distribution. This would be the case if the situation were faced a fixed number n of times, the probability of success was the same each time, and the outcome in one situation was independent of the outcomes in the other situations. In Chapter 5 we provided a formula for calculating the exact binomial probabilities.

Suppose we want to calculate the probability that at most 20 persons, among 100 selected at random from a large city, listened to a particular radio station in the past week. If only 20 persons in the sample did listen, then this probability would help us evaluate the station's claim that 30% of the people listen every week. Although computers can now evaluate binomial probabilities even with quite large values of n, later we will find it advantageous to have a normal approximation to the binomial probabilities.

When the success probability p is not too near 0 or 1 and the number of trials is large, the normal distribution serves as a good approximation to the binomial probabilities. Bypassing the mathematical proof, we concentrate on illustrating the manner in which this approximation works.

Figure 16 presents the binomial distribution for the number of trials $n = 5$, 12, and 25 when $p = .4$. Notice how the distribution begins to assume the distinctive bell shape for increasing n. Even though the binomial distributions with $p = .4$ are not symmetric, the lack of symmetry becomes negligible for large n.

We will sometimes use the normal distribution to approximate the probability that a binomial random variable X lies in an interval. This approximation to the binomial applies when the number of trials, n, is large and the probability of success, p, is not too close to 0 or 1. For large n, we ignore the small difference in the binomial probabilities of the intervals $[a \leq X \leq b]$ and $[a < X < b]$. We use the same normal approximation for both intervals.

The Normal Approximation to the Binomial

When np and $n(1 - p)$ are both large, say, greater than 15, the binomial distribution is well approximated by the normal distribution having mean $= np$ and sd $= \sqrt{np(1 - p)}$. That is,

$$Z = \frac{X - np}{\sqrt{np(1 - p)}} \quad \text{is approximately } N(0, 1)$$

Example 9   Normal approximation to binomial
Let X have a binomial distribution with $p = .6$ and $n = 150$. Approximate the probability that

(a) X is between 82 and 101, both inclusive.

(b) X is greater than 97.

SOLUTION AND DISCUSSION   (a)   We calculate the mean and standard deviation of X.

$$\text{mean} = np = 150(.6) = 90$$
$$\text{sd} = \sqrt{np(1 - p)} = \sqrt{150(.6)(.4)} = \sqrt{36} = 6$$

The standardized variable is

$$Z = \frac{X - 90}{6}$$

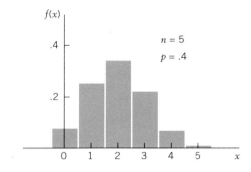

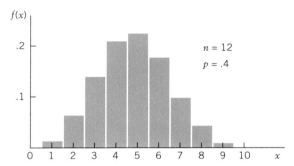

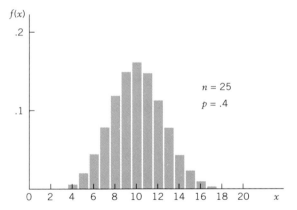

Figure 16. The binomial distributions for $p = .4$ and $n = 5, 12,$ and 25

We then approximate

$$P[82 \leq X \leq 101] = P\left[\frac{82 - 90}{6} \leq \frac{X - 90}{6} \leq \frac{101 - 90}{6}\right]$$
$$\approx P[-1.33 \leq Z \leq 1.83]$$

where Z has a standard normal distribution. From the normal table,

$$P[-1.33 \le Z \le 1.83] = .9664 - .0918 = .8746$$

That is, we approximate the binomial probability $P[82 \le X \le 101]$ by the normal probability .8746.

(b)   We approximate the binomial probability of $[X > 97]$ by the normal probability

$$P[X > 97] = P\left[\frac{X - 90}{6} > \frac{97 - 90}{6}\right]$$
$$\approx P[Z > 1.17] = 1 - .8790 = .1210$$

The normal approximation to the binomial gives $P[X > 97] \approx .1210$.   ■

Example 10   Probability calculations for a survey

A large-scale survey conducted five years ago revealed that 30% of the adult population were regular users of alcoholic beverages. If this is still the current rate, what is the probability that in a random sample of 1000 adults the number of users of alcoholic beverages will be (a) less than 280? (b) 316 or more?

SOLUTION AND DISCUSSION   Let X denote the number of users in a random sample of 1000 adults. Under the assumption that the proportion of the population who are users is .30, X has a binomial distribution with $n = 1000$ and $p = .3$. Since

$$np = 300 \qquad \sqrt{np(1 - p)} = \sqrt{210} = 14.49$$

the distribution of X is approximately $N(300, 14.49)$.

(a)   We approximate the binomial probability of $[X < 280]$ by the normal probability

$$P[X < 280] \approx P\left[Z < \frac{280 - 300}{14.49}\right]$$
$$= P[Z < -1.38]$$
$$= .0838$$

(b)
$$P[X \ge 316] \approx P\left[Z \ge \frac{316 - 300}{14.49}\right]$$
$$= P[Z \ge 1.10]$$
$$= 1 - .8643 = .1357$$ ■

Remark: If the object is to calculate binomial probabilities, today the best practice is to evaluate them directly using an established statistical computing package. The numerical details need not concern us. However, the fact that

$$\frac{X - np}{\sqrt{np(1-p)}} \quad \text{is approximately normal}$$

when np and $n(1-p)$ are both large remains important. We will use it in later chapters when discussing inferences about proportions.

Statistical Reasoning

Probability and Overbooking

Full flights mean profits for airlines. Because not every ticket purchaser shows up for any particular flight, airlines regularly resort to selling more tickets than the plane has seats. It is not always possible to seat all ticket holders who show up. However, this event will occur only infrequently if the probability of accommodating everyone on the flight is high enough.

How can a decision be made on the number of tickets to sell? The answer is based on a probability calculation. From its extensive data base, suppose that an airline knows that the probability is .9 that a ticketed passenger will actually take their seat on a particular flight. If the plane holds 400 passengers, then the airline could consider selling a number n greater than 400 tickets. However, the airline wants to keep a probability of at least .95 that no ticketed passengers will be turned away. Under this probability requirement, passengers would be bumped from only about 1 in 20 flights.

To proceed with our calculation, let

n = Number of customers who will be booked

X = Number of customers who actually show up to board the plane

David McGlynn/FPG International

Then X has a binomial distribution with n trials and probability of success $p =$ .9. The probability is reduced to finding the largest number of tickets n satisfying

$$P[X \le 400] \text{ is at least } .95$$

Because n is at least 400, $n(1 - p) = n(.1)$ is at least 40, and we will use the normal approximation to the binomial.

$$P[X \le 400] = P\left[Z \le \frac{400 - np}{\sqrt{np(1 - p)}}\right] = P\left[Z \le \frac{400 - n(.9)}{\sqrt{n(.9)(.1)}}\right]$$

We first try $n = 425$:

$$z = \frac{400 - 425(.9)}{\sqrt{425(.9)(.1)}} = 2.83 \quad \text{and} \quad P[Z \le 2.83] = .9977$$

Even when 25 extra tickets are sold, the probability is very high that all passengers who show up will get a seat.

After a few more trial values for n, we find

$$\text{for } n = 433: \quad z = \frac{400 - 433(.9)}{\sqrt{433(.9)(.1)}} = 1.65 \quad \text{and} \quad P[Z \le 1.65] = .9505$$

$$\text{for } n = 434: \quad z = \frac{400 - 434(.9)}{\sqrt{434(.9)(.1)}} = 1.50 \quad \text{and} \quad P[Z \le 1.50] = .9332$$

That is, the airline can book 433 passengers and still have probability greater than .95 that at most 400 will show up for the flight. Passengers traveling together do not make independent choices. Slightly fewer tickets should be sold.

Overbooking allows the airlines to have fewer empty seats on popular flights. Can you think of other applications similar to this example?

Exercises

6.1   A researcher, interested in the spread of the common cold, believes that 35% of the population will get a cold during the winter. She will collect data from 20 families of size 4 and count the number of persons X who get a cold during the winter. Can X be considered as arising from 80 trials and approximate probabilities be calculated by treating

$$\frac{X - np}{\sqrt{np(1 - p)}} = \frac{X - 80(.35)}{\sqrt{80(.35)(.65)}} = \frac{X - 28}{\sqrt{18.2}}$$

as a standard normal variable? Explain.

6.2   Home gardeners are concerned about bugs eating the tomato plants. They intend to inspect all of the tomato plants in six randomly selected home gardens. The count of the number of plants infested with bugs X, among the total of 90 plants in the six gardens, will be recorded.

It is conjectured that 17% of the plants are infested with bugs. Can X be considered as arising from 90 trials and approximate probabilities be calculated by treating

$$\frac{X - np}{\sqrt{np(1-p)}} = \frac{X - 90(.17)}{\sqrt{90(.17)(.83)}} = \frac{X - 15.3}{\sqrt{12.699}}$$

as a standard normal variable? Explain.

6.3 Let the number of successes X have a binomial distribution with $p = .25$ and $n = 300$. Approximate the probability of (a) $X \leq 65$ and (b) $68 \leq X \leq 89$.

6.4 Let the number of successes X have a binomial distribution with $n = 200$ and $p = .75$. Use the normal distribution to approximate the probability of

(a) $X \leq 160$
(b) $137 \leq X \leq 162$

6.5 State whether the normal approximation to the binomial is appropriate in each of the following situations.

(a) $n = 90$, $p = .23$
(b) $n = 100$, $p = .02$
(c) $n = 71$, $p = .4$
(d) $n = 120$, $p = .97$

6.6 State whether the normal approximation to the binomial is appropriate in each of the following situations.

(a) $n = 500$, $p = .33$ (b) $n = 10$, $p = .5$
(c) $n = 400$, $p = .01$ (d) $n = 200$, $p = .98$
(e) $n = 100$, $p = .61$

6.7 Copy Figure 16 and, underneath the x-axis, add the standard score scale $z = (x - np)/\sqrt{np(1-p)}$ for $n = 5$, 12, and 25. Notice how the distributions center on zero and most of the probability lies between $z = -2$ and $z = 2$.

6.8 The median age of residents of the United States is 32.3 years. If a survey of 200 residents is taken, approximate the probability that at least 110 will be under 32.3 years of age.

6.9 The unemployment rate in a city is 7.9%. A sample of 300 persons is selected from the labor force. Approximate the probability that

(a) Fewer than 18 unemployed persons are in the sample
(b) More than 30 unemployed persons are in the sample

6.10   A survey reports that 96% of the people think that violence has increased in the past five years. Out of a random sample of 50 persons, 48 expressed the opinion that citizens have become more violent in the past five years. Does the normal approximation seem appropriate for X = the number of persons who express the opinion that citizens have become more violent in the past five years? Explain.

6.11   Of the customers visiting the stereo section of a large electronics store, only 25% make a purchase. If 70 customers visit the stereo section tomorrow, find the probability that more than 20 will make a purchase.

6.12   The weekly amount spent by a company for travel has approximately a normal distribution with mean = $550 and standard deviation = $40. What is the probability that the actual expenses will exceed $570 in 20 or more weeks during the next year?

6.13   With reference to Exercise 6.12, calculate the probability that the actual expenses would exceed $580 for between 10 and 16 weeks, inclusive, during the next year.

6.14   In a large midwestern university, 30% of the students live in apartments. If 200 students are randomly selected, find the probability that the number of them living in apartments will be between 50 and 75 inclusive.

6.15   In the Statistical Reasoning example about overbooking, the probability that a ticketed passenger will take a seat on a particular flight is .9. Answer the same question in the example when that probability is now changed to .8.

7. CHECKING NORMALITY

Does a normal distribution serve as a reasonable population model for the observations? This question needs to be addressed because many of the most common statistical procedures require the population to be nearly normal. If the normal distribution is tentatively assumed as a plausible model, this assumption must be checked once the data are in hand.

Although they involve some judgment, graphical procedures are useful in detecting departures from normality. Stem-and-leaf plots, histograms, dot diagrams, and boxplots can be inspected for lack of symmetry and possible outliers. In addition, the bell-shaped appearance of the normal density can be checked by comparing the proportion of observations in the intervals $\bar{x} \pm s$, $\bar{x} \pm 2s$, and $\bar{x} \pm 3s$ with the normal probability proportions, .68, .95, and .997, respectively.

A more effective graphical procedure for checking normality is the **normal-scores plot**. It provides an assessment of the assumption of normality for moderate sample sizes that are too small for constructing a histogram. As a practical matter, 15 to 20 observations are required for a normal-scores plot to obtain a meaningful pattern.

The term **normal scores** refers to an idealized sample from a standard normal distribution. Basically, the standard normal scores divide the standard normal distribution into intervals of equal probability. Those near 0, the center, will be close together and those in the tail will be farther apart. We do not delve deeply into the calculation. Since different computer software programs use slight variants of normal scores, we will just use the scores provided.

To construct a normal-scores plot, first order the data from smallest to largest. The smallest observation is paired with the smallest normal score and the point is plotted. The next largest values are paired and the point plotted, and so on until the largest observation is paired with the largest normal score and the point plotted.

If the observations are generated from a normal distribution, the pattern in the normal-scores plot should resemble a straight line.

A straight-line pattern in a normal-scores plot supports the plausibility of a normal model. A curve appearance indicates a departure from normality.

The normal-scores plot of the accounting students' computer anxiety scores, from Table D.4 of the Data Bank, is shown in Figure 17. The pattern is nearly a straight line and we have no reason to doubt the tentative assumption that the population distribution is normal.

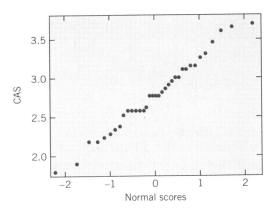

Figure 17. Normal-scores plot of the computer anxiety scores

Figure 18(a) on page 284 gives a normal-scores plot for the volume of timber, in cords, in 49 different plots within a large forest. These data, given in Table 2, were collected by a forester who randomly selected the plots to estimate the total volume of timber available in the forest.

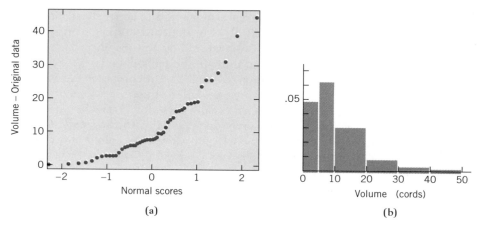

Figure 18. Timber data: (a) Normal-scores plot; (b) Histogram

TABLE 2   Volume of Timber in Cords

39.3	14.8	6.3	.9	6.5
3.5	8.3	10.0	1.3	7.1
6.0	17.1	16.8	.7	7.9
2.7	26.2	24.3	17.7	3.2
7.4	6.6	5.2	8.3	5.9
3.5	8.3	44.8	8.3	13.4
19.4	19.0	14.1	1.9	12.0
19.7	10.3	3.4	16.7	4.3
1.0	7.6	28.3	26.2	31.7
8.7	18.9	3.4	10.0	

Courtesy of Professor Alan Ek.

There is a definite curved pattern in this normal-scores plot. The largest values of lumber are too large—they lie above a straight line you can imagine passing through the lower points. The assumption that the data came from a normal distribution clearly is not valid. The histogram in Figure 18(b) also shows the long tail to the right.

Often, data that are far from normal can be transformed to more nearly conform to the assumption of a normal distribution. The largest observations on lumber need to be pulled down farther than those in the middle. We first tried taking the square root of the observations so the transformed values were $\sqrt{x_i}$. This resulted in an improved normal-scores plot but still not a straight-line pattern. After some trial and error with other powers, it was decided to take a square root a second time, or the fourth root of the original observations. The normal-scores plot, shown in Figure 19(a), arguably has a straight-line pattern. The as-

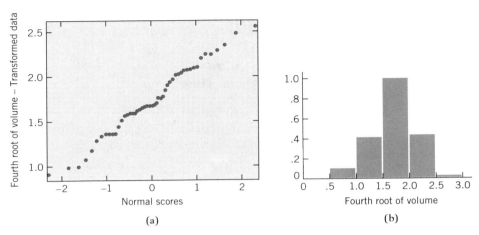

Figure 19. Fourth root of timber data: (a) Normal-scores plot; (b) Histogram

sumption of a normal distribution for the fourth root of lumber is tenable. The histogram in Figure 19(b) has a quite symmetric appearance.

For any data set that does not have a symmetric histogram, we consider a variety of transformations to improve the approximation to normality.

Some Useful Transformations

Make large values larger: Make large values smaller:

$$x^3, \quad x^2 \qquad\qquad \sqrt{x}, \quad \sqrt[4]{x}, \quad \log_e x, \quad \frac{1}{x}$$

You may recall that $\log_e x$ is the natural logarithm.

KEY IDEAS

The probability distribution for a continuous random variable X is specified by a probability density curve.

The probability that X lies in an interval from a to b is determined by the area under the probability density curve between a and b. The total area under the curve is 1, and the curve is never negative.

The normal distribution has a symmetric bell-shaped curve centered at the mean. The intervals of one, two, and three standard deviations around the mean contain the probabilities .683, .954, and .997, respectively.

If X is normally distributed with mean $= \mu$ and sd $= \sigma$, then

$$Z = \frac{X - \mu}{\sigma}$$

has the **standard normal distribution.**

When the number of trials n is large and the success probability p is not too near 0 or 1, the binomial distribution is well approximated by a normal distribution with mean $= np$ and sd $= \sqrt{np(1-p)}$. Specifically, the probabilities for a binomial variable X can be approximately calculated by treating

$$Z = \frac{X - np}{\sqrt{np(1-p)}}$$

as standard normal.

The **normal-scores plot** of a data set provides a diagnostic check for possible departure from a normal distribution.

Transformation of the measurement scale often helps to convert a long-tailed distribution to one that resembles a normal distribution.

8. REVIEW EXERCISES

8.1   Determine the median for the distribution shown in the following illustration.

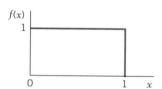

8.2   For X having the density in Exercise 8.1, find (a) $P[X > .8]$ (b) $P[.5 \le X \le .8]$ and (c) $P[.5 < X < .8]$.

8.3   Describe the reasoning that leads from a histogram to the concept of a probability density curve. (Think of successive histograms based on 100 birth heights, 5000 heights, 500,000 heights, and then an unlimited number.)

8.4   For a standard normal random variable Z, find:

    (a) $P[Z < 1.31]$ (b) $P[Z > 1.20]$

    (c) $P[.67 < Z < 1.98]$ (d) $P[-1.32 < Z < 1.05]$

8.5   For the standard normal distribution, find the value z such that the

    (a)   Area to its left is .0869   (b)   Area to its left is .12

    (c)   Area to its right is .2578   (d)   Area to its right is .25

8.6   Find the 20th and 80th percentiles of the standard normal distribution. How are these two related?

8.7   If Z is a standard normal random variable, what is the probability that
(a) Z exceeds $-.72$?
(b) Z lies in the interval $(-1.50, 1.50)$?
(c) $|Z|$ exceeds 2.0?
(d) $|Z|$ is less than 1.0?

8.8   The distribution of raw scores in a college qualification test has mean $=$ 582 and standard deviation $=$ 75.
(a)   If a student's raw score is 696, what is the corresponding standardized score?
(b)   If the standardized score is $-.8$, what is the raw score?
(c)   Find the interval of standardized scores corresponding to the raw scores of 380 to 560.
(d)   Find the interval of the raw scores corresponding to the standardized scores of -1.2 to 1.2.

8.9   If X is normally distributed with $\mu = 100$ and $\sigma = 8$, find
(a) $P[X < 107]$ (b) $P[X < 97]$
(c) $P[X > 110]$ (d) $P[X > 90]$
(e) $P[95 < X < 106]$ (f) $P[103 < X < 114]$
(g) $P[88 < X < 100]$ (h) $P[60 < X < 108]$

8.10 If X has a normal distribution with $\mu = 200$ and $\sigma = 5$, find b such that
(a) $P[X < b] = .6700$
(b) $P[X > b] = .0110$
(c) $P[|X - 200| < b] = .966$

8.11 Suppose that a student's verbal score X from next year's Graduate Record Exam can be considered an observation from a normal population having mean 497 and standard deviation 120. Find
(a) $P[X > 600]$
(b)   Probability that the student scores below 400

8.12 The lifting capacities of a class of industrial workers are normally distributed with mean $=$ 65 pounds and sd $=$ 10 pounds. What proportion of these workers can lift an 80-pound load?

8.13 The bonding strength of a drop of plastic glue is normally distributed with mean $=$ 100 pounds and sd $=$ 8 pounds. A broken plastic strip is repaired with a drop of this glue and then subjected to a test load of 98 pounds. What is the probability that the bonding will fail?

8.14   *Grading on a curve.*   The scores on an examination are normally distributed with mean $\mu = 70$ and standard deviation $\sigma = 8$. Suppose that the instructor decides to assign letter grades according to the following scheme (left endpoint included).

Scores	Grade
Less than 58	F
58 to 66	D
66 to 74	C
74 to 82	B
82 and above	A

Find the percentage of students in each grade category.

8.15   Suppose the duration of trouble-free operation of a new vacuum cleaner is normally distributed with mean = 530 days and sd = 100 days.

(a)   What is the probability that the vacuum cleaner will work for at least two years without trouble?

(b)   The company wishes to set the warranty period so that no more than 10% of the vacuum cleaners would need repair services while under warranty. How long a warranty period must be set?

8.16   An aptitude test administered to aircraft pilot trainees requires a series of operations to be performed in quick succession. Suppose that the time needed to complete the test is normally distributed with mean = 90 minutes and sd = 20 minutes.

(a)   To pass the test, a candidate must complete it within 80 minutes. What percentage of the candidates will pass the test?

(b)   If the top 5% of the candidates are to be given a certificate of commendation, how fast must a candidate complete the test to be eligible for a certificate?

8.17   It is known from past experience that 9% of the tax bills are paid late. If 20,000 tax bills are sent out, find the probability that

(a)   Fewer than 1750 are paid late.

(b)   2000 or more are paid late.

8.18   A particular program, say, program A, previously drew 30% of the television audience. To determine whether a recent rescheduling of the programs on a competing channel has adversely affected the audience of program A, a random sample of 400 viewers is to be asked whether they currently watch this program. If the percentage of viewers watching program A has not changed, what is the probability that fewer than 105 out of a sample of 400 will be found to watch the program?

8.19   The number of successes X has a binomial distribution. State whether the normal approximation is appropriate in each of the following situations: (a) $n = 400$, $p = .28$, (b) $n = 20$, $p = .04$, (c) $n = 90$, $p = .99$.

USING A COMPUTER

8.20   *Normal-scores plot.*   Use a computer program to make a normal-scores plot for the volume of timber data in Table 2. Comment on the departure from normality displayed by the normal-scores plot. With the data set in Column 1, the normal-scores plot is created by the MINITAB commands

```
NSCORE C1 SET IN C2
PLOT C1*C2
```

8.21   Use MINITAB or another package program to make a normal-scores plot of the fish growth data in Table D.6 of the Data Bank. Plot

(a)   The male freshwater growth

(b)   The male marine growth

(c)   The female freshwater growth

(d)   The female marine growth

8.22   *Transformations and normal-scores plots.*   The MINITAB computer language makes it possible to easily transform data. With the data already set in Column 1, the commands

```
LOGE C1 SET IN C4
SQRT C1 SET IN C5
```

place the $\log_e$ in Column C4 and square root in C5, respectively. Take the square root of the timber data in Table 2 and then do a normal-scores plot.

CHAPTER 7

Variation in Repeated Samples—Sampling Distributions

Chapter Objectives

After reading this chapter, you should be able to

▶ Distinguish between population parameters and statistics, which are based on samples.

▶ Discuss the sampling distribution of the mean $\overline{X}$.

▶ Discuss and use the central limit result to approximate probabilities for $\overline{X}$.

1. INTRODUCTION

When a random sample from the population of interest is available, it can be described by several different summary numbers, as discussed in Chapter 2. We can determine the sample mean, median, standard deviation, or the range. For a given sample, only *one* value is obtained for each sample summary.

In this chapter, we consider *all* of the possible values that a given sample summary can take as the sample varies. For a given sample summary, such as the sample mean, we want to know which ranges of its values have high probability, which have low probability, and simply, what is the pattern of variation. In other words, we study the **probability distribution** of the sample summary, which is also called its **sampling distribution.** The main emphasis here is on the distribution of possible values over different samples, not on the *single* number for the particular sample collected.

When we use any measuring device (for instance, a bathroom scale), we want to know:

What is the accuracy of the device?

Does the accuracy depend on the size of the object we are measuring?

Does it usually give a larger value than the true underlying value?

Think of the sample mean as a device for measuring the population mean. The same questions apply.

Most often, we use a sample summary as an estimate of the corresponding population value. Then, we need to know that this sample summary is generally on target as it varies according to the particular sample selected. That is, we need to know the properties of the sample summary to assess its reliability as an estimate of the corresponding population value.

The **sampling distribution** of a sample summary provides all of the information about these kinds of properties. For example, we may examine whether the majority of values of the sample summary, corresponding to various possible samples, highly concentrate around the population value to be estimated. If so, we know that the sample summary will produce a good estimate of the population value.

Of course, the sampling distribution of any sample summary is related to the characteristics of the population from which the sample is taken. Also, different types of sample summaries correspond to different forms of sampling distributions.

The sampling distribution of an appropriately chosen sample summary can therefore serve as a reference to determine whether the sample actually observed is consistent with assumptions made about the characteristics of the population. This use of the sampling distribution forms the basis for another type of statistical reasoning. It allows us *to take into account the uncertainty due to sampling* when testing whether or not certain hypotheses about the population characteristics hold.

PARAMETERS AND STATISTICS

The arguments outlined above concern the ideas of inference that are at the heart of statistics. They enable the investigator to argue from the particular sample to conclusions about the population. As indicated, these generalizations are founded on an understanding of the manner in which variation in the population is transmitted, by sampling, to variation in statistics like the sample mean. This key concept is the subject of this chapter.

Typically, we are interested in learning about some numerical feature of the population, such as the proportion possessing a stated characteristic, the mean and standard deviation of the population, or some other numerical measure of center or variability.

A numerical feature of a population is called a parameter.

The true value of a population parameter is an unknown constant. It can be correctly determined only by a complete study of the population. The concepts of statistical inference come into play whenever this is impossible or not practically feasible.

If we have access to only a sample from the population, our inferences about a parameter must then rest on an appropriate sample-based quantity. Whereas a parameter refers to some numerical characteristic of the population, a sample-based quantity is called a statistic.

A statistic is a numerical valued function of the sample observations.

For example, the sample mean

$$\overline{X} = \frac{X_1 + \cdots + X_n}{n}$$

is a statistic because its numerical value can be computed once the sample data, consisting of the values of $X_1, \ldots, X_n$, are available. Likewise, the sample median and the sample standard deviation are also sample-based quantities so each is a statistic.

A sample-based quantity (statistic) must serve as our source of information about the value of a parameter. Three points are crucial:

1. Because a sample is only a part of the population, the numerical value of a statistic cannot be expected to give us the exact value of the parameter.

2. The observed value of a statistic depends on the particular sample that happens to be selected.

3. There will be some variability in the values of a statistic over different occasions of sampling.

A brief example will help illustrate these important points. Suppose an urban planner wishes to study the average commuting distance of workers from their home to principal place of work. Here the statistical population consists of the commuting distances of all the workers in the city. The mean of this finite but vast and unrecorded set of numbers is called the population mean, which we denote by μ. We want to learn about the parameter μ by collecting data from a sample of workers. Suppose 80 workers are randomly selected and the (sample) mean of their commuting distances is found to be $\bar{x} = 8.3$ miles. Evidently, the population mean μ cannot be claimed to be exactly 8.3 miles. If one were to observe another random sample of 80 workers, would the sample mean again be 8.3 miles? Obviously, we do not expect the two results to be identical. Because the commuting distances do vary in the population of workers, the sample mean would also vary on different occasions of sampling. In practice, we observe only one sample and correspondingly a single value of the sample mean such as $\bar{x} = 8.3$.

In Section 3, we will see that, in repeated random samples, the values for $\overline{X}$ tend to concentrate in the neighborhood of the population mean μ. This fact gives support for using $\bar{x}$ as an estimate of μ. Section 2 discusses the important topic of sampling distributions in general settings.

2. THE SAMPLING DISTRIBUTION OF A STATISTIC

The fact that the value of the sample mean, or any other statistic, will vary as the sampling process is repeated is a key concept. Because any statistic, the sample mean in particular, varies from sample to sample, it is a random variable and has its own probability distribution. The variability of the statistic, in repeated sampling, is described by this probability distribution.

> The probability distribution of a statistic is called its sampling distribution.

The qualifier "sampling" indicates that the distribution is conceived in the context of repeated sampling from a population. We often drop the qualifier and simply say the distribution of a statistic.

Although in any given situation, we are limited to one sample and the corresponding single value for a statistic, over repeated samples from a population the statistic varies and has a sampling distribution. The sampling distribution of a statistic is determined from the distribution $f(x)$ that governs the population, and it also depends on the sample size n. Let us see how the distribution of $\overline{X}$ can be determined in a simple situation where the sample size is 2 and the population consists of 3 units.

Example 1   Illustration of a sampling distribution
A population consists of three housing units, where the value of X, the number of rooms for rent in each unit, is shown in the illustration.

Consider drawing a random sample of size 2 with replacement. That is, we select a unit at random, put it back, and then select another unit at random. Denote by X_1 and X_2 the observation of X obtained in the first and second drawing, respectively. Find the sampling distribution of $\overline{X} = (X_1 + X_2)/2$.

SOLUTION AND   The population distribution of X is given in Table 1, which simply formalizes the
DISCUSSION   fact that each of the X-values 2, 3, and 4 occurs in $\frac{1}{3}$ of the population of the housing units.

TABLE 1   The Population
Distribution

x	$f(x)$
2	$\dfrac{1}{3}$
3	$\dfrac{1}{3}$
4	$\dfrac{1}{3}$

Because each unit is equally likely to be selected, the observation X_1 from the first drawing has the same distribution as given in Table 1. Since the sampling is with replacement, the second observation X_2 also has this same distribution.
The possible samples (x_1, x_2) of size 2 and the corresponding values of $\overline{X}$ are

(x_1, x_2)	$(2, 2)$	$(2, 3)$	$(2, 4)$	$(3, 2)$	$(3, 3)$	$(3, 4)$	$(4, 2)$	$(4, 3)$	$(4, 4)$
$\overline{x} = \dfrac{x_1 + x_2}{2}$	2	2.5	3	2.5	3	3.5	3	3.5	4

The nine possible samples are equally likely so, for instance, $P[\overline{X} = 2.5] = \frac{2}{9}$. Continuing in this manner, we obtain the distribution of $\overline{X}$, which is given in Table 2.

TABLE 2 The Probability Distribution
of $\overline{X} = (X_1 + X_2)/2$

Value of $\overline{X}$	Probability
2	$\dfrac{1}{9}$
2.5	$\dfrac{2}{9}$
3	$\dfrac{3}{9}$
3.5	$\dfrac{2}{9}$
4	$\dfrac{1}{9}$

This sampling distribution pertains to repeated selection of random samples of size 2 with replacement. It tells us that if the random sampling is repeated a large number of times, then in about $\frac{1}{9}$ or 11% of the cases, the sample mean would be 2, and in $\frac{2}{9}$ or 22% of the cases, it would be 2.5, and so on.

Figure 1 shows the probability histograms of the distributions in Tables 1 and 2.

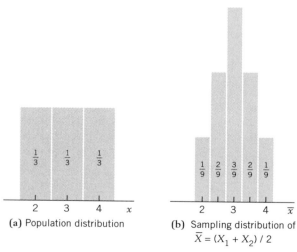

(a) Population distribution

(b) Sampling distribution of
$\overline{X} = (X_1 + X_2) / 2$

Figure 1. Idea of a sampling distribution

In the context of Example 1, suppose instead the population consists of 300 housing units, of which 100 units have 2 rooms, 100 units have 3 rooms, and 100 units have 4 rooms for rent. When we sample two units from this large population, it would make little difference whether or not we replace the unit after the first selection. Each observation would still have the same probability distribution—namely, $P[X = 2] = P[X = 3] = P[X = 4] = \frac{1}{3}$, which characterizes the population.

When the population is very large and the sample size relatively small, it is inconsequential whether or not a unit is replaced before the next unit is selected. Under these conditions, too, we refer to the observations as a random sample. What are the key conditions required for a sample to be random? The observations $X_1, X_2, \ldots, X_n$ are a **random sample of size n from the population distribution** if they result from independent selections, and each has the same distribution as the population.

Because of variation in the population, the random sample will vary and so will $\overline{X}$, the sample median, or any other statistic.

Example 1 discussed a sampling distribution when the underlying probability distribution describes an equally likely outcome situation. In contrast, Example 2 illustrates a situation where the outcomes have different probabilities of occurring. It also shows how the ideas from Chapters 4 and 5, concerning probability and random variables, can be used to construct the sampling distribution of a statistic. Notice that different statistics have different sampling distributions.

Example 2   The sample mean and median each have a sampling distribution
A large population is described by the probability distribution

x	$f(x)$
0	.2
3	.3
12	.5

Let X_1, X_2, X_3 be a random sample of size 3 from this distribution.

(a) List all the possible samples and determine their probabilities.

(b) Determine the sampling distribution of the sample mean.

(c) Determine the sampling distribution of the sample median.

SOLUTION AND DISCUSSION

(a) Taking a random sample means that each of the three observations X_1, X_2, X_3 has the same distribution as the population and they are independent. So, the probability of obtaining the sample 0, 3, 0 is $(.2) \times (.3) \times (.2) = .012$. The calculations for all $3 \times 3 \times 3 = 27$ possible samples are given in Table 3.

(b) The probabilities of all samples giving the same value $\bar{x}$ are added to obtain the sampling distribution in Table 3.

(c) The calculations and sampling distribution of the median are also given in Table 3. ∎

TABLE 3   Sampling Distributions

Population Distribution

x	$f(x)$
0	.2
3	.3
12	.5

Population mean: $E(X) = 0(.2) + 3(.3) + 12(.5) = 6.9 = \mu$

Pop. variance: $\text{Var}(X) = 0^2(.2) + 3^2(.3) + 12^2(.5) - 6.9^2$
$$= 27.09 = \sigma^2$$

	\multicolumn{3}{c}{Possible Samples x_1 x_2 x_3}	Sample Mean $\bar{x}$	Sample Median m	Probability
1	0	0	0	0	0	$(.2)(.2)(.2) = .008$
2	0	0	3	1	0	$(.2)(.2)(.3) = .012$
3	0	0	12	4	0	$(.2)(.2)(.5) = .020$
4	0	3	0	1	0	$(.2)(.3)(.2) = .012$
5	0	3	3	2	3	$(.2)(.3)(.3) = .018$
6	0	3	12	5	3	$(.2)(.3)(.5) = .030$
7	0	12	0	4	0	$(.2)(.5)(.2) = .020$
8	0	12	3	5	3	$(.2)(.5)(.3) = .030$
9	0	12	12	8	12	$(.2)(.5)(.5) = .050$
10	3	0	0	1	0	$(.3)(.2)(.2) = .012$
11	3	0	3	2	3	$(.3)(.2)(.3) = .018$
12	3	0	12	5	3	$(.3)(.2)(.5) = .030$
13	3	3	0	2	3	$(.3)(.3)(.2) = .018$
14	3	3	3	3	3	$(.3)(.3)(.3) = .027$
15	3	3	12	6	3	$(.3)(.3)(.5) = .045$
16	3	12	0	5	3	$(.3)(.5)(.2) = .030$
17	3	12	3	6	3	$(.3)(.5)(.3) = .045$
18	3	12	12	9	12	$(.3)(.5)(.5) = .075$
19	12	0	0	4	0	$(.5)(.2)(.2) = .020$
20	12	0	3	5	3	$(.5)(.2)(.3) = .030$
21	12	0	12	8	12	$(.5)(.2)(.5) = .050$
22	12	3	0	5	3	$(.5)(.3)(.2) = .030$
23	12	3	3	6	3	$(.5)(.3)(.3) = .045$
24	12	3	12	9	12	$(.5)(.3)(.5) = .075$
25	12	12	0	8	12	$(.5)(.5)(.2) = .050$
26	12	12	3	9	12	$(.5)(.5)(.3) = .075$
27	12	12	12	12	12	$(.5)(.5)(.5) = .125$
						Total = 1.000

Sampling Distribution of $\overline{X}$

$\overline{x}$	$f(\overline{x})$
0	.008
1	.036 = .012 + .012 + .012
2	.054 = .018 + .018 + .018
3	.027
4	.060 = .020 + .020 + .020
5	.180 = .030 + .030 + .030
	+ .030 + .030 + .030
6	.135 = .045 + .045 + .045
8	.150 = .050 + .050 + .050
9	.225 = .075 + .075 + .075
12	.125 = .125

$$E(\overline{X}) = \Sigma \overline{x} f(\overline{x}) = 0(.008) + 1(.036) + 2(.054) + 3(.027) + 4(.060)$$
$$+ 5(.180) + 6(.135) + 8(.150) + 9(.225) + 12(.125)$$
$$= 6.9 \text{ same as } E(X), \text{ pop. mean}$$

$$\text{Var}(\overline{X}) = \Sigma \overline{x}^2 f(\overline{x}) - \mu^2 = 0^2(.008) + 1^2(.036) + 2^2(.054)$$
$$+ 3^2(.027) + 4^2(.060) + 5^2(.180) + 6^2(.135)$$
$$+ 8^2(.150) + 9^2(.225) + 12^2(.125) - (6.9)^2$$
$$= 9.03 = \frac{27.09}{3} = \frac{\sigma^2}{3}$$

$\text{Var}(\overline{X})$ is one-third of the population variance.

Sampling Distribution of the Median m

m	$f(m)$
0	.104 = .008 + .012 + .020 + .012 + .020 + .012 + .020
3	.396 = .018 + .030 + .030 + .018 + .030 + .018 + .027 + .045 + .030 + .045 + .030 + .030 + .045
12	.500 = .050 + .075 + .050 + .075 + .050 + .075 + .125

Mean of the distribution of sample median
$= 0(.104) + 3(.396) + 12(.500) = 7.188 \neq 6.9 = \mu$
Different from the mean of the population distribution

Variance of the distribution of sample median
$= 0^2(.104) + 3^2(.396) + 12^2(.500) - (7.188)^2 = 23.897$
[*not* one-third of the population variance 27.09]

To illustrate the idea of a sampling distribution, we considered simple populations with only three possible values and small sample sizes $n = 2$ and $n = 3$. The calculation gets more tedious and extensive when a population has many values of X and n is large. However, the procedure remains the same. Once the population and sample size are specified,

1. List all possible samples of size n.
2. Calculate the value of the statistic for each sample.
3. List the distinct values of the statistic obtained in step 2. Calculate the corresponding probabilities by identifying all the samples that yield the same value of the statistic.

We leave the more complicated cases to statisticians who can sometimes use additional mathematical methods to derive exact sampling distributions.

Instead of a precise determination, one can turn to the computer to approximate a sampling distribution. The idea is to program the computer to actually draw a random sample and calculate the statistic. This procedure is then repeated a large number of times and a density histogram constructed from the values of the statistic. The resulting histogram will be an approximation to the sampling distribution. This approximation will be used in Example 4.

Exercises

2.1 Identify each of the following as either a parameter or a statistic.
   (a) Sample standard deviation
   (b) Sample range
   (c) Population 10th percentile
   (d) Sample first quartile
   (e) Population median

2.2 Identify the parameter, statistic, and population when they appear in each of the following statements.
   (a) There were 11 persons who served on the U.S. Supreme Court in the decade of the 1980s.
   (b) A survey of 1000 minority persons living in Chicago revealed that 185 were out of work.
   (c) Out of a sample of 100 dog owners who applied for dog licenses in northern Wisconsin, 18 had a Labrador retriever.

2.3 Data obtained from asking the wrong questions at the wrong time or in the wrong place can lead to misleading summary statistics. Explain why the following collection procedures are likely to produce useless data.

(a) To evaluate the number of students who are employed part time, the investigator interviews students who are hanging out at the student union.

(b) To study the pattern of spending of persons earning less than the minimum wage, a survey is taken during the first three weeks of December.

2.4 Explain why the following collection procedures are likely to produce data that fail to yield the desired information.

(a) To evaluate public opinion about import restrictions on automobiles, an interviewer asks persons, "Do you think that this unfair trade should be stopped?"

(b) To determine how eighth grade girls feel about having boys in the classroom, a random sample from a private girls' school is polled.

2.5 From the set of numbers $\{1, 3, 5\}$, a random sample of size 2 will be selected with replacement.

(a) List all possible samples and evaluate $\bar{x}$ for each.

(b) Determine the sampling distribution of $\bar{X}$.

2.6 A random sample of size 2 will be selected, with replacement, from the set of numbers $\{2, 4, 6\}$.

(a) List all possible samples and evaluate $\bar{x}$ and s^2 for each.

(b) Determine the sampling distribution of $\bar{X}$.

(c) Determine the sampling distribution of S^2.

2.7 A consumer wants to study the size of strawberries for sale at the market. If a sample of size 4 is taken from the top of a basket, will the berry sizes be a random sample? Explain.

2.8 To determine the time a cashier spends on a customer in the express lane, the manager decides to record the time to check-out for the customer who is being served at 10 past the hour, 20 past the hour, and so on. Will measurements collected in this manner be a random sample of the times a cashier spends on a customer?

2.9 Refer to Example 2. Find the sampling distribution of the Sample range = Maximum − Minimum.

2.10 Refer to Example 2. Replace the original probabilities .2, .3, .5 by .1, .5, .4, respectively. Determine the sampling distribution of the sample mean.

2.11 What is a statistic? What is the sampling distribution of a statistic? Explain why it is important to study sampling distributions.

3. DISTRIBUTION OF THE SAMPLE MEAN AND THE CENTRAL LIMIT THEOREM

Section 2 presented examples of sampling distributions for the mean and median. Statistical inference about a population mean is of prime importance in most practical studies.

Inferences about this parameter are based on the sample mean

$$\overline{X} = \frac{X_1 + X_2 + \cdots + X_n}{n}$$

and its sampling distribution. Consequently, we now explore the basic properties of the sampling distribution of $\overline{X}$ and explain the role of the normal distribution as a useful approximation.

In particular, we want to relate the sampling distribution of $\overline{X}$ to the population from which the random sample was selected. We denote the parameters of the population by

Population mean $= \mu$
Population standard deviation $= \sigma$

The sampling distribution of $\overline{X}$ also has a mean $E(\overline{X})$ and a standard deviation $sd(\overline{X})$. These can be expressed in terms of the population mean μ and standard deviation σ.

Mean and Standard Deviation of $\overline{X}$

The distribution of the sample mean, based on a random sample size of n, has

$$E(\overline{X}) = \mu \qquad (= \text{population mean})$$

$$\text{Var}(\overline{X}) = \frac{\sigma^2}{n} \qquad \left(= \frac{\text{population variance}}{\text{sample size}}\right)$$

$$sd(\overline{X}) = \frac{\sigma}{\sqrt{n}} \qquad \left(= \frac{\text{population standard deviation}}{\sqrt{\text{sample size}}}\right)$$

The first result shows that the distribution of $\overline{X}$ is centered at the population mean μ in the sense that expectation serves as a measure of center of a distribution. The last result states that the standard deviation of $\overline{X}$ equals the population standard deviation divided by the square root of the sample size. That is, the variability of the sample mean is governed by the two factors: the population variability σ and the sample size n. Large variability in the population induces large variability in $\overline{X}$, thus making the sample information about μ less dependable. However, this can be countered by choosing n large. For instance, with $n = 100$, the standard deviation of $\overline{X}$ is $\sigma/\sqrt{100} = \sigma/10$, a tenth of the population standard deviation. With increasing sample size, the standard deviation $\sigma/\sqrt{n}$ decreases and the distribution of $\overline{X}$ tends to become more concentrated around the population mean μ.

Example 3 The population mean and variance of $\overline{X}$

Calculate the mean and standard deviation for the population distribution given in Table 1 and for the distribution of $\overline{X}$ given in Table 2. Verify the relations $E(\overline{X}) = \mu$ and $\mathrm{sd}(\overline{X}) = \sigma/\sqrt{n}$.

SOLUTION AND DISCUSSION The calculations are performed in Table 4.

TABLE 4 Mean and Variance of $\overline{X} = (X_1 + X_2)/2$

Population Distribution				Distribution of $\overline{X} = (X_1 + X_2)/2$			
x	$f(x)$	$xf(x)$	$x^2f(x)$	$\overline{x}$	$f(\overline{x})$	$\overline{x}f(\overline{x})$	$\overline{x}^2f(\overline{x})$
2	$\frac{1}{3}$	$\frac{2}{3}$	$\frac{4}{3}$	2	$\frac{1}{9}$	$\frac{2}{9}$	$\frac{4}{9}$
3	$\frac{1}{3}$	$\frac{3}{3}$	$\frac{9}{3}$	2.5	$\frac{2}{9}$	$\frac{5}{9}$	$\frac{12.5}{9}$
4	$\frac{1}{3}$	$\frac{4}{3}$	$\frac{16}{3}$	3	$\frac{3}{9}$	$\frac{9}{9}$	$\frac{27}{9}$
				3.5	$\frac{2}{9}$	$\frac{7}{9}$	$\frac{24.5}{9}$
Total	1	3	$\frac{29}{3}$	4	$\frac{1}{9}$	$\frac{4}{9}$	$\frac{16}{9}$
				Total	1	3	$\frac{84}{9}$

$\mu = 3$

$\sigma^2 = \dfrac{29}{3} - (3)^2 = \dfrac{2}{3}$

$$E(\overline{X}) = 3 = \mu$$

$$\mathrm{Var}(\overline{X}) = \frac{84}{9} - (3)^2 = \frac{1}{3}$$

By direct calculation, $\mathrm{sd}(\overline{X}) = 1/\sqrt{3}$. This is confirmed by the relation

$$\mathrm{sd}(\overline{X}) = \frac{\sigma}{\sqrt{n}} = \frac{\sqrt{2/3}}{\sqrt{2}} = \frac{1}{\sqrt{3}}$$

See Example 2 (b) and page 299 for a similar illustration of $E(\overline{X}) = \mu$ and $\text{sd}(\overline{X}) = \sigma/\sqrt{n}$.

We now state two important results concerning the shape of the sampling distribution of $\overline{X}$. The first result gives the exact form of the distribution of $\overline{X}$ when the population distribution is normal:

$\overline{X}$ Is Normal When Sampling from a Normal Population

In random sampling from a normal population with mean μ and standard deviation σ, the sample mean $\overline{X}$ has the normal distribution with mean μ and standard deviation $\sigma/\sqrt{n}$.

When sampling from a nonnormal population, the distribution of $\overline{X}$ depends on the particular form of the population distribution that prevails.

In Example 2, the population is not normal, as its probability distribution is discrete, not continuous. The sampling distribution of $\overline{X}$ in Example 2 is also nonnormal. The probabilities are rather small for small $\bar{x}$ values, while larger probabilities are attached to the larger values of $\bar{x}$. That is, the distribution has a longer right-hand tail; it is skewed as opposed to being symmetric like a normal distribution.

If we change the values .2, .3, .5 in the probability distribution in Example 2, the sampling distribution of $\overline{X}$ will change (see Exercise 2.10). That is, the sampling distribution of $\overline{X}$ depends on the underlying probability distribution of the individual observations.

A surprising result, known as the central limit theorem, states that when the sample size n is large, the distribution of $\overline{X}$ is approximately normal, regardless of the shape of the population distribution. In practice, the normal approximation is usually adequate when n is greater than 30.

Central Limit Theorem

Whatever the population, the distribution of $\overline{X}$ is approximately normal when n is large.

In random sampling from an arbitrary population with mean μ and standard deviation σ, when n is large, the distribution of $\overline{X}$ is approximately normal with mean μ and standard deviation $\sigma/\sqrt{n}$. Consequently,

$$Z = \frac{\overline{X} - \mu}{\sigma/\sqrt{n}} \quad \text{is approximately } N(0, 1)$$

Whether the population distribution is continuous, discrete, symmetric, or asymmetric, the central limit theorem asserts that the distribution of the sample mean $\overline{X}$ is nearly normal[1] if the sample size is large. In this sense, the normal distribution plays a central role in the development of statistical procedures. Although a proof of the theorem requires higher mathematics, we can empirically demonstrate how this result works.

Example 4   Demonstration of the central limit theorem
Consider a population having a discrete uniform distribution that places a probability of .1 on each of the integers, 0, 1, . . . , 9. This may be an appropriate model for the distribution of the last digit in telephone numbers or the first overflow digit in computer calculations. The line diagram of this distribution appears in Figure 2. The population has $\mu = 4.5$ and $\sigma = 2.872$.

Figure 2. Uniform distribution on the integers 0, 1, . . . , 9

By means of a computer, 100 random samples of size 5 were generated from this distribution, and $\overline{x}$ was computed for each sample. The results of this repeated random sampling are presented in Table 5. The relative frequency histogram in Figure 3 is constructed from the 100 observed values of $\overline{x}$. Although the popu-

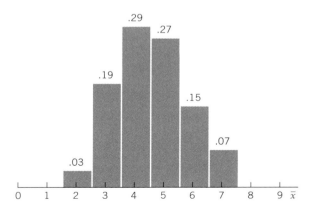

Figure 3. Relative frequency histogram of the $\overline{x}$ values recorded in Table 5

---

[1]Technically, the variance must be finite. This is the case for all populations in this book.

TABLE 5 Samples of Size 5 from a Discrete Uniform Distribution

Sample Number	Observations	Sum	Mean $\bar{x}$	Sample Number	Observations	Sum	Mean $\bar{x}$
1	4, 7, 9, 0, 6	26	5.2	51	4, 7, 3, 8, 8	30	6.0
2	7, 3, 7, 7, 4	28	5.6	52	2, 0, 3, 3, 2	10	2.0
3	0, 4, 6, 9, 2	21	4.2	53	4, 4, 2, 6, 3	19	3.8
4	7, 6, 1, 9, 1	24	4.8	54	1, 6, 4, 0, 6	17	3.4
5	9, 0, 2, 9, 4	24	4.8	55	2, 4, 5, 8, 9	28	5.6
6	9, 4, 9, 4, 2	28	5.6	56	1, 5, 5, 4, 0	15	3.0
7	7, 4, 2, 1, 6	20	4.0	57	3, 7, 5, 4, 3	22	4.4
8	4, 4, 7, 7, 9	31	6.2	58	3, 7, 0, 7, 6	23	4.6
9	8, 7, 6, 0, 5	26	5.2	59	4, 8, 9, 5, 9	35	7.0
10	7, 9, 1, 0, 6	23	4.6	60	6, 7, 8, 2, 9	32	6.4
11	1, 3, 6, 5, 7	22	4.4	61	7, 3, 6, 3, 6	25	5.0
12	3, 7, 5, 3, 2	20	4.0	62	7, 4, 6, 0, 1	18	3.6
13	5, 6, 6, 5, 0	22	4.4	63	7, 9, 9, 7, 5	37	7.4
14	9, 9, 6, 4, 1	29	5.8	64	8, 0, 6, 2, 7	23	4.6
15	0, 0, 9, 5, 7	21	4.2	65	6, 5, 3, 6, 2	22	4.4
16	4, 9, 1, 1, 6	21	4.2	66	5, 0, 5, 2, 9	21	4.2
17	9, 4, 1, 1, 4	19	3.8	67	2, 9, 4, 9, 1	25	5.0
18	6, 4, 2, 7, 3	22	4.4	68	9, 5, 2, 2, 6	24	4.8
19	9, 4, 4, 1, 8	26	5.2	69	0, 1, 4, 4, 4	13	2.6
20	8, 4, 6, 8, 3	29	5.8	70	5, 4, 0, 5, 2	16	3.2
21	5, 2, 2, 6, 1	16	3.2	71	1, 1, 4, 2, 0	8	1.6
22	2, 2, 9, 1, 0	14	2.8	72	9, 5, 4, 5, 9	32	6.4
23	1, 4, 5, 8, 8	26	5.2	73	7, 1, 6, 6, 9	29	5.8
24	8, 1, 6, 3, 7	25	5.0	74	3, 5, 0, 0, 5	13	2.6
25	1, 2, 0, 9, 6	18	3.6	75	3, 7, 7, 3, 5	25	5.0
26	8, 5, 3, 0, 0	16	3.2	76	7, 4, 7, 6, 2	26	5.2
27	9, 5, 8, 5, 0	27	5.4	77	8, 1, 0, 9, 1	19	3.8
28	8, 9, 1, 1, 8	27	5.4	78	6, 4, 7, 9, 3	29	5.8
29	8, 0, 7, 4, 0	19	3.8	79	7, 7, 6, 9, 7	36	7.2
30	6, 5, 5, 3, 0	19	3.8	80	9, 4, 2, 9, 9	33	6.6
31	4, 6, 4, 2, 1	17	3.4	81	3, 3, 3, 3, 3	15	3.0
32	7, 8, 3, 6, 5	29	5.8	82	8, 7, 7, 0, 3	25	5.0
33	4, 2, 8, 5, 2	21	4.2	83	5, 3, 2, 1, 1	12	2.4
34	7, 1, 9, 0, 9	26	5.2	84	0, 4, 5, 2, 6	17	3.4
35	5, 8, 4, 1, 4	22	4.4	85	3, 7, 5, 4, 1	20	4.0
36	6, 4, 4, 5, 1	20	4.0	86	7, 4, 5, 9, 8	33	6.6
37	4, 2, 1, 1, 6	14	2.8	87	3, 2, 9, 0, 5	19	3.8
38	4, 7, 5, 5, 7	28	5.6	88	4, 6, 6, 3, 3	22	4.4
39	9, 0, 5, 9, 2	25	5.0	89	1, 0, 9, 3, 7	20	4.0
40	3, 1, 5, 4, 5	18	3.6	90	2, 9, 6, 8, 5	30	6.0
41	9, 8, 6, 3, 2	28	5.6	91	4, 8, 0, 7, 6	25	5.0
42	9, 4, 2, 2, 8	25	5.0	92	5, 6, 7, 6, 3	27	5.4
43	8, 4, 7, 2, 2	23	4.6	93	3, 6, 2, 5, 6	22	4.4
44	0, 7, 3, 4, 9	23	4.6	94	0, 1, 1, 8, 4	14	2.8
45	0, 2, 7, 5, 2	16	3.2	95	3, 6, 6, 4, 5	24	4.8
46	7, 1, 9, 9, 9	35	7.0	96	9, 2, 9, 8, 6	34	6.8
47	4, 0, 5, 9, 4	22	4.4	97	2, 0, 0, 6, 8	16	3.2
48	5, 8, 6, 3, 3	25	5.0	98	0, 4, 5, 0, 5	14	2.8
49	4, 5, 0, 5, 3	17	3.4	99	0, 3, 7, 3, 9	22	4.4
50	7, 7, 2, 0, 1	17	3.4	100	2, 5, 0, 0, 7	14	2.8

lation distribution (Figure 2) is far from normal, the top of the histogram of the $\bar{x}$ values (Figure 3) has the appearance of a bell-shaped curve, even for the small sample size of 5. For larger sample sizes, the normal distribution would give an even closer approximation.

Calculating from the 100 simulated $\bar{x}$ values in Table 5, we find the sample mean and standard deviation to be 4.54 and 1.215, respectively. These are in close agreement with the theoretical values for the mean and standard deviation of $\bar{X}$: $\mu = 4.5$ and $\sigma/\sqrt{n} = 2.872/\sqrt{5} = 1.284$.   ■

It might be interesting for the reader to collect similar samples by reading the last digits of numbers from a telephone directory and then to construct a histogram of the $\bar{x}$ values.

Another graphic example of the central limit theorem appears in Figure 4, where the population distribution represented by the solid curve is a continuous asymmetric distribution with $\mu = 2$ and $\sigma = 1.41$. The distributions of the sample mean $\bar{X}$ for sample sizes $n = 3$ and $n = 10$ are plotted as dashed curves on the graph. These indicate that with increasing n, the distributions become more concentrated around μ and look more like the normal distribution.

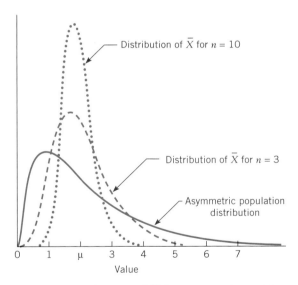

Figure 4. Distributions of $\bar{X}$ for $n = 3$ and $n = 10$ in sampling from an asymmetric population

Example 5   Probability calculations for $\bar{X}$—Large sample size
Consider a population with mean = 82 and standard deviation = 12.

(a)   If a random sample size of size 64 is selected, what is the probability that the sample mean will lie between 80.8 and 83.2?

(b) With a random sample of size 100, what is the probability that the sample mean will lie between 80.8 and 83.2?

SOLUTION AND DISCUSSION

(a) We have $\mu = 82$ and $\sigma = 12$. Since $n = 64$ is large, the central limit theorem tells us that the distribution of $\overline{X}$ is approximately normal with

$$\text{mean} = \mu = 82$$

$$\text{standard deviation} = \frac{\sigma}{\sqrt{n}} = \frac{12}{\sqrt{64}} = 1.5$$

To calculate $P[\,80.8 < \overline{X} < 83.2\,]$, we convert to the standardized variable

$$Z = \frac{\overline{X} - \mu}{\sigma/\sqrt{n}} = \frac{\overline{X} - 82}{1.5}$$

The z-values corresponding to 80.8 and 83.2 are

$$\frac{80.8 - 82}{1.5} = -.8 \quad \text{and} \quad \frac{83.2 - 82}{1.5} = .8$$

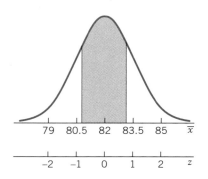

Consequently,

$$
\begin{aligned}
P[\,80.8 < \overline{X} < 83.2\,] &= P[\,-.8 < Z < .8\,] \\
&= .7881 - .2119 \quad (\text{using the normal table}) \\
&= .5762
\end{aligned}
$$

(b) We now have $n = 100$, so $\sigma/\sqrt{n} = 12/\sqrt{100} = 1.2$, and

$$Z = \frac{\overline{X} - 82}{1.2}$$

Therefore,

$$P[\,80.8 < \overline{X} < 83.2\,] = P\left[\frac{80.8 - 82}{1.2} < Z < \frac{83.2 - 82}{1.2}\right]$$
$$= P[\,-1.0 < Z < 1.0\,]$$
$$= .8413 - .1587$$
$$= .6826$$

Note that the interval $(80.8, 83.2)$ is centered at $\mu = 82$. The probability that $\overline{X}$ will lie in this interval is larger for $n = 100$ than for $n = 64$.　　　　　　　　　　　　　　　　　　　　　　　　　　■

Example 6　Probability calculations for $\overline{X}$, mean gripping strength
Suppose that the population distribution of the gripping strengths of industrial workers is known to have a mean of 110 and standard deviation of 10. For a random sample of 75 workers, what is the probability that the sample mean gripping strength will be

　(a)　Between 109 and 112?
　(b)　Greater than 111?

SOLUTION AND　Here the population mean and the standard deviation are $\mu = 110$ and $\sigma = 10$,
DISCUSSION　respectively. The sample size $n = 75$ is large, so the central limit theorem ensures that the distribution of $\overline{X}$ is approximately normal with

$$\text{mean} = 110$$
$$\text{standard deviation} = \frac{\sigma}{\sqrt{n}} = \frac{10}{\sqrt{75}} = 1.155$$

　(a)　To find $P[\,109 < \overline{X} < 112\,]$, we convert to the standardized variable

$$Z = \frac{\overline{X} - 110}{1.155}$$

and calculate the z-values

$$\frac{109 - 110}{1.155} = -.866, \quad \frac{112 - 110}{1.155} = 1.732$$

which round to $-.87$ and 1.73. The required probability is

$$P[\,109 < \overline{X} < 112\,] = P[\,-.87 < Z < 1.73\,]$$
$$= .9582 - .1922$$
$$= .7660$$

(b)

$$P[\overline{X} > 111] = P\left[Z > \frac{111 - 110}{1.155}\right]$$
$$= P[Z > .87]$$
$$= 1 - P[Z \le .87]$$
$$= 1 - .8078$$
$$= .1922 \qquad \blacksquare$$

A natural question that arises is how large should n be for the normal approximation to be used for the distribution of $\overline{X}$? The nature of the approximation depends on the extent to which the population distribution deviates from a normal form. If the population distribution is normal, then $\overline{X}$ is exactly normally distributed for all n, small or large. As the population distribution increasingly departs from normality, larger values of n are required for a good approximation. Ordinarily, $n > 30$ provides a satisfactory approximation.

STATISTICAL REASONING

Planning Refreshments for a Social Event

When social or interest groups meet, usually coffee, punch, and other refreshments are served. Statistical distributions can be the basis for an efficient approach to planning the amount needed. This is illustrated in the context of the number of cookies needed for an event that will be attended by 100 students.

From his extensive personal experience with a campus social group, Alex specifies the following distribution for the number of cookies taken by an individual.

Michael Tamborrino/FPG International

Probability Distribution of the Number of
Cookies Taken by an Individual

x	0	1	2	3	4
$f(x)$	.05	.2	.4	.3	.05

According to this probability distribution, it is most likely that a person will take
2 cookies. Consequently, in a large group, more persons would take exactly 2
cookies than would take 1 cookie or some other specific number.

Alex needs to find the number of cookies required for the 100 persons who
will attend the gathering. How many cookies should Alex prepare if he wants to
have probability .975 that there will be enough for the event?

From the probability distribution above, Alex obtains the mean and variance
of the number of cookies taken by an individual:

$$\mu = 0(.05) + 1(.2) + 2(.4) + 3(.3) + 4(.05) = 2.1$$
$$\sigma^2 = [0^2(.05) + 1^2(.2) + 2^2(.4) + 3^2(.3) + 4^2(.05)] - \mu^2$$
$$= 0 + .2 + 1.6 + 2.7 + .8 - (2.1)^2$$
$$= 5.3 - 4.41 = .89$$

Therefore, the standard deviation is $\sqrt{.89} = .9434$.

Alex next uses his knowledge about sampling distributions. Let $\overline{X}$ be the
average number of cookies taken by the 100 persons who will attend the event.
Then, by the central limit theorem, the sampling probability distribution of $\overline{X}$ is
approximately normal with mean 2.1 and standard deviation $.9434/\sqrt{100} =$
.09434.

Alex needs to find b, the number of cookies to prepare per person, so that,
with probability .975, $\overline{X}$ will not exceed b. That is, he requires that b satisfy

$$P[\overline{X} \le b] = .975$$

From the standard normal table, the upper 97.5th percentile is $z = 1.96$. Since
$\overline{X}$ has a normal distribution, the probability requirement gives

$$\frac{b - \text{mean}(\overline{X})}{\text{standard deviation}(\overline{X})} = 1.96$$

or

$$b = \text{mean}(\overline{X}) + 1.96 \text{ standard deviation}(\overline{X})$$
$$= 2.1 + 1.96(.09434)$$
$$= 2.2849$$

Consequently, when 100 persons attend, the required number of cookies is

$$100(2.2849) \approx 229$$

Note that if the probability .975 of having enough is replaced by .99, then we simply replace the standard normal percentile 1.96 by 2.33 and the number of cookies required becomes

$$100[\,2.1 + 2.33(\,.09434\,)\,] \approx 232$$

Let us compare Alex's approach with the following very conservative procedure. Since 4 is the maximum number of cookies taken by an individual, then $100(4) = 400$ is the maximum number that would be needed for the 100 persons attending the event. According to the probability calculations, for all practical purposes, 250 cookies is more than enough. Planning for everyone to take the maximum results in a great many leftover cookies!

Similar calculations can be applied to determine the amount of coffee to prepare. One need only replace the distribution of cookies for an individual by the distribution of cups of coffee.

One of the authors witnessed a considerable amount of food and coffee wasted at meetings held in his home. Apparently the student organizers had not yet learned enough statistics to plan effectively.

Exercises

3.1 A population has mean 99 and standard deviation 7. Calculate $E(\overline{X})$ and $sd(\overline{X})$ for a random sample of size (a) 4 and (b) 25.

3.2 A population has mean $= 250$ and standard deviation $= 12$. Calculate the expected value and standard deviation of $\overline{X}$ for a random sample of size (a) 3 (b) 16.

3.3 A population has standard deviation 10. What is the standard deviation of $\overline{X}$ for a random sample of size (a) $n = 25$ (b) $n = 100$ (c) $n = 400$?

3.4 A population has a standard deviation $= 84$. What is the standard deviation of $\overline{X}$ for a random sample of size (a) 36 (b) 144?

3.5 Using the sampling distribution determined for $\overline{X} = (X_1 + X_2)/2$ in Exercise 2.5, verify that $E(\overline{X}) = \mu$ and $sd(\overline{X}) = \sigma/\sqrt{2}$.

3.6 Using the sampling distribution determined for $\overline{X} = (X_1 + X_2)/2$ in Exercise 2.6, verify that $E(\overline{X}) = \mu$ and $sd(\overline{X}) = \sigma/\sqrt{2}$.

3.7 A normal population has $\mu = 27$ and $\sigma = 3$. For sample size $n = 4$, determine the (a) mean of $\overline{X}$, (b) standard deviation of $\overline{X}$, and (c) distribution of $\overline{X}$.

3.8 A normal population has mean $= 20$ and standard deviation $= 5$. For a random sample of size $n = 6$, determine the

(a)   mean of $\overline{X}$
(b)   standard deviation of $\overline{X}$
(c)   distribution of $\overline{X}$

3.9   The amount of sulfur in the daily emissions from a power plant has a normal distribution with a mean of 134 pounds and a standard deviation of 22 pounds. For a random sample of 5 days, find the probability that the total amount of sulfur emissions will exceed 700 pounds. (*Hint:*   A total of 5 measurements exceeding 700 pounds means that their average exceeds 140.)

3.10   To avoid difficulties with governmental consumer protection agencies, a manufacturer must make reasonably certain that its bags of potato chips actually contain 16 ounces of chips. One recognized way of monitoring production is to take a random sample of a few bags from each hour's production. The mean $\overline{X}$ of the contents of the bags then provides important information on whether or not the process is meeting the weight requirement. Records for one packaging machine indicate that its fill weights are nearly normally distributed with standard deviation = .122 ounce. If this machine is set so that the mean fill weight is 16.08 ounces per bag:

(a)   What is the probability that the sample mean contents of 9 bags, selected at random, will be less than 16.0 ounces?

(b)   In the long run, what proportion of the bags filled by this machine will contain less than 16.0 ounces of potato chips?

3.11   Refer to the Statistical Reasoning example concerning refreshments. Suppose the number of cups of coffee taken by an individual has the distribution

x	$f(x)$
0	.3
1	.4
2	.2
3	.1

Find b, the number of cups to prepare per person, so that the probability is .95 that $\overline{X}$ will not exceed b.

3.12   Suppose a certain type of airplane has a capacity of 400 passengers but that 350 people will take a particular flight. Can you specify the general form of the approximate probability distribution of the total weight of all the passengers? Explain.

3.13   A random sample of size 100 is taken from a population having a mean of 20 and a standard deviation of 5. The shape of the population distribution is unknown.

(a)   What can you say about the probability distribution of the sample mean $\overline{X}$?

(b)   Find the probability that $\overline{X}$ will exceed 20.75.

3.14   The lengths of the trout fry in a pond at the fish hatchery are approximately normally distributed with mean = 3.4 inches and standard deviation = .8 inch. A dozen fry will be netted and their lengths measured.

(a)   What is probability that the sample mean length of the 12 netted trout fry will be less than 3.0 inches?

(b)   Why might the fish in the net not represent a random sample of trout fry in the pond?

3.15   The heights of male students at a university have a nearly normal distribution with mean = 70 inches and standard deviation = 2.8 inches. If 5 male students are randomly selected to make up an intramural basketball team, what is the probability that the heights of the team will average over 72.0 inches?

3.16   According to the growth chart that doctors use as a reference, the heights of two-year-old boys are normally distributed with mean = 34.5 inches and standard deviation = 1.3 inches. For a random sample of 6 two-year-old boys, find the probability that the sample mean will be between 34.1 and 35.2 inches.

3.17   The weight of an almond is normally distributed with mean = .05 ounce and standard deviation = .015 ounce. Find the probability that a package of 100 almonds will weigh between 4.8 and 5.3 ounces. That is, find the probability that $\overline{X}$ will be between .048 and .053 ounce.

3.18   Refer to Table 5.

(a)   Calculate the sample median for each sample.

(b)   Construct a frequency table and make a histogram.

(c)   Compare the histogram for the median with that given in Figure 3 for the sample mean. Does your comparison suggest that the sampling distribution of the mean or median has the smaller variance?

3.19   How does the central limit theorem express the fact that the probability distribution of $\overline{X}$ concentrates more and more probability near μ as n increases?

KEY IDEAS

A **parameter** is a numerical characteristic of the population. It is a constant although its value is typically unknown to us. The object of a statistical analysis of sample data is to learn about the parameter.

A numerical characteristic of a sample is called a **statistic**. The value of a statistic varies in repeated sampling.

Random sampling from a population refers to independent selections where each observation has the same distribution as the population.

When random sampling from a population, a statistic is a random variable. The probability distribution of a statistic is called its **sampling distribution**.

The sampling distribution of $\overline{X}$ has mean μ and standard deviation $\sigma/\sqrt{n}$, where μ = population mean, σ = population standard deviation, and n = sample size.

With increasing n, the distribution of $\overline{X}$ is more concentrated around μ.

If the population distribution is normal $N(\mu, \sigma)$, the distribution of $\overline{X}$ is $N(\mu, \sigma/\sqrt{n})$.

Regardless of the shape of the population distribution, the distribution of $\overline{X}$ is approximately $N(\mu, \sigma/\sqrt{n})$, provided that n is large. This result is called the **central limit theorem**.

4. REVIEW EXERCISES

4.1   A population consists of the four numbers {2, 4, 6, 8}. Consider drawing a random sample of size 2 with replacement.

(a)   List all possible samples and evaluate $\bar{x}$ for each.

(b)   Determine the sampling distribution of $\overline{X}$.

(c)   Write down the population distribution and calculate its mean μ and standard deviation σ.

(d)   Calculate the mean and standard deviation of the sampling distribution of $\overline{X}$ obtained in part (b), and verify that these agree with μ and $\sigma/\sqrt{2}$, respectively.

4.2   Refer to Exercise 4.1 and, instead of $\overline{X}$ consider the statistic

Sample range R = Largest observation − Smallest observation

For instance, if the sample observations are (2, 6), the range is $6 - 2 = 4$.

(a)   Calculate the sample range for all possible samples.

(b)   Determine the sampling distribution of R.

4.3   Consider random sampling from a population that has mean = 550 and standard deviation = 70. Find the mean and standard deviation of $\overline{X}$ for

(a) Sample size 16

(b) Sample size 160

4.4 What sample size is required in order that the standard deviation of $\overline{X}$ be

(a) $\frac{1}{4}$ of the population standard deviation?

(b) $\frac{1}{8}$ of the population standard deviation?

(c) 15% of the population standard deviation?

4.5 Suppose a population distribution is normal with mean = 80 and standard deviation = 10. For a random sample of size $n = 9$,

(a) What are the mean and standard deviation of $\overline{X}$?

(b) What is the distribution of $\overline{X}$? Is this distribution exact or approximate?

(c) Find the probability that $\overline{X}$ lies between 76 and 84.

4.6 The weights of pears in an orchard are normally distributed with mean = .32 pound and standard deviation = .08 pound.

(a) If one pear is selected at random, what is the probability that its weight will be between .28 and .34 pound?

(b) If $\overline{X}$ denotes the average weight of a random sample of 4 pears, what is the probability that $\overline{X}$ will be between .28 and .34 pound?

4.7 Suppose that the size of pebbles in a river bed is normally distributed with mean = 12.1 mm and standard deviation = 3.2 mm. A random sample of 9 pebbles will be measured. Let $\overline{X}$ denote the average size of the sampled pebbles.

(a) What is the distribution of $\overline{X}$?

(b) What is the probability that $\overline{X}$ is smaller than 10?

(c) What percentage of the pebbles in the river bed are of size smaller than 10?

4.8 A random sample of size 150 is taken from a population that has mean = 60 and standard deviation = 8. The population distribution is not normal.

(a) Is it reasonable to assume a normal distribution for the sample mean $\overline{X}$? Why or why not?

(b) Find the probability that $\overline{X}$ lies between 59 and 61.

(c) Find the probability that $\overline{X}$ exceeds 62.

4.9 The distribution for the time it takes a student to complete the fall class registration has a mean of 94 minutes and a standard deviation of 10 minutes. For a random sample of 81 students,

(a) Determine the mean and standard deviation of $\overline{X}$.

(b) What can you say about the distribution of $\overline{X}$?

4.10   Refer to Exercise 4.9. Evaluate (a) $P[\overline{X} > 96]$ (b) $P[92.3 < \overline{X} < 96]$ and (c) $P[\overline{X} < 95]$.

4.11   The mean and standard deviation of the strength of a packaging material are 55 and 7 pounds, respectively. If 40 specimens of this material are tested,

(a)   What is the probability that the sample mean strength $\overline{X}$ will be between 54 and 56 pounds?

(b)   Find the interval centered at 55, where $\overline{X}$ will lie with probability .95.

4.12   Consider a random sample of size $n = 100$ from a population that has a standard deviation of $\sigma = 20$.

(a)   Find the probability that the sample mean $\overline{X}$ will lie within 2 units of the population mean—that is, $P[-2 \leq \overline{X} - \mu \leq 2]$.

(b)   Find the number k so that $P[-k \leq \overline{X} - \mu \leq k] = .90$.

(c)   What is the probability that $\overline{X}$ will differ from μ by more than 4 units?

4.13   The distribution of the diameter of hail hitting the ground during a storm has mean = .5 inch and standard deviation = .1 inch. What is the probability that the average diameter of 40 randomly selected hailstones will lie between .48 and .53 inch?

4.14   The time that customers take to complete their transaction at a money machine is a random variable with mean = 2 minutes and standard deviation = .6 minute. Find the probability that a random sample of 50 customers will take between 90 and 112.5 minutes to complete all their transactions. That is, find the probability that $\overline{X}$ will be between 1.8 and 2.25 minutes.

4.15   Refer to Example 2. Replace the original probabilities .2, .3, .5 by .3, .5, .2, respectively. Determine the sampling distribution of the sample mean.

4.16   Many cities in the United States must monitor the daily water consumption per household to ensure that no shortage in water supply is imminent. Consider a specific household where both parents work outside of the home and there are a son and daughter of high school age.

(a)   What is the likely shape of the probability distribution of the daily water consumption by the specific household?

(b)   Does the distribution remain the same for all days of the week? Why?

(c)   Does the distribution remain the same in all four seasons? Why?

STUDENT PROJECTS

1. (a) Count the number X of occupants including the driver in each of 20 passing cars. Calculate the mean $\bar{x}$ of your sample.
   (b) Repeat part (a) 10 times.
   (c) Collect the data sets of the individual car counts x from the entire class and construct a relative frequency histogram.
   (d) Collect the $\bar{x}$ values from the entire class (10 from each student) and construct a relative frequency histogram for $\bar{x}$, choosing appropriate class intervals.
   (e) Plot the two relative frequency histograms and comment on the closeness of their shapes to the normal distribution.

2. (a) Collect a sample of size 7 and compute $\bar{x}$ and the sample median.
   (b) Repeat part (a) 30 times.
   (c) Plot dot diagrams for the values of the two statistics in part (a). These plots reflect the individual sampling distributions.
   (d) Compare the amount of variation in $\overline{X}$ and the median.

   In this exercise, you might record weekly soft-drink consumptions, sentence lengths, or hours of sleep for different students.

COMPUTER PROJECT

Conduct a simulation experiment on the computer to verify the central limit theorem. Generate $n = 6$ observations from the continuous distribution that is uniform on 0 to 1. Calculate $\overline{X}$. Repeat 150 times. Make a histogram of the $\overline{X}$ values and a normal-scores plot. Does the distribution of $\overline{X}$ appear to be normal for $n = 6$? You may wish to repeat with $n = 20$.

If MINITAB is available, you could use the commands

```
RANDOM 150 C1-C6;
UNIFORM.
ADD C1-C6 SET C10
LET C11=C10/6.0
```

The 150 values of $\bar{x}$ in C11 can then be described by the commands

```
HISTOGRAM C11
DESCRIBE C11
```

or through the dialog box sequence

Stat > Basic Statistics > Descriptive Statistics

CHAPTER 8

Inferences About Means— Large Samples

Chapter Objectives

After reading this chapter, you should be able to

▶ Discuss the concept of making a statistical inference about a population parameter.
▶ Distinguish between point and interval estimators.
▶ Calculate a point estimate and its standard error.
▶ Construct a confidence interval for a population mean.
▶ Understand the steps in testing statistical hypotheses.
▶ Carry out a test of hypotheses concerning a population mean.
▶ Calculate *p*-values.

1. INTRODUCTION

Inferences are generalizations about a population that are made on the basis of a sample collected from the population. For instance, persons interested in fine cuisine and a little danger and excitement might consider beekeeping as a hobby. To learn more about this activity, you could take a survey of, say, 40 beekeepers. In fact, a company marketing a new honey-based product did take such a survey. One response requested was the number of times the beekeeper had been stung in the past year. These data could be described by the methods of Chapter 2 and they would tell us about the particular 40 persons in the sample. But, would the data apply to you if you were to take up this activity? That is, there can be a wider purpose of the study: to learn about the typical number of bee stings per year for all beekeepers. Attention is not confined to just the 40 beekeepers in the sample but to all beekeepers or even potential beekeepers. Having collected the 40 numbers of bee stings from participants in the survey, we need to generalize to the population of all potential numbers of stings. Typically, this generalization will be a statement about the population mean or some other feature of the population. We call these generalizations *statistical inferences* or just *inferences*.

> Statistical inference deals with drawing conclusions about population parameters from an analysis of the sample data.

In this chapter, we introduce two basic types of inference. First we discuss *estimation* or guessing the value of the mean, or some other feature, on the basis of a sample. Second, we discuss *hypothesis testing* or checking whether a specified value for the population quantity (a hypothesis) is compatible with the sample evidence. Later, we will see that the two forms of inference are related.

Companies are increasingly relying on temporary employees to meet peak staffing demands. A large mail order firm collects data on the number of calls handled per shift by individual temporary employees. This provides information on the activities of a typical worker. We could *estimate* the mean number of calls per worker. Coupled with historical records on the number of calls received in similar weeks, the data could be used to make decisions about the total staffing requirements for the next week.

New cable technology allows advertisers to determine the relative effectiveness of two different versions of a commercial. The first version can be shown to roughly half of the homes and the second version to the other half. After two hours, a random sample of viewers are questioned about the key point of the commercial. The object of this exercise is to determine whether, on average, viewers of the first version of the commercial remember the key message better than viewers of the second version of the commercial (*a test of hypotheses*).

A clear statement of purpose, as described in Chapter 1, can reveal the choice of inferences required to answer the important questions. To review, the two most important types of inference are

1. **estimation of parameter(s)**
2. **testing of statistical hypotheses**

The true value of any population parameter is an unknown constant. It can be correctly ascertained only by an exhaustive examination of the population, if indeed that is possible. Consequently, any inference about a population parameter will involve some uncertainty because it is based on a sample rather than the entire population.

The next example illustrates the estimation of parameters and testing hypotheses.

Example 1   Types of inference: Point estimation, interval estimation, and testing hypotheses

Beekeeping may be a good activity for some people interested in fine cuisine and a little excitement. A random sample of current beekeepers could provide information about what to expect. The question

How many times were you stung in the past year?

was asked by a company launching a new honey-based product. Table 1 gives the data on the number of bee stings in the past year for 40 beekeepers.

TABLE 1   Number of Bee Stings Per Year Reported by Beekeepers[1]

3	90	4	4	80	200	2	700
2	1000	200	3000	3	150	100	1
400	40	5	30	1000	100	200	200
1	100	150	1000	3	800	3	200
100	20	400	10	50	2	3	2000

[1]Courtesy of Jacob Leinenkugel Brewing Company.

Following the procedures in Chapter 2, we obtain a descriptive summary of the bee sting data:

sample mean $\bar{x} = 308.9$ sample standard deviation $s = 597.8$

sample median $= 95$ first quartile $= 3.5$ third quartile $= 200$

However, our interest lies not just in this particular set of 40 observations but in the vast population of all beekeepers and even potential beekeepers. The population distribution of number of stings per year received by a beekeeper is unknown to us and so are the population mean μ and population standard deviation σ.

We take the point of view that the 40 observations are a random sample from the population of numbers of bee stings. One purpose of this study could be "to learn about μ." More specifically, we may wish to make one, two, or all three of the following types of inference.

1. Use a single value to estimate the unknown mean μ (**point estimation**). We estimate the mean number of stings for all beekeepers to be 308.9. An alternative point estimate would be 95, which is the sample median.

2. Determine an interval of plausible values for μ (**interval estimation**). The population mean number of stings is between 123.6 and 494.2. (The calculation of intervals of plausible values is explained in Section 3.)

3. Decide whether the mean number of stings is 260, corresponding to five times a week (**testing hypotheses**). Some might consider this to be an acceptable number.   ■

2. POINT ESTIMATION OF A POPULATION MEAN

The object of point estimation is to calculate, from the sample data, a single number that is likely to be close to the unknown value of the parameter. The available information is assumed to be in the form of a random sample X_1, $X_2, \ldots, X_n$ of size n taken from the population. We wish to formulate a statistic such that its value computed from the sample data would reflect the value of the population parameter as closely as possible.

A statistic intended for estimating a parameter is called a **point estimator,** or simply an **estimator.** The standard deviation of an estimator is called its **standard error: S.E.**

When we estimate a population mean from a random sample, perhaps the most intuitive estimator is the sample mean,

$$\overline{X} = \frac{X_1 + X_2 + \cdots + X_n}{n}$$

For instance, to estimate the population mean number of bee stings in Example 1, we would naturally compute the sample mean of the 40 measurements. Employing the estimator $\overline{X}$, with the data of Table 1, we get the result $\overline{x} = 308.9$ stings, which we call a **point estimate** or simply an **estimate** of μ.

Recall the discussion in Chapter 7 regarding the sampling distribution of the sample mean $\overline{X}$. In particular, two cases resulted in a normal sampling distribution:

1. The distribution of $\overline{X}$ is *normal* whenever the underlying population of individual observations has a normal distribution.

2. The distribution of $\overline{X}$ is *approximately normal* provided that the sample size is *large* (at least 30). The underlying population need not be normal.

In this chapter, we develop statistical procedures for making inferences in the context of case 2, the large sample size case. Chapter 9 presents the corresponding results for case 1.

Without an assessment of accuracy, a single number quoted as an estimate may not serve a very useful purpose. We must indicate the extent of variability in the distribution of the estimator. The standard deviation, alternatively called the standard error of the estimator, provides information about its variability.

To study the properties of the sample mean $\overline{X}$ as an estimator of the population mean μ, let us review the results from Chapter 7.

1. $E(\overline{X}) = \mu$.

2. $\mathrm{sd}(\overline{X}) = \dfrac{\sigma}{\sqrt{n}}$ so S.E.$(\overline{X}) = \dfrac{\sigma}{\sqrt{n}}$.

3. With large n, $\overline{X}$ is nearly normally distributed with mean μ and standard deviation $\sigma/\sqrt{n}$.

The first two results show that the distribution of $\overline{X}$ is centered around μ and its standard error is $\sigma/\sqrt{n}$, where σ is the population standard deviation and n the sample size.

To understand how closely $\overline{X}$ is expected to estimate μ, we now examine the third result, which is depicted in Figure 1. Recall that, in a normal distribution, the interval running two standard deviations on each side of the mean contains probability .954. Thus, prior to sampling, the probability is .954 that the estimator $\overline{X}$ will be within a distance $2\sigma/\sqrt{n}$ from the true parameter value μ. This probability statement can be rephrased by saying that when we are estimating μ by $\overline{X}$ the 95.4% error margin is $2\sigma/\sqrt{n}$.

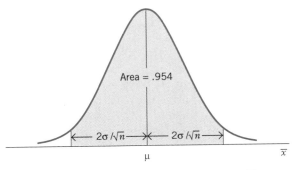

Area = .954

$2\sigma/\sqrt{n}$ — $2\sigma/\sqrt{n}$

μ $\overline{x}$

Figure 1. Approximate normal distribution of $\overline{X}$

Use of the probability .954, which corresponds to the multiplier 2 of the standard error, is by no means universal. The following notation will facilitate our writing of an expression for the $100(1 - \alpha)\%$ error margin, where $1 - \alpha$ denotes the desired high probability such as .95 or .90.

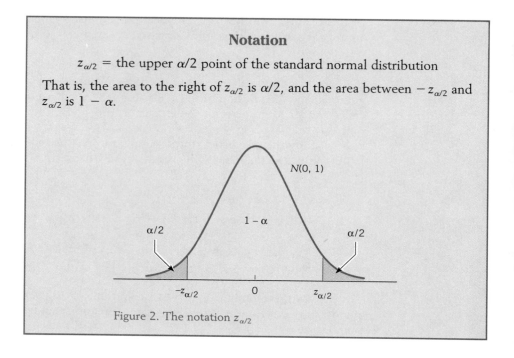

Figure 2. The notation $z_{\alpha/2}$

A few values of $z_{\alpha/2}$ obtained from the normal table appear in Table 2 for easy reference.

TABLE 2 Values of $z_{\alpha/2}$

$1 - \alpha$	.80	.85	.90	.95	.99
$z_{\alpha/2}$	1.28	1.44	1.645	1.96	2.58

To illustrate the notation, suppose we want to determine the 90% error margin. We then set $1 - \alpha = .90$ so $\alpha/2 = .05$ and we have $z_{.05} = 1.645$. Therefore, when estimating μ by $\overline{X}$, the 90% error margin is $1.645\sigma/\sqrt{n}$.

A minor difficulty remains in computing the standard error of $\overline{X}$. The expression involves the unknown population standard deviation σ, but we can estimate σ by the sample standard deviation

$$S = \sqrt{\dfrac{\sum\limits_{i=1}^{n} (X_i - \overline{X})^2}{n - 1}}$$

When n is large, the effect of estimating the standard error $\sigma/\sqrt{n}$ by $S/\sqrt{n}$ can be neglected. We now summarize.

Point Estimation of the Mean

Parameter: Population mean μ

Data: $X_1, \ldots, X_n$ (a random sample of size n)

Estimator: $\overline{X}$ (sample mean)

$$\text{S.E.}(\overline{X}) = \frac{\sigma}{\sqrt{n}}, \qquad \text{estimated S.E.}(\overline{X}) = \frac{S}{\sqrt{n}}$$

For large n, the $100(1-\alpha)\%$ error margin is $z_{\alpha/2}\sigma/\sqrt{n}$.
(If σ is unknown, use S in place of σ.)

Example 2   Point estimation of mean number of bee stings
Refer to the data of Example 1 consisting of 40 observations on the number of bee stings per year. Give a point estimate of the population mean number of stings and state a 95% error margin.

SOLUTION AND   The sample mean and standard deviation computed from the 40 observations in
DISCUSSION   Table 1 are

$$\bar{x} = \frac{\Sigma\, x}{n} = 308.9$$

$$s = \sqrt{\frac{\Sigma(x-\bar{x})^2}{39}} = \sqrt{357{,}393.5} = 597.8$$

To calculate the 95% error margin, we set $1-\alpha = .95$ so that $\alpha/2 = .025$ and $z_{\alpha/2} = 1.96$. Therefore, the 95% error margin is

$$1.96\,\frac{s}{\sqrt{n}} = 1.96\,\frac{597.8}{\sqrt{40}} = 185.3$$

Our point estimate of the population mean is 308.9 stings. We do not expect the population mean to be exactly this value so we attach an error of plus or minus 185.3.   ■

Caution: (a) Standard error should not be interpreted as the "typical" error in a problem of estimation as the word "standard" may suggest. For instance, when $\text{S.E.}(\overline{X}) = .3$, we should not think that the error $(\overline{X} - \mu)$ is likely to be .3, but rather, prior to observing the data, the probability is approximately .954 that the error will be within $\pm 2(\text{S.E.}) = \pm .6$.

(b) An estimate and its variability are often reported in either of the forms: estimate $\pm$ S.E. or estimate $\pm 2(\text{S.E.})$. In reporting a numerical result such as 53.4 ± 4.6, we must specify whether 4.6 represents S.E., $2(\text{S.E.})$, or some other multiple of the standard error.

DETERMINING THE SAMPLE SIZE

During the planning stage of an investigation, it is important to address the question of sample size. Because sampling is costly and time-consuming, the investigator needs to know beforehand the sample size required to give the desired precision.

To determine how large a sample is needed for estimating a population mean, we must specify

$$d = \text{the desired error margin}$$

and

$$1 - \alpha = \text{the probability associated with the error margin}$$

Referring to the expression for a $100(1 - \alpha)\%$ error margin, we then equate:

$$z_{\alpha/2} \frac{\sigma}{\sqrt{n}} = d$$

This gives an equation in which n is unknown. Solving for n, we obtain

$$n = \left[\frac{z_{\alpha/2}\sigma}{d} \right]^2$$

which determines the required sample size. Of course, the solution is rounded to the next higher integer, because a sample size cannot be fractional.

This determination of sample size is valid provided $n > 30$, so that the normal approximation to $\overline{X}$ is satisfactory.

To be $100(1 - \alpha)\%$ sure that the error of estimation $|\overline{X} - \mu|$ does not exceed d, the required sample size is

$$n = \left[\frac{z_{\alpha/2}\sigma}{d} \right]^2$$

If σ is completely unknown, a small-scale preliminary sampling is necessary to obtain an estimate of σ to be used in the formula to compute n.

Example 3   Determining the number of water samples

A limnologist wishes to estimate the mean phosphate content per unit volume of lake water. It is known from studies in previous years that the standard deviation has a fairly stable value of $\sigma = 4$. How many water samples must the limnologist analyze to be 90% certain that the error of estimation does not exceed .8 milligram?

SOLUTION AND DISCUSSION Here $\sigma = 4$ and $1 - \alpha = .90$, so $\alpha/2 = .05$. The upper .05 point of the $N(0, 1)$ distribution is $z_{.05} = 1.645$. The tolerable error is $d = .8$. Computing

$$n = \left[\frac{1.645 \times 4}{.8} \right]^2 = 67.65$$

we determine that the required sample size is $n = 68$. The limnologist must take care that the sampling sites are spread out and representative of the lake. ■

STATISTICAL REASONING

Sample the Correct Population—Otherwise Estimates Can Be Totally Misleading

Michael Krasowitz/FPG International

In essence, the headline of the state capital's city newspaper read

STATE INCOMES UP
BY 12.5% LAST YEAR

This conclusion, including the point estimate 12.5%, was obtained by Census Bureau from salary data collected from a sample of 700 households. Although the standard error was not reported, it would likely be less than one percent. The numerical work was correct and this study seemed to have produced an accurate estimate for the increase in median household income.

But something seemed very wrong. The announced increase of 12.5% for the state occurred during a period when the average increase for the whole United States was 2.7%. The big increase sounded too good to be true.

Remember the advice in Chapter 1—question data. Fortunately, someone in state government asked questions. It turned out that the agency responsible for producing the estimate was not given a budget to conduct a separate survey of salaries. Instead, they decided to use salary data that were already collected as part of a larger study that emphasized minorities including Native Americans. That was a fatal blunder.

Would you expect the salary history of Native Americans, most of whom live on reservation lands, to be representative of all workers in the state? No. Convenient samples cannot be substituted for samples from the population targeted in the study. Even samples of large size cannot overcome the mistake of not sampling from the target population.

How did the excessively high estimate 12.5% arise? The data were collected shortly after the Native Americans opened the only legal gambling casinos in the state. This activity brought many new jobs to the reservations and large profits for the casinos that spilled over to the local economy. In fact, one purpose of the original survey was to study the household incomes of minorities and the data were fine for that purpose.

Once the blunder was identified, a revised estimate of average increase was produced that was very near the national average 2.7%. The revised estimate is not even close to 12.5%, which was totally misleading.

The random sample must be taken from the target population, the one of interest. Otherwise, even though the sample size is large and the numerical calculations correct, an estimate can be totally misleading.

Exercises

2.1   Which of the following are point estimates?

   (a)   You are standing in line at a theme park and the sign says it takes 15 minutes to reach the front of the line from here.

   (b)   A microwave popcorn package reads "cook for 3 to 5 minutes."

   (c)   A report says the average dorm resident consumes 24 gallons of water per day.

2.2   Which of the following are point estimates?

   (a)   A package containing a microwave macaroni and cheese dinner reads "cook for 12 minutes."

   (b)   When calling an 800 number for information, you are placed on hold and are told by an electronic voice that the wait should be about 3 minutes.

   (c)   The package of a frozen pizza reads "cook for 10–12 minutes."

2.3   You have just purchased a bag of your favorite potato chips and the package reads "8 oz."

(a)   Interpret this as a point estimate. What is the population?

(b)   What is the quantity, or feature of bags, that is being estimated?

(c)   Explain how you would have more information if a 95% error margin were also printed on the bag.

2.4   You have just purchased your favorite chocolate bar and the package reads "6 oz."

(a)   Interpret this as a point estimate. What is the population?

(b)   What is the quantity, or feature of bars, that is being estimated?

(c)   Explain how you would have more information if a 95% error margin were also printed on the package.

2.5   For estimating a population mean with the sample mean $\overline{X}$, find (i) the standard error of $\overline{X}$ and (ii) $100(1 - \alpha)\%$ error margin in each case.

(a) $n = 138,$ $\sigma = 22,$ $1 - \alpha = .95$

(b) $n = 65,$ $\sigma = 8.2,$ $1 - \alpha = .99$

(c) $n = 320,$ $\sigma = 56,$ $1 - \alpha = .92$

2.6   Determine the point estimate of the population mean μ and its $100(1 - \alpha)\%$ margin of error in each case.

(a) $n = 150,$ $\overline{x} = 86.2,$ $s = 9.56,$ $1 - \alpha = .975$

(b) $n = 220,$ $\overline{x} = 925,$ $s = 87,$ $1 - \alpha = .88$

(c) $n = 1100,$ $\overline{x} = .728,$ $s = .085,$ $1 - \alpha = .90$

2.7   Consider the problem of estimating a population mean μ based on a random sample of size n from the population. Compute a point estimate of μ and the estimated standard error in each of the following cases.

(a) $n = 70,$ $\Sigma x_i = 852,$ $\Sigma (x_i - \overline{x})^2 = 215$

(b) $n = 140,$ $\Sigma x_i = 1653,$ $\Sigma (x_i - \overline{x})^2 = 464$

(c) $n = 160,$ $\Sigma x_i = 1985,$ $\Sigma (x_i - \overline{x})^2 = 475$

2.8   Determine a 95.4% error margin for the estimation of μ in each of the three cases in Exercise 2.7.

2.9   Data on the average weekly earnings were obtained from a survey of 50 nonsupervisory production workers in the mining industry. The sample mean and standard deviation were found to be $630 and $35, respectively. Estimate the true mean weekly earnings and determine the 95% error margin.

2.10   Fifty-eight trout caught in a lake had average weight $= 4.37$ pounds and standard deviation $= 1.61$ pounds. From these data, estimate the mean weight of catchable trout in this lake and give a 90% error margin.

2.11   For each case, determine the sample size n that is required for estimating the population mean. The population standard deviation σ and the desired error margin are specified.

   (a) $\sigma = 3.8$,   95% error margin = .75
   (b) $\sigma = 125$,   80% error margin = 4.5
   (c) $\sigma = .092$,   98% error margin = .025

2.12   Referring to Exercise 2.9, suppose that the survey of 50 workers was, in fact, a pilot study intended to give an idea of the population standard deviation. Assuming $\sigma = \$35$, determine the sample size that is needed for estimating the population mean weekly earnings with a 98% error margin of $3.50.

2.13   Assume that the standard deviation of the heights of five-year-old boys is 3.5 inches. How many five-year-old boys need to be sampled if we want to be 90% sure that the population mean height is estimated within .5 inch?

2.14   Let the abbreviation PSLT stand for the percent of the gross family income that goes into paying state and local taxes. Suppose one wants to estimate the mean PSLT for the population of all families in New York City with gross incomes in the range $35,000 to $40,000. If $\sigma = 2.5$, how many such families should be surveyed if one wants to be 90% sure of being able to estimate the true mean PSLT within .5?

3. CONFIDENCE INTERVAL FOR A POPULATION MEAN

For point estimation, a single number lies in the forefront even though a standard error is attached. Instead, it is often more desirable to produce an interval of values that is likely to contain the true value of the parameter.

Ideally, we would like to be able to collect a sample and then use it to calculate an interval that would definitely contain the true value of the parameter. This goal, however, is not achievable because of sample-to-sample variation. Instead, we insist that, before sampling, the proposed interval will contain the true value with a specified high probability. This probability, called the **level of confidence**, is typically taken as .90, .95, or .99.

To develop this concept, we first confine our attention to the construction of a confidence interval for a population mean μ, assuming that the population is normal and the standard deviation σ is *known*. This restriction helps to simplify the initial presentation of the concept of a confidence interval. Later on, we will treat the more realistic case where σ is also unknown.

A probability statement about $\overline{X}$ based on the normal distribution provides the cornerstone for the development of a confidence interval. From Chapter 7, recall that when the population is normal, the distribution of $\overline{X}$ is also normal. It has mean μ and standard deviation $\sigma/\sqrt{n}$. Here μ is unknown, but $\sigma/\sqrt{n}$ is a known number because the sample size n is known and we have assumed that σ is known.

The normal table shows that the probability is .05 that a normal random variable will lie within 1.96 standard deviations from its mean. For $\overline{X}$, we then have

$$P\left[\mu - 1.96\,\frac{\sigma}{\sqrt{n}} < \overline{X} < \mu + 1.96\,\frac{\sigma}{\sqrt{n}}\right] = .95$$

as shown in Figure 3.

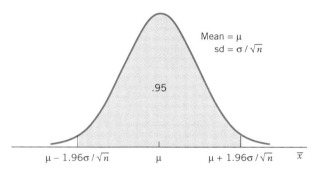

Figure 3. Normal distribution of $\overline{X}$

Now, the relation

$$\mu - 1.96\,\frac{\sigma}{\sqrt{n}} < \overline{X} \qquad \text{is the same as} \qquad \mu < \overline{X} + 1.96\,\frac{\sigma}{\sqrt{n}}$$

and

$$\overline{X} < \mu + 1.96\,\frac{\sigma}{\sqrt{n}} \qquad \text{is the same as} \qquad \overline{X} - 1.96\,\frac{\sigma}{\sqrt{n}} < \mu$$

as we can see by transposing $1.96\sigma/\sqrt{n}$ from one side of each inequality to the other. Therefore, the event

$$\left[\mu - 1.96\,\frac{\sigma}{\sqrt{n}} < \overline{X} < \mu + 1.96\,\frac{\sigma}{\sqrt{n}}\right]$$

is equivalent to

$$\left[\overline{X} - 1.96\,\frac{\sigma}{\sqrt{n}} < \mu < \overline{X} + 1.96\,\frac{\sigma}{\sqrt{n}}\right]$$

In essence, both events state that the difference $(\overline{X} - \mu)$ lies between $-1.96\sigma/\sqrt{n}$ and $1.96\sigma/\sqrt{n}$. Thus, the probability statement

$$P\left[\mu - 1.96 \frac{\sigma}{\sqrt{n}} < \overline{X} < \mu + 1.96 \frac{\sigma}{\sqrt{n}} \right] = .95$$

can also be expressed as

$$P\left[\overline{X} - 1.96 \frac{\sigma}{\sqrt{n}} < \mu < \overline{X} + 1.96 \frac{\sigma}{\sqrt{n}} \right] = .95$$

This second form tells us that, before we sample, the random interval from $\overline{X} - 1.96\sigma/\sqrt{n}$ to $\overline{X} + 1.96\sigma/\sqrt{n}$ will include the unknown parameter μ with a probability of .95. Because σ is assumed to be known, both the upper and lower endpoints can be computed as soon as the sample data are available. Guided by the above reasonings, we say that the interval

$$\left(\overline{X} - 1.96 \frac{\sigma}{\sqrt{n}}, \quad \overline{X} + 1.96 \frac{\sigma}{\sqrt{n}} \right)$$

or its realization $(\overline{x} - 1.96\sigma/\sqrt{n}, \overline{x} + 1.96\sigma/\sqrt{n})$ is a **95% confidence interval** for μ when the population is normal and σ known.

Example 4   Calculation of a confidence interval
Given a random sample of 25 observations from a normal population for which μ is unknown and $\sigma = 8$, the sample mean is found to be $\overline{x} = 42.7$. Construct a 95% confidence interval for μ.

SOLUTION AND DISCUSSION   The population is normal, and the observed value of the sample mean is $\overline{x} = 42.7$.

$$\left(42.7 - 1.96 \frac{8}{\sqrt{25}}, \quad 42.7 + 1.96 \frac{8}{\sqrt{25}} \right) = (39.6, \quad 45.8)$$

is a 95% confidence interval for μ. Since μ is unknown, we do not know if μ lies in this interval. ∎

Referring to the confidence interval obtained in Example 4, we must **not** speak of the probability of the fixed interval $(39.6, 45.8)$ covering the true mean μ. The particular interval $(39.6, 45.8)$ either does or does not cover μ, and we will never know which is the case.

We need not always tie our discussion of confidence intervals to the choice of a 95% level of confidence. An investigator may wish to specify a different high probability. We denote this probability by $1 - \alpha$ and speak of a $100(1 - \alpha)\%$ confidence interval. The only change is to replace 1.96 with $z_{\alpha/2}$, where $z_{\alpha/2}$ denotes the upper $\alpha/2$ point of the standard normal distribution (i.e., the area to the right of $z_{\alpha/2}$ is $\alpha/2$, as shown in Figure 2).

In summary, when the population is normal and σ is known, a $100(1 - \alpha)\%$ confidence interval for μ is given by

$$\left(\bar{X} - z_{\alpha/2} \frac{\sigma}{\sqrt{n}}, \qquad \bar{X} + z_{\alpha/2} \frac{\sigma}{\sqrt{n}} \right)$$

INTERPRETATION OF CONFIDENCE INTERVALS

To better understand the meaning of a confidence statement, we use the computer to generate repeated samples from a normal distribution with $\mu = 100$ and $\sigma = 10$. Ten samples of size 7 are selected, and a 95% confidence interval $\bar{x} \pm 1.96 \times 10/\sqrt{7}$ is computed from each. For the first sample, $\bar{x} = 104.3$ and the interval is 104.3 ± 7.4 or 96.9 to 111.7. This and the other intervals are illustrated in Figure 4, where each vertical line segment represents one confidence interval. The midpoint of a line is the observed value of $\bar{X}$ for that particular sample. Also note that all the intervals are of the same length $2 \times 1.96\sigma/\sqrt{n} = 14.8$. Of the 10 intervals shown, 9 cover the true value of μ. This is not surprising. The specified probability .95 represents the long-run relative frequency of the intervals $\bar{x} \pm 1.96 \times 10/\sqrt{7}$ covering the true $\mu = 100$.

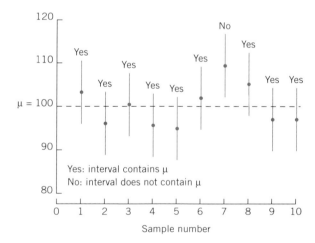

Figure 4. Interpretation of the confidence interval for μ

Because confidence interval statements are the most useful way to communicate information obtained from a sample, certain aspects of their formulation merit special emphasis. Stated in terms of a 95% confidence interval for μ, these are

1. Before we sample, a confidence interval $(\bar{X} - 1.96\sigma/\sqrt{n}, \bar{X} + 1.96\sigma/\sqrt{n})$ is a random interval that attempts to cover the true value of the parameter μ.

2. The probability

$$P\left[\overline{X} - 1.96\,\frac{\sigma}{\sqrt{n}} < \mu < \overline{X} + 1.96\,\frac{\sigma}{\sqrt{n}} \right] = .95$$

interpreted as the long-run relative frequency over many repetitions of sampling asserts that about 95% of the intervals will cover μ.

3. Once $\bar{x}$ is calculated from an observed sample, the interval

$$\left(\bar{x} - 1.96\,\frac{\sigma}{\sqrt{n}},\ \ \bar{x} + 1.96\,\frac{\sigma}{\sqrt{n}} \right),$$

which is a realization of the random interval, is presented as a 95% confidence interval for μ. A numerical interval having been determined, it is no longer sensible to speak about the probability of its covering a fixed quantity μ.

4. In any application we never know whether the 95% confidence interval covers the unknown mean μ. Relying on the long-run relative frequency of coverage in property 2, we adopt the terminology **confidence** once the interval is calculated.

At this point, one might protest, "I have only one sample and I am not really interested in repeated sampling." But if the confidence estimation techniques presented in this text are mastered and followed each time a problem of interval estimation arises, then over a lifetime approximately 95% of the intervals will cover the true parameter. Of course, this is contingent on the validity of the assumptions underlying the techniques. Here we require independent normal observations.

LARGE SAMPLE CONFIDENCE INTERVALS FOR μ

Having established the basic concepts underlying confidence interval statements, we now turn to the more realistic situation for which the population standard deviation σ is unknown. We require the sample size n to be large to dispense with the assumption of a normal population. The central limit theorem then tells us that $\overline{X}$ is nearly normal whatever the form of the population. Referring to the normal distribution of $\overline{X}$ in Figure 5 and the discussion accompanying Figure 2, we again have the probability statement

$$P\left[\overline{X} - z_{\alpha/2}\,\frac{\sigma}{\sqrt{n}} < \mu < \overline{X} + z_{\alpha/2}\,\frac{\sigma}{\sqrt{n}} \right] = 1 - \alpha$$

(Strictly speaking, this probability is approximately $1 - \alpha$ for a nonnormal population.) Even though the interval

$$\left(\overline{X} - z_{\alpha/2}\,\frac{\sigma}{\sqrt{n}},\ \ \ \overline{X} + z_{\alpha/2}\,\frac{\sigma}{\sqrt{n}} \right)$$

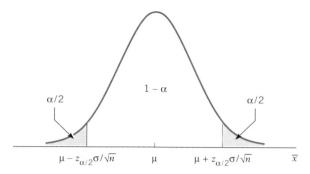

Figure 5. Normal distribution of $\overline{X}$

will include μ with the probability $1 - \alpha$, it does not serve as a confidence interval because it involves the unknown quantity σ. However, because n is large, replacing $\sigma/\sqrt{n}$ with its estimator $S/\sqrt{n}$ does not appreciably affect the probability statement. Summarizing, we find that the large sample confidence interval for μ has the form

$$\text{Estimate} \pm (\text{Tabled } z\text{-value})(\text{estimated standard error})$$

Large Sample Confidence Interval for μ

When n is large, a $100(1 - \alpha)\%$ confidence interval for μ is given by

$$\left(\overline{X} - z_{\alpha/2} \frac{S}{\sqrt{n}}, \quad \overline{X} + z_{\alpha/2} \frac{S}{\sqrt{n}} \right)$$

where S is the sample standard deviation.

Example 5   Confidence intervals for mean weekly earnings

To estimate the average weekly income of restaurant waiters and waitresses in a large city, an investigator collects weekly income data from a random sample of 75 restaurant workers. The mean and the standard deviation are found to be $227 and $15, respectively. Compute (a) 90% and (b) 80% confidence intervals for the mean weekly income.

SOLUTION AND   The sample size is $n = 75$ so a normal approximation for the distribution of the
DISCUSSION   sample mean $\overline{X}$ is appropriate. From the sample data, we know that

$$\bar{x} = \$227 \quad \text{and} \quad s = \$15$$

(a) With $1 - \alpha = .90$, we have $\alpha/2 = .05$, and $z_{\alpha/2} = 1.645$

$$1.645 \frac{s}{\sqrt{n}} = \frac{1.645 \times 15}{\sqrt{75}} = 2.85$$

Hence, a 90% confidence interval for the population mean μ becomes

$$\left(\bar{x} - 1.645 \frac{s}{\sqrt{n}}, \quad \bar{x} + 1.645 \frac{s}{\sqrt{n}} \right) = (227 - 2.85, \quad 227 + 2.85)$$

or approximately (224,   230)

This means that the investigator is 90% confident the mean income μ is in the interval of \$224 to \$230. That is, 90% of the time random samples of 75 waiters and waitresses would produce intervals $\bar{x} \pm 1.645 \, s/\sqrt{75}$ that contain μ.

(b) With $1 - \alpha = .80$, we have $\alpha/2 = .10$, and $z_{.10} = 1.28$

$$z_{.10} \frac{s}{\sqrt{n}} = \frac{1.28 \times 15}{\sqrt{75}} = 2.22$$

Hence, an 80% confidence interval for μ becomes

$$(227 - 2.22, \quad 227 + 2.22) \quad \text{or} \quad (225, \quad 229) \quad \text{dollars}$$

Comparing the two results, we note that the 80% confidence interval is shorter than the 90% interval. A shorter interval seems to give a more precise location for μ, but suffers from a lower long-run frequency of being correct. ■

Example 6   A confidence interval for mean time to complete test
In Madison, Wisconsin, recruits for the fire department must complete a timed test that simulates working conditions. It includes placing a ladder against a building, pulling out a section of fire hose, dragging a weighted object, and crawling in a simulated attic environment. The times, in seconds, for recruits to complete the test for Madison firefighters are

425	389	380	421	438	331	368	417	403	416	385	315
427	417	386	386	378	300	321	286	269	225	268	317
287	256	334	342	269	226	291	280	221	283	302	308
296	266	238	286	317	276	254	278	247	336	296	259
270	302	281	228	317	312	327	288	395	240	264	246
294	254	222	285	254	264	277	266	228	347	322	232
365	356	261	293	354	236	285	303	275	403	268	250
279	400	370	399	438	287	363	350	278	278	234	266
319	276	291	352	313	262	289	273	317	328	292	279
289	312	334	294	297	304	240	303	255	305	252	286
297	353	350	276	333	285	317	296	276	247	339	328
267	305	291	269	386	264	299	261	284	302	342	304
336	291	294	323	320	289	339	292	373	410	257	406
374	268										

Obtain a 95% confidence interval for the mean time of recruits who complete the test.

SOLUTION AND DISCUSSION A computer calculation gives

$$
\begin{array}{ll}
\text{SAMPLE SIZE} & 158 \\
\text{MEAN} & 307.77 \\
\text{STD DEV} & 51.852
\end{array}
$$

Since $1 - \alpha = .95$, $\alpha/2 = .025$, and $z_{.025} = 1.96$, the large sample 95% confidence interval for μ becomes

$$
\left(\bar{x} - z_{\alpha/2} \frac{s}{\sqrt{n}}, \quad \bar{x} + z_{\alpha/2} \frac{s}{\sqrt{n}} \right)
$$

$$
= \left(307.77 - 1.96 \frac{51.852}{\sqrt{158}}, \quad 307.77 + 1.96 \frac{51.852}{\sqrt{158}} \right)
$$

or

$$
(299.69, \quad 315.86) \text{ seconds}
$$

When the sample size is large, the sample also contains information on the shape of distribution that can be elicited by graphical displays. Figure 6 gives the stem-and-leaf display, with the data rounded to two places, accompanied by the boxplot. The confidence interval pertains to the mean of a population with a long right-hand tail.

STEM–AND–LEAF OF TIME N = 158
LEAF UNIT = 10

```
2 | 2222223333
2 | 44444555555555
2 | 666666666666666677777777777777
2 | 88888888888888899999999999999999
3 | 000000000001111111111
3 | 222222233333333
3 | 444555555
3 | 6667777
3 | 88888899
4 | 00001111
4 | 22233
```

Figure 6. A stem-and-leaf display and boxplot give more information about the form of the population.   ∎

Exercises

3.1 Determine a 90% confidence interval for μ if $n = 48$, $\bar{x} = 86.5$, and $s = 7.9$.

3.2 Determine a 99% confidence interval for μ if $n = 120$, $\bar{x} = .816$, and $s = .061$.

3.3 An experimenter always calculates 90% confidence intervals for a mean. After 200 applications, about how many of these intervals would actually cover the respective means? Explain.

3.4 A forester measures 100 needles off a pine tree and finds $\bar{x} = 3.1$ centimeters and $s = .7$ centimeter. She reports that a 95% confidence interval for the mean needle length is

$$3.1 - 1.96 \frac{.7}{\sqrt{100}} \quad \text{to} \quad 3.1 + 1.96 \frac{.7}{\sqrt{100}} \quad \text{or} \quad (2.96, \quad 3.24)$$

(a) Is the statement correct?

(b) Does the interval $(2.96, 3.24)$ cover the true mean? Explain.

3.5 In a study on the nutritional qualities of fast foods, the amount of fat was measured for a random sample of 35 hamburgers of a particular restaurant chain. The sample mean and standard deviation were found to be 30.2 and 3.8 grams, respectively. Use these data to construct a 95% confidence interval for the mean fat content in hamburgers served in these restaurants.

3.6 In the same study described in Exercise 3.5, the sodium content was also measured for the sampled hamburgers, and the sample mean and standard deviation were 658 and 47 milligrams, respectively. Determine a 98% confidence interval for the true mean sodium content.

3.7 From a random sample of 70 high school seniors in a large school district, the mean and standard deviation of the verbal scores in the Scholastic Assessment Test (SAT) are found to be 433 and 47, respectively. Based on this sample, construct a 98% confidence interval for the mean verbal score in the SAT for the population of all seniors in this school district.

3.8 Refer to Exercise 3.7. The sample mean and standard deviation of the math scores in the SAT are found to be 496 and 75, respectively. Determine a 95% confidence interval for the mean math score of all seniors in the school district.

3.9 Based on a survey of 170 employed persons in a city, the mean and standard deviation of the commuting distances between home and the principal place of business are found to be 9.8 and 4.3 miles, respectively. Determine a 90% confidence interval for the mean commuting distance for the population of all employed persons in the city.

3.10 Referring to Exercise 2.9, determine a 98% confidence interval for the mean weekly earnings of nonsupervisory production workers in the mining industry.

3.11 Referring to Exercise 2.10, determine a 95% confidence interval for the mean weight of trout.

3.12 In a study to determine whether a certain stimulant produces hyperactivity, 55 mice were injected with 10 micrograms of the stimulant. Afterward, each mouse is given a hyperactivity rating score. The mean score was $\bar{x} = 14.9$ and $s = 2.8$. Give a 95% confidence interval for μ = population mean score.

3.13 Refer to the 40 numbers of bee stings given in Table 1 and their summary statistics reported in Example 1. Calculate a 99% confidence interval for the population mean number of stings per year.

3.14 Radiation measurements on a sample of 65 microwave ovens produced $\bar{x} = .11$ and $s = .06$. Determine a 95% confidence interval for the mean radiation.

3.15 With a random sample of size $n = 144$, someone proposes

$$(\bar{X} - .12\,S, \qquad \bar{X} + .12\,S)$$

to be a confidence interval for μ. What then is the level of confidence?

3.16 By measuring the heights of 62 six-year-old girls selected at random, someone has determined that a 95% confidence interval for the population mean height μ of six-year-old girls is (42.2 inches, 46.1 inches).
  Answer the following questions with "Yes," "No," or "Can't tell" and justify your answer.
  (a) Does the population mean lie in the interval (42.2, 46.1)?
  (b) Does the sample mean lie in the interval (42.2, 46.1)?
  (c) For a future sample of 62 six-year-old girls, will the sample mean lie in the interval (42.1, 46.1)?
  (d) Do 95% of the sample data lie in the interval (42.2, 46.1)?
  (e) For a greater confidence, say, 99%, will the confidence interval calculation from the same data produce an interval narrower than (42.2, 46.1)?

4. TESTING HYPOTHESES CONCERNING A POPULATION MEAN

Broadly speaking, the goal of testing statistical hypotheses is to determine whether a claim or conjecture about some feature of the population, a parameter, is strongly supported by the information obtained from the sample data. Here we illustrate the testing of hypotheses concerning a population mean μ. The available data will be assumed to be a random sample of size n from the population of interest. Further, the sample size n will be large ($n > 30$ for a rule of thumb).

  The formulation of a hypotheses testing problem and then the steps for solving it require a number of definitions and concepts. We will introduce these key statistical concepts:

Null hypothesis and the alternative hypothesis
Type I and Type II errors
Level of significance
Rejection region
P-value

in the context of a specific problem to help integrate them with intuitive reasoning.

PROBLEM: *Can an upgrade reduce the mean transaction time at automated teller machines?* At peak periods, customers are subject to unreasonably long waits before receiving cash. To help alleviate this difficulty, the bank wants to reduce the time it takes a customer to complete a transaction. From extensive records, it is found that the transaction times have a distribution with mean = 270 and standard deviation = 24 seconds. The teller machine vendor suggests that a new software and hardware upgrade will reduce the mean time for a customer to complete a transaction. For experimental verification, a random sample of 38 transaction times will be taken at a machine with the upgrade and the sample mean $\overline{X}$ calculated. How should the result be used toward a statistical validation of the claim that the true (population) mean transaction time is less than 270 seconds?

Whenever we seek to establish a claim or conjecture on the basis of strong support from sample data, the problem is called one of hypothesis testing.

FORMULATING THE HYPOTHESES

In the language of statistics, the claim or the research hypothesis that we wish to establish is called the alternative hypothesis H_1. The opposite statement, one that nullifies the research hypothesis, is called the null hypothesis H_0. The word "null" in this context means that the assertion we are seeking to establish is actually void.

Formulation of H_0 and H_1

When our goal is to establish an assertion with substantive support obtained from the sample, the negation of the assertion is taken to be the null hypothesis H_0 and the assertion itself is taken to be the alternative hypothesis H_1.

Our initial question, "Is there strong evidence in support of the claim?" now translates to, "Is there strong evidence for rejecting H_0?" The first version typically appears in the statement of a practical problem, whereas the second version is ingrained in the conduct of a statistical test. It is crucial to understand the correspondence between the two formulations of a question.

Before claiming that a statement is established statistically, adequate evidence from data must be produced to support it. A close analogy can be made to a court trial where the jury clings to the null hypothesis of "not guilty" unless there is convincing evidence of guilt. The intent of the hearings is to establish the assertion that the accused is guilty, rather than to prove that he or she is innocent.

	Court Trial	Testing Statistical Hypothesis
Requires strong evidence to establish:	Guilt	Conjecture (research hypothesis)
Null hypothesis (H_0):	Not guilty	Conjecture is false.
Alternative hypothesis (H_1):	Guilty	Conjecture is true.
Attitude:	Uphold "not guilty" unless there is a strong evidence of guilt	Retain the null hypothesis unless it makes the sample data very unlikely to happen.

False rejection of H_0 is a more serious error than failing to reject H_0 when H_1 is true.

Once H_0 and H_1 are formulated, our goal is to analyze the sample data to choose between them.

A decision rule, or a test of the null hypothesis, specifies a course of action by stating what sample information is to be used and how it is to be used in making a decision. Bear in mind that we are to make one of the following two decisions:

Decisions

Either

       reject H_0 and conclude that H_1 is substantiated

or

       retain H_0 and conclude that H_1 fails to be substantiated

---

Rejection of H_0 amounts to saying that H_1 is substantiated, whereas nonrejection or retention of H_0 means that H_1 fails to be substantiated. A key point is that a decision to reject H_0 must be based on strong evidence. Otherwise, the claim H_1 could not be established beyond a reasonable doubt.

In our problem of evaluating the upgraded teller machine, let μ be the population mean transaction time. Because μ is claimed to be lower than 270 seconds, we formulate the alternative hypothesis as $H_1: \mu < 270$. According to the description of the problem, the researcher does not care to distinguish between the situations that $\mu = 270$ and $\mu > 270$ for the claim is false in either case. For this reason, it is customary to write the null hypothesis simply as a statement of no difference. Accordingly, we formulate the

Testing Problem
Test $H_0: \mu = 270$ versus $H_1: \mu < 270$

TEST CRITERION AND REJECTION REGION

Naturally, the sample mean $\overline{X}$, calculated from the measurements of $n = 38$ randomly selected transaction times, ought to be the basis for rejecting H_0 or not. The question now is: For what sort of values of $\overline{X}$ should we reject H_0? Because the claim states that μ is low (a left-sided alternative), only low values of $\overline{X}$ can contradict H_0 in favor of H_1. Therefore, a reasonable decision rule should be of the form

$$\text{Reject } H_0 \text{ if } \overline{X} \le c$$
$$\text{Retain } H_0 \text{ if } \overline{X} > c$$

This decision rule is conveniently expressed as $R: \overline{X} \le c$, where R stands for the rejection of H_0. Further, $R: \overline{X} \le c$ is the set of outcomes consisting of the observed values $\overline{x}$ where H_0 is rejected. This set is called the rejection region or critical region, and the cutoff point c is called the critical value.

The cutoff point c must be specified to fully describe a decision rule. To this end, we consider the case when H_0 holds, that is $\mu = 270$. Rejection of H_0 would then be a wrong decision, amounting to a false acceptance of the claim—a serious error. For an adequate protection against this kind of error, we must ensure that $P[\overline{X} \leq c]$ is very small when $\mu = 270$. For example, suppose that we wish to hold a low probability of $\alpha = .05$ for a wrong rejection of H_0. Then our task is to find the c that makes

$$P[\overline{X} \leq c] = .05 \quad \text{when } \mu = 270$$

We know that, for large n, the distribution of $\overline{X}$ is approximately normal with mean μ and standard deviation $\sigma/\sqrt{n}$, whatever the form of the underlying population. Here $n = 38$ is large, and we initially assume that σ is known. Specifically, we assume that $\sigma = 24$ seconds, the same standard deviation as with the original money machines. Then, when $\mu = 270$, the distribution of $\overline{X}$ is $N(270, 24/\sqrt{38})$ so

$$Z = \frac{\overline{X} - 270}{24/\sqrt{38}}$$

has the $N(0, 1)$ distribution.

Because $P[Z \leq -1.645] = .05$, the cutoff c on the $\overline{X}$-scale must be 1.645 standard deviations below $\mu_0 = 270$, the value of the mean specified by the null hypothesis. Therefore,

$$c = 270 - 1.645\left(\frac{24}{\sqrt{38}}\right) = 270 - 6.40 = 263.60$$

Our decision rule is now completely specified by the rejection region (see Figure 7)

$$R: \quad \overline{X} \leq 263.6$$

that has $\alpha = .05$ as the probability of wrongly rejecting H_0.

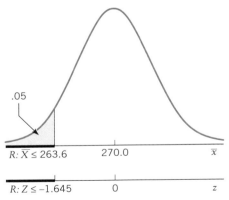

.05

$R: \overline{X} \leq 263.6$ 270.0 $\overline{x}$

$R: Z \leq -1.645$ 0 z

Figure 7. Rejection region with the cutoff $c = 263.6$

Instead of locating the rejection region on the scale of $\overline{X}$, we can cast the decision criterion on the standardized scale as well:

$$Z = \frac{\overline{X} - \mu_0}{\sigma/\sqrt{n}} = \frac{\overline{X} - 270}{24/\sqrt{38}}$$

and set the rejection region as $R: Z \le -1.645$ (see Figure 7). This form is more convenient because the cutoff -1.645 is directly read off the normal table, whereas the determination of c involves additional numerical work.

The random variable $\overline{X}$ whose value serves to determine the action is called the **test statistic**.

A **test of the null hypothesis** is a course of action specifying the set of values of a test statistic $\overline{X}$, for which H_0 is to be rejected.

This set is called the **rejection region** of the test.

A test is completely specified by a test statistic and the rejection region.

TWO TYPES OF ERROR AND THEIR PROBABILITIES

Up to this point we only considered the probability of rejecting H_0 when, in fact, H_0 is true and illustrated how a decision rule is determined by setting this probability equal to .05. The following table shows all the consequences that might arise from the use of a decision rule.

Decision Based on Sample	Unknown True Situation	
	H_0 True $\mu = 270$	H_1 True $\mu < 270$
Reject H_0	Wrong rejection of H_0 (Type I error)	Correct decision
Retain H_0	Correct decision	Wrong retention of H_0 (Type II error)

In particular, when our sample based decision is to reject H_0, we either have a correct decision (if H_1 is true) or we commit a Type I error (if H_0 is true). On the other hand, a decision to retain H_0 either constitutes a correct decision (if H_0 is true) or leads to a Type II error.

The probability of rejecting H_0, when H_0 is true, is called the **level of significance**. It is denoted by α.

Two Types of Error

Type I error: Rejection of H_0 when H_0 is true

Type II error: Nonrejection of H_0 when H_1 is true

α = Probability of making a Type I error
(also called the level of significance)

β = Probability of making a Type II error

In our problem of evaluating the upgraded teller machine, the rejection region is of the form $R: \overline{X} \leq c$; thus,

$$\alpha = P[\,\overline{X} \leq c\,] \quad \text{when } \mu = 270 \quad (H_0 \; true)$$
$$\beta = P[\,\overline{X} > c\,] \quad \text{when } \mu < 270 \quad (H_1 \; true)$$

Of course, the probability β depends on the numerical value of μ that prevails under H_1. Figure 8 shows the Type I error probability α as the shaded area under the normal curve that has $\mu = 270$ and the Type II error probability β as the shaded area under the normal curve that has $\mu = 262$, a case of H_1 being true.

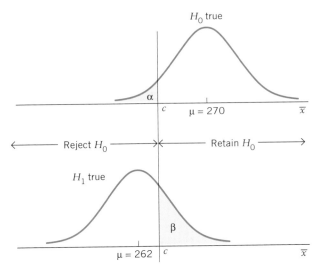

Figure 8. The error probabilities α and β

From Figure 8, it is apparent that no choice of the cutoff c can minimize both the error probabilities α and β. If c is moved to the left, α gets smaller but β gets larger, and if c is moved to the right, just the opposite effects take place. In view of this dilemma and the fact that a wrong rejection of H_0 is the more serious error, we hold α at a predetermined low level such as .10, .05, or .01 when choosing a rejection region. We will not pursue the evaluation of β, but we do note that if β turns out to be uncomfortably large, the sample size must be increased.

PERFORMING A TEST

When determining the rejection region for this example, we assumed that $\sigma = 24$ seconds, the same standard deviation as with the original money machines. Then, when $\mu = 270$, the distribution of $\overline{X}$ is $N(270, 24/\sqrt{38})$ and the rejection region $R: \overline{X} \leq 263.6$ was arrived at by fixing $\alpha = .05$ and referring

$$Z = \frac{\overline{X} - 270}{24/\sqrt{38}}$$

to the standard normal distribution.

In practice, we are usually not sure about the assumption that $\sigma = 24$, the standard deviation of the transaction times using the upgraded machine is the same as with the original teller machine. But that does not cause any problem as long as the sample size is large. When n is large ($n > 30$), the normal approximation for $\overline{X}$ remains valid even if σ is estimated by the sample standard deviation S. Therefore, for testing $H_0: \mu = \mu_0$ versus $H_1: \mu < \mu_0$ with level of significance α, we employ the test statistic

$$Z = \frac{\overline{X} - \mu_0}{S/\sqrt{n}}$$

and set the rejection region $R: Z \leq -z_\alpha$. This test is commonly called a **large sample normal test** or a **Z-test**.

To increase familiarity, we review the notation and elements for testing hypotheses. The test statistic is denoted by a capital letter to remind us that it is a random variable before the sample is chosen. Once the particular values of the observations in the sample are substituted into the formula for the test statistic, the resulting value of the test statistic is denoted by the corresponding lowercase letter. For example, when testing $H_0: \mu = \mu_0$ versus $H_1: \mu > \mu_0$ using the test statistic

$$Z = \frac{\overline{X} - \mu_0}{S/\sqrt{n}}$$

we reject H_0 in favor of H_1 for samples that result in a large observed value of Z. In other words, the set consisting of large z-values, $R: Z \geq z_{.05}$, is the rejection region for a test of level $\alpha = .05$.

Example 7   Conducting a test of hypotheses about mean transaction times

Referring to the automated teller machine transaction times, suppose that, from the measurements of a random sample of 38 transaction times, the sample mean and standard deviation are found to be 261 and 22 seconds, respectively. Test the null hypothesis $H_0: \mu = 270$ versus $H_1: \mu < 270$ using a 2.5% level of significance and state whether or not the claim ($\mu < 270$) is substantiated.

SOLUTION AND DISCUSSION Because $n = 38$ and the null hypothesis specifies that μ has the value $\mu_0 = 270$, we employ the test statistic

$$Z = \frac{\overline{X} - 270}{S/\sqrt{38}}$$

The rejection region should consist of small values of Z because H_1 is left-sided. For a 2.5% level of significance, we take $\alpha = .025$, and since $z_{.025} = 1.96$, the rejection region is (see Figure 9) $R: Z \leq -1.96$.

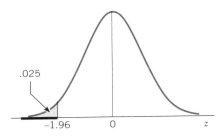

.025

−1.96   0 z

Figure 9. Rejection region for Z

With the observed values $\overline{x} = 261$ and $s = 22$, we calculate the test statistic

$$z = \frac{261 - 270}{22/\sqrt{38}} = -2.52$$

Because this observed z is in R, the null hypothesis is rejected at the level of significance $\alpha = .025$. We conclude that the claim of a reduction in the mean transaction time is strongly supported by the data. ∎

P-VALUE: HOW STRONG IS A REJECTION OF H_0?

Our test in Example 7 was based on the fixed level of significance $\alpha = .025$, and we rejected H_0 because the observed $z = -2.52$ fell in the rejection region $R: Z \leq -1.96$. A strong evidence against H_0 emerged due to the fact that a small α was used. The natural question at this point is: How small an α could we use and still arrive at the conclusion of rejecting H_0? To answer this question, we consider the observed $z = -2.52$ itself as the cutoff point (critical value) and calculate the rejection probability

$$P[Z \leq -2.52] = .0059$$

The smallest possible α that would permit rejection of H_0, on the basis of the observed $z = -2.52$, is therefore .0059 (see Figure 10). It is called the observed significance level or P-value of the observed z. This very small P-value, .0059, signifies a strong rejection of H_0 or that the result is highly statistically significant.

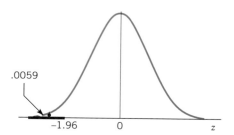

.0059

−1.96   0 z

Figure 10. P-value with left-sided rejection region

P-value is the probability, calculated under H_0, that the test statistic takes a value equal to or more extreme than the value actually observed. The P-value serves as a measure of the strength of evidence against H_0. A small P-value means that the null hypothesis is strongly rejected or the result is highly statistically significant.

Our illustrations of the basic concepts of tests of hypotheses thus far focused on a problem where the alternative hypothesis is of the form $H_1 : \mu < \mu_0$, called a left-sided alternative. If the alternative hypothesis in a problem states that the true μ is larger than its null hypothesis value of μ_0, we formulate the right-sided alternative $H_1 : \mu > \mu_0$ and use a right-sided rejection region $R: Z \geq z_\alpha$.

We illustrate the right-sided case in an example after summarizing the main steps that are involved in the conduct of a statistical test.

The Steps for Testing Hypotheses

1. **Hypotheses.**

    (a) Define the symbols you are using for parameters, in the context of the problem.

    (b) Specify the null hypothesis H_0 and the alternative hypothesis H_1 in terms of these parameters. The assertion to be established is taken as the alternative hypothesis.

    (c) Notice whether the alternative is one-sided or two-sided.

2. **Level of significance.** Use the specified level of significance α. If none is given, take $\alpha = .05$ or $.01$.

(Continued)

3. ***Test statistic.***   Specify the test statistic.
4. ***Rejection region.***
    (a) Determine the general form of the rejection region in terms of large, small, or both large and small values of the test statistic. Make sure this determination is consistent with the form of the alternative hypothesis—either two-sided or one-sided.
    (b) For the specified α, write down the rejection region. Determine the boundary value for the rejection region by using the distribution of the test statistic.
5. ***Calculation.***   Calculate the test statistic from the available data.
6. ***Conclusion.***
    (a) Determine whether or not the value of the test statistic lies in the rejection region. Then state whether or not H_0 is rejected at the specified level of significance.
    (b) Also calculate the P-value, or observed significance level, to strengthen the conclusion when H_0 is rejected.
    (c) Express the conclusion in the context of the problem, using common English.

Example 8   Evaluating a weight loss diet—Calculation of P-value

A brochure inviting subscriptions for a new diet program states that the participants are expected to lose over 22 pounds in five weeks. Suppose that, from the data of the five-week weight losses of 56 participants, the sample mean and standard deviation are found to be 23.5 and 10.2 pounds, respectively. Could the statement in the brochure be substantiated on the basis of these findings? Test, with $\alpha = .05$. Also calculate the P-value and interpret the result.

SOLUTION AND DISCUSSION

1. *Hypotheses.*   Let μ denote the mean five-week weight loss from participation in the program. Our aim is to substantiate the assertion that, on average, over 22 pounds are lost, $\mu > 22$. The assertion is taken as the alternative hypothesis so we formulate the hypotheses

$$\text{Null hypothesis } H_0: \quad \mu = 22 = \mu_0$$
$$\text{Alternative hypothesis } H_1: \quad \mu > 22$$

2. *Level of significance.*   We are given $\alpha = .05$.
3. *Test statistic.*   The sample size $n = 56$ is large, and $\mu_0 = 22$, so the test statistic is

$$Z = \frac{\overline{X} - \mu_0}{S/\sqrt{n}} = \frac{\overline{X} - 22}{S/\sqrt{56}}$$

4. *Rejection region.* The alternative hypothesis is one-sided so we reject H_0 if Z is large. For the specified $\alpha = .05$, from the normal table, $z_{.05} = 1.645$ so the one-sided rejection region is (see Figure 11)

$$R: \quad Z = \frac{\overline{X} - 22}{S / \sqrt{56}} \geq 1.645$$

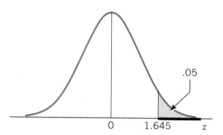

Figure 11. Right-sided rejection region with $\alpha = .05$

5. *Calculation.* We observe $\overline{x} = 23.5$ pounds and $s = 10.2$ pounds, so the observed value of the test statistic is

$$Z = \frac{\overline{X} - 22}{S / \sqrt{56}} = \frac{23.5 - 22}{10.2 / \sqrt{56}} = 1.10$$

6. *Conclusion.* The observed value of the test statistic, 1.10, does not fall in the rejection range R.

At level $\alpha = .05$, we fail to reject the null hypothesis in favor of the alternative that the mean weight loss is greater than 22 pounds.

The observed value of Z is 1.10 and larger values are more extreme evidence in favor of the alternative hypothesis. Therefore, the observed level of significance is

$$P\text{-value} = P(Z > 1.10) = .1357 \quad (\text{from the normal table})$$

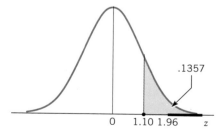

Figure 12. *P*-value with right-sided rejection region

That is, .1357 is the smallest α at which H_0 would be rejected. This is not ordinarily considered a negligible chance so we conclude that the data do not provide a strong basis for rejection of H_0.

The study data have failed to provide convincing evidence that the mean weight loss, after five weeks under the diet, is greater than 22 pounds. ■

The preceding hypotheses are called one-sided hypotheses, because the values of the parameter μ under the alternative hypothesis lie on one side of those under the null hypothesis. The corresponding tests are called one-sided tests or one-tailed tests. By contrast, we can have a problem of testing the null hypothesis

$$H_0: \quad \mu = \mu_0$$

versus the two-sided alternative

$$H_1: \quad \mu \neq \mu_0$$

Here H_0 is to be rejected if $\overline{X}$ is too far away from μ_0 in either direction—that is, if Z is too small or too large. For a level α test, we divide the rejection probability α equally between the two tails and construct the rejection region

$$R: \quad Z \leq -z_{\alpha/2} \quad \text{or} \quad Z \geq z_{\alpha/2}$$

which can also be expressed in the more compact notation

$$R: \quad |Z| \geq z_{\alpha/2}$$

Example 9   Testing hypotheses about the mean number of bee stings
Refer to the data of Example 1 consisting of 40 observations on the number of bee stings per year. A mean of 10 stings a week amounts to 520 a year. Do these data indicate that the population mean number of stings is different from 520 stings per year? Take $\alpha = .05$.

SOLUTION AND DISCUSSION

1. *Hypotheses.* Let μ denote the mean number of stings per year for the population of beekeepers. We are seeking support for $\mu \neq 520$ so the hypotheses are formulated as

$$\text{Null hypothesis } H_0: \quad \mu = 520 = \mu_0$$
$$\text{Alternative hypothesis } H_1: \quad \mu \neq 520$$

2. *Level of significance.* We are given $\alpha = .05$.

3. *Test statistic.* The sample size $n = 40$ is large, and $\mu_0 = 520$, so the test statistic is

$$Z = \frac{\overline{X} - \mu_0}{S/\sqrt{n}} = \frac{\overline{X} - 520}{S/\sqrt{40}}$$

4. *Rejection region.* The two-sided form of the alternative hypothesis dictates that the rejection region must also be two-sided. We reject H_0 if Z is large, $Z > c$, or if Z is small, $Z < -c$.

   The specified level of significance is $\alpha = .05$ so $\alpha/2 = .025$ and, from the normal table, $z_{.025} = 1.96$. The two-sided rejection region is (see Figure 13)

$$R: \quad Z = \frac{\overline{X} - 520}{S/\sqrt{40}} \geq 1.96 \quad \text{or} \quad Z = \frac{\overline{X} - 520}{S/\sqrt{40}} \leq -1.96$$

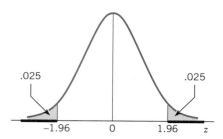

Figure 13. Two-sided rejection region with $\alpha = .05$

5. *Calculation.* From Example 1, $\overline{x} = 308.9$ stings and $s = 597.8$ stings so the observed value of the test statistic is

$$z = \frac{\overline{x} - 520}{s/\sqrt{40}} = \frac{308.9 - 520}{597.8/\sqrt{40}} = -2.23$$

6. *Conclusion.* The observed value of Z is less than -1.96 so it lies in the rejection region R. At level $\alpha = .05$, we reject the null hypothesis in favor of the alternative hypothesis $\mu \neq 520$ stings.

   In fact, the observed value $z = -2.23$ provides even stronger evidence for the rejection of H_0 than is indicated by $\alpha = .05$. How small an α could we have set and still reject H_0? This is precisely the idea behind the P-value.

   The observed value of Z is -2.23 and smaller values, as well as those greater than 2.23, are more extreme evidence in favor of the alternative hypothesis (see Figure 14). Consequently, the observed level of significance is

$$P\text{-value} = P[Z < -2.23] + P[Z > 2.23] = 2 \times .0129 = .0258$$

   With α as small as .0258, H_0 would still be rejected. This small P-value gives stronger support for rejection of H_0.

   The data provide convincing evidence that the mean number of stings per year received by beekeepers is different from 520.

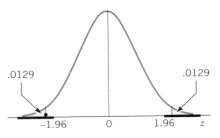

Figure 14. *P*-value with two-sided rejection region

In summary:

Large Sample Tests for μ

When the sample size is large, a Z-test of the null hypothesis $H_0 : \mu = \mu_0$ is based on the normal test statistic

$$Z = \frac{\overline{X} - \mu_0}{S/\sqrt{n}}$$

The rejection region is one- or two-sided depending on the alternative hypothesis. Specifically,

$$
\begin{aligned}
H_1 : \quad &\mu > \mu_0 &\text{requires} \quad &R: \quad Z \geq z_\alpha \\
H_1 : \quad &\mu < \mu_0 & &R: \quad Z \leq -z_\alpha \\
H_1 : \quad &\mu \neq \mu_0 & &R: \quad Z \geq z_{\alpha/2} \text{ or } Z \leq -z_{\alpha/2}
\end{aligned}
$$

Because the central limit theorem prevails for large n, no assumption is required as to the shape of the population distribution.

Exercises

4.1 Stated here are some claims or research hypotheses that are to be substantiated by sample data. In each case, identify the null hypothesis H_0 and the alternative hypothesis H_1 in terms of the population mean μ.

(a) The average mathematics score of the college-bound students in Milwaukee who participated in the American College Testing (ACT) program in 1996 was higher than 17.2.

(b)   The mean time for an airline passenger to obtain his or her luggage, once luggage starts coming out on the conveyer belt, is less than 210 seconds.

(c)   The content of fat in a name-brand chocolate ice cream is more than 4%, the amount printed on the label.

(d)   The average weight of a brand of motors is different from the manufacturer's target of 6 pounds.

4.2   From an analysis of the sample data, suppose that the decision has been to reject the null hypothesis. In the context of each part (a)–(d) of Exercise 4.1, answer the following questions:

In what circumstance is it a correct decision?

When is it a wrong decision, and what type of error is then made?

4.3   From an analysis of the sample data, suppose that the decision has been made to retain the null hypothesis. In the context of each part (a)–(d) of Exercise 4.1, answer the following questions.

In what circumstance is it a correct decision?

When is it a wrong decision, and what type of error is then made?

4.4   For each situation (a)–(d) in Exercise 4.1, state which of the following three forms of the rejection range is appropriate when σ is known.

$$R: \overline{X} \leq c \quad (\text{left-sided})$$
$$R: \overline{X} \geq c \quad (\text{right-sided})$$
$$R: \overline{X} - \mu_0 \leq -c \quad \text{or} \quad \overline{X} - \mu_0 \geq c \quad (\text{two-sided})$$

4.5   Each part of this problem gives the population standard deviation σ, the statement of a claim about μ, the sample size n, and the desired level of significance α. Formulate (i) the hypotheses, (ii) the test statistic Z, and (iii) the rejection region. [The answers to part (a) are provided.]

(a) $\sigma = 2$,   claim: $\mu > 30$, $n = 55$, $\alpha = .05$

[Answers: (i) $H_0: \mu = 30$, $H_1: \mu > 30$

$$(\text{ii})\ Z = \frac{\overline{X} - 30}{2/\sqrt{55}}$$

(iii) $R: Z \geq 1.645$]

(b) $\sigma = .085$,   claim: $\mu < .15$, $n = 150$, $\alpha = .025$
(c) $\sigma = 8.6$,   claim: $\mu \neq 75$, $n = 38$, $\alpha = .01$
(d) $\sigma = 1.62$,   claim: $\mu \neq 0$, $n = 40$, $\alpha = .06$

4.6   Suppose that the observed values of the sample mean in the contexts of parts (a)–(d) of Exercise 4.5 are given as follows. Calculate the test statistic Z and state the conclusion with the specified α.

(a) $\bar{x} = 30.54$ (b) $\bar{x} = .142$ (c) $\bar{x} = 72.35$ (d) $\bar{x} = -.79$

4.7 Suppose you are to verify the claim that $\mu > 20$ on the basis of a random sample of size 70, and you know that $\sigma = 5.6$.

(a) If you set the rejection region to be $R: \overline{X} \geq 21.31$, what is the level of significance of your test?

(b) Find the numerical value of c so that the test $R: \overline{X} \geq c$ has a 5% level of significance.

4.8 A sample of 40 sales receipts from a university bookstore has $\bar{x} = \$121$ and $s = \$10.2$. Use these values to perform a test of $H_0: \mu = 125$ against $H_1: \mu < 125$ with $\alpha = .05$.

4.9 Use the values in Exercise 4.8 to test

$$H_0: \quad \mu = 125 \quad \text{versus} \quad H_1: \quad \mu \neq 125$$

with $\alpha = .05$.

4.10 A sample of 80 measurements was taken to test the null hypothesis that the population mean equals 1100 against the alternative that it is different from 1100. The sample mean and standard deviation were found to be 1060 and 210, respectively.

(a) Perform the hypothesis test at the 1% level of significance.

(b) Calculate the P-value and interpret the result.

4.11 In a random sample of 250 observations, the mean and standard deviation are found to be 169.8 and 31.6, respectively. Is the claim that μ is larger than 169.0 substantiated by these data at the 10% level of significance?

4.12 It is claimed that a new treatment is more effective than the standard treatment for prolonging the lives of terminal cancer patients. The standard treatment has been in use for a long time, and from records in medical journals, the mean survival period is known to be 4.2 years. The new treatment is administered to 80 patients and their duration of survival recorded. The sample mean and the standard deviation are found to be 4.5 and 1.1 years, respectively. Is the claim supported by these results? Test at $\alpha = .05$. Also calculate the P-value.

4.13 Calculating from a random sample of 36 observations, one obtains the results $\bar{x} = 80.4$ and $s = 16.2$. In the context of each of the following hypothesis testing problems, determine the P-value of these results and state whether or not it signifies a strong rejection of H_0.

(a) Test $H_0: \mu = 74$ versus $H_1: \mu > 74$

(b) Test $H_0: \mu = 85$ versus $H_1: \mu < 85$

(c) Test $H_0: \mu = 76$ versus $H_1: \mu \neq 76$

4.14 In a given situation, suppose H_0 was not rejected at $\alpha = .02$. Answer the following questions as "yes," "no," or "can't tell" as the case may be.

(a) Would H_0 also be retained at $\alpha = .01$?

(b) Would H_0 also be retained at $\alpha = .05$?

(c) Is the P-value smaller than .02?

4.15   From extensive records, it is known that the duration of treating a disease by a standard therapy has a mean of 15 days. It is claimed that a new therapy can reduce the treatment time. To test this claim, the new therapy is tried on 70 patients, and from the data of their times to recovery, the sample mean and standard deviation are found to be 14.6 and 3.0 days, respectively.

(a)   Perform the hypothesis test using a 2.5% level of significance.

(b)   Calculate the P-value and interpret the result.

4.16   A company's mixed nuts are sold in cans and the label says that 25% of the contents are cashews. Suspecting that this might be an overstatement, an inspector takes a random sample of 35 cans and measures the percent weight of cashews [i.e., 100(weight of cashews / weight of all nuts)] in each can. The mean and standard deviation of these measurements are found to be 23.5 and 3.1, respectively. Do these results constitute strong evidence in support of the inspector's belief? (Answer by calculating and interpreting the P-value.)

4.17   Biological Oxygen Demand (BOD) is an index of pollution that is monitored in the treated effluent of paper mills on a regular basis. From 43 determinations of BOD (in pounds per day) at a particular paper mill, the mean and standard deviation were found to be 3246 and 757, respectively. The company had set the target that the mean BOD should be 3000 pounds per day. Do the sample data indicate that the actual amount of BOD is significantly off the target? (Use $\alpha = .05$.)

4.18   Refer to Exercise 4.17. Along with the determinations of BOD, the discharge of suspended solids (SS) was also monitored at the same site. The mean and standard deviation of the 43 determinations of SS were found to be 5710 and 1720 pounds per day, respectively. Do these results strongly support the company's claim that the true mean SS is lower than 6000 pounds per day? (Answer by calculating and interpreting the P-value.)

KEY IDEAS AND FORMULAS

The concepts of statistical inference are useful when we wish to make generalizations about a population on the basis of sample data.

Two basic forms of inference are (1) **estimation** of a population parameter and (2) **testing statistical hypotheses**.

A parameter can be estimated in two ways: by quoting (1) a single numerical value (**point estimation**), or (2) an interval of plausible values (**interval estimation**).

The standard deviation of a point estimator is also called its standard error. To be meaningful, a point estimate must be accompanied by an evaluation of its error margin.

A $100(1 - \alpha)\%$ confidence interval is an interval that, before sampling, will cover the true value of the parameter with probability $1 - \alpha$. The interval must be computable from the sample data.

If random samples are repeatedly drawn from a population and a $100(1 - \alpha)\%$ confidence interval is calculated from each, then approximately $100(1 - \alpha)\%$ of those intervals will include the true value of the parameter. We never know what happens in a single application. Our confidence draws from the success rate of $100(1 - \alpha)\%$ over many applications.

A statistical hypothesis is a statement about a population parameter.

A statement or claim that is to be established with strong support from the sample data is formulated as the alternative hypothesis (H_1). The null hypothesis (H_0) says that the claim is void.

A test is a decision rule that tells us when to reject H_0 and when not to reject H_0. A test is specified by a test statistic and a rejection region.

A wrong decision may occur in one of the two ways:

A false rejection of H_0 (Type I error)

Failure to reject H_0 when H_1 is true   (Type II error)

Errors cannot always be prevented when making a decision based on a sample. It is their probabilities that we attempt to keep small.

A Type I error is considered to be more serious. The maximum Type I error probability of a test is called its level of significance and is denoted by α.

The observed significance level or P-value of an observed test statistic is the smallest α for which this observation leads to the rejection of H_0.

Main Steps in Testing Statistical Hypotheses

1. Formulate the null hypothesis H_0 and the alternative hypothesis H_1.
2. Use the specified level or take $\alpha = .05$ or $.01$.
3. State the test statistic.
4. With a specified α, determine the rejection region.
5. Calculate the test statistic from the data.
6. Draw a conclusion: State whether or not H_0 is rejected at the specified α and interpret the conclusion in the context of the problem. Also, it is a good statistical practice to calculate the P-value and strengthen the conclusion.

The Type II error probability is denoted by β.

Inferences About a Population Mean When *n* Is Large

When n is large, we need not be concerned about the shape of the population distribution. The central limit theorem tells us that the sample mean $\overline{X}$ is nearly normally distributed with mean μ and standard deviation $\sigma/\sqrt{n}$. Moreover, $\sigma/\sqrt{n}$ can be estimated by $S/\sqrt{n}$.

Parameter of interest is

$$\mu = \text{population mean}$$

Inferences are based on

$$\overline{X} = \text{sample mean}$$

1. A point estimator of μ is the sample mean $\overline{X}$.

$$\text{Estimated standard error} = \frac{S}{\sqrt{n}}$$

$$\text{Approximate } 100(1 - \alpha)\% \text{ error margin} = z_{\alpha/2}\frac{S}{\sqrt{n}}$$

2. A $100(1 - \alpha)\%$ confidence interval for μ is

$$\left(\overline{X} - z_{\alpha/2}\frac{S}{\sqrt{n}}, \quad \overline{X} + z_{\alpha/2}\frac{S}{\sqrt{n}} \right)$$

3. To test hypotheses about μ, when $H_0: \mu = \mu_0$, the test statistic is

$$Z = \frac{\overline{X} - \mu_0}{S/\sqrt{n}}$$

where μ_0 is the value of μ that marks the boundary between H_0 and H_1. Given a level of significance α,

Reject $H_0: \mu = \mu_0$ in favor of $H_1: \mu > \mu_0$ if $Z \geq z_\alpha$
Reject $H_0: \mu = \mu_0$ in favor of $H_1: \mu < \mu_0$ if $Z \leq -z_\alpha$
Reject $H_0: \mu = \mu_0$ in favor of $H_1: \mu \neq \mu_0$ if $Z \geq z_{\alpha/2}$ or $Z \leq -z_{\alpha/2}$

5. REVIEW EXERCISES

5.1 For estimating a population mean with the sample mean $\overline{X}$, calculate (i) the standard error of $\overline{X}$ and (ii) $100(1 - \alpha)\%$ error margin in each case.

(a) $n = 43$, $\sigma = .825$, $1 - \alpha = .98$
(b) $n = 122$, $\sigma = 98$, $1 - \alpha = .80$
(c) $n = 980$, $\sigma = 55$, $1 - \alpha = .90$

5.2 Consider the problem of estimating μ based on a random sample of size n. Compute a point estimate of μ and the estimated standard error when

(a) $n = 80$, $\Sigma x_i = 752$, $\Sigma(x_i - \overline{x})^2 = 345$
(b) $n = 169$, $\Sigma x_i = 1290$, $\Sigma(x_i - \overline{x})^2 = 842$

5.3   The time it takes for a taxi to drive from the office to the airport was recorded on 40 occasions. It was found that $\bar{x} = 47$ minutes and $s = 5$ minutes. Give

(a)   An estimate of $\mu = $ population mean time to drive

(b)   An approximate 95.4% error margin

5.4   By what factor should the sample size be increased to reduce the standard error of $\bar{X}$ to

(a) $\frac{1}{2}$ its original value?

(b) $\frac{1}{4}$ its original value?

5.5   For each case, determine the sample size that is required for estimating the population mean. The population standard deviation σ and the desired error margin are specified.

(a) $\sigma = .025$,   90% error margin $= .005$

(b) $\sigma = 108$,   99% error margin $= 22$

(c) $\sigma = 6.8$,   95.4% error margin $= .5$

5.6   An investigator, interested in estimating a population mean, wants to be 95% certain that the error of estimation does not exceed 2.5. What sample size should she use if $\sigma = 18$?

5.7   A food service manager wants to be 95% certain that the error in the estimate of the mean number of sandwiches dispensed over the lunch hour is 10 or less. What sample size should be selected if a preliminary sample suggests $\sigma = 40$?

5.8   What sample size is required if $\sigma = 80$ in Exercise 5.7?

5.9   A zoologist wishes to estimate the mean blood sugar level of a species of animal when injected with a specified dosage of adrenaline. A sample of 55 animals of a common breed are injected with adrenaline, and their blood sugar measurements are recorded in units of milligrams per 100 milliliters of blood. The mean and the standard deviation of these measurements are found to be 126.9 and 10.5, respectively.

(a)   Give a point estimate of the population mean and find a 95.4% error margin.

(b)   Determine a 90% confidence interval for the population mean.

5.10   Determine a $100(1 - \alpha)\%$ confidence interval for the population mean in each of the following cases:

(a) $n = 60$, $\Sigma x_i = 752$, $\Sigma (x_i - \bar{x})^2 = 426$, $1 - \alpha = .95$

(b) $n = 150$, $\Sigma x_i = 2562$, $\Sigma (x_i - \bar{x})^2 = 3722$, $1 - \alpha = .90$

5.11   A sample of 64 measurements provide the sample mean $\bar{x} = 8.76$ and the sample standard deviation $s = 1.84$. For the population mean, construct a

(a) 90% confidence interval

(b) 99% confidence interval

5.12 After feeding a special diet to 80 mice, the scientist measures the weight in grams and obtains $\bar{x} = 35$ grams and $s = 4$ grams. He states that a 90% confidence interval for μ is given by

$$\left(35 - 1.645 \frac{4}{\sqrt{80}}, \quad 35 + 1.65 \frac{4}{\sqrt{80}} \right) \qquad \text{or} \qquad (34.26, \quad 35.74)$$

(a) Was the confidence interval calculated correctly? If not, provide the correct result.

(b) Does the interval $(34.26, 35.74)$ cover the true mean? Explain your answer.

5.13 In each case, identify the null hypothesis (H_0) and the alternative hypothesis (H_1) using the appropriate symbol for the parameter of interest.

(a) A consumer group plans to test-drive several cars of a new model to document that its average highway mileage is less than 50 miles per gallon.

(b) Subsoil water specimens will be analyzed to determine whether there is convincing evidence that the mean concentration of a chemical agent has exceeded .008.

(c) The setting of an automatic dispenser needs adjustment when the mean fill differs from the intended amount of 16 ounces. Several fills will be accurately measured to decide whether there is a need for resetting.

5.14 Suppose you are to verify the claim that $\mu < -6.5$ with a sample of size 40 and you know $\sigma = 2.2$.

(a) If you set the rejection region to be $R: \bar{X} \le -7.1$, what is the level of significance of your test?

(b) Find the numerical value of c so that the test $R: \bar{X} \le c$ has $\alpha = .10$.

5.15 Suppose you are to verify the claim that $\mu \ne 52$ with a sample of size 100 and you know $\sigma = 10.6$.

(a) If you set the rejection to be $R: \bar{X} - 52 \le -2.47$ or $\bar{X} - 52 \ge 2.47$, what is the numerical value of α?

(b) If you want the test $R: \bar{X} - 52 \le -c$ or $\bar{X} - 52 \ge c$ to have $\alpha = .05$, what must the numerical value of c be?

5.16 In a given situation, suppose H_0 was rejected at $\alpha = .05$. Answer the following questions as "yes," "no," or "can't tell" as the case may be.

(a) Would H_0 also be rejected at $\alpha = .02$?

(b) Would H_0 also be rejected at $\alpha = .10$?

(c) Is the P-value larger than .05?

5.17   Assuming that n is large in each case, write the test statistic and determine the rejection region at the given level of significance.

(a) $H_0: \mu = 20$ versus $H_1: \mu \neq 20, \quad \alpha = .05$

(b) $H_0: \mu = 30$ versus $H_1: \mu < 30, \quad \alpha = .02$

5.18   In a problem of testing $H_0: \mu = 75$ versus $H_1: \mu > 75$, the following sample quantities are recorded.

$$n = 56, \qquad \bar{x} = 77.04, \qquad s = 6.80$$

(a)   State the test statistic and find the rejection region with $\alpha = .05$.

(b)   Calculate the test statistic and draw a conclusion with $\alpha = .05$.

(c)   Find the P-value and interpret the result.

5.19   A sample of 42 measurements was taken to test the null hypothesis that the population mean equals 8.5 against the alternative that it is different from 8.5. The sample mean and standard deviation were found to be 8.79 and 1.27, respectively.

(a)   Perform the hypothesis test using .10 as the level of significance.

(b)   Calculate the P-value and interpret the result.

5.20   In a large-scale, cost-of-living survey undertaken last January, weekly grocery expenses for families with one or two children were found to have a mean of $98 and a standard deviation of $15. To investigate the current situation, a random sample of families with one or two children is to be chosen and their last week's grocery expenses are to be recorded.

(a)   How large a sample should be taken if one wants to be 95% sure that the error of estimation of the population mean grocery expenses per week for families with one or two children does not exceed $2? (Use the previous s as an estimate of the current σ.)

(b)   A random sample of 100 families is actually chosen, and from the data of their last week's grocery bills, the mean and the standard deviation are found to be $103 and $12, respectively. Construct a 98% confidence interval for the current mean grocery expense per week for the population of families with one or two children.

USING A COMPUTER

5.21   The calculation of confidence intervals and tests about μ can be conveniently done by computer. We illustrate with the blood cholesterol data given in Exercise 7.17 in Chapter 2. The data disk file C2P7-17.DAT contains the cholesterol readings. The sequence of choices,

> Stat > Basic Statistics > Descriptive Statistics

followed by

> Type *Choleste* in **Variables**. Click **Graphs**. Click **Graphical Summary**. Click **OK**. Click **OK**.

produces a rather complete summary, which includes the following.

Descriptive Statistics

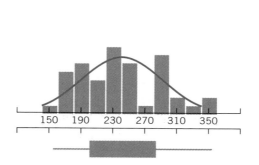

Variable:	Choleste
Mean	239.605
St. Dev	50.987
Variance	2599.67
N	43
Minimum	154.000
1st Quartile	196.000
Median	233.000
3rd Quartile	282.000
Maximum	357.000

Use the output for the cholesterol data to

(a) Determine the large sample 95% confidence interval for μ.

(b) Determine the P-value for the two-sized Z-test of $H_0 : \mu = 250$ versus $H_1 : \mu \neq 250$.

5.22 Refer to the data on number of bee stings per year, in Example 1. Use MINITAB (or some other package program) to obtain $\bar{x}$ and s.

(a) Find a 97% percent confidence interval for the mean number of stings.

(b) Test $H_0 : \mu = 550$ versus $H_1 : \mu \neq 550$ stings with $\alpha = .03$.

ACTIVITIES TO IMPROVE UNDERSTANDING: STUDENT PROJECTS

Collect a large sample of data, consisting of at least 40 observations. Obtain one data set of counts and one set of continuous measurements. For instance,

1. *Count data.*   The number of movies viewed in the past week, the number of books read in the previous month, or the number of e-mail messages received per day
2. *Continuous data.*   Length of time spent playing video games last week, time to walk to class, number of hours of study in a day

Once you formulate a question of interest concerning a population mean, you should develop a statement of purpose and continue according to the following plan.

Statement of Purpose

Plan for Data Collection

      Variable(s)

      Variable 1 _____ *discrete / continuous* _____

      Variable 2 _____ *discrete / continuous* _____

      Variable 3 _____ *discrete / continuous* _____

      Population   _____

      Sample   Specify how the sample is collected.

Calculations

Determine confidence intervals and conduct any appropriate test of hypotheses.

Summary of Findings

Interpret your tests and confidence intervals. Relate findings to the purpose of the study.

CHAPTER 9

Small Sample Inferences for Normal Populations

Chapter Objectives

After reading this chapter, you should be able to

▶ Construct a confidence interval for a normal population mean.
▶ Carry out a test of hypotheses concerning a normal population mean.
▶ Describe the relation between confidence intervals and two-sided tests of hypotheses.

1. INTRODUCTION

In Chapter 8, we discussed inferences about a population mean when a large sample is available. Those methods are deeply rooted in the central limit theorem, which guarantees that the distribution of $\overline{X}$ is approximately normal. By the versatility of the central limit theorem, we did not need to know the specific form of the population distribution.

Many investigations, especially those involving costly experiments, require statistical inferences to be drawn from small samples ($n \leq 30$, as a rule of thumb). Since the sample mean $\overline{X}$ will still be used for inferences about μ, we must address the question, "What is the sampling distribution of $\overline{X}$ when n is not large?" Unlike the large sample situation, here we do not have an unqualified answer. In fact, when n is small, the distribution of $\overline{X}$ does depend to a considerable extent on the form of the population distribution. With the central limit theorem no longer applicable, more information concerning the population is required for the development of statistical procedures. In other words, the appropriate methods of inference depend on the restrictions met by the population distribution.

In this chapter, we describe how to set confidence intervals and test hypotheses when it is reasonable to assume that the population distribution is normal.

We begin with inferences about the mean μ of a normal population.

2. STUDENT'S *t* DISTRIBUTION

When $\overline{X}$ is based on a random sample of size n from a normal $N(\mu, \sigma)$ population, we know that $\overline{X}$ is exactly distributed as $N(\mu, \sigma/\sqrt{n})$. Consequently, the standardized variable

$$Z = \frac{\overline{X} - \mu}{\sigma/\sqrt{n}}$$

has the standard normal distribution.

Because σ is typically unknown, an intuitive approach is to estimate σ by the sample standard deviation S. Just as we did in the large sample situation, we consider the ratio

$$T = \frac{\overline{X} - \mu}{S/\sqrt{n}}$$

Its probability density function is still symmetric about zero. Although estimating σ with S does not appreciably alter the distribution in large samples, it does make a substantial difference if the sample is small. The new notation T is required to distinguish it from the standard normal variable Z. In fact, this ratio is no longer standardized. Replacing σ by the sample quantity S introduces more variability in the ratio, making its standard deviation larger than 1.

The distribution of the ratio T is known in statistical literature as "Student's *t* distribution." This distribution was first studied by the British chemist W. S. Gosset, who published his work in 1908 under the pseudonym "Student." The brewery for which he worked apparently did not want the competition to know that it was using statistical techniques to better understand and improve its fermentation process.

Student's *t* Distribution

If $X_1, \ldots, X_n$ is a random sample from a normal population $N(\mu, \sigma)$ and

$$\overline{X} = \frac{1}{n} \sum X_i \quad \text{and} \quad S^2 = \frac{\sum (X_i - \overline{X})^2}{n - 1}$$

then the distribution of

$$T = \frac{\overline{X} - \mu}{S/\sqrt{n}}$$

is called **Student's *t* distribution** with $n - 1$ degrees of freedom.

The qualification "with $n - 1$ degrees of freedom" is necessary, because with each different sample size or value of $n - 1$, there is a different *t* distribution. The choice $n - 1$ coincides with the divisor for the estimator S^2 that is based on $n - 1$ degrees of freedom.

The *t* distributions are all symmetric around 0, but have tails that are more spread out than the $N(0, 1)$ distribution. However, with increasing degrees of freedom, the *t* distributions tend to look more like the $N(0, 1)$ distribution. This

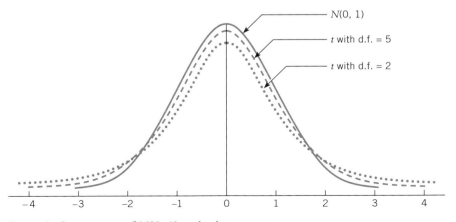

Figure 1. Comparison of $N(0, 1)$ and *t* density curves

agrees with our previous remark that for large n, the ratio

$$\frac{\overline{X} - \mu}{S/\sqrt{n}}$$

is approximately standard normal. The density curves for t with 2 and 5 degrees of freedom are plotted in Figure 1 along with the $N(0, 1)$ curve.

Appendix B, Table 4, gives the upper α points t_α for some selected values of α and the degrees of freedom (abbreviated d.f.).

The curve is symmetric about zero so the lower α point is simply $-t_\alpha$. The entries in the last row marked "d.f. = infinity" in Appendix B, Table 4, are exactly the percentage points of the $N(0, 1)$ distribution.

Example 1   Obtaining percentage points of t distributions
Using Appendix B, Table 4, determine the upper .10 point (90th percentile) of the t distribution with 5 degrees of freedom. Also find the lower .10 point (10th percentile).

SOLUTION AND   Locate .10 on the top margin of the table and the degrees of freedom on the side
DISCUSSION   margin. With d.f. = 5, the upper .10 point of the t distribution is found from Appendix B, Table 4, to be $t_{.10} = 1.476$. Since the curve is symmetric about 0, the lower .10 point is simply $-t_{.10} = -1.476$. See Figure 2.

Percentage Points of
t Distributions

d.f. $\diagdown$ α	...	.10
.	.	
.	.	
.	.	
5		$\dashrightarrow$ 1.476

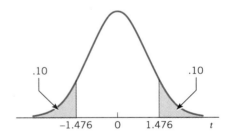

Figure 2. The upper and lower .10 points of the t distribution with d.f. = 5

Example 2   Determining a central interval having probability .9
For the t distribution with d.f. = 9, find the number b such that
$P[-b < T < b] = .90$.

SOLUTION AND   For the probability in the interval $(-b, b)$ to be .90, we must have a probability
DISCUSSION   of .05 to the right of b and, correspondingly, a probability of .05 to the left of $-b$ (see Figure 3). Thus, the number b is the upper $\alpha = .05$ point of the t distribution. Reading Appendix B, Table 4, at $\alpha = .05$ and d.f. = 9, we find $t_{.05} = 1.833$, so $b = 1.833$.

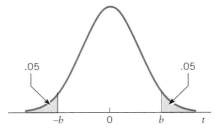

Figure 3. Finding the 5th percentile   ■

Exercises

2.1 Using the table for the *t* distributions, find
  (a) The upper .05 point when d.f. = 5
  (b) The lower .025 point when d.f. = 18
  (c) The lower .01 point when d.f. = 11
  (d) The upper .10 point when d.f. = 16

2.2 Name the *t* percentiles shown and find their values from Appendix B, Table 4.

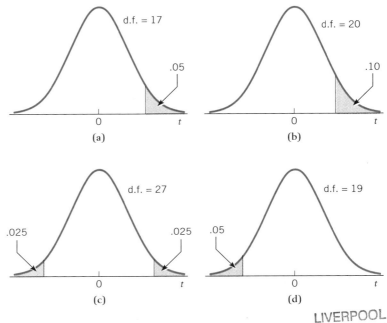

2.3 Using the table for the t distributions, find
   (a) The 90th percentile of the t distribution when d.f. = 11
   (b) The 99th percentile of the t distribution when d.f. = 3
   (c) The 5th percentile of the t distribution when d.f. = 26
   (d) The lower and upper quartiles of the t distribution when d.f. = 19

2.4 Find the probability of
   (a) $T < -1.740$ when d.f. = 17
   (b) $T > 3.143$ or $T < -3.143$ when d.f. = 6
   (c) $-1.330 < T < 1.330$ when d.f. = 18
   (d) $-1.372 < T < 2.764$ when d.f. = 10

2.5 In each case, find the number b so that
   (a) $P[T < b] = .95$ when d.f. = 6
   (b) $P[-b < T < b] = .95$ when d.f. = 15
   (c) $P[T > b] = .01$ when d.f. = 8
   (d) $P[T > b] = .99$ when d.f. = 11

2.6 Record the $t_{.05}$ values for d.f. = 5, 10, 15, 20, and 29. Does this percentile increase or decrease with increasing degrees of freedom?

2.7 Using the table for the t distributions, make an assessment for the probability of the stated event. [The answer to part (a) is provided.]
   (a) $T > 2.6$ when d.f. = 7. (Answer: $P[T > 2.6]$ is between .01 and .025 because 2.6 lies between $t_{.025} = 2.365$ and $t_{.01} = 2.998$.)
   (b) $T > 1.9$ when d.f. = 15
   (c) $T < -1.5$ when d.f. = 17
   (d) $T > 1.9$ or $T < -1.9$ when d.f. = 15
   (e) $-2.8 < T < 2.8$ when d.f. = 20
   (You can find the exact values using a computer or a statistics calculator.)

3. INFERENCES ABOUT THE MEAN—SMALL SAMPLE SIZE

CONFIDENCE INTERVAL FOR μ

The distribution of

$$T = \frac{\overline{X} - \mu}{S/\sqrt{n}}$$

provides the key for determining a confidence interval for the mean of a normal population. For a $100(1 - \alpha)\%$ confidence interval, we consult the t table (Appendix B, Table 4) and find $t_{\alpha/2}$, the upper $\alpha/2$ point of the t distribution with $n - 1$ degrees of freedom (see Figure 4).

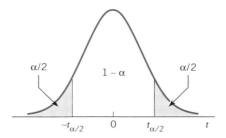

Figure 4. $t_{\alpha/2}$ and the probabilities

Since $\dfrac{\overline{X} - \mu}{S/\sqrt{n}}$ has the t distribution with d.f. $= n - 1$, we have

$$P\left[-t_{\alpha/2} < \frac{\overline{X} - \mu}{S/\sqrt{n}} < t_{\alpha/2} \right] = 1 - \alpha$$

To obtain a confidence interval, let us rearrange the terms inside the brackets so that only the parameter μ remains in the center. The above probability statement then becomes

$$P\left[\overline{X} - t_{\alpha/2}\frac{S}{\sqrt{n}} < \mu < \overline{X} + t_{\alpha/2}\frac{S}{\sqrt{n}} \right] = 1 - \alpha$$

which is precisely in the form required for a confidence statement about μ. The probability is $1 - \alpha$ that the random interval $\overline{X} - t_{\alpha/2} S/\sqrt{n}$ to $\overline{X} + t_{\alpha/2} S/\sqrt{n}$ will cover the true population mean μ. This argument is virtually the same as in Section 3 of Chapter 8. Only now the unknown σ is replaced by the sample standard deviation S, and the t percentage point is used instead of the standard normal percentage point.

A 100$(1 - \alpha)$% Confidence Interval for a Normal Population Mean

$$\left(\overline{X} - t_{\alpha/2}\frac{S}{\sqrt{n}}, \quad \overline{X} + t_{\alpha/2}\frac{S}{\sqrt{n}} \right)$$

where $t_{\alpha/2}$ is the upper $\alpha/2$ point of the t distribution with d.f. $= n - 1$.

Let us review the meaning of a confidence interval in the present context. Imagine that random samples of size n are repeatedly drawn from a normal population and the interval $(\bar{x} - t_{\alpha/2} s/\sqrt{n}, \bar{x} + t_{\alpha/2} s/\sqrt{n})$ calculated in each case. The interval is centered at $\bar{x}$ so the center varies from sample to sample. The length of an interval, $2t_{\alpha/2} \, s/\sqrt{n}$, also varies from sample to sample because it is a multiple of the sample standard deviation s. (This is unlike the fixed length situation illustrated in Figure 4 of Chapter 8, which was concerned with a known σ.) Thus, in repeated sampling, the intervals have variable centers and variable lengths. However, our confidence statement means that if the sampling is repeated many times, about $100(1 - \alpha)\%$ of the resulting intervals would cover the true population mean μ.

Figure 5 shows the results of drawing 10 samples of size $n = 7$ from the normal population with $\mu = 100$ and $\sigma = 10$. Selecting $\alpha = .05$, we find that the value of $t_{.025}$ with 6 d.f. is 2.447, so the 95% confidence interval is $\overline{X} \pm 2.447S/\sqrt{7}$. In the first sample, $\bar{x} = 103.88$ and $s = 7.96$, so the interval is $(96.52, 111.24)$. The 95% confidence intervals are shown by the vertical line segments.

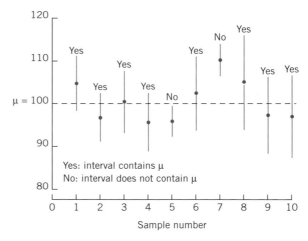

Figure 5. Behavior of confidence intervals based on the t distribution

Example 3   Interpreting a confidence interval

A new alloy has been devised for use in a space vehicle. Tensile strength measurements are made on 15 pieces of the alloy, and the mean and the standard deviation of these measurements are found to be 39.3 and 2.6, respectively.

(a)   Find a 90% confidence interval for the mean tensile strength of the alloy.

(b)   Is μ included in this interval?

SOLUTION AND DISCUSSION

(a)   If we assume that strength measurements are normally distributed, a 90% confidence interval for the mean μ is given by

$$\left(\overline{X} - t_{.05} \frac{S}{\sqrt{n}}, \quad \overline{X} + t_{.05} \frac{S}{\sqrt{n}} \right)$$

where $n = 15$ and, consequently, d.f. $= 14$. Consulting the t table, we find $t_{.05} = 1.761$. Hence, a 90% confidence interval for μ computed from the observed sample is

$$39.3 \pm 1.761 \times \frac{2.6}{\sqrt{15}} = 39.3 \pm 1.18 \quad \text{or} \quad (38.12, 40.48)$$

We are 90% confident the mean tensile strength is between 38.12 and 40.48, because 90% of the intervals calculated in this manner will contain the true mean tensile strength.

(b) We will never know whether a single realization of the confidence interval, such as $(38.12, 40.48)$, covers the unknown μ. Our confidence in the method is based on the high percentage of times μ is covered by intervals constructed from repeated samples. ∎

Remark: When repeated independent measurements are made on the same material and any variation in the measurements is basically due to experimental error (possibly compounded by nonhomogeneous materials), the normal model is often found to be appropriate. It is still necessary to graph the individual data points (not given here) in a dot diagram and normal-scores plot to reveal any wild observations or serious departures from normality. In all small sample situations, it is important to remember that the validity of a confidence interval rests on the reasonableness of the normal distribution assumed for the population. *Some departure from normality is alright, especially with moderate sample sizes.*

Recall from the previous chapter that the length of a $100(1 - \alpha)\%$ confidence interval for a normal μ is $2z_{\alpha/2}\, \sigma/\sqrt{n}$ when σ is known, whereas it is $2t_{\alpha/2}\, S/\sqrt{n}$ when σ is unknown. Given a small sample size n and consequently a small number of degrees of freedom $(n - 1)$, the extra variability caused by estimating σ with S makes the t percentage point $t_{\alpha/2}$ much larger than the normal percentage point $z_{\alpha/2}$. For instance, with d.f. $= 4$, we have $t_{.025} = 2.776$, which is considerably larger than $z_{.025} = 1.96$. Thus, when σ is unknown, the confidence estimation of μ based on a very small sample size is expected to produce a much less precise inference (namely, a wide confidence interval) compared to the situation when σ is known. With increasing n, σ can be more closely estimated by S and the difference between $t_{\alpha/2}$ and $z_{\alpha/2}$ tends to diminish.

HYPOTHESES TESTS FOR μ

The steps for conducting a test of hypotheses concerning a population mean were presented in the previous chapter. If the sample size is small, basically the same procedure can be followed provided it is reasonable to assume that the population distribution is normal. However, in the small sample situation, our test statistic

$$T = \frac{\overline{X} - \mu_0}{S/\sqrt{n}}$$

has Student's t distribution with $n - 1$ degrees of freedom.

The t table (Appendix B, Table 4) is used to determine the rejection region.

Hypotheses Tests for μ—Small Samples

To test $H_0: \mu = \mu_0$ concerning the mean of a normal population, the test statistic is

$$T = \frac{\overline{X} - \mu_0}{S/\sqrt{n}}$$

which has Student's t distribution with $n - 1$ degrees of freedom:

$$
\begin{array}{ll}
H_1: \mu > \mu_0 & R: \ T \geq t_\alpha \\
H_1: \mu < \mu_0 & R: \ T \leq -t_\alpha \\
H_1: \mu \neq \mu_0 & R: \ T \geq t_{\alpha/2} \ \text{ or } \ T \leq -t_{\alpha/2}
\end{array}
$$

The test is called a Student's t-test, or simply a t-test.

Example 4   A Student's t-test—Is the water safe?
A city health department wishes to determine whether the mean bacteria count per unit volume of water at a lake beach is within the safety level of 200. A researcher collected 10 water samples of unit volume and found the bacteria counts to be

$$175 \quad 190 \quad 207 \quad 193 \quad 184 \quad 204 \quad 205 \quad 193 \quad 196 \quad 180$$

Do the data strongly indicate that there is no cause for concern? Test with $\alpha = .05$.

SOLUTION AND DISCUSSION

1. *Hypotheses.* Let μ denote the current (population) mean bacteria count per unit volume of water. Then, the statement "no cause for concern" translates to $\mu < 200$, and the researcher is seeking strong evidence in support of this hypothesis. So the formulation of the null and alternative hypotheses should be

$$H_0: \ \mu = 200 \qquad \text{versus} \qquad H_1: \ \mu < 200$$

2. *Level of significance.* We are given $\alpha = .05$.

3. *Test statistic.* Since the counts are spread over a wide range, an approximation by a continuous distribution is not unrealistic for inference about the mean. Assuming further that the measurements constitute a sample from a normal population, we employ the *t*-test with

$$T = \frac{\overline{X} - 200}{S/\sqrt{10}}, \quad \text{d.f.} = 9$$

4. *Rejection region.* Let us perform the test at the level of significance $\alpha = .05$. Since H_1 is left-sided, we set the rejection region $T \leq -t_{.05}$. From the *t* table we find that $t_{.05}$ with d.f. $= 9$ is 1.833, so our rejection region is $R: T \leq -1.833$ (see Figure 6).

5. *Calculation.* Computations from the sample data yield

$$\overline{x} = 192.7$$
$$s = 10.81$$
$$t = \frac{192.7 - 200}{10.81/\sqrt{10}} = \frac{-7.3}{3.418} = -2.14$$

6. *Conclusion.* The observed value of T is -2.14, and smaller values are more extreme evidence in favor of the alternative hypothesis. Since, with 9 degrees of freedom, $t_{.05} = 1.833$ and $t_{.025} = 2.262$, the *P*-value is between .05 and .025. The accompanying computer output gives

$$P\text{-value} = P(T \leq -2.14) = .031 \quad (\text{see Figure 7})$$

There is strong evidence that the mean bacteria count is within the safety level.

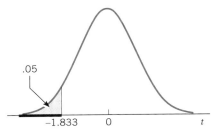

Figure 6. Left-sided rejection region $T \leq -1.833$

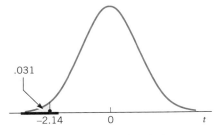

Figure 7. *P*-value for left-sided rejection region, *T* statistic

Computer Solution to Example 4

(see Exercise 3.21 for the MINITAB COMMANDS)

T-Test of the Mean

Test of mu = 200.00 vs mu < 200.00

Variable	N	Mean	StDev	SE Mean	T	P
Bac Coun	10	192.70	10.81	3.42	−2.14	0.031

Statistical Reasoning

Checking Drugstore Scanners—Practical Significance and Statistical Significance

Grocery stores, drugstores, and large superstores all use scanners to calculate a customer's bill. Scanners should be as accurate as possible. A state agency regularly monitors stores by randomly selecting a large number of items and comparing the shelf price with the checkout scanner price. One question concerns fairness. Are the overcharges balanced by the undercharges, or is the mean overcharge of all different items in the store positive?

Bill Losh/FPG International

During one check by the agency, 16 items were found to be incorrectly scanned. The amounts of overcharge were

2.00	−.99	1.00	−.50	.40	−.60	.20	.30
.50	3.00	−1.20	1.00	.50	.30	−.70	.40

A negative sign indicates an undercharge—the scanner price was below the shelf price.

We test the null hypothesis that the mean overcharge is 0 versus the alternative that it is not 0. The computer output

T-Test of the Mean

```
Test of mu = 0.000 vs mu not = 0.000

Variable   N   Mean   StDev   SE Mean      T      P
overchg    16  0.351  1.083    0.271    1.30   0.21
```

Confidence Intervals

```
Variable    N   Mean   StDev   SE Mean       95.0 % CI
overchg    16  0.351  1.083    0.271  ( −0.226,    0.928)
```

shows that the null hypothesis cannot be rejected at any reasonable level of significance.

Since the sample size is not large, our conclusion is based on the assumption that the underlying population is normal. The normal-scores plot, as described in Section 7 of Chapter 6, has a slight bend (see Figure 8), but the evidence against the normal distribution is not strong.

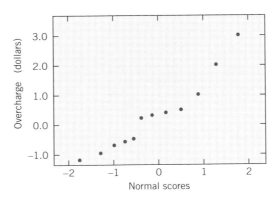

Figure 8. Normal-scores plot of overcharges

We did not reject the null hypothesis of zero mean but this does not imply we established that it is correct. The 95% confidence interval contains a mean overcharge of 90 cents, which is substantial in this application. According to the test of hypotheses, the mean overcharge is not **statistically significant** or different from zero. However, 90 cents is **practically significant**. The two concepts of significance are different.

Larger samples can detect smaller departures from the null hypothesis. With a very large sample size, we would likely find that the mean is statistically different from zero and yet the amount of difference can be ignored for all practical purposes.

Statistical significance is different from practical significance. In some situations, confidence intervals help us assess the practical significance of departures from a null hypothesis.

Exercises

3.1   A random sample of $n = 20$ from a normal population gives the sample mean = 140 and the sample standard deviation = 8.

(a)   Construct a 98% confidence interval for the population mean.

(b)   What is the length of this confidence interval? What is its center?

(c)   If a 98% confidence interval were calculated from another random sample of $n = 20$, would it have the same length as that found in part (b)? Why or why not?

3.2   Recorded here are the germination times (number of days) for seven seeds of a new strain of snap bean.

12   16   15   20   17   11   18

Stating any assumptions that you make, determine a 95% confidence interval for the true mean germination time for this strain.

3.3   The daily number of returns handled by one employee at a large mail order company were recorded for ten days. These data are given here:

28   35   32   19   48   29   30   43   36   21

Stating any assumptions you make, obtain a 95% confidence interval for the population mean number of returns.

3.4   A national sales firm monitors operators taking orders by phone. The average time for a phone call, in seconds, is used as one summary of the activity for the shift. The average talk times, for 15 persons, are

195   223   230   237   271   239   275   262
226   275   179   214   176   208   189

Stating any assumptions you make, obtain a 95% confidence interval for the population mean.

3.5   An experimenter studying the feasibility of extracting protein from seaweed to use in animal feed makes 18 determinations of the protein extract, each based on a different 50-kilogram sample of seaweed. The sample mean and the standard deviation are found to be 3.6 and .8 kilograms, respectively. Determine a 95% confidence interval for the mean yield of protein extract per 50 kilograms of seaweed.

3.6 In a lake pollution study, the concentration of lead in the upper sedimentary layer of a lake bottom is measured from 25 sediment samples of 1000 cubic centimeters each. The sample mean and the standard deviation of the measurements are found to be .38 and .06, respectively. Compute a 99% confidence interval for the mean concentration of lead per 1000 cubic centimeters of sediment in the lake bottom.

3.7 From a random sample of size 12, one has calculated the 95% confidence interval for μ and obtained the result (18.6, 26.2).

   (a) What were the $\bar{x}$ and s for that sample?

   (b) Calculate a 98% confidence interval for μ.

3.8 Suppose that with a random sample of size 18 from a normal population, one has calculated the 90% confidence interval for the population mean μ and obtained the result (122, 146). Using this result, obtain

   (a) A point estimate of μ and its 90% margin of error

   (b) A 95% confidence interval for μ

3.9 A random sample of size 20 from a normal population has $\bar{x} = 182$ and $s = 2.3$. Test $H_0: \mu = 181$ against $H_1: \mu > 181$ with $\alpha = .05$.

3.10 Referring to Exercise 3.6, test the null hypothesis $H_0: \mu = .34$ versus $H_1: \mu \neq .34$ at $\alpha = .01$, where μ denotes the population mean concentration of lead per 1000 cubic centimeters of sediment.

3.11 Do the data in Exercise 3.6 provide strong evidence that the true mean concentration of lead is higher than .35? (Answer by means of the P-value.)

3.12 Heights were measured for 12 plants grown under the treatment of a particular nutrient. The sample mean and standard deviation of those measurements were 182 and 10 inches, respectively.

   (a) Do these results substantiate the claim that the population mean height is more than 176 inches? Use $\alpha = .05$.

   (b) Construct a 95% confidence interval for the population mean.

3.13 The following measurements of the diameters (in feet) of Indian mounds in southern Wisconsin were gathered by examining reports in *Wisconsin Archeologist* (courtesy of J. Williams).

    22   24   24   30   22   20   28   30   24   34   36   15   37

   Do these data substantiate the conjecture that the population mean diameter is larger than 21 feet? Test at $\alpha = .01$.

3.14 Referring to Exercise 3.13, determine a 90% confidence interval for the population mean diameter of Indian mounds.

3.15 Measurements of the acidity (pH) of rain samples were recorded at 13 sites in an industrial region.

3.5	5.1	5.0	3.6	4.8	3.6	4.7
4.3	4.2	4.5	4.9	4.7	4.8	

Determine a 95% confidence interval for the mean acidity of rain in that region.

3.16   District court records provided data on sentencing for 19 criminals convicted of negligent homicide. The mean and standard deviation of the sentences were found to be 72.7 and 10.2 months, respectively. Determine a 95% confidence interval for the mean sentence for this crime.

3.17   Five years ago, the average size of farms in a state was 160 acres. From a recent survey of 27 farms, the mean and standard deviation were found to be 180 and 36 acres, respectively.

   (a)   Is there strong evidence that the average farm size is larger than what it was 5 years ago?

   (b)   Give a 98% confidence interval for the current average size.

3.18   The mean drying time of a brand of spray paint is known to be 90 seconds. The research division of the company that produces this paint contemplates that adding a new chemical ingredient to the paint will accelerate the drying process. To investigate this conjecture, the paint with the chemical additive is sprayed on 15 surfaces and the drying times are recorded. The mean and the standard deviation computed from these measurements are 86 and 4.5 seconds, respectively.

   (a)   Do these data provide strong evidence that the mean drying time is reduced by the addition of the new chemical?

   (b)   Construct a 98% confidence interval for the mean drying time of the paint with the chemical additive.

3.19   In a metropolitan area, the concentrations of cadmium (Cd) and zinc (Zn) in leaf lettuce were measured at six representative gardens where sewage sludge was used for fertilizer. The following measurements (in milligrams/kilogram of dry weight) were obtained.

Cd:	21	38	12	15	14	8
Zn:	140	190	130	150	160	140

   (a)   Obtain 95% confidence intervals for the mean concentrations of Cd and Zn.

   (b)   Is there strong evidence that the mean concentration of Cd is higher than 12?

3.20   A few years ago, noon bicycle traffic past a busy section of campus had a mean of $\mu = 300$ in ten minutes. To see whether any change in traffic has occurred, counts were taken for a sample of 19 weekdays. It was found that $\bar{x} = 340$ and $s = 30$.

(a) Construct an $\alpha = .05$ test of $H_0 : \mu = 300$ against the alternative that some change has occurred.

(b) Obtain a 95% confidence interval for μ.

USING A COMPUTER

3.21 The calculation of confidence intervals and tests about μ can be conveniently done by computer. We illustrate this with a small sample 2, 7, 3, 6. With MINITAB, the commands and the outputs are as follows.

Data:

C1: 2 7 3 6

Dialog box:	Session command
Stat > Basic Statistics > 1-Sample t	MTB > TINTERVAL 95 PERCENT C1
Type *C1* in **Variables**. Click **OK**.	

Output:

Confidence Intervals

	N	Mean	StDev	SE Mean	95.0 % CI
C1	4	4.50	2.38	1.19	(0.71, 8.29)

Dialog Box:	Session command
Stat > Basic Statistics > 1-Sample t	MTB > TTEST MU=2.5 C1
Type *C1* in **Variables** and Click **Test Mean**	
Type 2.5. Select Alternative **not equal** Click **OK**.	

Output:

T-Test of the Mean

Test of mu = 2.50 vs mu not = 2.50

	N	Mean	StDev	SE Mean	T	P
C1	4	4.50	2.38	1.19	1.68	0.19

(a) What is the conclusion of testing $H_0 : \mu = 2.5$ versus $H_1 : \mu \neq 2.5$ at $\alpha = .10$?

(b) Using the data in Exercise 3.4, find a 95% confidence interval for the mean talk time.

3.22 Conduct a simulation experiment to verify the long-run coverage property of the confidence intervals $\bar{x} - t_{\alpha/2} s/\sqrt{n}$ to $\bar{x} + t_{\alpha/2} s/\sqrt{n}$. Generate $n = 7$ normal observations having mean 100 and standard deviation 8. Calculate the 95% confidence interval $\bar{x} - t_{.025} s/\sqrt{7}$ to $\bar{x} + t_{.025} s/\sqrt{7}$. Repeat a large number of times. Students may combine their results. Graph the intervals as in Figure 5 and find the proportion of intervals that cover the true mean $\mu = 100$. Notice how the lengths vary.

With MINITAB, the commands

```
RANDOM 7 OBSERVATIONS IN C1-C12;
NORMAL MU = 100 AND SIGMA = 8.
TINTERVAL 95 PERCENT C1-C12
```

provide the realization of 12 confidence intervals.

	N	MEAN	STDEV	SE MEAN	95.0 PERCENT C.I.
C1	7	101.87	12.64	4.78	(90.17, 113.56)
C2	7	107.12	7.03	2.66	(100.62, 113.62)
C3	7	103.39	6.91	2.61	(96.99, 109.78)
C4	7	101.95	9.13	3.45	(93.50, 110.40)
C5	7	99.08	2.77	1.05	(96.52, 101.65)
C6	7	104.95	11.23	4.25	(94.56, 115.35)
C7	7	97.12	7.49	2.83	(90.19, 104.05)
C8	7	100.15	10.29	3.89	(90.63, 109.67)
C9	7	96.64	10.56	3.99	(86.86, 106.41)
C10	7	100.63	4.89	1.85	(96.10, 105.15)
C11	7	95.14	3.94	1.49	(91.49, 98.78)
C12	7	94.90	3.50	1.32	(91.66, 98.14)

4. THE RELATION BETWEEN TESTS AND CONFIDENCE INTERVALS

By now the careful reader should have observed a similarity between the formulas we use in testing hypotheses and in estimation by a confidence interval. To clarify the link between these two concepts, let us consider again the inferences about the mean μ of a normal population.

A $100(1 - \alpha)\%$ confidence interval for μ is

$$\left(\bar{X} - t_{\alpha/2} \frac{S}{\sqrt{n}}, \quad \bar{X} + t_{\alpha/2} \frac{S}{\sqrt{n}} \right)$$

because before the sample is taken, the probability that

$$\bar{X} - t_{\alpha/2} \frac{S}{\sqrt{n}} < \mu < \bar{X} + t_{\alpha/2} \frac{S}{\sqrt{n}}$$

is $1 - \alpha$. On the other hand, the rejection region of a level α test for $H_0: \mu = \mu_0$ versus the two-sided alternative $H_1: \mu \neq \mu_0$ is

$$R: \quad \frac{\overline{X} - \mu_0}{S/\sqrt{n}} \leq -t_{\alpha/2} \quad \text{or} \quad \frac{\overline{X} - \mu_0}{S/\sqrt{n}} \geq t_{\alpha/2}$$

Let us use the name "acceptance region" to mean the opposite (or complement) of the rejection region. Reversing the inequalities in R, we obtain

Acceptance region

$$-t_{\alpha/2} < \frac{\overline{X} - \mu_0}{S/\sqrt{n}} < t_{\alpha/2}$$

which can also be written as

Acceptance region

$$\overline{X} - t_{\alpha/2} \frac{S}{\sqrt{n}} < \mu_0 < \overline{X} + t_{\alpha/2} \frac{S}{\sqrt{n}}$$

The latter expression shows that any given null hypothesis value μ_0 will be accepted (more precisely, will not be rejected) at level α if μ_0 lies within the $100(1 - \alpha)\%$ confidence interval. Thus, having established a $100(1 - \alpha)\%$ confidence interval for μ, we know at once that all possible null hypothesis values μ_0 lying outside this interval will be rejected at level of significance α and all those lying inside will not be rejected.

Example 5   Relation between 95% confidence interval and $\alpha = .05$ test
A random sample of size $n = 9$ from a normal population produced the mean $\overline{x} = 8.3$ and the standard deviation $s = 1.2$. Obtain a 95% confidence interval for μ and also test $H_0: \mu = 8.5$ versus $H_1: \mu \neq 8.5$ with $\alpha = .05$.

SOLUTION AND DISCUSSION   A 95% confidence interval has the form

$$\left(\overline{X} - t_{.025} \frac{S}{\sqrt{n}}, \quad \overline{X} + t_{.025} \frac{S}{\sqrt{n}} \right)$$

where $t_{.025} = 2.306$ corresponds to $n - 1 = 8$ degrees of freedom. Here $\overline{x} = 8.3$ and $s = 1.2$, so the interval becomes

$$\left(8.3 - 2.306 \frac{1.2}{\sqrt{9}}, \quad 8.3 + 2.306 \frac{1.2}{\sqrt{9}} \right) = (7.4, \quad 9.2)$$

Turning now to the problem of testing $H_0: \mu = 8.5$, we observe that the value 8.5 lies in the 95% confidence interval we have just calculated. Using the correspondence between confidence interval and acceptance region, we can at once conclude that $H_0: \mu = 8.5$ should not be rejected at $\alpha = .05$. Alternatively, a formal step-by-step solution can be based on the test statistic

$$T = \frac{\overline{X} - 8.5}{S/\sqrt{n}}$$

The rejection region consists of both large and small values

Rejection region

$$\frac{\overline{X} - 8.5}{S/\sqrt{n}} \leq -2.306 \quad \text{or} \quad \frac{\overline{X} - 8.5}{S/\sqrt{n}} \geq t_{.025} = 2.306$$

Now the observed value $t = \sqrt{9}(8.3 - 8.5)/1.2 = -.5$ does not fall in the rejection region, so the null hypothesis $H_0: \mu = 8.5$ is not rejected at $\alpha = .05$. This conclusion agrees with the one we arrived at from the confidence interval. ∎

This relationship indicates how confidence estimation and tests of hypotheses with two-sided alternatives are really integrated in a common framework. A confidence interval statement is regarded as a more comprehensive inference procedure than testing a single null hypothesis, because a confidence interval statement in effect tests many null hypotheses at the same time.

Exercises

4.1 Based on a random sample of size 18 from a normal distribution, an investigator calculates the 95% confidence interval and gets the result (17.1, 29.3).

    (a) What is the conclusion of the t-test for $H_0: \mu = 19$ versus $H_1: \mu \neq 19$ at level $\alpha = .05$?

    (b) What is the conclusion if $H_0: \mu = 29.8$?

4.2 In Example 3, the 90% confidence interval for the mean tensile strength of an alloy was found to be (38.12, 40.48).

    (a) What is the conclusion of testing $H_0: \mu = 39$ versus $H_1: \mu \neq 39$ at level $\alpha = .10$?

    (b) What is the conclusion if $H_0: \mu = 42$?

4.3 Suppose that from a random sample, a 90% confidence interval for the population mean has been found to be (26.8, 29.6). Answer each question "yes," "no," or "can't tell" and justify your answer. On the basis of the same sample,

    (a) Would $H_0: \mu = 27$ be rejected in favor of $H_1: \mu \neq 27$ at $\alpha = .10$?

    (b) Would $H_0: \mu = 30$ be rejected in favor of $H_1: \mu \neq 30$ at $\alpha = .10$?

4.4 Recorded here are the amounts of decrease in percent body fat for eight participants in an exercise program over three weeks.

1.8	10.6	− 1.2	12.9	15.1	− 2.0	6.2	10.8

(a) Construct a 95% confidence interval for the population mean amount μ of decrease in percent body fat over the three-week program.

(b) If you were to test $H_0: \mu = 15$ versus $H_1: \mu \neq 15$ at $\alpha = .05$, what would you conclude from your result in part (a)?

(c) Perform the hypothesis test indicated in part (b) and confirm your conclusion.

4.5 Refer to the data in Exercise 4.4.

(a) Construct a 90% confidence interval for μ.

(b) If you were to test $H_0: \mu = 10$ versus $H_1: \mu \neq 10$ at $\alpha = .10$, what would you conclude from your result in part (a)? Why?

(c) Perform the hypothesis test indicated in part (b) and confirm your conclusion.

4.6 Establish the connection between the large sample Z-test, which rejects $H_0: \mu = \mu_0$ in favor of $H_1: \mu \neq \mu_0$, at $\alpha = .05$, if

$$Z = \frac{\overline{X} - \mu_0}{S/\sqrt{n}} \geq 1.96 \qquad \text{or} \qquad \frac{\overline{X} - \mu_0}{S/\sqrt{n}} \leq -1.96$$

and the 95% confidence interval

$$\overline{X} - 1.96 \frac{S}{\sqrt{n}} \qquad \text{to} \qquad \overline{X} + 1.96 \frac{S}{\sqrt{n}}$$

5. INFERENCES ABOUT THE STANDARD DEVIATION (THE CHI-SQUARE DISTRIBUTION)

Aside from inferences about the population mean, the population variability may also be of interest. Apart from the record of a baseball player's batting average, information on the variability of the player's performance from one game to the next may be an indicator of reliability. Uniformity is often a criterion of production quality for a manufacturing process. The quality control engineer must ensure that the variability of the measurements does not exceed a specified limit. It may also be important to ensure sufficient uniformity of the raw material for trouble-free operation of the machines.

In this section, we consider inferences for the standard deviation σ of a population under the assumption that the population distribution is normal. In contrast to the inference procedures concerning the population mean μ, the usefulness of the methods to be presented here is extremely limited when this normal assumption is violated.

To make inferences about σ^2, the natural choice of a statistic is its sample analog, which is the sample variance,

$$S^2 = \frac{\sum\limits_{i=1}^{n} (X_i - \overline{X})^2}{n - 1}$$

We take S^2 as the point estimator of σ^2 and its square root S as the point estimator of σ. To estimate by confidence intervals and to test hypotheses, we must consider the sampling distribution of S^2. To do this, we introduce a new distribution, called the χ^2 distribution (read "chi-square distribution"), whose form depends on $n - 1$.

χ^2 **Distribution**

Let $X_1, \ldots, X_n$ be a random sample from a normal population $N(\mu, \sigma)$. Then the distribution of

$$\chi^2 = \frac{\sum\limits_{i=1}^{n} (X_i - \overline{X})^2}{\sigma^2} = \frac{(n - 1)S^2}{\sigma^2}$$

is called the χ^2 distribution with $n - 1$ degrees of freedom.

Unlike the normal or t distribution, the probability density curve of a χ^2 distribution is an asymmetric curve stretching over the positive side of the line and having a long right tail. The form of the curve depends on the value of the degrees of freedom. A typical χ^2 curve is illustrated in Figure 9.

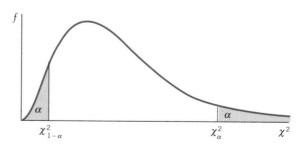

Figure 9. Probability density curve of a χ^2 distribution

Appendix B, Table 5, provides the upper α points of χ^2 distributions for various values of α and the degrees of freedom. As in both the cases of the t and the normal distributions, the upper α point χ_α^2 denotes the χ^2 value such that the area to the right is α. The lower α point or 100αth percentile, read from the column $\chi_{1-\alpha}^2$ in the table, has an area $1 - \alpha$ to the right. For example, the lower .05 point is obtained from the table by reading the $\chi_{.95}^2$ column, whereas the upper .05 point is obtained by reading the column $\chi_{.05}^2$.

Example 6   Finding percentage points of the chi-square distribution
Find the upper .05 point of the χ^2 distribution with 17 degrees of freedom. Also find the lower .05 point.

SOLUTION AND
DISCUSSION

Percentage Points of the χ^2 Distributions
(Appendix B, Table 5)

α d.f.	$\cdots$	.95	$\cdots$	.05
. . .				
17	$\dashrightarrow$	8.67	$\dashrightarrow$	27.59

The upper .05 point is read from the column labeled $\alpha = .05$. We find $\chi_{.05}^2 = 27.59$ for 17 d.f. The lower .05 point is read from the column $\alpha = .95$. We find $\chi_{.95}^2 = 8.67$, as the lower .05 point.   ∎

The χ^2 is the basic distribution for constructing confidence intervals for σ^2 or σ. We outline the steps in terms of a 95% confidence interval for σ^2. Dividing the probability $\alpha = .05$ equally between the two tails of the χ^2 distribution and using the notation just explained, we have

$$P\left[\chi_{.975}^2 < \frac{(n-1)S^2}{\sigma^2} < \chi_{.025}^2 \right] = .95$$

where the percentage points are read from the χ^2 table at d.f. $= n - 1$. Because

$$\frac{(n-1)S^2}{\sigma^2} < \chi_{.025}^2 \quad \text{is equivalent to} \quad \frac{(n-1)S^2}{\chi_{.025}^2} < \sigma^2$$

and

$$\chi_{.975}^2 < \frac{(n-1)S^2}{\sigma^2} \quad \text{is equivalent to} \quad \sigma^2 < \frac{(n-1)S^2}{\chi_{.975}^2}$$

we have

$$P\left[\frac{(n-1)S^2}{\chi_{.025}^2} < \sigma^2 < \frac{(n-1)S^2}{\chi_{.975}^2} \right] = .95$$

This last statement, concerning a random interval covering σ^2, provides a 95% confidence interval for σ^2.

A confidence interval for σ can be obtained by taking the square root of the endpoints of the interval. For a confidence level .95, the interval for σ becomes

$$\left(S\sqrt{\frac{n-1}{\chi^2_{.025}}}, \quad S\sqrt{\frac{n-1}{\chi^2_{.975}}} \right)$$

Example 7 A confidence interval for the standard deviation

A precision watchmaker wishes to learn about the variability of his product. To do this, he decides to obtain a confidence interval for σ based on a random sample of 10 watches selected from a much larger number of watches that have passed the final quality check. The deviations of these 10 watches from a standard clock are recorded at the end of one month and the following statistics calculated

$$\bar{x} = .7 \text{ second}, \qquad s = .4 \text{ second}$$

Assuming that the distribution of the measurements can be modeled as a normal distribution, find a 90% confidence interval for σ.

SOLUTION AND DISCUSSION

Here $n = 10$, so d.f. $= n - 1 = 9$. The χ^2 table gives $\chi^2_{.95} = 3.33$ and $\chi^2_{.05} = 16.92$. Using the preceding formula, we determine that a 90% confidence interval for σ^2 is

$$\left(\frac{9 \times (.4)^2}{16.92}, \quad \frac{9 \times (.4)^2}{3.33} \right) = (.085, \quad .432)$$

and the corresponding interval for σ is $(\sqrt{.085}, \sqrt{.432}) = (.29, .66)$. The watchmaker can be 90% confident that σ is between .29 and .66 second, because 90% of the intervals calculated by this procedure in repeated samples will cover the true σ. ∎

It is instructive to note that the midpoint of the confidence interval for σ^2 in Example 7 is not $s^2 = .16$, which is the best point estimate. This is in sharp contrast to the confidence intervals for μ, and it serves to accent the difference in logic between interval and point estimation.

For a test of the null hypothesis $H_0: \sigma^2 = \sigma_0^2$ it is natural to employ the statistic S^2. If the alternative hypothesis is one-sided, say $H_1: \sigma^2 > \sigma_0^2$, then the rejection region should consist of large values of S^2 or, alternatively, large values of the

$$\text{Test statistic} \quad \chi^2 = \frac{(n-1)S^2}{\sigma_0^2}, \quad \text{d.f.} = n - 1$$

The rejection region of a level α test is, therefore,

$$R: \quad \frac{(n-1)S^2}{\sigma_0^2} \geq \chi^2_\alpha, \quad \text{d.f.} = n - 1$$

For a two-sided alternative $H_1 : \sigma^2 \neq \sigma_0^2$, a level α rejection region is

$$R: \quad \frac{(n-1)S^2}{\sigma_0^2} \leq \chi_{1-\alpha/2}^2 \quad \text{or} \quad \frac{(n-1)S^2}{\sigma_0^2} \geq \chi_{\alpha/2}^2$$

Once again, we remind the reader that the inference procedures for σ presented in this section are extremely sensitive to departures from a normal population.

Exercises

5.1 Using the table for the χ^2 distributions, find
   (a) The upper 5% point when d.f. = 8
   (b) The upper 1% point when d.f. = 17
   (c) The lower 2.5% point when d.f. = 11
   (d) The lower 1% point when d.f. = 23

5.2 Name the χ^2 percentiles shown and find their values from Appendix B, Table 5.

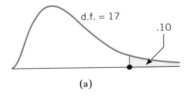

(a)

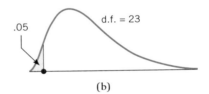

(b)

   (c) Find the percentile in figure (a) if d.f. = 40.
   (d) Find the percentile in figure (b) if d.f. = 7.

5.3 Find the probability of
   (a) $\chi^2 > 30.14$ when d.f. = 19
   (b) $\chi^2 < .83$ when d.f. = 5
   (c) $3.24 < \chi^2 < 15.99$ when d.f. = 10
   (d) $3.49 < \chi^2 < 17.53$ when d.f. = 8

5.4 Given the sample data

$$12 \quad 18 \quad 9 \quad 15 \quad 14$$

   (a) Obtain a point estimate of the population standard deviation σ.
   (b) Construct a 95% confidence interval for σ.
   (c) Examine whether or not your point estimate is located at the center of the confidence interval.

5.5   Plastic sheets produced by a machine are periodically monitored for possible fluctuations in thickness. Uncontrollable heterogeneity in the viscosity of the liquid mold makes some variation in thickness measurements unavoidable. However, if the true standard deviation of thickness exceeds 1.5 millimeters, there is cause to be concerned about the product quality. Thickness measurements (in millimeters) of 10 specimens produced on a particular shift resulted in the following data.

226   228   226   225   232   228   227   229   225   230

Do the data substantiate the suspicion that the process variability exceeded the stated level on this particular shift? (Test at $\alpha = .05$.) State the assumption you make about the population distribution.

5.6   Refer to Exercise 5.5. Construct a 95% confidence interval for the true standard deviation of the thickness of sheets produced on this shift.

5.7   From a data set of $n = 10$ observations, one has calculated the 95% confidence interval for σ and obtained the result $(.81, 2.15)$.
   (a)   What was the standard deviation s for the sample? (*Hint:*   Examine how s enters the formula of a confidence interval.)
   (b)   Calculate a 90% confidence interval for σ.

5.8   Suppose that, with a random sample of 19 observations from a normal population, one has found the 95% confidence interval for σ to be $(3.28, 6.32)$. Using this result,
   (a)   Find the 90% confidence interval for σ.
   (b)   State the conclusion for the test of $H_0: \sigma = 9$ versus $H_1: \sigma \neq 9$ at $\alpha = .05$.

5.9   Referring to Exercise 3.13, construct a 95% confidence interval for the population standard deviation of the diameters of Indian mounds.

5.10   Do the data in Exercise 3.15 substantiate the conjecture that the true standard deviation of the acidity measurements is larger than .4? Test at $\alpha = .05$.

5.11   Refer to the data in Exercise 3.19.
   (a)   Obtain a 95% confidence interval for the true standard deviation of the concentration of cadmium.
   (b)   Is there strong evidence that the true standard deviation of the concentration of zinc is less than 30? Test at $\alpha = .10$.

KEY IDEAS AND FORMULAS

When the sample size is small, additional conditions need to be imposed on the population. In this chapter, we assume normal populations.

Inferences about the mean of a normal population are based on

$$T = \frac{\overline{X} - \mu}{S / \sqrt{n}}$$

which is distributed as Student's t with $n - 1$ degrees of freedom.

Inferences about the standard deviation of a normal population are based on $(n - 1)S^2/\sigma^2$, which has a χ^2 distribution with $n - 1$ degrees of freedom.

Moderate departures from a normal population distribution do not seriously affect inferences based on t. These procedures are robust.

Nonnormality can seriously affect inferences about σ.

Inferences About a Normal Population Mean

When n is small, we assume that the population is approximately normal. Inference procedures are derived from Student's t sampling distribution of $T = \dfrac{\overline{X} - \mu}{S / \sqrt{n}}$.

1. A $100(1 - \alpha)\%$ confidence interval for μ is

$$\left(\overline{X} - t_{\alpha/2} \frac{S}{\sqrt{n}}, \quad \overline{X} + t_{\alpha/2} \frac{S}{\sqrt{n}} \right)$$

2. To test hypotheses about μ, the test statistic is

$$T = \frac{\overline{X} - \mu_0}{S / \sqrt{n}}$$

Given a level of significance α:

Reject $H_0 : \mu = \mu_0$ in favor of $H_1 : \mu > \mu_0$ if $T \geq t_\alpha$

Reject $H_0 : \mu = \mu_0$ in favor of $H_1 : \mu < \mu_0$ if $T \leq -t_\alpha$

Reject $H_0 : \mu = \mu_0$ in favor of $H_1 : \mu \neq \mu_0$ if $T \leq -t_{\alpha/2}$ or $T \geq t_{\alpha/2}$

Inferences About a Normal Population Standard Deviation

Inferences are derived from the χ^2 distribution for $(n - 1)S^2/\sigma^2$.

1. Point estimator of σ is the sample standard deviation S.
2. A 95% confidence interval for σ is

$$\left(S \sqrt{\frac{n - 1}{\chi^2_{.025}}}, \quad S \sqrt{\frac{n - 1}{\chi^2_{.975}}} \right)$$

3. To test hypotheses about σ, the test statistic is

$$\frac{(n - 1)S^2}{\sigma_0^2}$$

Given a level of significance α,

$$\left.\begin{array}{l} \text{Reject } H_0 : \sigma = \sigma_0 \\ \text{in favor of } H_1 : \sigma < \sigma_0 \end{array}\right\} \text{ if } \quad \frac{(n-1)S^2}{\sigma_0^2} \leq \chi^2_{1-\alpha}$$

$$\left.\begin{array}{l} \text{Reject } H_0 : \sigma = \sigma_0 \\ \text{in favor of } H_1 : \sigma > \sigma_0 \end{array}\right\} \text{ if } \quad \frac{(n-1)S^2}{\sigma_0^2} \geq \chi^2_{\alpha}$$

$$\left.\begin{array}{l} \text{Reject } H_0 : \sigma = \sigma_0 \\ \text{in favor of } H_1 : \sigma \neq \sigma_0 \end{array}\right\} \begin{array}{l} \text{if } \quad \frac{(n-1)S^2}{\sigma_0^2} \leq \chi^2_{1-\alpha/2} \\ \\ \text{or } \quad \frac{(n-1)S^2}{\sigma_0^2} \geq \chi^2_{\alpha/2} \end{array}$$

6. REVIEW EXERCISES

6.1 Using the table of percentage points for the t distributions, find
   (a) $t_{.05}$ when d.f. = 4
   (b) $t_{.025}$ when d.f. = 13
   (c) The lower .05 point when d.f. = 4
   (d) The lower .05 point when d.f. = 13

6.2 Using the table for the t distributions, find the probability of
   (a) $T > 2.710$ when d.f. = 23
   (b) $T < 3.355$ when d.f. = 8
   (c) $-2.110 < T < 2.110$ when d.f. = 17
   (d) $-.703 < T < 2.262$ when d.f. = 9

6.3 A t distribution assigns more probability to large values than the standard normal.
   (a) Find $t_{.05}$ for d.f. = 15 and then evaluate $P[Z > t_{.05}]$. Verify that $P[T > t_{.05}]$ is greater than $P[Z > t_{.05}]$.
   (b) Examine the relation for d.f. = 5 and 20, and comment.

6.4 Measurements of the amount of suspended solids in river water, on 15 Monday mornings, yield $\bar{x} = 47$ and $s = 9.4$. Obtain a 95% confidence interval for the mean amount of suspended solids. State any assumption you make about the population.

6.5 Determine a 99% confidence interval for μ using the data in Exercise 6.4.

6.6 The time to blossom of 21 plants has $\bar{x} = 39$ days and $s = 5.1$ days. Give a 95% confidence interval for the mean time to blossom.

6.7   Refer to Exercise 6.4. The water quality is acceptable if the mean amount of suspended solids is less than 49. Construct an $\alpha = .05$ test to establish that the quality is acceptable.

(a)   Specify H_0 and H_1 .

(b)   State the test statistic.

(c)   What does the test conclude?

6.8   Refer to Exercise 6.6. Do these data provide strong evidence that the mean time to blossom is less than 42 days? Test with $\alpha = .01$.

6.9   A random sample of 16 observations provided $\bar{x} = 182$ and $s = 12$.

(a)   Test $H_0: \mu = 190$ versus $H_1: \mu < 190$ at $\alpha = .05$. State your assumption about the population distribution.

(b)   What can you say about the P-value of the test statistic calculated in part (a)?

6.10   The supplier of a particular brand of vitamin pills claims that the average potency of these pills after a certain exposure to heat and humidity is at least 65. Before buying these pills, a distributor wants to verify the supplier's claim. To this end, the distributor will choose a random sample of 9 pills from a batch and measure their potency after the specified exposure.

(a)   Formulate the hypotheses about the mean potency μ.

(b)   Determine the rejection region of the test with $\alpha = .05$. State any assumption you make about the population.

(c)   The data are 63, 72, 64, 69, 59, 65, 66, 64, 65. Apply the test and state your conclusion.

6.11   A weight loss program advertises "LOSE 40 POUNDS IN 4 MONTHS." A random sample of $n = 25$ customers has $\bar{x} = 32$ pounds lost and $s = 12$. Test $H_0: \mu = 40$ against $H_1: \mu < 40$ with $\alpha = .05$.

6.12   A car advertisement asserts that with the new collapsible bumper system, the average body repair cost for the damages sustained in a collision impact of 10 miles per hour does not exceed $1500. To test the validity of this claim, 5 cars are crashed into a stone barrier at an impact force of 10 miles per hour and their subsequent body repair costs are recorded. The mean and the standard deviation are found to be $1620 and $90, respectively. Do these data strongly contradict the advertiser's claim?

6.13   Combustion efficiency measurements were recorded for 10 home heating furnaces of a new model. The sample mean and standard deviation were found to be 73.2 and 2.74, respectively. Do these results provide strong evidence that the average efficiency of the new model is higher than 70? (Test at $\alpha = .05$. Comment also on the P-value.)

6.14   The number of days to maturity was recorded for 25 plants grown from seeds of a single stock. The mean and standard deviation were $\bar{x} = 68.4$ days and $s = 6.5$ days.

(a) Do these results contradict the claim that the average maturity time is 65 days for this stock?

(b) Construct a 95% confidence interval for the mean maturity time.

(c) Construct a 90% confidence interval for σ.

6.15 Using the table of percentage points of the χ^2 distributions, find

(a) $\chi^2_{.05}$ with d.f. = 4

(b) $\chi^2_{.025}$ with d.f. = 25

(c) Lower .05 point with d.f. = 4

(d) Lower .025 point with d.f. = 25

6.16 Using the table for the χ^2 distributions, find

(a) The 90th percentile of χ^2 when d.f. = 12

(b) The 10th percentile of χ^2 when d.f. = 9

(c) The median of χ^2 when d.f. = 22

(d) The 5th percentile of χ^2 when d.f. = 50

6.17 Test $H_0 : \sigma = 10$ versus $H_1 : \sigma > 10$ with $\alpha = .05$ in each case.

(a) $n = 25$, $\Sigma(x_i - \bar{x})^2 = 4016$

(b) $n = 15$, $s = 12$

(c) $n = 6$ and the sample observations are 110, 126, 131, 149, 156, 165

6.18 Refer to the data of Exercise 6.13. Is there strong evidence that the standard deviation for the efficiency of the new model is below 3.0?

6.19 From a random sample of size 25 from a normal population, the 95% confidence interval for the population mean has been found to be (106.8, 115.2). Using this result,

(a) Find the value of the sample standard deviation.

(b) With $\alpha = .05$, test the null hypothesis that the population standard deviation equals 3.0 against the alternative that it exceeds 3.0.

6.20 From a random sample, a 90% confidence interval for the population standard deviation σ was found to be (8.6, 15.3). With the same data, what would be the conclusion of testing $H_0 : \sigma = 7$ versus $H_1 : \sigma \neq 7$ at $\alpha = .10$?

ACTIVITIES TO IMPROVE UNDERSTANDING: STUDENT PROJECTS

Formulate a question of interest concerning a population mean in a situation where you expect the distribution of the individual observations to follow a normal distribution.

Collect a sample consisting of at least 20 observations. The variable should preferably be continuous, although a count that is spread over many values can often reasonably be approximated by a normal distribution.

You can follow the outline of student projects from the previous chapter, except that you must add a step to check the assumption that the population is normal. This assumption is less crucial when the sample size is even moderately large. You may have to consider a transformation as discussed in Section 7 of Chapter 6.

Statement of Purpose

Plan for Data Collection

Variable _____

Population _____

Sample   Specify how the sample is collected.

Checking Normality

Normal-scores plot _____ State findings.

Check for outliers _____ State findings.

Calculations

Determine confidence intervals and conduct any appropriate test of hypotheses.

Summary of Findings

Interpret your tests and confidence intervals. Relate findings to the purpose of the study.

CHAPTER 10

Comparing Two Treatments

Chapter Objectives

After reading this chapter, you should be able to

▶ Discuss matched pair and independent sample designs.
▶ Explain how to randomize treatments.
▶ Construct a confidence interval for the trifference in two means using independent random samples.
▶ Carry out tests of hypotheses for the difference in two means using independent random samples.
▶ Construct confidence intervals and conduct tests for the mean difference using paired observations.

1. INTRODUCTION

In virtually every area of human activity, new procedures are invented and existing techniques revised. Advances occur whenever a new technique proves to be better than the old. To compare them, we conduct experiments, collect data about their performance, and then draw conclusions from statistical analyses. The manner in which sample data are collected, called an experimental design or sampling design, is crucial to an investigation. In this chapter, we introduce two experimental designs that are most fundamental to a comparative study: (1) independent samples and (2) matched pair sample.

As we shall see, the methods of analyzing the data are quite different for these two processes of sampling. First, we outline a few illustrative situations where a comparison of two methods requires statistical analysis of data.

Example 1 Experimental design issues for comparative experiments: Agricultural field trials

To ascertain whether a new strain of seeds actually produces a higher yield per acre compared to a current major variety, field trials must be performed by planting each variety under appropriate farming conditions. A record of crop yields from the field trials will form the database for making a comparison between the two varieties. It may also be desirable to compare the varieties at several geographic locations representing different climate or soil conditions. ∎

Example 2 Subject selection issues in comparative experiments: Drug evaluation

Pharmaceutical researchers strive to synthesize chemicals to improve their efficiency in curing diseases. New chemicals may result from educated guesses concerning potential biological reactions, but evaluations must be based on their effects on diseased animals or human beings. To compare the effectiveness of two drugs in controlling tumors in mice, several mice of an identical breed may be taken as experimental subjects. After infecting them with cancer cells, some will be subsequently treated with drug 1 and others with drug 2. The data of tumor sizes for the two groups will then provide a basis for comparing the drugs. When testing the drugs on human subjects, the experiment takes a different form. Artificially infecting them with cancer cells is absurd! In fact, it would be criminal. Instead, the drugs will be administered to cancer patients who are available for the study. In contrast with a pool of mice of an "identical breed," here the available subjects may be of varying conditions of general health, prognosis of the disease, and other factors. ∎

When discussing a comparative study, the common statistical term treatment is used to refer to the things that are being compared. The basic units that are exposed to one treatment or another are called experimental units or experimental subjects, and the characteristic that is recorded after the application of a treatment to a subject is called the **response**. For instance, the two treatments in Example 1 are the two varieties of seeds, the experimental subjects are the agricultural plots, and the response is crop yield.

The term **experimental design** refers to the manner in which subjects are chosen and assigned to treatments. For comparing two treatments, the two basic types of design are:

1. Independent samples (complete randomization)
2. Matched pair sample (randomization within each matched pair)

The case of independent samples arises when the subjects are randomly divided into two groups, one group is assigned to treatment 1 and the other to treatment 2. The response measurements for the two treatments are then unrelated because they arise from separate and unrelated groups of subjects. Consequently, each set of response measurements can be considered a sample from a population, and we can speak in terms of a comparison between two population distributions.

With the matched pair design, the experimental subjects are chosen in pairs so that the members in each pair are alike, whereas those in different pairs may be substantially dissimilar. One member of each pair receives treatment 1 and the other treatment 2. Example 3 illustrates these ideas.

Example 3   Matched pair versus independent samples: Comparing two drugs
To compare the effectiveness of two drugs in curing a disease, suppose that 8 patients are included in a clinical study. Here, the time to cure is the response of interest. Figure 1(a), page 400, portrays a design of independent samples where the 8 patients are randomly split into groups of 4, one group is treated with drug 1, and the other with drug 2. The observations for drug 1 will have no relation to those for drug 2 because the selection of patients in the two groups is left completely to chance.

To conduct a matched pair design, one would first select the patients in pairs. The two patients in each pair should be as alike as possible in regard to their physiological conditions; for instance, they should be of the same gender and age group and have about the same severity of the disease. These preexisting conditions may be different from one pair to another. Having paired the subjects, we randomly select one member from each pair to be treated with drug 1 and the other with drug 2. Figure 1(b) shows this matched pair design.

In contrast with the situation of Figure 1(a), here we would expect the responses within each pair to be dependent for the reason that they are governed by the same preexisting conditions of the subjects.

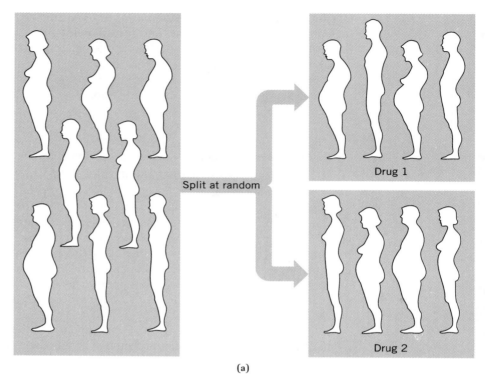

(a)

Figure 1(a). Independent samples, each of size 4

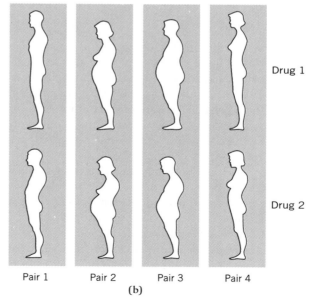

(b)

Figure 1(b). Matched pair design with four pairs of subjects

In summary, a carefully planned experimental design is crucial to a successful comparative study. The design determines the structure of the data. In turn, the design provides the key to selecting an appropriate analysis.

Exercises

1.1 Al, Bob, Carol, Dennis, and Ellen are available as subjects. Make a list of all possible ways to split them into two groups with the first group having two subjects and the second three subjects.

1.2 Six mice—Alpha, Tau, Omega, Pi, Beta, and Phi—are to serve as subjects. List all possible ways to split them into two groups with the first having 4 mice and the second 2 mice.

1.3 Six students in a psychology course have volunteered to serve as subjects in a matched pair experiment.

Name	Age	Gender
Tom	18	M
Sue	20	F
Erik	18	M
Grace	20	F
Chris	18	F
Roger	18	M

(a) List all possible sets of pairings if subjects are paired by age.

(b) If subjects are paired by gender, how many pairs are available for the experiment?

1.4 Identify the following as either matched pair or independent samples. Also identify the experimental units, treatments, and response in each case.

(a) Twelve persons are given a high-potency vitamin C capsule once a day. Another twelve do not take extra vitamin C. Investigators will record the number of colds in 5 winter months.

(b) One self-fertilized plant and one cross-fertilized plant are grown in each of 7 pots. Their heights will be measured after 3 months.

(c) Ten newly married couples will be interviewed. Both the husband and wife will respond to the question, "How many children would you like to have?"

(d) Learning times will be recorded for 5 dogs trained by a reward method and 3 dogs trained by a reward–punishment method.

2. INDEPENDENT RANDOM SAMPLES FROM TWO POPULATIONS

Here we discuss the methods of statistical inference for comparing two treatments or two populations on the basis of independent samples. Recall that with the design of independent samples, a collection of $n_1 + n_2$ subjects is randomly divided into two groups and the responses are recorded. We conceptualize population 1 as the collection of responses that would result if a vast number of subjects were given treatment 1. Similarly, population 2 refers to the population of responses under treatment 2. The design of independent samples can then be viewed as one that produces unrelated random samples from two populations (see Figure 2). In other situations, the populations to be compared may be quite real entities. For instance, one may wish to compare the residential property values in the east suburb of a city to those in the west suburb. Here the issue of assigning experimental subjects to treatments does not arise. The collection of all residential properties in each suburb constitutes a population from which a sample will be drawn at random.

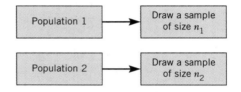

Figure 2. Independent random samples

With the design of independent samples, we obtain

Sample	Summary Statistics	
$X_1, X_2, \ldots, X_{n_1}$ from population 1	$\overline{X} = \dfrac{1}{n_1} \displaystyle\sum_{i=1}^{n_1} X_i$	$S_1^2 = \dfrac{\displaystyle\sum_{i=1}^{n_1} (X_i - \overline{X})^2}{n_1 - 1}$
$Y_1, Y_2, \ldots, Y_{n_2}$ from population 2	$\overline{Y} = \dfrac{1}{n_2} \displaystyle\sum_{i=1}^{n_2} Y_i$	$S_2^2 = \dfrac{\displaystyle\sum_{i=1}^{n_2} (Y_i - \overline{Y})^2}{n_2 - 1}$

To make confidence statements or to test hypotheses, we specify a statistical model for the data.

Statistical Model: Independent Random Samples

1. $X_1, X_2, \ldots, X_{n_1}$ is a random sample of size n_1 from population 1 whose mean is denoted by μ_1 and standard deviation by σ_1.

2. $Y_1, Y_2, \ldots, Y_{n_2}$ is a random sample of size n_2 from population 2 whose mean is denoted by μ_2 and standard deviation by σ_2.

3. The samples are independent. In other words, the response measurements under one treatment are unrelated to the response measurements under the other treatment.

We now set our goal toward drawing a comparison between the mean responses of the two treatments or populations. In statistical language, we are interested in making inferences about the parameter

$$\mu_1 - \mu_2 = (\text{Mean of population 1}) - (\text{Mean of population 2})$$

INFERENCES FROM LARGE SAMPLES

Inferences about the difference $\mu_1 - \mu_2$ are naturally based on its estimate $\overline{X} - \overline{Y}$, the difference between the sample means. When both sample sizes n_1 and n_2 are large (say, greater than 30), $\overline{X}$ and $\overline{Y}$ are each approximately normal and their difference $\overline{X} - \overline{Y}$ is approximately normal with

Mean

$$E(\overline{X} - \overline{Y}) = \mu_1 - \mu_2$$

Variance

$$\text{Var}(\overline{X} - \overline{Y}) = \frac{\sigma_1^2}{n_1} + \frac{\sigma_2^2}{n_2}$$

Standard Error

$$\text{S.E.}(\overline{X} - \overline{Y}) = \sqrt{\frac{\sigma_1^2}{n_1} + \frac{\sigma_2^2}{n_2}}$$

Note: Because the entities $\overline{X}$ and $\overline{Y}$ vary in repeated sampling and independently of each other, the distance between them becomes more variable than the individual members. This explains the mathematical fact that the variance of the difference $(\overline{X} - \overline{Y})$ equals the *sum* of the variances of $\overline{X}$ and $\overline{Y}$.

When n_1 and n_2 are both large, the normal approximation remains valid if σ_1^2 and σ_2^2 are replaced by their estimators

$$S_1^2 = \frac{\sum\limits_{i=1}^{n_1} (X_i - \overline{X})^2}{n_1 - 1} \quad \text{and} \quad S_2^2 = \frac{\sum\limits_{i=1}^{n_2} (Y_i - \overline{Y})^2}{n_2 - 1}$$

We conclude that, when the sample sizes n_1 and n_2 are large,

$$Z = \frac{(\overline{X} - \overline{Y}) - (\mu_1 - \mu_2)}{\sqrt{\dfrac{S_1^2}{n_1} + \dfrac{S_2^2}{n_2}}} \quad \text{is approximately } N(0, 1)$$

A confidence interval for $\mu_1 - \mu_2$ is constructed from this sampling distribution. As we did for the single sample problem, we obtain a confidence interval of the form

Estimate of parameter $\pm$ (Tabled z-value)(Estimated standard error)

Large Sample Confidence Interval for $\mu_1 - \mu_2$

When n_1 and n_2 are greater than 30, an approximate $100(1 - \alpha)\%$ confidence interval for $\mu_1 - \mu_2$ is given by

$$\left(\overline{X} - \overline{Y} - z_{\alpha/2}\sqrt{\frac{S_1^2}{n_1} + \frac{S_2^2}{n_2}}, \quad \overline{X} - \overline{Y} + z_{\alpha/2}\sqrt{\frac{S_1^2}{n_1} + \frac{S_2^2}{n_2}} \right)$$

where $z_{\alpha/2}$ is the upper $\alpha/2$ point of $N(0, 1)$.

Example 4   A confidence interval for difference of mean age at marriage
To compare the age at first marriage of females in two ethnic groups, A and B, a random sample of 100 ever-married females is taken from each group and the ages at first marriage are recorded. The means and the standard deviations are found to be

	A	B
Mean	20.7	18.5
Standard deviation	6.3	5.8

Find a 95% confidence interval for the difference in means of age at first marriage.

SOLUTION AND DISCUSSION   Let μ_1 be the mean age at first marriage for ethnic group A and let μ_2 be the mean age at first marriage for ethnic group B.

We have

$$n_1 = 100, \qquad \bar{x} = 20.7, \qquad s_1 = 6.3$$
$$n_2 = 100, \qquad \bar{y} = 18.5, \qquad s_2 = 5.8$$

$$\sqrt{\frac{s_1^2}{n_1} + \frac{s_2^2}{n_2}} = \sqrt{\frac{(6.3)^2}{100} + \frac{(5.8)^2}{100}} = .8563$$

For a 95% confidence interval, we use $z_{.025} = 1.96$. We calculate

$$\bar{x} - \bar{y} = 20.7 - 18.5 = 2.2$$

$$z_{.025}\sqrt{\frac{s_1^2}{n_1} + \frac{s_2^2}{n_2}} = 1.96 \times .8563 = 1.68$$

Therefore, a 95% confidence interval for $\mu_1 - \mu_2$ is given by

$$2.2 \pm 1.68 \qquad \text{or} \qquad (.52, 3.88)$$

Females in ethnic group B tend, on the average, to marry .52 year to 3.88 years younger than those in ethnic group A. ∎

A test of the null hypothesis that the two population means are the same, $H_0 : \mu_1 - \mu_2 = 0$, employs the test statistic

$$Z = \frac{\bar{X} - \bar{Y}}{\sqrt{\dfrac{S_1^2}{n_1} + \dfrac{S_2^2}{n_2}}}$$

which is approximately $N(0, 1)$ when $\mu_1 - \mu_2 = 0$.

Example 5 Testing for a difference in mean age at marriage
Do the data in Example 4 provide strong evidence that the mean age at first marriage of females is different for the two groups? Test at $\alpha = .02$.

SOLUTION AND DISCUSSION

1. *Hypotheses.* Let μ_1 be the mean age at first marriage for ethnic group A and let μ_2 be the mean age at first marriage for ethnic group B. Because our goal is to establish that the population means μ_1 and μ_2 are different, we formulate the hypotheses

$$H_0 : \quad \mu_1 = \mu_2 \qquad \text{versus} \qquad H_1 : \quad \mu_1 \neq \mu_2$$

2. *Level of significance.* We are given $\alpha = .02$.
3. *Test statistic.* We use the test statistic

$$Z = \frac{\bar{X} - \bar{Y}}{\sqrt{\dfrac{S_1^2}{n_1} + \dfrac{S_2^2}{n_2}}}$$

4. *Rejection region.* Since the alternative hypothesis is two-sided, we set a two-sided rejection region. Specifically, with $\alpha = .02$, we have $\alpha/2 = .01$ and $z_{\alpha/2} = 2.33$, so the rejection region is $R: Z \geq 2.33$ or $Z \leq -2.33$.

5. *Calculation.* Using the sample data given in Example 4, we calculate

$$z = \frac{20.7 - 18.5}{\sqrt{\dfrac{(6.3)^2}{100} + \dfrac{(5.8)^2}{100}}} = \frac{2.2}{.8563} = 2.57$$

6. *Conclusion.* The observed value $Z = 2.57$ lies in the rejection region, so we reject the null hypothesis in favor of H_1. The P-value is

$$P[Z > 2.57] + P[Z < -2.57] = .0102 \quad (\text{see Figure 3})$$

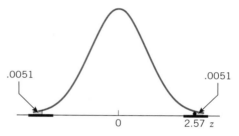

Figure 3. P-value with two-sided rejection region

The mean ages at first marriage are significantly different for the two ethnic groups. ∎

Statistical Reasoning

The Importance of a Control Group

The military was interested in determining whether large doses of vitamin C and B-complex vitamins would keep personnel healthy. They conducted a test under harsh winter conditions in the mountains. Participants slept in housing that was drafty and were awakened early for outside hiking and exercise. In general, the conditions were intended to cause colds and sickness. Vitamin supplements were given to the group. It was found that they experienced significantly fewer health problems than a typical member of the military and their physical fitness improved. It looked as if the vitamin supplement was helpful.

Fortunately, this study was designed with a control group. That is, another group of participants also went through the same hardship conditions, but they

Jim Cummins/FPG International

did not receive the extra vitamin supplements. Surprisingly, this group also showed an increase in health and physical fitness.

Without a control group undergoing identical conditions, except for the vitamin treatment, there would not be a proper baseline. Here the health and fitness of a typical member of the military is not the proper baseline.

Any study of a new treatment should include a control group.

Example 6   Testing for lower levels of chlorine

In June 1993, chemical analyses were made of 85 water samples (each of unit volume) taken from various parts of a city lake, and the measurements of chlorine content were recorded. During the next two winters, the use of road salt was substantially reduced in the catchment areas of the lake. In June 1995, 110 water samples were analyzed and their chlorine contents recorded. Calculations of the mean and the standard deviation for the two sets of data, along with the sample sizes, are given in the table.

	Chlorine Content	
	1993	1995
Sample size	85	110
Mean	18.3	17.8
Standard deviation	1.2	1.8

Test the claim that lower salt usage has reduced the amount of chlorine in the lake. Use $\alpha = .05$.

SOLUTION AND DISCUSSION

1. *Hypotheses.* Let μ_1 = mean chlorine content in 1993 and μ_2 = mean chlorine content in 1995. Because the claim is that μ_2 is less than μ_1, we formulate the hypotheses

$$H_0: \mu_1 - \mu_2 = 0 \quad \text{versus} \quad H_1: \mu_1 - \mu_2 > 0$$

2. *Level of significance.* We are given $\alpha = .05$.

3. *Test statistic.* We use the test statistic

$$Z = \frac{\bar{X} - \bar{Y}}{\sqrt{\dfrac{S_1^2}{n_1} + \dfrac{S_2^2}{n_2}}}$$

4. *Rejection region.* The rejection region should be of the form $R: Z \geq c$ because H_1 is right-sided. Since $\alpha = .05$, we reject H_0 for $Z \geq 1.645$.

5. *Calculation.* Using the data

$$n_1 = 85, \quad \bar{x} = 18.3, \quad s_1 = 1.2$$
$$n_2 = 110, \quad \bar{y} = 17.8, \quad s_2 = 1.8$$

we calculate

$$z = \frac{18.3 - 17.8}{\sqrt{\dfrac{(1.2)^2}{85} + \dfrac{(1.8)^2}{110}}} = \frac{.5}{.2154} = 2.32$$

6. *Conclusion.* The observed value $z = 2.32$ lies in the rejection region, so we reject the null hypothesis in favor of H_1. The P-value is

$$P[Z > 2.32] = .0102 \quad (\text{see Figure 4})$$

Because the P-value is very small, we conclude that there is strong evidence in support of H_1. The mean chlorine content is lower following the reduction in use of road salt.

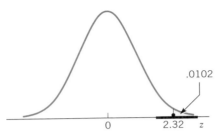

Figure 4. P-value = .0102

We summarize the procedure for testing $\mu_1 - \mu_2 = \delta_0$, where δ_0 is specified under the null hypothesis. The case $\mu_1 = \mu_2$ corresponds to $\delta_0 = 0$.

Testing $H_0: \mu_1 - \mu_2 = \delta_0$ with Large Samples

Test statistic:

$$Z = \frac{\overline{X} - \overline{Y} - \delta_0}{\sqrt{\dfrac{S_1^2}{n_1} + \dfrac{S_2^2}{n_2}}}$$

Alternative Hypothesis	Level α Rejection Region
$H_1: \ \mu_1 - \mu_2 > \delta_0$	R: $Z \geq z_\alpha$
$H_1: \ \mu_1 - \mu_2 < \delta_0$	R: $Z \leq -z_\alpha$
$H_1: \ \mu_1 - \mu_2 \neq \delta_0$	R: $Z \geq z_{\alpha/2}$ or $Z \leq -z_{\alpha/2}$

Example 7 A statistical analysis of a fast change or evolution

Darwin's finches, living on a small isolated island in the Galápagos archipelago, appeared to undergo a change in size over a period of just two to three years. Because evolutionary changes usually take thousands of times longer, this was big news.[1]

The island supports a breeding population of four species of Darwin's finches but an 18-month-long major drought from the middle of 1976 to the beginning of 1978 prevented breeding. Food was scarce and the typical variety was not available. Bigger birds with larger beaks were better able to find and eat what was available. They survived better than the smaller birds. Would first-generation birds after the drought be larger?

[1]This study, led by P. Grant and R. Grant, was documented on a *NOVA* program on public television. The summary statistics are from their paper in *Evolution* **49**(2), 241–251, 1995.

We summarize the findings for the bill depth of the medium ground finch. The findings were similar for other beak measurements and body size.

Mean Bill Depth (mm) of Medium Ground Finch

Population	Sample Size	Sample Mean	Estimated Standard Error
Before drought (1976)	$n_1 = 634$	$\bar{x} = 9.21$	$\dfrac{s_1}{\sqrt{n_1}} = \dfrac{s_1}{\sqrt{634}} = .03$
Next generation (1978)	$n_2 = 135$	$\bar{y} = 9.70$	$\dfrac{s_2}{\sqrt{n_2}} = \dfrac{s_2}{\sqrt{135}} = .06$

(a) Test, with $\alpha = .05$, whether the mean bill depth of the next generation is more than .3 mm larger.

(b) Obtain a 95% confidence interval for the difference in mean bill depth before and after the drought.

SOLUTION AND DISCUSSION

(a) 1. *Hypotheses.* Let μ_1 be the mean bill depth for finches in 1976 and μ_2 be the mean bill depth for the next generation. If the mean bill depth of the next generation is .3 mm larger, then $\mu_1 - \mu_2 = -.3$. We want to show that the mean bill depth of the next generation is more than .3 mm larger, so this assertion is taken as the alternative hypothesis.

$$\text{Null hypothesis } H_0: \quad \mu_1 - \mu_2 = -.3$$
$$\text{Alternative hypothesis } H_1: \quad \mu_1 - \mu_2 < -.3$$

2. *Level of significance.* We are given $\alpha = .05$.

3. *Test statistic.* The sample sizes are both large. We employ the large sample Z statistic with $\delta_0 = -.3$.

$$Z = \frac{\bar{X} - \bar{Y} - (-.3)}{\sqrt{\dfrac{S_1^2}{n_1} + \dfrac{S_2^2}{n_2}}}$$

4. *Rejection region.* We reject the null hypothesis for small values since the alternative hypothesis is left-sided. With $\alpha = .05$, we find $z_{.05} = 1.645$ so the rejection region is $R: Z < -1.645$.

5. *Calculation.* We first use the estimated standard error to calculate

$$\frac{s_1^2}{n_1} + \frac{s_2^2}{n_2} = (.03)^2 + (.06)^2 = .0045$$

so the estimated standard deviation is

$$\sqrt{\frac{s_1^2}{n_1} + \frac{s_2^2}{n_2}} = \sqrt{.0045} = .0671$$

The test statistic takes the value

$$Z = \frac{\bar{x} - \bar{y} + .3}{\sqrt{\frac{s_1^2}{n_1} + \frac{s_2^2}{n_2}}} = \frac{9.21 - 9.70 + .3}{.0671} = -2.83$$

6. *Conclusion.*   The observed value $z = -2.83$ lies in the rejection region. Consequently, we reject the null hypothesis in favor of H_1. The *P*-value is

$$P[Z < -2.83] = .0026 \quad (\text{see Figure 5})$$

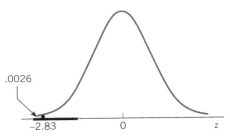

Figure 5. *P*-value = .0026

The evidence against the null hypothesis is extremely strong. We conclude that the mean bill depth of the next generation is more than .3 mm larger.

(b) The 95% confidence interval uses the tabled value $z_{.025} = 1.96$. The confidence interval becomes

$$\bar{x} - \bar{y} \pm 1.96 \sqrt{\frac{s_1^2}{n_1} + \frac{s_2^2}{n_2}}$$

$$= (9.21 - 9.70) \pm 1.96 \times (.0671) \quad \text{or} \quad (-.62, \ -.36)$$

An interesting thing happened when another 18-month drought occurred from 1984 to 1986. This time a different food supply favored finches with smaller bills, and the next generation had smaller bills. ∎

INFERENCES FROM SMALL SAMPLES

Not surprisingly, more distributional structure is required to formulate appropriate inference procedures for small samples. Here we introduce the small sample inference procedures that are valid under the following assumptions about the population distributions. Naturally, the usefulness of such procedures depends on how closely these assumptions are realized.

Additional Assumptions When the Sample Sizes Are Small

1. Both populations are normal.
2. The population standard deviations σ_1 and σ_2 are equal.

A restriction to normal populations is not new. It was previously introduced for inferences about the mean of a single population. The second assumption, requiring equal variability of the populations, is somewhat artificial but we reserve comment until later. Letting σ denote the common standard deviation, we summarize:

Small Sample Assumptions

1. $X_1, X_2, \ldots, X_{n_1}$ is a random sample from $N(\mu_1, \sigma)$.
2. $Y_1, Y_2, \ldots, Y_{n_2}$ is a random sample from $N(\mu_2, \sigma)$.
   (*Note:* σ is the same for both distributions.)
3. $X_1, X_2, \ldots, X_{n_1}$ and $Y_1, Y_2, \ldots, Y_{n_2}$ are independent.

Again, $\overline{X} - \overline{Y}$ is our choice for a statistic.

$$\text{Mean of } (\overline{X} - \overline{Y}) = E(\overline{X} - \overline{Y}) = \mu_1 - \mu_2$$

$$\text{Var}(\overline{X} - \overline{Y}) = \frac{\sigma^2}{n_1} + \frac{\sigma^2}{n_2} = \sigma^2 \left(\frac{1}{n_1} + \frac{1}{n_2} \right)$$

The common variance σ^2 can be estimated by combining information provided by both samples. Specifically, the sum $\Sigma_{i=1}^{n_1} (X_i - \overline{X})^2$ incorporates $n_1 - 1$ pieces of information about σ^2, in view of the constraint that the deviations $X_i - \overline{X}$ sum to zero. Independently of this, $\Sigma_{i=1}^{n_2} (Y_i - \overline{Y})^2$ contains $n_2 - 1$ pieces of information about σ^2. These two quantities can then be combined

$$\sum (X_i - \overline{X})^2 + \sum (Y_i - \overline{Y})^2$$

to obtain a pooled estimate of the common σ^2. The proper divisor is the sum of the component degrees of freedom, or $(n_1 - 1) + (n_2 - 1) = n_1 + n_2 - 2$.

Pooled Estimator of the Common σ^2

$$S^2_{\text{pooled}} = \frac{\sum_{i=1}^{n_1}(X_i - \overline{X})^2 + \sum_{i=1}^{n_2}(Y_i - \overline{Y})^2}{n_1 + n_2 - 2}$$

$$= \frac{(n_1 - 1)S_1^2 + (n_2 - 1)S_2^2}{n_1 + n_2 - 2}$$

Example 8   Calculation of the pooled estimate of σ^2
Calculate s^2_{pooled} from these two samples.

$$\text{Sample from population 1:} \quad 8 \quad 5 \quad 7 \quad 6 \quad 9 \quad 7$$
$$\text{Sample from population 2:} \quad 2 \quad 6 \quad 4 \quad 7 \quad 6$$

SOLUTION AND DISCUSSION   The sample means are

$$\overline{x} = \frac{\sum x_i}{6} = \frac{42}{6} = 7$$

$$\overline{y} = \frac{\sum y_i}{5} = \frac{25}{5} = 5$$

Furthermore,

$$(6 - 1)s_1^2 = \sum (x_i - \overline{x})^2$$
$$= (8 - 7)^2 + (5 - 7)^2 + (7 - 7)^2 + (6 - 7)^2 + (9 - 7)^2 + (7 - 7)^2$$
$$= 10$$
$$(5 - 1)s_2^2 = \sum (y_i - \overline{y})^2$$
$$= (2 - 5)^2 + (6 - 5)^2 + (4 - 5)^2 + (7 - 5)^2 + (6 - 5)^2$$
$$= 16$$

Thus, $s_1^2 = 2$, $s_2^2 = 4$, and the pooled variance is

$$s^2_{\text{pooled}} = \frac{\sum (x_i - \overline{x})^2 + \sum (y_i - \overline{y})^2}{n_1 + n_2 - 2} = \frac{10 + 16}{6 + 5 - 2} = 2.89$$

The pooled variance is closer to 2 than 4 because the first sample size is larger.
   These arithmetic details serve to demonstrate the concept of pooling. Using a calculator with a "standard deviation" key, one can directly get the sample standard deviations $s_1 = 1.414$ and $s_2 = 2.000$. Noting that $n_1 = 6$ and $n_2 = 5$, we can then calculate the pooled variance as

$$s^2_{\text{pooled}} = \frac{5(1.414)^2 + 4(2.000)^2}{9} = 2.89$$ ∎

Employing the pooled estimator $\sqrt{S_{\text{pooled}}^2}$ for the common σ, we obtain a Student's t variable that is basic to inferences about $\mu_1 - \mu_2$.

$$T = \frac{(\overline{X} - \overline{Y}) - (\mu_1 - \mu_2)}{S_{\text{pooled}}\sqrt{\dfrac{1}{n_1} + \dfrac{1}{n_2}}}$$

has Student's t distribution with $n_1 + n_2 - 2$ degrees of freedom.

We can now obtain confidence intervals for $(\mu_1 - \mu_2)$, which are of the form

Estimate of parameter $\pm$ (Tabled t-value) $\times$ (Estimated standard error)

Confidence Interval for $\mu_1 - \mu_2$
Small Samples

A $100(1 - \alpha)\%$ confidence interval for $\mu_1 - \mu_2$ is given by

$$\overline{X} - \overline{Y} \pm t_{\alpha/2}\, S_{\text{pooled}}\sqrt{\frac{1}{n_1} + \frac{1}{n_2}}$$

where

$$S_{\text{pooled}}^2 = \frac{(n_1 - 1)S_1^2 + (n_2 - 1)S_2^2}{n_1 + n_2 - 2}$$

and $t_{\alpha/2}$ is the upper $\alpha/2$ point of the t distribution with d.f. $= n_1 + n_2 - 2$.

Example 9   Inspecting plots for agreement with assumptions
A feeding test is conducted on a herd of 25 milking cows to compare two diets, one of dewatered alfalfa and the other of field-wilted alfalfa. A sample of 12 cows randomly selected from the herd are fed dewatered alfalfa; the remaining 13 cows are fed field-wilted alfalfa. From observations made over a three-week period, the average daily milk production is recorded for each cow.

TABLE 1   Milk Yields Under Two Diets

						Milk Yield (Pounds)						
Field-wilted alfalfa	44	44	56	46	47	38	58	53	49	35	46	30	41
Dewatered alfalfa	35	47	55	29	40	39	32	41	42	57	51	39	

Obtain a 95% confidence interval for the difference in mean daily milk yield per cow between the two diets.

SOLUTION AND The dot diagrams of these data, plotted in Figure 6, give the appearance of ap-
DISCUSSION proximately equal amounts of variation.

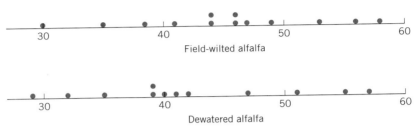

Figure 6. Dot diagrams of milk-yield data in Example 9

We assume that the milk-yield data for both field-wilted and dewatered alfalfa are random samples from normal populations with means μ_1 and μ_2, respectively, and a common standard deviation σ. Computations from these data provide the summary statistics

Field-wilted alfalfa

$n_1 = 13 \qquad \bar{x} = 45.15 \qquad s_1 = 7.998$

Dewatered alfalfa

$n_2 = 12 \qquad \bar{y} = 42.25 \qquad s_2 = 8.740$

We calculate

$$s_{\text{pooled}} = \sqrt{\frac{12(7.998)^2 + 11(8.740)^2}{23}} = 8.36$$

With a 95% confidence level, $\alpha/2 = .025$ and consulting the t table, we find that $t_{.025} = 2.069$ with d.f. $= n_1 + n_2 - 2 = 23$. Thus, a 95% confidence interval for $\mu_1 - \mu_2$ becomes

$$\bar{x} - \bar{y} \pm t_{.025}\, s_{\text{pooled}} \sqrt{\frac{1}{n_1} + \frac{1}{n_2}}$$

$$= 45.15 - 42.25 \pm 2.069 \times 8.36 \sqrt{\frac{1}{13} + \frac{1}{12}}$$

$$= 2.90 \pm 6.92 \qquad \text{or} \qquad (-4.02, 9.82)$$

We can be 95% confident that the mean yield from field-wilted alfalfa can be anywhere from 4.02 pounds lower to 9.82 pounds higher than for dewatered alfalfa. ∎

Testing $H_0: \mu_1 - \mu_2 = \delta_0$ with Small Samples

Test statistic:

$$T = \frac{\overline{X} - \overline{Y} - \delta_0}{S_{\text{pooled}} \sqrt{\dfrac{1}{n_1} + \dfrac{1}{n_2}}} \qquad \text{d.f.} = n_1 + n_2 - 2$$

Alternative Hypothesis	Level α Rejection Region
$H_1: \ \mu_1 - \mu_2 > \delta_0$	R: $\ T \geq t_\alpha$
$H_1: \ \mu_1 - \mu_2 < \delta_0$	R: $\ T \leq -t_\alpha$
$H_1: \ \mu_1 - \mu_2 \neq \delta_0$	R: $\ T \geq t_{\alpha/2}$ or $T \leq -t_{\alpha/2}$

Example 10   Testing for a difference in mean milk yield
Refer to the feeding test on cows described in Example 9 and the summary statistics of the data of milk yield:

Field-wilted alfalfa

$n_1 = 13 \qquad \overline{x} = 45.15 \qquad s_1 = 7.998$

Dewatered alfalfa

$n_2 = 12 \qquad \overline{y} = 42.25 \qquad s_2 = 8.740$

Do these data strongly indicate that the milk yield is less with dewatered alfalfa than field-wilted alfalfa? Test at $\alpha = .05$.

SOLUTION AND DISCUSSION

1. *Hypotheses.*   We are seeking strong evidence in support of the claim that the mean milk yield with dewatered alfalfa (μ_2) is less than the mean milk yield with field-wilted alfalfa (μ_1). Therefore, the alternative hypothesis should be taken as $H_1: \mu_1 > \mu_2$ or $H_1: \mu_1 - \mu_2 > 0$, and our problem then is to test

$$H_0: \ \mu_1 - \mu_2 = 0 \qquad \text{versus} \qquad H_1: \ \mu_1 - \mu_2 > 0$$

2. *Level of significance.*   We are given $\alpha = .05$.

3. *Test statistic.*   We employ the test statistic

$$T = \frac{\overline{X} - \overline{Y}}{S_{\text{pooled}} \sqrt{\dfrac{1}{n_1} + \dfrac{1}{n_2}}} \qquad \text{d.f.} = n_1 + n_2 - 2$$

4. *Rejection region.*   With $\alpha = .05$, we set the right-sided rejection region $R: T \geq t_{.05}$. For d.f. $= n_1 + n_2 - 2 = 23$, the tabled value is $t_{.05} = 1.714$, so the rejection region is $R: T \geq 1.714$.

5. *Calculation.* With $s_{pooled} = 8.36$ already calculated in Example 9, the observed value of T is

$$t = \frac{45.15 - 42.25}{8.36 \sqrt{\dfrac{1}{13} + \dfrac{1}{12}}} = \frac{2.90}{3.347} = .87$$

6. *Conclusion.* The observed value $t = .87$ is not in the rejection region; nor is it even close to R. At level $\alpha = .05$, we cannot reject H_0.

   The accompanying computer output includes the *P*-value $= .20$. The data fail to establish that the mean milk yield is lower with dewatered alfalfa.

MINITAB Output for Example 10
See Exercise 2.22 for the commands.

Two Sample T-Test and Confidence Interval

```
        Two sample T for Fieldwlt vs Dewater
                 N       Mean      StDev    SE Mean
    Fieldwlt    13      45.15       8.00       2.2
    Dewater     12      42.25       8.74       2.5

95% CI for mu Fieldwlt − mu Dewater: ( −4.0, 9.8)
T-Test mu Fieldwlt = mu Dewater (vs >): T = 0.87  P = 0.20  DF = 23
Both use Pooled StDev = 8.36
```

DECIDING WHETHER OR NOT TO POOL

Our preceding discussion of large sample and small sample inferences raises a few questions:

For small sample inference, why do we assume the population standard deviations to be equal when no such assumption was needed in the large sample case?

When should we be wary about this assumption, and what procedure should we use when the assumption is not reasonable?

Learning statistics would be a step simpler if the ratio

$$\frac{(\overline{X} - \overline{Y}) - (\mu_1 - \mu_2)}{\sqrt{\dfrac{S_1^2}{n_1} + \dfrac{S_2^2}{n_2}}}$$

had a t distribution for small samples from normal populations. Unfortunately, statistical theory proves it otherwise. The distribution of this ratio is *not* a t and, worse yet, it depends on the unknown quantity σ_1/σ_2. The assumption $\sigma_1 = \sigma_2$ and the change of the denominator to $S_{pooled} \sqrt{1/n_1 + 1/n_2}$ allow the t-based inferences to be valid. However, the $\sigma_1 = \sigma_2$ restriction and accompanying pooling are not needed in large samples where a normal approximation holds.

With regard to the second question, the relative magnitude of the two sample standard deviations s_1 and s_2 should be the prime consideration. The assumption $\sigma_1 = \sigma_2$ is reasonable if s_1/s_2 is not very much different from 1. As a working rule, the range of values $\frac{1}{2} \leq s_1/s_2 \leq 2$ may be taken as reasonable cases for making the assumption $\sigma_1 = \sigma_2$ and hence for pooling. If s_1/s_2 is seen to be smaller than $\frac{1}{2}$ or larger than 2, the assumption $\sigma_1 = \sigma_2$ would be suspect. In that case, some approximate methods of inference about $\mu_1 - \mu_2$ are available, but those will not be discussed here because of their complex forms. Instead, we outline a simple though conservative procedure, which treats the ratio

$$T^* = \frac{(\overline{X} - \overline{Y}) - (\mu_1 - \mu_2)}{\sqrt{\dfrac{S_1^2}{n_1} + \dfrac{S_2^2}{n_2}}}$$

as a t-variable with d.f. = smaller of $n_1 - 1$ and $n_2 - 1$.

Small Sample Inferences for $\mu_1 - \mu_2$ When the Populations Are Normal, but σ_1 and σ_2 Are Not Assumed to Be Equal

A $100(1 - \alpha)\%$ conservative confidence interval for $\mu_1 - \mu_2$ is given by

$$\overline{X} - \overline{Y} \pm t^*_{\alpha/2} \sqrt{\frac{S_1^2}{n_1} + \frac{S_2^2}{n_2}}$$

where $t^*_{\alpha/2}$ denotes the upper $\alpha/2$ point of the t distribution with d.f. = smaller of $n_1 - 1$ and $n_2 - 1$.

The null hypothesis $H_0: \mu_1 - \mu_2 = \delta_0$ is tested using the test statistic

$$T^* = \frac{\overline{X} - \overline{Y} - \delta_0}{\sqrt{\dfrac{S_1^2}{n_1} + \dfrac{S_2^2}{n_2}}} \qquad \text{d.f. = smaller of } n_1 - 1 \text{ and } n_2 - 1$$

Here, the confidence interval is conservative in the sense that the actual level of confidence is at least $(1 - \alpha)$. Likewise, the level α test is conservative in the sense that the actual Type I error probability is no more than α.

Example 11   Testing the difference of two means when the variances are not equal

A new method of storing snap beans is believed to retain more ascorbic acid than an old method. In an experiment, snap beans were harvested under uniform conditions and frozen in 18 equal-size packages. Nine of these packages were randomly selected and stored according to the new method, and the other 9 packages were stored by the old method. Subsequently, ascorbic acid determinations (in mg/kg) were made, and the following summary statistics were calculated.

	New Method	Old Method
Mean ascorbic acid	449	410
Standard deviation	19	45

Do these data substantiate the claim that more ascorbic acid is retained under the new method of storing? Test at $\alpha = .05$.

SOLUTION AND DISCUSSION

1. *Hypotheses.* Let μ_1 denote the population mean ascorbic acid under the new method of storing and μ_2 that under the old method. The problem concerns substantiation of the claim that μ_1 is larger than μ_2. Therefore, we formulate the testing problem as

$$H_0: \quad \mu_1 - \mu_2 = 0 \quad \text{versus} \quad H_1: \quad \mu_1 - \mu_2 > 0$$

2. *Level of significance.* We are given $\alpha = .05$.

3. *Test statistic.* We assume that both population distributions are normal. Looking at the observed sample standard deviations, we note that s_2 is more than twice s_1 so the assumption $\sigma_1 = \sigma_2$ is suspect. We therefore use the conservative test based on the test statistic:

$$T^* = \frac{\overline{X} - \overline{Y}}{\sqrt{\dfrac{S_1^2}{n_1} + \dfrac{S_2^2}{n_2}}} \qquad \begin{array}{l} \text{d.f.} = \text{smaller of } n_1 - 1 \text{ and } n_2 - 1 \\ \quad = 8 \end{array}$$

4. *Rejection region.* With $\alpha = .05$, we reject H_0 for large values of T^*. For d.f. $= 8$, the tabled value is $t_{.05} = 1.860$, so we set the rejection region $R: T^* \geq 1.860$.

5. *Calculation.* The summary statistics are

$$n_1 = 9, \qquad \overline{x} = 449, \qquad s_1 = 19$$
$$n_2 = 9, \qquad \overline{y} = 410, \qquad s_2 = 45$$

and the observed value of the test statistic is

$$t^* = \frac{449 - 410}{\sqrt{\dfrac{(19)^2}{9} + \dfrac{(45)^2}{9}}} = 2.40$$

6. *Conclusion.* The observed value $t^* = 2.40$ is in the rejection region R. At level $\alpha = .05$, H_0 is rejected in favor of H_1. Since $t_{.025} = 2.306$, the P-value is smaller than .025.

   The evidence against H_0 is quite strong. The mean ascorbic acid content is higher under the new method of storage. ∎

Finally, we would like to emphasize that with large samples, we can also learn about other differences between the two populations.

Example 12   Graphical displays provide additional information

Natural resource managers have attempted to use the Satellite Landsat Multispectral Scanner data for improved landcover classification. When the satellite was flying over country known to consist of forest, the following intensities were recorded on the near-infrared band of a thermatic mapper. The sample has already been ordered.

77	77	78	78	81	81	82	82	82	82	82	83
83	84	84	84	84	85	86	86	86	86	86	87
87	87	87	87	87	87	89	89	89	89	89	89
89	90	90	90	91	91	91	91	91	91	91	91
91	91	93	93	93	93	93	93	94	94	94	94
94	94	94	94	94	94	94	94	95	95	95	95
95	96	96	96	96	96	96	97	97	97	97	97
97	97	97	97	98	99	100	100	100	100	100	100
100	100	100	101	101	101	101	101	101	102	102	102
102	102	102	103	103	104	104	104	105	107		

When the satellite was flying over urban areas, the intensity of reflected light on the same near-infrared band was

71	72	73	74	75	77	78	79	79	79
79	80	80	80	81	81	81	82	82	82
82	84	84	84	84	84	84	85	85	85
85	85	85	86	86	87	88	90	91	94

If the means are different, the readings could be used to tell urban from forest areas. Obtain a 95% confidence interval for the difference in mean radiance levels.

SOLUTION AND DISCUSSION   Computer calculations give

	Forest	Urban
Number	118	40
Mean	92.932	82.075
Std. dev.	6.9328	4.9789

and, for large sample sizes, the approximate 95% confidence interval for $\mu_1 - \mu_2$ is given by

$$\left(\overline{X} - \overline{Y} - z_{.025} \sqrt{\frac{S_1^2}{n_1} + \frac{S_2^2}{n_2}}, \quad \overline{X} - \overline{Y} + z_{.025} \sqrt{\frac{S_1^2}{n_1} + \frac{S_2^2}{n_2}} \right)$$

Since $z_{.025} = 1.96$, the 95% confidence interval is calculated as

$$92.932 - 82.075 \pm 1.96 \sqrt{\frac{(6.9328)^2}{118} + \frac{(4.9789)^2}{40}} \quad \text{or} \quad (8.87, \quad 12.84)$$

The mean for the forest is 8.87 to 12.84 levels of radiance higher than the mean for the urban areas.

Because the sample sizes are large, we can also learn about other differences between the two populations. The stem-and-leaf displays and boxplots in Figure 7 reveal that there is some difference in the standard deviation as well as the means. The graphs further indicate a range of high readings that are more likely to come from forests than urban areas. This feature has proven helpful in discriminating between forest and urban areas on the basis of near-infrared readings.

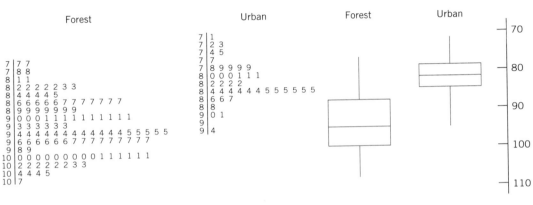

Figure 7. Stem-and-leaf displays and boxplots give additional information about population differences.

Exercises

2.1 Independent random samples from two populations have provided the summary statistics

Sample 1	Sample 2
$n_1 = 52$	$n_2 = 42$
$\bar{x} = 75$	$\bar{y} = 66$
$s_1^2 = 155$	$s_2^2 = 142$

(a) Obtain a point estimate of $\mu_1 - \mu_2$ and calculate the estimated standard error.

(b) Construct a 95% confidence interval for $\mu_1 - \mu_2$.

2.2   Rural and urban students are to be compared on the basis of their scores on a nationwide musical aptitude test. Two random samples of sizes 90 and 100 are selected from rural and urban seventh grade students. The summary statistics from the test scores are

	Rural	Urban
Sample size	90	100
Mean	76.4	81.2
Standard deviation	8.2	7.6

Establish a 98% confidence interval for the difference in population mean scores between urban and rural students.

2.3   Construct a test to determine whether there is a significant difference between the population mean scores in Exercise 2.2. Use $\alpha = .05$.

2.4   A group of 141 subjects is used in an experiment to compare two treatments. Treatment 1 is given to 79 subjects selected at random, and the remaining 62 are given treatment 2. The means and standard deviations of the responses are

	Treatment 1	Treatment 2
Mean	109	128
Standard deviation	46.2	53.4

Determine a 98% confidence interval for the mean difference of the treatment effects.

2.5   Refer to the data in Exercise 2.4. Suppose the investigator wishes to establish that treatment 2 has a higher mean response than treatment 1.
(a)   Formulate H_0 and H_1.
(b)   State the test statistic and the rejection region with $\alpha = .05$.
(c)   Perform the test at $\alpha = .05$. Also, find the P-value and comment.

2.6   In one part of a larger study on child development and social factors, interest was focused on comparing the verbal skills of white first graders whose parents were not high school graduates (Group 1) and those whose parents had some college education (Group 2). Random samples of 66 and 38 children were taken from Groups 1 and 2, respectively. Calculations from the data of their verbal skills, as measured by the scores in the California Achievement Test (CAT)–Verbal, provided the following summary statistics (based on D. R. Entwisle and K. L. Alexander, "Beginning School Math Competence: Minority and Majority Comparison," *Child Development*, **61** (1990), pp. 454–471).

	Group 1	Group 2
Mean	273	311
Standard deviation	29	40

(a) Is there strong evidence of a difference in verbal skills between the two groups? (Test at $\alpha = .01$).

(b) Determine a 98% confidence interval for the difference between the population mean scores for the two groups of children.

2.7 In a study of interspousal aggression and its possible effect on child behavior, the behavior problem checklist (BPC) scores were recorded for 47 children whose parents were classified as aggressive. The sample mean and standard deviation were 7.92 and 3.45, respectively. For a sample of 38 children whose parents were classified as nonaggressive, the mean and standard deviation of the BPC scores were 5.80 and 2.87, respectively. Do these observations substantiate the conjecture that the children of aggressive families have a higher mean BPC than those of nonaggressive families? (Answer by calculating the *P*-value.)

2.8 Given that $n_1 = 15$, $\bar{x} = 20$, $\Sigma (x_i - \bar{x})^2 = 28$, and $n_2 = 12$, $\bar{y} = 17$, $\Sigma (y_i - \bar{y})^2 = 22$,

(a) Obtain s_{pooled}^2.

(b) Test $H_0: \mu_1 = \mu_2$ against $H_1: \mu_1 > \mu_2$ with $\alpha = .05$.

(c) Determine a 95% confidence interval for $\mu_1 - \mu_2$.

2.9 The following summary statistics are recorded for independent random samples from two populations.

Sample 1	Sample 2
$n_1 = 9$	$n_2 = 6$
$\bar{x} = 16.18$	$\bar{y} = 4.22$
$s_1 = 1.54$	$s_2 = 1.37$

Stating any assumptions that you make,

(a) Calculate a 95% confidence interval for $\mu_1 - \mu_2$.

(b) Test the null hypothesis $\mu_1 - \mu_2 = 10$ versus the alternative $\mu_1 - \mu_2 > 10$ with $\alpha = .01$.

2.10   Suppose that for a random sample of size 5 from population 1, the sample mean and standard deviation were 10.6 and 3.62, respectively, while a random sample of size 8 from population 2 yielded a sample mean of 14.9 and a sample standard deviation of 4.17.

   (a)   Determine a 90% confidence interval for the difference between the population means.

   (b)   Test the null hypothesis $\mu_1 = \mu_2$ versus the alternative $\mu_1 < \mu_2$ at $\alpha = .025$.

2.11   The following summary statistics are recorded for independent random samples from two populations.

Sample 1	Sample 2
$n_1 = 7$	$n_2 = 8$
$\bar{x} = 86.2$	$\bar{y} = 74.7$
$s_1 = 14.2$	$s_2 = 5.5$

Stating any assumptions that you make,

   (a)   Determine a 98% confidence interval for $\mu_1 - \mu_2$.

   (b)   Test the null hypothesis $\mu_1 - \mu_2 = 20$ versus the alternative $\mu_1 - \mu_2 < 20$ with $\alpha = .05$.

2.12   Given here are the sample sizes and the sample standard deviations for independent random samples from two populations. For each case, state which of the three tests you would use in testing hypotheses about $\mu_1 - \mu_2$: (1) Z-test, (2) t-test with pooling, (3) conservative t-test without pooling. Also, state any assumptions you would make about the population distributions.

   (a) $n_1 = 45, \quad s_1 = 12.2$
         $n_2 = 53, \quad s_2 = 8.6$

   (b) $n_1 = 9, \quad s_1 = .86$
         $n_2 = 7, \quad s_2 = 1.12$

   (c) $n_1 = 8 \quad s_1 = 1.65$
         $n_2 = 12, \quad s_2 = 5.22$

   (d) $n_1 = 70, \quad s_1 = 6.2$
         $n_2 = 62, \quad s_2 = 2.3$

2.13   Psychologists have made extensive studies on the relationship between child abuse and later criminal behavior. The processes of data collection and interpretation of the statistical findings are quite intricate, as discussed in C. S. Widom, "The Cycle of Violence," *Science*, **244** (1989), pp. 160–165. Consider a study that consisted of the follow-ups of 52 boys who were abused in their preschool years and 67 boys who were not abused. The data of the number of criminal offenses of those boys in their teens yielded the following summary statistics.

	Abused	Nonabused
Mean	2.52	1.63
Standard deviation	1.84	1.22

Is the mean number of criminal offenses significantly higher for the abused group than for the nonabused group? Answer by calculating the *P*-value.

2.14 Referring to the data in Exercise 2.13, determine a 99% confidence interval for the difference between the true means for the two groups.

2.15 The peak oxygen intake per unit of body weight, called the "aerobic capacity," of an individual performing a strenuous activity is a measure of work capacity. For a comparative study, measurements of aerobic capacities are recorded for a group of 20 Peruvian Highland natives and a group of 10 U.S. lowlanders acclimatized as adults in high altitudes. The following summary statistics were obtained from the data [*Source:* A. R. Frisancho, *Science*, **187** (1975), p. 317].

	Peruvian Natives	U.S. Subjects Acclimatized
Mean	46.3	38.5
Standard deviation	5.0	5.8

Construct a 98% confidence interval for the mean difference in aerobic capacity between the two groups.

2.16 Do the data in Exercise 2.15 provide a strong indication of a difference in mean aerobic capacity between the highland natives and the acclimatized lowlanders? Test with $\alpha = .05$.

2.17 To compare two programs for training industrial workers to perform a skilled job, 20 workers are included in an experiment. Of these, 10 are selected at random and trained by method 1; the remaining 10 workers are trained by method 2. After completion of training, all the workers are subjected to a time-and-motion test that records the speed of performance of a skilled job. The following data are obtained.

	Time (minutes)									
Method 1	15	20	11	23	16	21	18	16	27	24
Method 2	23	31	13	19	23	17	28	26	25	28

(a) Can you conclude from the data that the mean job time is significantly less after training with method 1 than after training with
method 2? (Test with $\alpha = .05$.)

(b) State the assumptions you make for the population distributions.

(c) Construct a 95% confidence interval for the population mean difference in job times between the two methods.

2.18 Reading level can be quantified in different ways. One simple measure is
sentence length. To overcome difficulties with different typefaces and
sizes, an investigator just counted the number of letters and punctuation
signs. Random samples of 20 sentences were selected from *The New Yorker*
and *Sports Illustrated*.

New Yorker	37	18	69	155	49	43	68	33	164	125
	74	92	37	72	127	191	111	231	134	59
Sports Illustrated	134	23	135	45	111	51	132	146	81	65
	115	33	46	178	132	207	95	117	37	74

(a) Test for equality of the mean sentence lengths (number of characters) using $\alpha = .05$. Comment.

(b) Determine a 97% confidence interval for the difference in means.

(c) Suppose any sentence was chosen by randomly selecting a page and
then haphazardly pointing with a finger to a position on the page.
Explain why this selection procedure would lead to a sample where
the sentence lengths (number of characters) are biased toward large
values.

2.19 Suppose that you are asked to design an experiment to study the effect of
a hormone injection on the weight gain of pregnant rats during gestation.
You have decided to inject 6 of the 12 rats available for the experiment
and retain the other 6 as controls.

(a) Briefly explain why it is important to randomly divide the rats into
the two groups. What might be wrong with the experimental results
if you choose to give the hormone treatment to 6 rats that are easy
to grab from their cages?

(b) Suppose that the 12 rats are tagged with serial numbers from 1
through 12 and 12 marbles identical in appearance are also numbered from 1 through 12. How can you use these marbles to randomly select the rats in the treatment and control groups?

2.20 To compare the effectiveness of isometric and isotonic exercise methods,
20 potbellied business executives are included in an experiment: 10 are
selected at random and assigned to one exercise method; the remaining 10

are assigned to the other exercise method. After five weeks, the reductions in abdomen measurements are recorded in centimeters, and the following results obtained.

	Isometric Method 1	Isotonic Method 2
Mean	2.5	3.1
Standard deviation	.8	1.0

(a) Do these data support the claim that the isotonic method is more effective?

(b) Construct a 95% confidence interval for $\mu_2 - \mu_1$.

2.21 Refer to Exercise 2.20.

(a) Aside from the type of exercise method, identify a few other factors that are likely to have an important effect on the amount of reduction accomplished in a five-week period.

(b) What role does randomization play in achieving a valid comparison between the two exercise methods?

(c) If you were to design this experiment, describe how you would divide the 20 business executives into two groups.

USING A COMPUTER

2.22 A two-sample analysis can be done using MINITAB as illustrated here.

Data: C1: 4 6 2 7 C2: 7 9 5	
Dialog box:	**Session command**
Stat > Basic Statistics > 2-Sample t	MTB> TWOSAMPLE 95 C1 C2;
Click **Samples in different columns**	POOLED.
Type *C1* in **First** *C2* in **Second.**	
Click **OK.**	
Click **Assume equal variances**	

Output:

Two Sample T-Test and Confidence Interval

Two sample T for C1 vs C2

	N	Mean	StDev	SE Mean
C1	4	4.75	2.22	1.1
C2	3	7.00	2.00	1.2

95% CI for mu C1 − mu C2: (−6.7, 2.2)
T-Test mu C1 = mu C2 (vs not =): T = −1.41 P = 0.23 DF = 4

Find a 97% confidence interval for the difference of means in Exercise 2.17. The modified steps are:

Type .97 in **Confidence level.**   MTB> TWOSAMPLE 97 C1 C2;

3. RANDOMIZATION AND ITS ROLE IN INFERENCE

We have presented the methods of drawing inferences about the difference be-tween two population means. Let us now turn to some important questions re-garding the design of the experiment or data collection procedure. The manner in which experimental subjects are chosen for the two treatment groups can be crucial. For example, suppose that a remedial-reading instructor has developed a new teaching technique and is permitted to use the new method to instruct half the pupils in the class. The instructor might choose the most alert or the students who are more promising in some other way, leaving the weaker students to be taught in the conventional manner. Clearly, a comparison between the reading achievements of these two groups would not just be a comparison of two teaching methods. A similar fallacy can result in comparing the nutritional quality of a new lunch package if the new diet is given to a group of children suffering from mal-nutrition and the conventional diet is given to a group of children who are already in good health.

When the assignment of treatments to experimental units is under our con-trol, steps can be taken to ensure a valid comparison between the two treatments. At the core lies the principle of impartial selection, or *randomization.* The choice of the experimental units for one treatment or the other must be made by a chance mechanism that does not favor one particular selection over any other. It must not be left to the discretion of the experimenters because, even uncon-sciously, they may be partial to one treatment.

Suppose that a comparative experiment is to be run with N experimental units, of which n_1 units are to be assigned to treatment 1 and the remaining $n_2 = N - n_1$ units are to be assigned to treatment 2. The principle of randomi-zation tells us that the n_1 units for treatment 1 must be chosen at random from the available collection of N units—that is, in a manner such that all $\binom{N}{n_1}$ possible choices are equally likely to be selected.

Randomization Procedure for Comparing Two Treatments

From the available $N = n_1 + n_2$ experimental units, choose n_1 units at random to receive treatment 1 and assign the remaining n_2 units to treatment 2. The random choice entails that all $\binom{N}{n_1}$ possible selections are equally likely to be chosen.

As a practical method of random selection, we can label the available units from 1 to N. Then N identical cards, marked from 1 to N, can be shuffled thoroughly and n_1 cards can be drawn blindfolded. These n_1 experimental units receive treatment 1 and the remaining units receive treatment 2. For a quicker and more efficient means of random sampling, one can use the computer or a table of random digits.

For instance, suppose you want to compare the taste of steaks grilled inside the house with those cooked on the outside grill. Of 7 steaks available, three will be cooked on the inside grill and four on the outside grill. Number the steaks 1 to 7, or just set them in a pile. Then, using Appendix B, Table 1, we decide to start at column 8, row 14, and read to the right to obtain

$$6 \quad 2 \quad 2 \quad 1$$

We ignore the second 2. Steaks 6, 2, and 1 would then be grilled inside the house and the other steaks outside.

Although randomization is not a difficult concept, it is one of the most fundamental principles of a good experimental design. It guarantees that uncontrolled sources of variation have the same chance of favoring the response of treatment 1 as they do of favoring the response of treatment 2. Any systematic effects of uncontrolled variables, such as age, strength, resistance, or intelligence, are chopped up or confused in their attempt to influence the treatment response.

Randomization prevents uncontrolled sources of variation from influencing the responses in a systematic manner.

Of course, in many cases, the investigator does not have the luxury of randomization. Consider comparing crime rates of cities before and after a new law. Aside from a package of criminal laws, other factors such as poverty, inflation, and unemployment play a significant role in the prevalence of crime. As long as these contingent factors cannot be regulated during the observation period, caution should be exercised in crediting the new law if a decline in the crime rate is observed or discrediting the new law if an increase in the crime rate is observed.

When randomization cannot be performed, extreme caution must be exercised in crediting an apparent difference in means to a difference in the treatments. The differences may well be due to another factor.

Exercises

3.1   Randomly allocate 2 subjects from among

Al, Bob, Carol, Dennis, Ellen

to be in the control group. The others will receive a treatment.

3.2   Randomly allocate 3 subjects from among 6 mice,

Alpha, Tau, Omega, Pi, Beta, Phi,

to group 1.

3.3   Observations on 10 mothers who nursed their babies and 8 who did not revealed that nursing mothers felt warmer toward their babies. Can we conclude that nursing affects a mother's feelings toward her child?

3.4   Early studies showed a disproportionate number of heavy smokers among lung cancer patients. One scientist theorized that the presence of a particular gene could tend to make a person want to smoke and be susceptible to lung cancer.

(a)   How would randomization settle this question?

(b)   Would such a randomization be ethical with human subjects?

4. MATCHED PAIR COMPARISONS

In comparing two treatments, it is desirable that the experimental units or subjects be as alike as possible, so that a difference in responses between the two groups can be attributed to differences in treatments. If some identifiable conditions vary over the units in an uncontrolled manner, they could introduce a large variability in the measurements. In turn, this could obscure a real difference in treatment effects. On the other hand, the requirement that all subjects be alike may impose a severe limitation on the number of subjects available for a comparative experiment. To compare two analgesics, for example, it would be impractical to look for a sizable number of patients who are of the same sex, age, and general health condition and who have the same severity of pain. Aside from the question of practicality, we would rarely want to confine a comparison to such a narrow group. A broader scope of inference can be attained by applying the treatments on a variety of patients of both sexes and different age groups and health conditions.

The concept of **matching** or **blocking** is fundamental to providing a compromise between the two conflicting requirements that the experimental units be alike and also of different kinds. The procedure consists of choosing units in pairs or blocks so that the units in each block are similar and those in different blocks are dissimilar. One of the units in each block is assigned to treatment 1, the other to treatment 2. This process preserves the effectiveness of a comparison within each block and permits a diversity of conditions to exist in different blocks. Of course, the treatments must be allotted to each pair randomly to avoid selection bias. This design is called **sampling by matched pairs**. For example, in studying how two different environments influence the learning capacities of preschoolers, it is desirable to remove the effect of heredity: Ideally, this is accomplished by working with twins who were raised apart.

Matched Pair Design

Units in each pair are alike, whereas units in different pairs may be dissimilar. In each pair, a unit is chosen at random to receive treatment 1, the other unit receives treatment 2.

In a matched pair design, the response of an experimental unit is influenced by:

1. The conditions prevailing in the block (pair)
2. A treatment effect

By taking the difference between the two observations in a block, we can filter out the common block effect. These differences then permit us to focus on the effects of treatments that are freed from undesirable sources of variation.

Pairing (or Blocking)

Pairing like experimental units according to some identifiable characteristic(s) serves to remove this source of variation from the experiment.

The structure of the observations in a paired comparison is given below, where X and Y denote the responses to treatments 1 and 2, respectively. The difference between the responses in each pair is recorded in the last column, and the summary statistics are also presented.

Structure of Data for a Matched Pair Comparison

Pair	Treatment 1	Treatment 2	Difference
1	X_1	Y_1	$D_1 = X_1 - Y_1$
2	X_2	Y_2	$D_2 = X_2 - Y_2$
.	.	.	.
.	.	.	.
.	.	.	.
n	X_n	Y_n	$D_n = X_n - Y_n$

The differences $D_1, D_2, \ldots D_n$ are a random sample.

Summary statistics:

$$\overline{D} = \frac{1}{n}\sum_{i=1}^{n} D_i \qquad S_D^2 = \frac{\sum_{i=1}^{n}(D_i - \overline{D})^2}{n-1}$$

Although the pairs (X_i, Y_i) are independent of one another, X_i and Y_i within the ith pair will usually be dependent. In fact, if the pairing of experimental units is effective, we would expect X_i and Y_i to be relatively large or small together. Expressed in another way, we would expect (X_i, Y_i), to have a high positive correlation. Because the differences $D_i = X_i - Y_i$, $i = 1, 2, \ldots, n$, are freed from the block effects, it is reasonable to assume that they constitute a random sample from a population with mean $= \delta$ and variance $= \sigma_D^2$, where δ represents the mean difference of the treatment effects. In other words,

$$E(D_i) = \delta$$
$$\text{Var}(D_i) = \sigma_D^2 \qquad i = 1, \ldots, n$$

If the mean difference δ is zero, then the two treatments can be considered equivalent. A positive δ signifies that treatment 1 has a higher mean response than treatment 2. Considering $D_1, \ldots, D_n$ to be a single random sample from a population, we can immediately apply the techniques discussed in Chapters 8 and 9 to learn about the population mean δ.

As we learned in Chapter 8, the assumption of an underlying normal distribution can be relaxed when the sample size is large. The central limit theorem applied to the differences $D_1, \ldots, D_n$ suggests that when n is large, say, greater than 30,

$$\frac{\overline{D} - \delta}{S_D / \sqrt{n}} \quad \text{is approximately } N(0, 1)$$

The inferences can then be based on the percentage points of the $N(0, 1)$ distribution or, equivalently, those of the t-distribution, with the degrees of freedom marked "infinity."

When the sample size is not large, we make the additional assumption that the distribution of the differences is normal.

In summary,

Small Sample Inferences About the Mean Difference δ

Assume that the differences $D_i = X_i - Y_i$ are a random sample from $N(\delta, \sigma_D)$ distribution. Let

$$\overline{D} = \frac{\sum_{i=1}^{n} D_i}{n} \quad \text{and} \quad S_D = \sqrt{\frac{\sum_{i=1}^{n} (D_i - \overline{D})^2}{n - 1}}$$

Then

1. A $100(1 - \alpha)\%$ confidence interval for δ is given by

$$\left(\overline{D} - t_{\alpha/2} \frac{S_D}{\sqrt{n}}, \quad \overline{D} + t_{\alpha/2} \frac{S_D}{\sqrt{n}} \right)$$

    where $t_{\alpha/2}$ is based on $n - 1$ degrees of freedom.

2. A test of $H_0: \delta = \delta_0$ is based on the test statistic

$$T = \frac{\overline{D} - \delta_0}{S_D / \sqrt{n}}, \quad \text{d.f.} = n - 1$$

Example 13   A one-sided paired t-test: Do birth control pills reduce blood pressure? A medical researcher wishes to determine whether a birth control pill has the undesirable side effect of reducing the blood pressure of the user. The study involves recording the initial blood pressures of 15 college-age women. After they use the pill regularly for six months, their blood pressures are again recorded. The researcher wishes to draw inferences about the effect of the pill on blood pressure from the observations given in Table 2.

(a)   Calculate a 95% confidence interval for the mean reduction in blood pressure.

(b)   Do the data substantiate the claim that use of the pill reduces blood pressure? Test at $\alpha = .01$.

TABLE 2   Blood-Pressure Measurements Before and After Use of Pill

	Subject														
	1	2	3	4	5	6	7	8	9	10	11	12	13	14	15
Before (x)	70	80	72	76	76	76	72	78	82	64	74	92	74	68	84
After (y)	68	72	62	70	58	66	68	52	64	72	74	60	74	72	74
$d = (x - y)$	2	8	10	6	18	10	4	26	18	-8	0	32	0	-4	10

Courtesy of a family planning clinic.

SOLUTION AND DISCUSSION

(a)   Here each subject represents a block generating a pair of measurements: one before using the pill and the other after using the pill. The paired differences $d_i = x_i - y_i$ are computed in the last row of Table 2, and we calculate the summary statistics

$$\bar{d} = \frac{\sum d_i}{15} = 8.80, \qquad s_D = \sqrt{\frac{\sum (d_i - \bar{d})^2}{14}} = 10.98$$

If we assume that the paired differences constitute a random sample from a normal population $N(\delta, \sigma_D)$, a 95% confidence interval for the mean difference δ is given by

$$\bar{D} \pm t_{.025} \frac{s_D}{\sqrt{15}}$$

where $t_{.025}$ is based on d.f. $= 14$. From the t table, we find $t_{.025} = 2.145$. The 95% confidence interval is then computed as

$$8.80 \pm 2.145 \times \frac{10.98}{\sqrt{15}} = 8.80 \pm 6.08 \qquad \text{or} \qquad (2.72, \quad 14.88)$$

This means that we are 95% confident the mean reduction of blood pressure is between 2.72 and 14.88.

(b)   1.   *Hypotheses.*   Let δ be the mean difference in blood pressure. Because the claim is that $\delta > 0$, we formulate

$$H_0: \quad \delta = 0 \qquad \text{versus} \qquad H_1: \quad \delta > 0$$

  2.   *Level of significance.*   We are given $\alpha = .01$.

  3.   *Test statistic.*   We employ the test statistic

$$T = \frac{\bar{D}}{s_D / \sqrt{n}} \qquad \text{d.f.} = 14$$

  4.   *Rejection region.*   We reject H_0 for large values of T since the alternative is right-sided. With $\alpha = .01$ and d.f. $= 14$, we find $t_{.01} = 2.624$, so the rejection region is $R: T \geq 2.624$.

5. *Calculation.* The observed value of the test statistic is

$$t = \frac{\bar{d}}{s_D/\sqrt{n}} = \frac{8.80}{10.98/\sqrt{15}} = \frac{8.80}{2.84} = 3.10$$

6. *Conclusion.* The observed value $t = 3.10$ falls in the rejection region R. At level $\alpha = .01$, H_0 is rejected in favor of H_1. Since $t_{.005} = 2.977$, the P-value is smaller than .005.
   The accompanying computer output gives

$$P\text{-value} = .0039 \quad (\text{see Figure 8})$$

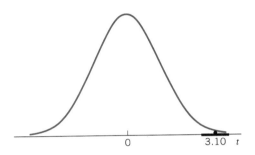

Figure 8. P-value $= .0039$

The evidence against H_0 is very strong. We conclude that a reduction in mean blood pressure, following use of the pill, is strongly supported by the data.

Note: To be more solidly convinced that the birth control pill causes the reduction in blood pressure, it is advisable to measure the blood pressures of the same subjects once again after they have stopped using the pill for a period of time. This amounts to performing the experiment in reverse order to check the findings of the first stage.

MINITAB Output for Example 13
See Exercise 4.12 for the commands.

T-Test of the Mean

Test of mu $= 0.00$ vs mu > 0.00

Variable	N	Mean	StDev	SE Mean	T	P
Diff	15	8.80	10.98	2.83	3.11	0.0039

Confidence Intervals

Variable	N	Mean	StDev	SE Mean	95.0 % CI	
Diff	15	8.80	10.98	2.83	(2.72,	14.88) ∎

Example 13 is a typical before–after situation. Data gathered to determine the effectiveness of a safety program or an exercise program would have the same structure. In such cases, there is really no way to choose how to order the experiments within a pair. The before situation must precede the after situation. If something other than the institution of the program causes performance to improve, the improvement will be incorrectly credited to the program. However, when the order of the application of treatments can be determined by the investigator, something can be done about such systematic influences. Suppose that a coin is flipped to select the treatment for the first unit in each pair. Then the other treatment is applied to the second unit. Because the coin is flipped again for each new pair, any uncontrolled variable has an equal chance of helping either the performance of treatment 1 or treatment 2. After eliminating an identified source of variation by pairing, we return to randomization in an attempt to reduce the systematic effects of any uncontrolled sources of variation.

Randomization with Pairing

After pairing, the assignment of treatments should be randomized for each pair.

Randomization within each pair chops up or diffuses any systematic influences that we are unable to control.

Exercises

4.1 Given the following paired sample data,

x	y
6	7
10	9
8	11
13	11

(a) Evaluate the t statistic $= \dfrac{\bar{d}}{s_D / \sqrt{n}}$.

(b) How many degrees of freedom does this t have?

4.2 Given the following paired sample data,

x	y
5	4
3	2
7	4
5	6
8	8
7	9

(a) Evaluate the t statistic $= \dfrac{\bar{d}}{s_D / \sqrt{n}}$.

(b) How many degrees of freedom does this t have?

4.3   To compare two treatments, a matched pair experiment was conducted with 12 pairs of subjects and the paired differences, response to treatment A minus the response to treatment B, were recorded.

$$2 \quad 5 \quad 6 \quad 8 \quad -6 \quad 4 \quad 18 \quad -12 \quad 17 \quad -7 \quad 16 \quad 12$$

(a)   Is there strong evidence that treatment A produces a larger mean response than treatment B? Test at $\alpha = .05$.

(b)   Construct a 95% confidence interval for the mean difference of the responses to the two treatments.

4.4   A manufacturer claims his boot waterproofing is better than the major brand. Five pairs of shoes are available for a test.

(a)   Explain how you would conduct a paired sample test.

(b)   Write down your assignment of waterproofing to each shoe. How did you randomize?

4.5   It is claimed that an industrial safety program is effective in reducing the loss of working hours due to factory accidents. The following data are collected concerning the weekly loss of working hours due to accidents in six plants both before and after the safety program is instituted.

	Plant					
	1	2	3	4	5	6
Before	12	29	16	37	28	15
After	10	28	17	35	25	16

Do the data substantiate the claim? Use $\alpha = .05$.

4.6   A food scientist wants to study whether quality differences exist between yogurt made from skim milk with and without the preculture of a particular type of bacteria, called Psychrotrops (PC). Samples of skim milk are procured from seven dairy farms. One-half of the milk sampled from each farm is inoculated with PC, and the other half is not. After yogurt is made with these milk samples, the firmness of the curd is measured, and those measurements are given below.

Curd Firmness	Dairy Farm						
	A	B	C	D	E	F	G
With PC	68	75	62	86	52	46	72
Without PC	61	69	64	76	52	38	68

Do these data substantiate the conjecture that the treatment of PC results in a higher degree of curd firmness? Test at $\alpha = .05$.

4.7   Referring to Exercise 4.6, determine a 90% confidence interval for the mean increase of curd firmness due to the PC treatment.

4.8   A study is to be made of the relative effectiveness of two kinds of cough medicines in increasing sleep. Six people with colds are given medicine A the first night and medicine B the second night. Their hours of sleep each night are recorded.

	\multicolumn{6}{c}{Subject}
	1	2	3	4	5	6
Medicine A	4.8	4.1	5.8	4.9	5.3	7.4
Medicine B	3.9	4.2	5.0	4.9	5.4	7.1

(a)   Establish a 95% confidence interval for the mean change in hours of sleep when switching from medicine A to medicine B.

(b)   How and what would you randomize in this study? Briefly explain your reason for randomization.

4.9   Two methods of memorizing difficult material are being tested to determine whether one produces better retention. Nine pairs of students are included in the study. The students in each pair are matched according to IQ and academic background and then assigned to the two methods at random. A memorization test is given to all the students, and the following scores are obtained:

	\multicolumn{9}{c}{Pair}
	1	2	3	4	5	6	7	8	9
Method A	90	86	72	65	44	52	46	38	43
Method B	85	87	70	62	44	53	42	35	46

At $\alpha = .05$, test to determine whether there is a significant difference in the effectiveness of the two methods.

4.10   In an experiment conducted to see whether electrical pricing policies can affect consumer behavior, 10 homeowners in Wisconsin had to pay a premium for power use during the peak hours. They were offered lower off-peak rates. For each home, the July on-peak usage (kilowatt hours) under the pricing experiment was compared to the previous July usage.

	Year	
Previous	Experimental	
200	160	
180	175	
240	210	
425	370	
120	110	
333	298	
418	368	
380	250	
340	305	
516	477	

(a) Find a 95% confidence interval for the mean decrease.

(b) Test $H_0: \delta = 0$ against $H_1: \delta \neq 0$ at level $\alpha = .05$.

(c) Comment on the feasibility of randomization of treatments.

(d) Without randomization, in what way could the results in (a) and (b) be misleading? (*Hint:* What if air conditioner use is a prime factor, and the July with experimental pricing was cooler than the previous July?)

4.11 Measurements of the left- and right-hand gripping strengths of 10 left-handed writers are recorded.

	Person									
	1	2	3	4	5	6	7	8	9	10
Left hand	140	90	125	130	95	121	85	97	131	110
Right hand	138	87	110	132	96	120	86	90	129	100

(a) Do the data provide strong evidence that people who write with the left hand have a greater gripping strength in the left hand than in the right hand?

(b) Construct a 90% confidence interval for the mean difference.

USING A COMPUTER

4.12 The analysis of a matched pair sample using the MINITAB software package is illustrated on page 440 with reference to the data in Example 13.

DataC10T2.DAT:

C1: 70 80 72 · · · 68 84
C2: 68 72 62 · · · 72 74

Dialog box:	Session command
Calc > Calculator	MTB > NAME C3 'Diff'
Type *Diff* in **Store result in variable**	MTB > LET 'DIFF' = C1 - C2
Type *C1 - C2* in the **Expression** box. Click **OK**.	
Dialog box:	Session command
Stat > Basic Statistics > 1-Sample t	MTB > TINTERVAL C3
Type *Diff* in **Variables**. Click **OK**. Click **OK**.	

To test the null hypothesis of zero mean difference, use

Click **Test mean**. Click **OK**.	MTB > TTEST C3

Output:

Confidence Intervals

```
Variable         N       Mean     StDev   SE Mean       95.0 % CI
Diff            15       8.80     10.98      2.83   (   2.72,   14.88)
```

T-Test of the Mean

Test of mu = 0.00 vs mu not = 0.00

```
Variable         N      Mean      StDev    SE Mean         T          P
Diff            15      8.80      10.98       2.83      3.11     0.0077
```

(a) Do Exercise 4.10 on a computer.
(b) Do Exercise 4.11 on a computer.

Summary of Confidence Intervals and Tests

Table 3 summarizes the major statistical procedures thus far. Notice the common structure of confidence intervals and tests.

TABLE 3 General Formulas for Inferences about a Mean (μ), Difference of Two Means ($\mu_1 - \mu_2$)

Confidence interval = Point estimator ± (Tabled value) (Estimated or true std. dev.)

$$\textbf{Test statistic} = \frac{\text{Point estimator} - \text{Parameter value at } H_0 \text{ (null hypothesis)}}{\text{(Estimated or true) std. dev. of point estimator}}$$

	Ch. 8	Ch. 9	Ch. 10	Independent Samples		Matched Samples
	General	Normal with unknown σ	Normal $N(\mu_1, \sigma_1), N(\mu_2, \sigma_2)$ $\sigma_1 = \sigma_2 = \sigma$	Normal $N(\mu_1, \sigma_1), N(\mu_2, \sigma_2)$ $\sigma_1 \neq \sigma_2$	General	
Population(s)					Normal $N(\mu_1, \sigma_1), N(\mu_2, \sigma_2)$	Normal for the difference $D_i = X_i - Y_i$
Inference on	Mean μ	Mean μ	$\mu_1 - \mu_2$	$\mu_1 - \mu_2$	$\mu_1 - \mu_2$	$\delta = \mu_1 - \mu_2$
Sample(s)	$X_1, \ldots, X_n$	$X_1, \ldots, X_n$	$X_1, \ldots, X_{n_1}$ $Y_1, \ldots, Y_{n_2}$	$X_1, \ldots, X_{n_1}$ $Y_1, \ldots, Y_{n_2}$	$X_1, \ldots, X_{n_1}$ $Y_1, \ldots, Y_{n_2}$	$\begin{matrix} X_1 & Y_1 & D_1 = X_1 - Y_1 \\ X_2 & Y_2 & D_2 = X_2 - Y_2 \\ \vdots & \vdots & \vdots \\ X_n & Y_n & D_n = X_n - Y_n \end{matrix}$
Sample size n	Large $n > 30$	$n \geq 2$	$n_1 \geq 2$ $n_2 \geq 2$	$n_1 \geq 2$ $n_2 \geq 2$	$n_1 > 30$ $n_2 > 30$	$n \geq 2$
Point estimator	$\bar{X}$	$\bar{X}$	$\bar{X} - \bar{Y}$	$\bar{X} - \bar{Y}$	$\bar{X} - \bar{Y}$	$\bar{D} = \bar{X} - \bar{Y}$
Variance of point estimator	$\dfrac{\sigma^2}{n}$	$\dfrac{\sigma^2}{n}$	$\sigma^2\left(\dfrac{1}{n_1} + \dfrac{1}{n_2}\right)$	$\dfrac{\sigma_1^2}{n_1} + \dfrac{\sigma_2^2}{n_2}$	$\dfrac{\sigma_1^2}{n_1} + \dfrac{\sigma_2^2}{n_2}$	$\dfrac{\sigma_D^2}{n}$
Std. dev. of point estimator	$\dfrac{\sigma}{\sqrt{n}}$	$\dfrac{\sigma}{\sqrt{n}}$	$\sigma\sqrt{\dfrac{1}{n_1} + \dfrac{1}{n_2}}$	$\sqrt{\dfrac{\sigma_1^2}{n_1} + \dfrac{\sigma_2^2}{n_2}}$	$\sqrt{\dfrac{\sigma_1^2}{n_1} + \dfrac{\sigma_2^2}{n_2}}$	$\dfrac{\sigma_D}{\sqrt{n}}$
Estimated std. dev.	$\dfrac{S}{\sqrt{n}}$	$\dfrac{S}{\sqrt{n}}$	$S_{\text{pooled}}\sqrt{\dfrac{1}{n_1} + \dfrac{1}{n_2}}$	$\sqrt{\dfrac{S_1^2}{n_1} + \dfrac{S_2^2}{n_2}}$	$\sqrt{\dfrac{S_1^2}{n_1} + \dfrac{S_2^2}{n_2}}$	$\dfrac{S_D}{\sqrt{n}}$
Distribution	Normal	t with d.f. = $n - 1$	t with d.f. = $n_1 + n_2 - 2$	t with d.f. = smaller of $n_1 - 1$ and $n_2 - 1$	Normal	t with d.f. = $n - 1$
Test statistic	$\dfrac{\bar{X} - \mu_0}{S/\sqrt{n}}$	$\dfrac{\bar{X} - \mu_0}{S/\sqrt{n}}$	$\dfrac{(\bar{X} - \bar{Y}) - \delta_0}{S_{\text{pooled}}\sqrt{\dfrac{1}{n_1} + \dfrac{1}{n_2}}}$ $S_{\text{pooled}}^2 = \dfrac{(n_1 - 1)S_1^2 + (n_2 - 1)S_2^2}{n_1 + n_2 - 2}$	$\dfrac{(\bar{X} - \bar{Y}) - \delta_0}{\sqrt{\dfrac{S_1^2}{n_1} + \dfrac{S_2^2}{n_2}}}$	Same as $\leftarrow$	$\dfrac{\bar{D} - \delta_0}{S_D/\sqrt{n}}$ S_D = sample std. dev. of the D_i's

KEY IDEAS AND FORMULAS

A carefully designed experiment is fundamental to the success of a comparative study. The appropriate methods of statistical inference depend on the sampling scheme used to collect data.

The most basic experimental designs to compare two treatments are **independent samples** and **matched pair sample.**

The design of independent samples requires the subjects to be randomly selected for assignment to each treatment. **Randomization** prevents uncontrolled factors from systematically favoring one treatment over the other.

With a matched pair design, subjects in each pair are alike, while those in different pairs may be dissimilar. For each pair, the two treatments should be randomly allocated to the members.

Pairing subjects according to some feature prevents that source of variation from interfering with treatment comparisons. By contrast, random allocation of subjects according to the independent random sampling design spreads these variations between the two treatments.

Inferences with Two Independent Random Samples

1. *Large samples.* When n_1 and n_2 are both greater than 30, inferences about $\mu_1 - \mu_2$ are based on the fact that

$$\frac{(\overline{X} - \overline{Y}) - (\mu_1 - \mu_2)}{\sqrt{\dfrac{S_1^2}{n_1} + \dfrac{S_2^2}{n_2}}} \qquad \text{is approximately } N(0, 1)$$

A $100(1 - \alpha)\%$ confidence interval for $(\mu_1 - \mu_2)$ is

$$(\overline{X} - \overline{Y}) \pm z_{\alpha/2} \sqrt{\frac{S_1^2}{n_1} + \frac{S_2^2}{n_2}}$$

To test $H_0: \mu_1 - \mu_2 = \delta_0$, we use the normal test statistic

$$Z = \frac{(\overline{X} - \overline{Y}) - \delta_0}{\sqrt{\dfrac{S_1^2}{n_1} + \dfrac{S_2^2}{n_2}}}$$

No assumptions are needed in regard to the shape of the population distributions.

2. *Small samples.* When n_1 and n_2 are small, inferences using the t distribution require the assumptions:

(a) Both populations are normal.

(b) $\sigma_1 = \sigma_2$

The common σ^2 is estimated by

$$S^2_{pooled} = \frac{(n_1 - 1)S_1^2 + (n_2 - 1)S_2^2}{n_1 + n_2 - 2}$$

Inferences about $\mu_1 - \mu_2$ are based on

$$T = \frac{(\bar{X} - \bar{Y}) - (\mu_1 - \mu_2)}{S_{pooled}\sqrt{\dfrac{1}{n_1} + \dfrac{1}{n_2}}}, \qquad \text{d.f.} = n_1 + n_2 - 2$$

A $100(1 - \alpha)\%$ confidence interval for $(\mu_1 - \mu_2)$ is

$$(\bar{X} - \bar{Y}) \pm t_{\alpha/2}\, S_{pooled}\sqrt{\dfrac{1}{n_1} + \dfrac{1}{n_2}}$$

To test $H_0: \mu_1 - \mu_2 = \delta_0$, the test statistic is

$$T = \frac{(\bar{X} - \bar{Y}) - \delta_0}{S_{pooled}\sqrt{\dfrac{1}{n_1} + \dfrac{1}{n_2}}}, \qquad \text{d.f.} = n_1 + n_2 - 2$$

Inferences with a Matched Pair Sample

With a paired sample $(X_1, Y_1), \ldots, (X_n, Y_n)$, the first step is to calculate the differences $D_i = X_i - Y_i$, their mean $\bar{D}$, and standard deviation S_D.

If n is small, we assume that the D_i's are normally distributed $N(\delta, \sigma_D)$. Inferences about δ are based on

$$T = \frac{\bar{D} - \delta}{S_D/\sqrt{n}}, \qquad \text{d.f.} = n - 1$$

A $100(1 - \alpha)\%$ confidence interval for δ is

$$\bar{D} \pm t_{\alpha/2}\, S_D/\sqrt{n}$$

The test of $H_0: \delta = \delta_0$ is performed with the test statistic:

$$T = \frac{\bar{D} - \delta_0}{S_D/\sqrt{n}}, \qquad \text{d.f.} = n - 1$$

If n is large, the assumption of normal distribution for the D_i's is not needed. Inferences are based on the fact that

$$Z = \frac{\bar{D} - \delta_0}{S_D/\sqrt{n}} \qquad \text{is approximately } N(0, 1)$$

5. REVIEW EXERCISES

5.1 The following summary is recorded for independent samples from two populations.

	Sample 1	Sample 2
	$n_1 = 65$	$n_2 = 60$
	$\bar{x} = 194$	$\bar{y} = 165$
	$s_1^2 = 86$	$s_2^2 = 137$

(a) Construct a 98% confidence interval for $\mu_1 - \mu_2$.

(b) Test $H_0 : \mu_1 - \mu_2 = 35$ versus $H_1 : \mu_1 - \mu_2 \neq 35$ with $\alpha = .02$.

(c) Test $H_0 : \mu_1 - \mu_2 = 35$ versus $H_1 : \mu_1 - \mu_2 < 35$ with $\alpha = .05$.

5.2 Given here are two sets of values for the standard deviations of independent random samples, each of size 52, from two populations. For each case, determine whether or not an observed difference of 8 between the sample means is statistically significant at $\alpha = .05$.

(a) $s_1 = 21$, $s_2 = 29$

(b) $s_1 = 11$, $s_2 = 14$

5.3 A group of 88 subjects is used in an experiment to compare two treatments. From this group, 40 subjects are randomly selected to be assigned to treatment 1 and the remaining 48 subjects are assigned to treatment 2. The means and standard deviations of the responses are

	Treatment 1	Treatment 2
Mean	15.62	27.25
Standard deviation	2.88	4.32

Determine a 95% confidence interval for the mean difference of the treatment effects.

5.4 Refer to the data in Exercise 5.3. Suppose the investigator wishes to establish that the mean response of treatment 2 is larger than that of treatment 1 by more than 10 units.

(a) Formulate the null hypothesis and the alternative hypothesis.

(b) State the test statistic and the rejection region with $\alpha = .10$.

(c) Perform the test at $\alpha = .10$. Also, find the P-value and comment.

5.5   A study of postoperative pain relief is conducted to determine whether drug A has a significantly longer duration of pain relief than drug B. Observations of the hours of pain relief are recorded for 55 patients given drug A and 58 patients given drug B. The summary statistics are

	A	B
Mean	5.64	5.03
Standard deviation	1.25	1.82

(a)   Formulate H_0 and H_1.
(b)   State the test statistic and the rejection region with $\alpha = .10$.
(c)   State the conclusion of your test with $\alpha = .10$. Also, find the P-value and comment.

5.6   Consider the data of Exercise 5.5.
(a)   Construct a 90% confidence interval for $\mu_A - \mu_B$.
(b)   Give a 95% confidence interval for μ_A using the data of drug A alone. (Note: Refer to Chapter 8.)

5.7   Given the following two samples,

$$8, \quad 10, \quad 5, \quad 9, \quad 8 \qquad \text{and} \qquad 6, \quad 2, \quad 4, \quad 8$$

obtain (a) s^2_{pooled} and (b) value of the t statistic for testing H_0: $\mu_1 - \mu_2 = 2$. State the d.f. of the t.

5.8   Two work designs are being considered for possible adoption in an assembly plant. A time study is conducted with 10 workers using design 1 and 12 workers using design 2. The means and standard deviations of their assembly times (in minutes) are

	Design 1	Design 2
Mean	78.3	85.6
Standard deviation	4.8	6.5

Is the mean assembly time significantly higher for design 2?

5.9   Refer to the data of Exercise 5.8. Give 95% confidence intervals for the mean assembly times for design 1 and design 2 individually (see Chapter 9).

5.10   A fruit grower wishes to evaluate a new spray that is claimed to reduce the loss due to damage by insects. To this end, he performs an experiment with 27 trees in his orchard by treating 12 of those trees with the new spray and the other 15 trees with the standard spray. From the data of fruit yield (in pounds) of those trees, the following summary statistics were found.

	New Spray	Standard Spray
Mean yield	249	233
Standard deviation	19	45

Do these data substantiate the claim that a higher yield should result from the use of the new spray? State the assumptions you make and test at $\alpha = .05$.

5.11   Referring to Exercise 5.10, construct a conservative 95% confidence interval for the difference in mean yields between the new spray and the standard spray.

5.12   Referring to Exercise 5.10, construct 90% confidence intervals for the mean yields under the use of the new spray and the standard spray individually.

5.13   In the early 1970s, students started a phenomenon called "streaking." Within a two-week period following the first streaking sighted on campus, a standard psychological test was given to a group of 19 males who were admitted streakers and a control group of 19 males who were nonstreakers. S. Stoner and M. Watman reported the following data [*Source: Psychology*, 11 (4) (1975), pp.14–16], regarding scores on a test designed to determine extroversion:

Streakers	Nonstreakers
$\bar{x} = 15.26$	$\bar{y} = 13.90$
$s_1 = 2.62$	$s_2 = 4.11$

(a)   Construct a 95% confidence interval for the difference in population means. Does there appear to be a difference between the two groups?

(b)   It may be true that those who admit to streaking differ from those who do not admit to streaking. In light of this possibility, what criticism can be made of your analytical conclusion?

5.14 In each of the following cases, how would you select the experimental units and conduct the experiment?

   (a) Compare the mileage obtained from two gasolines; 16 cars are available for the experiment.

   (b) Test two varnishes; 12 birch boards are available for the experiment.

   (c) Compare two methods of teaching basic ice skating; 40 seven-year-old girls are available for the experiment.

5.15 To compare two treatments, a matched pair experiment was conducted with 9 pairs of subjects and the differences, response to treatment 1 minus the response to treatment 2, were recorded.

$$-1.26, \quad -.82, \quad 1.32, \quad 3.16, \quad 4.62, \quad .60, \quad -.55, \quad 1.54, \quad 2.18$$

   (a) Is there strong evidence of a difference between the mean responses of the two treatments? Test at $\alpha = .02$.

   (b) Construct a 90% confidence interval for the mean difference of the response to the two treatments.

5.16 An experiment is conducted to determine whether the use of a special chemical additive with a standard fertilizer accelerates plant growth. Ten locations are included in the study. At each location, two plants growing in close proximity are treated: One is given the standard fertilizer, the other the standard fertilizer with the chemical additive. Plant growth after four weeks is measured in centimeters.

   Do the following data substantiate the claim that use of the chemical additive accelerates plant growth? State the assumptions that you make and devise an appropriate test of the hypothesis. Take $\alpha = .05$.

	Location									
	1	2	3	4	5	6	7	8	9	10
Without additive	20	31	16	22	19	32	25	18	20	19
With additive	23	34	15	21	22	31	29	20	24	23

5.17 Obtain a 95% confidence interval for δ using the data in Exercise 5.16.

5.18 Referring to Exercise 5.16, suppose that the two plants at each location are situated in the east–west direction. In designing this experiment, you must decide which of the 2 plants at each location—the one in the east or the one in the west—is to be given the chemical additive.

   (a) Explain how by repeatedly tossing a coin, you can randomly allocate the treatments to the plants at the 10 locations.

   (b) Perform the randomization by actually tossing a coin 10 times.

5.19   A trucking firm wishes to choose between two alternate routes for trans-
porting merchandise from one depot to another. One major concern is the
travel time. In a study, 5 drivers were randomly selected from a group of
10 and assigned to route A, the other 5 were assigned to route B. The
following data were obtained.

	Travel Time (hours)				
Route A	18	24	30	21	32
Route B	22	29	34	25	35

(a)   Is there a significant difference between the mean travel times between the two routes? State the assumptions you have made in performing the test.

(b)   Suggest an alternative design for this study that would make a comparison more effective.

5.20   It is anticipated that a new instructional method will more effectively improve the reading ability of elementary school children than the standard
method currently in use. To test this conjecture, 16 children are divided
at random into two groups of 8 each. One group is instructed using the
standard method and the other group using the new method. The children's scores on a reading test are found to be

	Reading Test Scores							
Standard	65	70	76	63	72	71	68	68
New	75	80	72	77	69	81	71	78

Use both hypotheses testing and confidence interval methods to draw
inferences about the difference of the population mean scores. Take
$\alpha = .05$.

USING A COMPUTER

5.21   Refer to the alligator data in Table D.7 of the Data Bank. Using the data
on testosterone x_4 from the Lake Apopka alligators, find a 95% confidence
interval for the difference of means between males and females. There
should be a large difference for healthy alligators. Comment on your conclusion.

5.22   Refer to the alligator data in Table D.7 of the Data Bank. Using the data on testosterone x_4 for male alligators, compare the means for the two lake regions.

(a)   Should you pool the variances with these data?

(b)   Find a 90% confidence interval for the difference of means between the two lakes. Use the conservative procedure on page 418.

(c)   Which population has the higher mean and how much higher is it?

ACTIVITIES TO IMPROVE UNDERSTANDING: STUDENT PROJECTS

Formulate a question of interest concerning two treatments and conduct a comparative experiment. Report your activities.

Statement of Purpose

Plan for Data Collection

   Treatments

   Treatment 1 ————————————————

   Treatment 2 ————————————————

   Variable ————————————————

   Choice of Design
   Matched Pair or Independent Samples   Explain your choice.

   Population(s) ————————————————

   Randomization   Specify your randomization and how the sample is collected.

Calculations
Determine confidence intervals and conduct any appropriate tests of hypotheses.

Summary of Findings
Interpret your tests and confidence intervals. Relate findings to the purpose of the study.

CHAPTER 11

Analyzing Count Data

Chapter Objectives

At the end of this chapter, you should be able to

▶ Identify categorical data.
▶ Construct a confidence interval for a population proportion.
▶ Conduct tests of hypotheses involving a single population proportion.
▶ Construct confidence intervals and conduct tests for the difference of two population proportions.
▶ Interpret the chi-square statistic and test for comparing two populations from frequencies of a categorical variable.
▶ Interpret the chi-square statistic and test for the independence of two simultaneously observed categorical variables.

1. INTRODUCTION

When the outcomes for variables are categories, which is the case when we observe traits or characteristics of the experimental units, the basic data will be counts. Examples include the number of applications accepted and the number of serious complaints received. Whenever responses to survey questions are classified into categories or the responses are directly recorded on, say, a 5-point scale, the data again consist of the counts for each category.

Data that consist of counts associated with categories are called **categorical data.** Descriptive summaries of these count, or frequency, data were introduced in Chapter 3. In the simplest case, there are only two categories and any inference will concern the population proportion for a category. In terms of the basic models discussed in Chapter 4, the two category case corresponds to either the large urn or the spinner model.

Example 1   A single population proportion—A large urn model

A survey of 200 students revealed that 115 of them would rather E-mail their professor than appear in person during office hours. Does this provide convincing information that over half of all students at the university would prefer E-mail to office hour contacts? What are the plausible values for the population proportion of students who prefer E-mail? Statistical tests and confidence intervals can answer these questions.   ■

For purposes of comparison, samples can be taken from different populations.

Example 2   Comparing two population proportions

Before commercials are placed on national television, they undergo testing and modification. Technology has now made it easy to show one version of a commercial to half the audience and a second version to the other half. A follow-up telephone survey is conducted and the responses are categorized as

don't remember key point   or   *able to remember key point*

In one such situation, 80 viewers of Commercial A and 70 viewers of Commercial B were interviewed. The results are shown in Table 1.

TABLE 1   Remembering Key Point in a Commercial

	Don't Remember Key Point	Remember Key Point	Total
Commercial A	43	37	80
Commercial B	52	18	70
Total	95	55	150

A statistical test, based on these data, can answer the question of whether or not the two versions of the commercial are equally well remembered. This is a comparison of proportions from two populations. One population consists of all potential viewers of Commercial A and the other population consists of all potential viewers of Commercial B. ∎

2. INFERENCES ABOUT A PROPORTION

When a single response is classified into only two categories—for example, employed/not employed, male/female, defective/not defective—we can call the category of interest *success*, S, and the other *failure*, F.

Two basic models that we met in Chapter 4 are appropriate for the two category case. One is the large urn model where a random sample of n units is selected from an urn with a large number of units having a characteristic labeled as S or F. The number of S's is recorded and the population proportion of S's, p, is a quantity of interest.

The second model is the spinner model with two categories, S and F. A total of n spins are made and the number of S's recorded. The probability of success, again denoted by p, is the quantity of interest.

Under the spinner model, the responses for different spins are independent and the probability of success is the same for every trial. Independence and constant population proportion hold, approximately, for the responses from different units selected from the large urn. The conditions of Bernoulli trials described in Chapter 5 are satisfied. Consequently, under either the spinner or large urn model, the total number of successes, X, has the binomial distribution.

In summary, for the two category case, the data will consist of the count X of the number in the sample of size n that are labeled S.

Common sense suggests the **sample proportion**

$$\hat{p} = \frac{X}{n}$$

as an estimator of the **population proportion** p. The "hat" above p reminds us that this random quantity is a statistic that provides an estimate of the parameter p.

SAMPLING PROPERTIES OF THE SAMPLE PROPORTION $\hat{p}$

The outcome of a single trial is either a success, S, or a failure, F. We can represent the outcome as a random variable Y that takes the value 1 or 0 according to whether or not S occurs. Then Y takes the value 1 with probability p and 0 with probability $q = 1 - p$. We can determine the mean and variance of Y from its distribution:

Value y	Probability $f(y)$
0	$1 - p$
1	p

$$\text{Mean} = \mu_Y = 0 \cdot (1 - p) + 1 \cdot p = p$$
$$\text{Variance} = \sigma_Y^2 = E[Y^2] - \mu_Y^2$$
$$= 0^2 \cdot (1 - p) + 1^2 \cdot p - p^2 = p(1 - p)$$

so the standard deviation of Y is $\sigma_Y = \sqrt{pq}$.

Let $Y_i = 1$ if the ith trial is an S and $Y_i = 0$ otherwise. Each Y_i has the same distribution as Y and the Y_i are independent. Further, the total number of successes X is the sum $X = Y_1 + Y_2 + \cdots + Y_n$, and $\hat{p} = X/n = \overline{Y}$. Because $\hat{p}$ is actually a sample mean, we know that its sampling distribution has

$$\text{Mean of } \hat{p} = \mu_Y = p$$

and

$$\text{Standard deviation of } \hat{p} = \frac{\sigma_Y}{\sqrt{n}} = \sqrt{\frac{pq}{n}}$$

Further, for large sample size, we can appeal to the central limit result to conclude that **the sampling distribution of $\hat{p}$ is approximately normal with mean p and standard deviation $\sqrt{pq/n}$** so

$$Z = \frac{\hat{p} - p}{\sqrt{pq/n}} = \frac{\overline{Y} - \mu_Y}{\sigma_Y/\sqrt{n}}$$

is nearly standard normal. This fact, illustrated in Figure 1, is crucial to making inferences about the population proportion p.

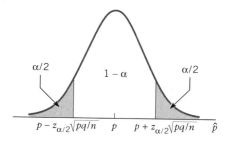

Figure 1. Approximate normal distribution for $\hat{p}$: Large n

POINT ESTIMATION OF p

From our discussion of the sampling distribution above, we know that when the count X has a binomial distribution for n trials, the sample proportion $\hat{p} = X/n$ has mean and standard deviation

$$\mu_{\hat{p}} = p$$

$$\sigma_{\hat{p}} = \sqrt{\frac{pq}{n}}$$

The first result states that the sampling distribution of $\hat{p}$ has the population proportion p as its mean. The second result says that the standard error of the estimator $\hat{p}$ is

$$\text{S.E.}(\hat{p}) = \sqrt{\frac{pq}{n}}$$

If $\hat{p}$ is used as an estimate of p, its estimated standard error is obtained by replacing p by $\hat{p}$ and q by $\hat{q} = 1 - \hat{p}$ in the formula for standard error.

$$\text{Estimated S.E.}(\hat{p}) = \sqrt{\frac{\hat{p}\hat{q}}{n}}$$

When the sample size n is large, prior to sampling, the probability is approximately .954 that the error of estimation $|\hat{p} - p|$ will be less than $2 \times$ (estimated S.E.). That is, $2\sqrt{\hat{p}\hat{q}/n}$ is a 95.4% error margin.

Point Estimation of a Population Proportion

Parameter: Population proportion p

Data: $X =$ Count of number with characteristic in a random sample of size n

Estimator: $\hat{p} = \dfrac{X}{n}$

$$\text{S.E.}(\hat{p}) = \sqrt{\frac{pq}{n}} \quad \text{and} \quad \text{Estimated S.E.}(\hat{p}) = \sqrt{\frac{\hat{p}\hat{q}}{n}}$$

For large n, $z_{\alpha/2}\sqrt{\hat{p}\hat{q}/n}$ is an approximate $100(1 - \alpha)\%$ error margin.

Example 3   A point estimate for the proportion who prefer E-mail

A survey of 200 students revealed that 115 of them would rather E-mail their professor than appear in person during office hours.

Give a point estimate of the proportion of students that prefer E-mail to office hour visits. Also, give the approximate 95.4% error margin.

SOLUTION AND
DISCUSSION

The 200 students in the sample represent only a small fraction of the total enrollment at the university, so the count can be treated as if it were a binomial random variable.

Here $n = 200$ and $X = 115$, so the estimate of the population proportion is

$$\hat{p} = \frac{115}{200} = .575$$

$$\text{Estimated S.E.}(\hat{p}) = \sqrt{\frac{\hat{p}\hat{q}}{n}} = \sqrt{\frac{.575 \times .425}{200}} = .035$$

The estimate $\hat{p} = .575$ has approximate 95.4% error margin $2 \times .035 = .070$.   ■

Producers of breakfast cereals continually experiment with new products. Each promising new cereal must be market-tested on a sample of potential purchasers. An added twist here is that youngsters are a major component of the market. To elicit accurate information from young people, one firm developed a smiling face scale.

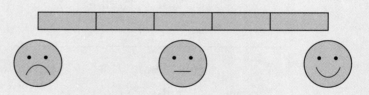

After tasting a new product, respondents are asked to check one box to rate the taste. A good product should have most of the youngsters responding in the top two boxes. Grouping these into a single top category and the lower three boxes into a lower category, we are in the situation of estimating the proportion of the market population that would rate taste in the top category.

Out of a sample of 42 youngsters, 30 rated a new cereal in the top category.

Example 4   Estimating the proportion of purchasers

A large mail-order club that offers monthly specials wishes to try out a new item. A trial mailing is sent to a random sample of 250 members selected from the list of over 9000 subscribers. Based on this sample mailing, 70 of the members decide to purchase the item. Give a point estimate of the proportion of club members that could be expected to purchase the item and attach a 95.4% error margin.

SOLUTION AND DISCUSSION The number in the sample represents only a small fraction of the total membership, so the count can be treated as if it were a binomial variable.

Here $n = 250$ and $X = 70$, so the estimate of the population proportion is

$$\hat{p} = \frac{70}{250} = .28$$

$$\text{Estimated S.E.}(\hat{p}) = \sqrt{\frac{\hat{p}\hat{q}}{n}} = \sqrt{\frac{.28 \times .72}{250}} = .028$$

$$95.4\% \text{ error margin} = 2 \times .028 = .056$$

Therefore, the estimated proportion is $\hat{p} = .28$, with a 95.4% error margin of .06 (rounded to two decimals).   ■

CONFIDENCE INTERVAL FOR p

A large sample confidence interval for a population proportion follows from the approximate normality of the sample proportion $\hat{p}$. Since $\hat{p}$ is nearly normal with mean p and standard deviation $\sqrt{pq/n}$, the random interval $\hat{p} \pm z_{\alpha/2} \sqrt{pq/n}$ is a candidate. However, the standard deviation involves the unknown parameter p, so we use the estimated standard deviation $\sqrt{\hat{p}\hat{q}/n}$ to set the endpoints of the confidence interval. Notice again that the common form of the confidence interval is

$$\text{Estimate} \pm (\text{Tabled } z\text{-value})(\text{Estimated standard error})$$

Large Sample Confidence Interval for p

For large n, a $100(1 - \alpha)\%$ confidence interval for p is given by

$$\left(\hat{p} - z_{\alpha/2}\sqrt{\frac{\hat{p}\hat{q}}{n}}, \quad \hat{p} + z_{\alpha/2}\sqrt{\frac{\hat{p}\hat{q}}{n}} \right)$$

Example 5   A confidence interval for the proportion unemployed
A local government agency wishes to investigate the prevailing rate of unemployment. It reasons correctly that this assessment could be made accurately and efficiently by sampling a small fraction of the labor force.

Among 500 randomly selected persons interviewed, 41 are found to be unemployed. Compute a 95% confidence interval for the rate of unemployment.

SOLUTION AND DISCUSSION   The sample size $n = 500$ is large, so a normal approximation to the distribution of the sample proportion $\hat{p}$ is justified. Since $(1 - \alpha) = .95$, we have $\alpha/2 = .025$ and $z_{.025} = 1.96$. The observed $\hat{p} = 41/500 = .082$, and $\hat{q} = 1 - .082 = .918$. We calculate

$$z_{.025}\sqrt{\frac{\hat{p}\hat{q}}{n}} = 1.96\sqrt{\frac{.082 \times .918}{500}} = 1.96 \times .012 = .024$$

Therefore, a 95% confidence interval for the rate of unemployment in the county is $.082 \pm .024 = (.058, .106)$, or $(5.8\%, 10.6\%)$ in percentages.

Because our procedure will produce true statements 95% of the time, we can be 95% confident that the rate of unemployment is between 5.8% and 10.6%.   ■

DETERMINING THE SAMPLE SIZE

Note that, prior to sampling, the numerical estimate $\hat{p}$ of p is not available. Therefore, for a $100(1 - \alpha)\%$ error margin for the estimation of p, we use the expression $z_{\alpha/2}\sqrt{pq/n}$. The required sample size is obtained by equating $z_{\alpha/2}\sqrt{pq/n} = d$ where d is the specified error margin. We then obtain

$$n = pq\left[\frac{z_{\alpha/2}}{d}\right]^2$$

If the value of p is known to be roughly in the neighborhood of a value p^*, then n can be determined from

$$n = p^*(1 - p^*)\left[\frac{z_{\alpha/2}}{d}\right]^2$$

Without prior knowledge of p, pq can be replaced by its maximum possible value $\frac{1}{4}$ and n determined from the relation

$$n = \frac{1}{4}\left[\frac{z_{\alpha/2}}{d}\right]^2$$

Example 6   **Sample size selection for estimating a proportion**
A public health survey is to be designed to estimate the proportion p of a population having defective vision. How many persons should be examined if the public health commissioner wishes to be 98% certain that the error of estimation is below .05 when

(a)   There is no knowledge about the value of p?

(b) p is known to be about .3?

SOLUTION AND DISCUSSION   The tolerable error is $d = .05$. Also $(1 - \alpha) = .98$, so $\alpha/2 = .01$. From the normal table, we know that $z_{.01} = 2.33$. Therefore,

(a) Since p is unknown, the conservative bound on n yields

$$\frac{1}{4}\left[\frac{2.33}{.05}\right]^2 = 542.9$$

A sample of size 543 would suffice.

(b) If $p^* = .3$, the required sample size is

$$n = (.3 \times .7)\left[\frac{2.33}{.05}\right]^2 = 456$$

The most dramatic point of these calculations is that a sample size of several hundred is required to obtain accurate estimates of a proportion.   ∎

LARGE SAMPLE TESTS ABOUT p

We consider testing $H_0: p = p_0$ versus $H_1: p \neq p_0$. With a large number of trials n, the sample proportion

$$\hat{p} = \frac{X}{n}$$

is approximately normally distributed. Under the null hypothesis, p has the specified value p_0 and the distribution of $\hat{p}$ is approximately $N(p_0, \sqrt{p_0 q_0/n})$. Consequently, the standardized statistic

$$Z = \frac{\hat{p} - p_0}{\sqrt{p_0 q_0/n}}$$

has the $N(0, 1)$ distribution. Since the alternative hypothesis is two-sided, the rejection region of a level α test is given by

$$R: \quad Z \geq z_{\alpha/2} \quad \text{or} \quad Z \leq -z_{\alpha/2}$$

For one-sided alternatives, we use a one-tailed test in exactly the same way we discussed in Chapter 8, Section 4, in connection with tests about μ.

Example 7   Test for a change in proportion below poverty level
A five-year-old census recorded that 20% of the families in a large community lived below the poverty level. To determine whether this percentage has changed, a random sample of 400 families is studied and 70 are found to be living below the poverty level. Does this finding indicate that the current percentage of families earning incomes below the poverty level has changed from what it was five years ago? Use $\alpha = .05$.

SOLUTION AND DISCUSSION   1. *Hypotheses.* Let p denote the current population proportion of families living below the poverty level. Because we are seeking evidence to determine whether p is *different* from .20, we wish to test

$$H_0: \quad p = .20 \quad \text{versus} \quad H_1: \quad p \neq .20$$

2. *Level of significance.* We are given $\alpha = .05$.

3. *Test statistic.* The sample size $n = 400$ being large, the Z-test is appropriate. The test statistic is

$$Z = \frac{\hat{p} - .2}{\sqrt{.2 \times .8/400}}$$

4. *Rejection region.* Since $\alpha = .05$, the rejection region is $R: Z \leq -1.96$ or $Z \geq 1.96$.

5. *Calculation.* From the sample data, $\hat{p} = 70/400 = .175$ so the computed value of Z is

$$z = \frac{(70/400) - .2}{\sqrt{.2 \times .8/400}} = \frac{.175 - .2}{.020} = -1.25$$

6. *Conclusion.* The observed value $z = -1.25$ is not in the rejection region. The null hypothesis is not rejected at level $\alpha = .05$.

The observed level of significance is

$$\begin{aligned} P\text{-value} &= P[\, Z \leq -1.25 \quad \text{or} \quad Z \geq 1.25 \,] \\ &= P[\, Z \leq -1.25 \,] + P[\, Z \geq 1.25 \,] \\ &= 2 \times .1056 = .2112 \quad (\text{see Figure 2}) \end{aligned}$$

We would have to inflate α to more than .21 to reject the null hypothesis. Thus, the evidence against H_0 is really weak.

The data have failed to establish that the proportion of families in the community living below the poverty level has changed.

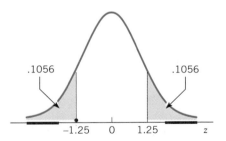

.1056   .1056

−1.25   0   1.25   z

Figure 2. *P*-value with two-sided rejection region ■

Exercises

2.1 To estimate a population proportion p, suppose that n units are randomly sampled and x of the sampled units are found to have the characteristic of interest. For each case, provide a point estimate of p and determine its 95% error margin.

(a) $n = \quad 50, \quad x = \quad 31$
(b) $n = \quad 460, \quad x = \quad 52$
(c) $n = 2500, \quad x = 1997$

2.2   For each case in Exercise 2.1, determine the 98% error margin of the estimate.

2.3   For each case in Exercise 2.1, calculate a 90% confidence interval for the population proportion.

2.4   According to a survey of persons applying for a driver's license, 1152 out of 1687 persons had used a designated driver at some time.[1]

(a)   Give a point estimate of the proportion of drivers who have used a designated driver at some time. Also give a 95% error margin.

(b)   If this proportion seems too high, is there some weakness in the sampling plan? Comment on the representativeness of age distribution of those applying for licenses as compared to that of all drivers.

2.5   It was found that 90% of the dogwoods of a standard strain die within three years of planting in a particular industrial belt. A new strain, believed to be resistant to air pollution, is currently under study. In an initial phase of the study, 45 seedlings of the new strain were planted in this region and 28 of them died in three years. Estimate the three-year mortality rate of the new strain and give an 80% error margin.

2.6   Manufacturers of food products reported that, in 1996, less than 2% of the coupons they circulated were ever redeemed. If nearly the same proportion holds today, select a sample size so we are 90% confident that the error of estimation is less than .003.

2.7   An automobile club, which pays for emergency road services (ERS) requested by its members, wishes to estimate the proportions of the different types of ERS requests. Upon examining a sample of 2927 ERS calls, it finds that 1499 calls related to starting problems, 849 calls involved serious mechanical failures requiring towing, 498 calls involved flat tires or lockouts, and 81 calls were for other reasons.

(a)   Estimate the true proportion of ERS calls that involve serious mechanical problems requiring towing and determine its 95% margin of error.

(b)   Calculate a 98% confidence interval for the true proportion of ERS calls that relate to starting problems.

[1]Wisconsin Department of Transportation, Traffic/Safety Study 1997.

2.8   In a survey of 750 high school graduates of 1998, it was found that 59.8% had enrolled in colleges, 4.2% in vocational institutions, while the other 36% did not pursue any further studies. Based on these findings,

(a)   Calculate a 95% confidence interval for the college enrollment rate of 1998 high school graduates.

(b)   Determine a 90% confidence interval for the true percent of the 1998 high school graduates enrolling in vocational institutions.

2.9   Refer to Example 1 and determine a 95% confidence interval for the proportion of all students who prefer E-mail.

2.10   In a psychological experiment, individuals are permitted to react to a stimulus in one of two ways, say, A or B. The experimenter wishes to estimate the proportion p of persons exhibiting reaction A. How many persons should be included in the experiment to be 90% confident that the error of estimation is within .04 if the experimenter

(a)   Knows that p is about .2?

(b)   Has no idea about the value of p?

2.11   A national safety council wishes to estimate the proportion of automobile accidents that involve pedestrians. How large a sample of accident records must be examined to be 98% certain that the estimate does not differ from the true proportion by more than .04? (The council believes that the true proportion is below .25.)

2.12   Manufacturers of food products reported that, in 1996, less than 2% of the coupons they circulated were ever redeemed. You wish to show the same is true for a coupon circulated on campus.

(a)   Determine H_0 and H_1.

(b)   Suppose 5000 coupons are placed in circulation on campus and 79 are redeemed. What does the level $\alpha = .05$ test conclude? Also give the P-value.

2.13   Identify the null and the alternative hypotheses in the following situations.

(a)   A university official believes that the proportion of students that currently hold part-time jobs has increased from the value .23 that prevailed four years ago.

(b)   A cable company claims that, because of improved procedures, the proportion of its cable subscribers that have complaints against the cable company is now less than .12.

(c)   Referring to part (b), suppose a consumer advocate believes the proportion of cable subscribers that have complaints against the cable company this year is greater than .12. She will conduct a survey to challenge the cable company's claim.

(d)   An inspector wants to establish that 2 × 4 lumber at a mill does not meet a specification that requires at most 5% break under a standard load.

2.14   Given here are the descriptive statements of some claims that one intends to establish on the basis of data. In each case, identify the null and the alternative hypotheses in terms of a population proportion p.

(a)   Of smokers who eventually quit smoking, less than 50% are able to do so in just one attempt.

(b)   On a particular freeway, over 20% of the cars that use a lane restricted exclusively to multipassenger cars use the lane illegally.

(c)   At a particular clinic, less than 10% of the patients wait over half an hour to see the doctor.

2.15   Each part of this problem specifies a claim about a population proportion, the sample size n, and the desired level of significance α. Formulate (i) the hypotheses, (ii) the test statistic, and (iii) the rejection region. [The answers to part (a) are provided for illustration.]

(a)   Claim: $p < .32$, $n = 120$, $\alpha = .05$

[Answers: (i) $H_0: p = .32$, $H_1: p < .32$

(ii) $Z = \dfrac{\hat{p} - .32}{\sqrt{.32 \times .68/120}} = \dfrac{\hat{p} - .32}{.0426}$

(iii)   R: $Z \leq -1.645$]

(b)   Claim: $p > .75$, $n = 248$, $\alpha = .02$
(c)   Claim: $p \neq .60$, $n = 88$, $\alpha = .02$
(d)   Claim: $p < .56$, $n = 96$, $\alpha = .10$

2.16   A concerned group of citizens wants to show that less than half the voters support the President's handling of a recent crisis. Let $p = $ proportion of voters who support the handling of the crisis.

(a)   Determine H_0 and H_1.

(b)   If a random sample of 500 voters gives 228 in support, what does the test conclude? Use $\alpha = .05$. Also evaluate the P-value.

2.17   Refer to Exercise 2.7. Perform a test of hypotheses to determine whether the proportion of ERS calls involving flat tires or lockouts was significantly smaller than .19, the true proportion for previous years. (Use a 5% level of significance.)

2.18   Refer to Exercise 2.8. Perform a test of hypotheses to determine whether the college enrollment rate in the population of 1998 high school graduates was significantly higher than 57%. (Calculate the P-value and interpret the result.)

2.19   From telephone interviews with 980 adults, it was found that 78% of those persons supported tougher legislation for antipollution measures. Does this poll substantiate the conjecture that more than 75% of the adult population are in favor of tougher legislation for antipollution measures? (Answer by calculating the *P*-value.)

2.20   Refer to the box with the smiling face scale for rating cereals. Using the data that 30 out of 42 youngsters in a sample rated a cereal in the top category, find an approximate 95% confidence interval for the corresponding population proportion.

3. COMPARING TWO POPULATION PROPORTIONS

We are often interested in comparing two populations with regard to the rate of incidence of a particular characteristic. Comparing the jobless rates in two cities, the percentages of female employees in two categories of jobs, and infant mortality in two ethnic groups are just a few examples. Let p_1 denote the proportion of members possessing the characteristic in Population 1, and p_2 that in Population 2. Our goals in this section are to construct confidence intervals for $p_1 - p_2$ and test $H_0: p_1 = p_2$, the null hypothesis that the rates are the same for two populations. The methods would also apply to the problems of comparison between two treatments, where the response of a subject falls into one of two possible categories that we may technically call "success" and "failure." The success rates for the two treatments can then be identified as the two population proportions p_1 and p_2.

The form of the data is displayed in Table 2, where X and Y denote the numbers of successes in independent random samples of sizes n_1 and n_2 taken from Population 1 and Population 2, respectively.

TABLE 2   Independent Samples from Two
Dichotomous Populations

	No. of Successes	No. of Failures	Sample Size
Population 1	X	$n_1 - X$	n_1
Population 2	Y	$n_2 - Y$	n_2

The population proportions of successes p_1 and p_2 are estimated by the corresponding sample proportions

$$\hat{p}_1 = \frac{X}{n_1} \quad \text{and} \quad \hat{p}_2 = \frac{Y}{n_2}$$

Naturally, $(\hat{p}_1 - \hat{p}_2)$ serves to estimate the difference of population proportions, $(p_1 - p_2)$. The hats again remind us that $\hat{p}_1$ and $\hat{p}_2$ are sample proportions or estimates.

SAMPLING PROPERTIES OF $\hat{p}_1 - \hat{p}_2$

By the independence of $\hat{p}_1$ and $\hat{p}_2$, the standard deviation, or standard error, of $\hat{p}_1 - \hat{p}_2$ is

$$\text{S.E.}(\hat{p}_1 - \hat{p}_2) = \sqrt{\frac{p_1 q_1}{n_1} + \frac{p_2 q_2}{n_2}}$$

where $q_1 = 1 - p_1$ and $q_2 = 1 - p_2$. This formula for the standard error stems from the fact that because $\hat{p}_1$ and $\hat{p}_2$ are based on independent samples, the variance of their difference equals the sum of their individual variances.

We can calculate the estimated standard error of $(\hat{p}_1 - \hat{p}_2)$ by using the above expression with the population proportions replaced by the corresponding sample proportions. Moreover, when n_1 and n_2 are large, the estimator $(\hat{p}_1 - \hat{p}_2)$ is approximately normally distributed. Specifically,

$$Z = \frac{(\hat{p}_1 - \hat{p}_2) - (p_1 - p_2)}{\text{Estimated standard error}} \qquad \text{is approximately } N(0, 1)$$

and this can be the basis for constructing confidence intervals for $(p_1 - p_2)$.

CONFIDENCE INTERVAL FOR $p_1 - p_2$

The normal distribution for $\hat{p}_1 - \hat{p}_2$ leads to a confidence interval for the difference of population proportions $p_1 - p_2$.

Large Sample Confidence Interval for $p_1 - p_2$

An approximate $100(1 - \alpha)\%$ confidence interval for $p_1 - p_2$ is

$$(\hat{p}_1 - \hat{p}_2) \pm z_{\alpha/2} \sqrt{\frac{\hat{p}_1(1 - \hat{p}_1)}{n_1} + \frac{\hat{p}_2(1 - \hat{p}_2)}{n_2}}$$

provided the sample sizes n_1 and n_2 are large.

Example 8   A confidence interval to show that medicated patches work
An investigation comparing a medicated patch with the unmedicated control patch for helping smokers quit the habit was discussed on page 105. At the end of the study, the number of persons in each group who were abstinent and who were smoking are repeated in Table 3 on page 466.

Determine a 95% confidence interval for the difference in success probabilities.

TABLE 3   Quitting Smoking

	Abstinent	Smoking	Total
Medicated patch	21	36	57
Unmedicated patch	11	44	55
Total	32	80	112

SOLUTION AND DISCUSSION   Let p_1 and p_2 denote the probabilities of quitting smoking with the medicated and unmedicated patches, respectively. We calculate

$$\hat{p}_1 = \frac{21}{57} = .3684, \qquad \hat{p}_2 = \frac{11}{55} = .2000$$

$$\hat{p}_1 - \hat{p}_2 = .1684$$

$$\sqrt{\frac{\hat{p}_1\hat{q}_1}{n_1} + \frac{\hat{p}_2\hat{q}_2}{n_2}} = \sqrt{\frac{.3684 \times .6316}{57} + \frac{.2000 \times .8000}{55}} = .0836$$

A 95% confidence interval for $p_1 - p_2$ is

$$(.1684 - 1.96 \times .0836, \quad .1684 + 1.96 \times .0836) \qquad \text{or} \qquad (.005, .332)$$

The confidence interval covers only positive values, so we conclude that the success rate with the medicated patch is .005 to .332 higher than for the control group that received the untreated patches. The lower value is so close to 0 that it is still plausible that the medicated patch is not very effective. ∎

Note: A confidence interval for each of the population proportions p_1 and p_2 can be determined by using the method described in Section 2. For instance, with the data of Example 8, a 90% confidence interval for p_1 is calculated as

$$\hat{p}_1 \pm 1.645 \sqrt{\frac{\hat{p}_1\hat{q}_1}{n_1}} = .3684 \pm 1.645 \sqrt{\frac{.3684 \times .6316}{57}}$$

$$= .3684 \pm .1051 \qquad \text{or} \qquad (.263, .474)$$

TESTING HYPOTHESES ABOUT TWO PROPORTIONS

To formulate a test of $H_0: p_1 = p_2$ when the sample sizes n_1 and n_2 are large, we again turn to the fact that $(\hat{p}_1 - \hat{p}_2)$ is approximately normally distributed. But now we note that under H_0, the mean of this normal distribution is $p_1 - p_2 = 0$ and the standard deviation, or standard error, is

$$\text{S.E.}(\hat{p}_1 - \hat{p}_2) = \sqrt{\frac{p_1 q_1}{n_1} + \frac{p_2 q_2}{n_2}} = \sqrt{pq}\sqrt{\frac{1}{n_1} + \frac{1}{n_2}}$$

where p stands for the common probability of success $p_1 = p_2$, and $q = 1 - p$. The unknown common p is estimated by pooling information from the two samples. The proportion of successes in the combined sample provides

$$\text{Pooled estimate } \hat{p} = \frac{X + Y}{n_1 + n_2}$$

or, alternatively,

$$\hat{p} = \frac{n_1 \hat{p}_1 + n_2 \hat{p}_2}{n_1 + n_2}$$

Also, $$\text{Estimated S.E.}(\hat{p}_1 - \hat{p}_2) = \sqrt{\hat{p}\hat{q}}\sqrt{\frac{1}{n_1} + \frac{1}{n_2}}$$

In summary,

Testing $H_0: p_1 = p_2$ with Large Samples

Test statistic:

$$Z = \frac{\hat{p}_1 - \hat{p}_2}{\sqrt{\hat{p}(1 - \hat{p})}\sqrt{\frac{1}{n_1} + \frac{1}{n_2}}}, \qquad \text{where } \hat{p} = \frac{X + Y}{n_1 + n_2}$$

The level α rejection region is $R: Z \leq -z_{\alpha/2}$ or $Z \geq z_{\alpha/2}$, $R: Z \leq -z_\alpha$, or $R: Z \geq z_\alpha$, according to whether the alternative hypothesis is $p_1 \neq p_2$, $p_1 < p_2$, or $p_1 > p_2$.

Example 9   Testing the equality of two proportions
A study (courtesy of R. Golubjatnikov) is undertaken to compare the rates of prevalence of CF antibody to Parainfluenza I virus among boys and girls in the age group 5 to 9 years. Among 113 boys tested, 34 are found to have the antibody; among 139 girls tested, 54 have the antibody. Do the data provide strong evidence that the rate of prevalence of the antibody is significantly higher in girls than boys? Use $\alpha = .05$. Also, find the P-value.

SOLUTION AND DISCUSSION

1. *Hypotheses.* Let p_1 denote the population proportion of boys who have the CF antibody and p_2 the population proportion of girls who have the CF antibody. Because we are looking for strong evidence in support of $p_1 < p_2$, we formulate the hypotheses as

$$H_0: \quad p_1 = p_2 \qquad \text{versus} \qquad H_1: \quad p_1 < p_2$$

or equivalently as

$$H_0: \quad p_1 - p_2 = 0 \qquad \text{versus} \qquad H_1: \quad p_1 - p_2 < 0$$

2.   *Level of significance.*   We are given $\alpha = .05$.

3.   *Test statistic.*   The sample sizes $n_1 = 113$ and $n_2 = 139$ being large, we will employ the test statistic

$$Z = \frac{\hat{p}_1 - \hat{p}_2}{\sqrt{\hat{p}\hat{q}}\sqrt{\dfrac{1}{n_1} + \dfrac{1}{n_2}}}$$

4.   *Rejection region.*   Because the alternative hypothesis is left-sided, the rejection region consists of small values for Z. Since $\alpha = .05$, the rejection region is $R: Z \le -1.645$.

5.   *Calculation.*   From the sample data, we calculate

$$\hat{p}_1 = \frac{34}{113} = .301, \qquad \hat{p}_2 = \frac{54}{139} = .388$$

$$\text{Pooled estimate } \hat{p} = \frac{34 + 54}{113 + 139} = .349$$

The observed value of the test statistic is then

$$z = \frac{.301 - .388}{\sqrt{.349 \times .651}\sqrt{\dfrac{1}{113} + \dfrac{1}{139}}} = -1.44$$

6.   *Conclusion.*   Because the observed value $z = -1.44$ is not in the rejection region, the null hypothesis H_0 is not rejected at level $\alpha = .05$. The observed level of significance is

$$\text{P-value} = P[\, Z \le -1.44 \,]$$
$$= .0749 \quad (\text{see Figure 3})$$

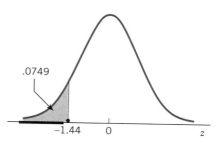

.0749

−1.44   0 z

Figure 3. *P*-value for left-sided rejection region

This means that we must allow an α of at least .0749 to consider the result significant. Consequently, the assertion that the girls have a higher rate of prevalence of the CF antibody than boys is not substantiated at the level of significance $\alpha = .05$. ∎

STATISTICAL REASONING

The Choice of Wording Can Greatly Influence the Response

When creating estimates of unemployment, the Bureau of Labor Statistics has clear-cut rules about who to include in the sample and who to exclude. For instance, persons who have not looked for a job in the past two months are excluded from the calculation. They are not considered to belong to the population of unemployed. The conditions of unemployed and employed need to be clearly defined and they are.

Stephen Simpson/FPG International

One year, there was interest in providing an estimate of unemployment for an area that included the city and some part of the surrounding county. A small company was given a contract to conduct the survey and come up with an estimate. They randomly selected names of a few hundred residents and mailed out a short questionnaire. On the basis of the responses, the unemployment rate was estimated at over 20%.

The paper announced this result but it was immediately questioned. The area of interest historically has the lowest unemployment in the state and this number was much higher than that for any other area.

It soon came to light that the estimate of unemployment was based on the answers to two questions. Essentially, people were asked

1. Are you working at this time?   *Yes/No*
2. Would you like to have a job?   *Yes/No*

Not surprisingly, many persons answered "yes" to the second question even though, prior to receiving the questionnaire, they were not even thinking about looking for jobs. A full-time student may want a 5–10 hour a week job. This wording was vague and biased toward a "yes" response.

Wording is important when data are collected via a survey. A poorly worded question can completely defeat the objective of the study.

Example 10   Comparing the effectiveness of two commercials
Refer to Example 2, where part of the TV audience was shown one version of a commercial and the other part was shown a second version. After two hours, a random sample of each audience was interviewed and the data in Table 4 were obtained.

TABLE 4   Remembering the Key Point in a Commercial

	Don't Remember Key Point	Remember Key Point	Total
Commercial A	43	37	80
Commercial B	52	18	70
Total	95	55	150

Test the null hypothesis that the two versions are equally effective versus the alternative that they are not. Use $\alpha = .05$.

SOLUTION AND DISCUSSION

1. *Hypotheses.*   Let p_1 and p_2 denote the probabilities of remembering the key point in Commercial A and Commercial B, respectively. Because the alternative is two-sided, not equally effective, we formulate the hypotheses as

$$H_0: \ p_1 = p_2 \quad \text{versus} \quad H_1: \ p_1 \neq p_2$$

or, equivalently,

$$H_0: \ p_1 - p_2 = 0 \quad \text{versus} \quad H_1: \ p_1 - p_2 \neq 0$$

2. *Level of significance.*   We are given $\alpha = .05$.
3. *Test statistic.*   Since the sample sizes $n_1 = 80$ and $n_2 = 70$ are large, we use the test statistic

$$Z = \frac{\hat{p}_1 - \hat{p}_2}{\sqrt{\hat{p}(1 - \hat{p})}\sqrt{\dfrac{1}{n_1} + \dfrac{1}{n_2}}}$$

4. *Rejection region.*   Because $\alpha = .05$ and the alternative hypothesis is two-sided, we set the two-sided rejection region $R: Z \le -1.96$ or $Z \ge 1.96$.

5. *Calculation.*   From the sample data, we first calculate

$$\hat{p}_1 = \frac{37}{80} = .463 \qquad \hat{p}_2 = \frac{18}{70} = .257$$

$$\text{Pooled estimate } \hat{p} = \frac{37 + 18}{80 + 70} = .367$$

The observed value of the test statistic is

$$z = \frac{.463 - .257}{\sqrt{.367(.633)}\sqrt{\dfrac{1}{80} + \dfrac{1}{70}}} = 2.61$$

6. *Conclusion.*   The value $z = 2.61$ lies in the rejection region R, so we reject the equality of the two proportions at the .05 level of significance. The observed level of significance is

$$P\text{-value} = P[Z \le -2.61] + P[Z \ge 2.61]$$
$$= .0045 + .0045 = .0090 \quad (\text{see Figure 4})$$

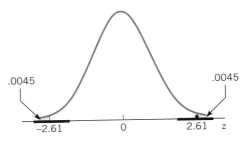

.0045   .0045

−2.61   0   2.61   z

Figure 4. *P*-value when $Z = 2.61$

We conclude that the evidence is very strong that the two commercials are not equally effective.   ∎

Exercises

3.1   Given the data $n_1 = 100, \qquad \hat{p}_1 = \dfrac{50}{100} = .5$

$n_2 = 200, \qquad \hat{p}_2 = \dfrac{140}{200} = .7$

(a)   Find a 95% confidence interval for $p_1 - p_2$.

(b) Perform the Z-test for testing the null hypothesis $H_0: p_1 = p_2$ versus $H_1: p_1 < p_2$.

3.2 A sample of 100 females was collected from ethnic group A and a sample of 100 from ethnic group B. Each female was asked, "Did you get married before you were 19?" The following counts were obtained:

	A	B
Yes	62	29
No	38	71

(a) Test for equality of two proportions against a two-sided alternative. Take $\alpha = .05$.

(b) Establish a 95% confidence interval for the difference $p_A - p_B$.

3.3 In a comparative study of two new drugs, A and B, 120 patients were treated with drug A and 150 patients with drug B, and the following results were obtained.

	Drug A	Drug B
Cured	52	88
Not cured	68	62
Total	120	150

(a) Do these results demonstrate a significantly higher cure rate with drug B than drug A? Test at $\alpha = .05$.

(b) Construct a 95% confidence interval for the difference in the cure rates of the two drugs.

3.4 Referring to Exercise 3.3, construct a 95% confidence interval for the cure rate of each of the two drugs.

3.5 In a study of the relationship between temperament and personality, 49 female high school students who had a high level of reactivity (HRL) and 54 students who had a low level of reactivity (LRL) were classified according to their attitude to group pressure with the following results.

Reactivity	Attitude		Total
	Submissive	Resistant	
HRL	33	16	49
LRL	12	42	54

Is resistance to group pressure significantly lower in the HRL group than the LRL group? Answer by calculating the *P*-value.

3.6   Refer to the data in Exercise 3.5. Determine a 99% confidence interval for the difference between the proportions of resistant females in the HRL and LRL populations.

3.7   According to a survey, 73 males out of 786 and 43 females out of 943 report that they usually drive 10 or more miles per hour over the speed limit in the city.[2]

   (a)   Is the proportion of male speeders higher than the proportion of female speeders? Answer by calculating the *P*-value.

   (b)   Obtain a 95% confidence interval for the difference between the proportions of males and females who usually drive 10 or more miles per hour over the speed limit.

   (c)   These data were collected by distributing a questionnaire to a sample of persons applying for a driver's license. Name a possible source of bias.

3.8   The popular disinfectant Listerine is named after Joseph Lister, a British physician who pioneered the use of antiseptics. Lister conjectured that human infections might have an organic origin and thus could be prevented by using a disinfectant. Over a period of several years, he performed 75 amputations: 40 using carbolic acid as a disinfectant and 35 without any disinfectant. The following results were obtained.

	Patient Survived	Patient Died	Total
With carbolic acid	34	6	40
Without carbolic acid	19	16	35

Are the survival rates significantly different between the two groups? Test at $\alpha = .05$ and calculate the *P*-value.

3.9   Referring to the data of Exercise 3.8, calculate a 95% confidence interval for the difference between the survival rates for the two groups.

3.10   Random samples of 250 persons in the 30–40 year age group and 250 persons in the 60–70 year age group are asked about the average number of hours they sleep per night, and the following summary data are recorded.

[2]Wisconsin Department of Transportation, Traffic/Safety Study 1997.

Age	Hours of Sleep ≤ 8	> 8	Total
30–40	172	78	250
60–70	120	130	250
Total	292	208	500

Do these data demonstrate that the proportion of persons who have ≤ 8 hours of sleep per night is significantly higher for the age group 30–40 than that for the age group 60–70? Answer by calculating the P-value.

3.11   Referring to Exercise 3.10, denote by p_1 and p_2 the population proportions in the two groups who have ≤ 8 hours of sleep per night. Construct a 95% confidence interval for $p_1 - p_2$.

3.12   Consider independent random samples of sizes n_1 and n_2 from two populations and suppose that the sample proportions of successes are $\hat{p}_1 = .62$ and $\hat{p}_2 = .51$. For each case, examine whether or not the observed difference $\hat{p}_1 - \hat{p}_2 = .11$ is statistically significant at the specified α.

(a) $n_1 = 50,$ $n_2 = 90,$ $\alpha = .05$
(b) $n_1 = 200,$ $n_2 = 210,$ $\alpha = .05$
(c) $n_1 = 200,$ $n_2 = 210,$ $\alpha = .01$

3.13   A major clinical trial of a new vaccine for type-B hepatitis was conducted with a high-risk group of 1083 male volunteers. From this group, 549 men were given the vaccine and the other 534 a placebo. A follow-up of all these individuals yielded the data (see Exercise 7.4, Chapter 3):

	Follow-up Got Hepatitis	Did Not Get Hepatitis	Total
Vaccine	11	538	549
Placebo	70	464	534

(a)   Do these observations testify that the vaccine is effective? Use $\alpha = .01$.

(b)   Construct a 95% confidence interval for the difference between the incidence rates of hepatitis among the vaccinated and nonvaccinated individuals in the high-risk group.

3.14   Records of drivers with a major medical condition (diabetes, heart condition, or epilepsy) and also a group of drivers with no known health conditions were retrieved from a motor vehicle department. Drivers in each group were classified according to their driving record in the last year.

Medical Condition	Traffic Violations		Total
	None	One or more	
Diabetes	119	41	160
Heart condition	121	39	160
Epilepsy	72	78	150
None (control)	157	43	200

Let p_D, p_H, p_E, and p_C denote the population proportions of drivers having one or more traffic violations in the last year for the four groups "diabetes," "heart condition," "epilepsy," and "control," respectively.

(a)   Test $H_0: p_D = p_C$ versus $H_1: p_D > p_C$ at $\alpha = .10$.

(b)   Is there strong evidence that p_E is higher than p_C? Answer by calculating the P-value.

3.15   Refer to Exercise 3.14.

(a)   Construct a 95% confidence interval for $p_E - p_H$.

(b)   Construct a 90% confidence interval for $p_H - p_C$.

4. COMPARING POPULATIONS WITH A CHI-SQUARE TEST

Categorical data arise when a trait observed for an individual sampling unit is recorded in some qualitative categories. Examples in the previous section are employed/unemployed and don't remember/remember a key point of a commercial. We are not limited to two categories or even two populations as in Section 3. We discussed arranging categorical or frequency data as two-way tables in Section 2 of Chapter 3.

From each of several populations, suppose we draw a random sample of fixed size and classify each response into a category. These data can be summarized as counts in the form of a two-way table. The columns refer to the categories and the rows to different populations. Our objective is to test whether the populations are alike or **homogeneous** with respect to the probabilities of the categories. That is, we need to determine whether the observed proportion for each category is nearly the same for all populations. The test is called a **test of homogeneity** and it applies to two-way **contingency tables** that have *one margin fixed*, the row total, corresponding to the fixed sample size for each population.

We pursue our analysis using the full data on commercials, which are given in Table 5. The data in Example 2 were obtained by combining the first two categories into a single "Don't remember key point" category.

Statistical inferences are based on models for variation. Here, each category in a row of the table has a probability or unknown population proportion. These are entered in Table 6.

TABLE 5   Remembering a Commercial

	Don't Remember	Remember Seeing	Remember Key Point	Total
Commercial A	19	24	37	80
Commercial B	24	28	18	70
Total	43	52	55	150

TABLE 6   A Model for No Difference Between the Commercials: The Probabilities for Each Cell of the Two-Way Table

	Don't Remember	Remember Seeing	Remember Key Point	Total
Commercial A	p_{A1}	p_{A2}	p_{A3}	1
Commercial B	p_{B1}	p_{B2}	p_{B3}	1

The null hypothesis of "no difference in," or "homogeneity," specifies that the probability for a category is the same for all rows or populations:

$$H_0: \quad p_{A1} = p_{B1}, \quad p_{A2} = p_{B2}, \quad p_{A3} = p_{B3}$$

Under the null hypothesis, let us denote these common probabilities by p_1, p_2, and p_3, respectively. Each of the positions within the table, such as *Commercial A—Remember Seeing* is called a **cell**. For any given row, the expected cell frequency is obtained by multiplying the probability of that cell by the sample size. Consequently, for the first row, the expected cell frequencies are $80p_1$, $80p_2$, and $80p_3$, while those for the second row are $70p_1$, $70p_2$, and $70p_3$. However, the common probabilities p_1, p_2, and p_3 are not specified under H_0 and must be estimated.

The column totals 43, 52, and 55 in Table 5 are the frequency counts for the categories in the combined sample of size $70 + 80 = 150$. Under the null hypothesis that the populations are alike, the estimated common probabilities are

$$\hat{p}_1 = \frac{43}{150}, \qquad \hat{p}_2 = \frac{52}{150}, \qquad \hat{p}_3 = \frac{55}{150}$$

Using these estimates, the expected frequencies for the first row are $80\hat{p}_1$, $80\hat{p}_2$, $80\hat{p}_3$ or

$$\frac{80 \times 43}{150}, \qquad \frac{80 \times 52}{150}, \qquad \frac{80 \times 55}{150}$$

with similar expressions for the second row using 70 rather than 80. Notice the pattern:

$$\text{Expected cell frequency} = \frac{\text{Row total} \times \text{Column total}}{\text{Grand total}}$$

where the grand total is the sum of all the cell frequencies.

Table 7(a) presents the observed frequencies (O) and the expected frequencies (E). The latter are given in parentheses. Table 7(b) computes the discrepancy measure

$$\frac{(O - E)^2}{E}$$

for each cell. These are then summed over all cells to obtain the **chi-square**, or χ^2, **statistic**

$$\chi^2 = \sum_{\text{cells}} \frac{(O - E)^2}{E}$$

which measures the overall departure from homogeneity.

TABLE 7(a) The Observed and Expected Values for Data on Commercials in Table 5

	Don't Remember	Remember Seeing	Remember Key Point
Commercial A	19 (22.93)	24 (27.73)	37 (29.33)
Commercial B	24 (20.07)	28 (24.27)	18 (25.67)

TABLE 7(b) The Values of $(O - E)^2/E$

	Don't Remember	Remember Seeing	Remember Key Point	
Commercial A	.675	.503	2.004	$6.816 = \chi^2$
Commercial B	.771	.574	2.290	

What values of the χ^2 statistic are considered to be too large if the distributions for the two rows are the same? Any test of significance for the equality relies on the answer. What is available is an approximate sampling distribution for the χ^2 statistic when H_0 prevails. This distribution is called the **chi-square distribution** and there is a different distribution for each number of **degrees of freedom.**

It was introduced in Section 5 of Chapter 9 in connection with inference about the variance of a normal population. The upper 100α percentile, χ_α^2, is the value of a chi-square variable that is exceeded with probability α. (See Figure 5.) The values of χ_α^2 have been tabulated in Table 5 for various values for α and degrees of freedom.

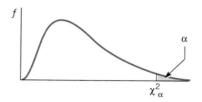

Figure 5. The chi-square χ_α^2 notation

In the context of two-way tables,

Number of degrees of freedom = (No. of rows $-$ 1) $\times$ (No. of columns $-$ 1)

or $(2 - 1)(3 - 1) = 2$ here. Consulting the table for the chi-square distribution, in the column for $\alpha = .05$ and the row for 2 degrees of freedom, we find that the upper 5th percentile is $\chi_{.05}^2 = 5.99$. Since the observed value $\chi^2 = 6.816$ exceeds the upper 5th percentile, the evidence against the populations being alike is strong.

A closer examination shows that the large contributions to χ^2 come from the "remember key point" category, where the relative frequency is $37/80 = .463$ for Commercial A and $18/70 = .257$ for Commercial B. These data suggest that Commercial A is better. The test itself concludes only that the two commercials are different.

The χ^2 Test of Homogeneity in a Contingency Table

Null hypothesis

In each response category, the probabilities are equal for all the populations.

Test statistic

$$\chi^2 = \sum_{\text{cells}} \frac{(O - E)^2}{E}, \quad \begin{cases} O = \text{Observed cell frequency} \\ E = \dfrac{\text{Row total} \times \text{Column total}}{\text{Grand total}} \end{cases}$$

d.f. = (No. of rows $-$ 1) $\times$ (No. of columns $-$ 1)

Rejection region

$$\chi^2 \geq \chi_\alpha^2$$

Example 11   Comparing several populations
A survey is undertaken to determine the incidence of alcoholism in different professional groups. Random samples of the clergy, educators, executives, and merchants are interviewed, and the observed frequency counts given in Table 8.

TABLE 8   Contingency Table of Alcoholism versus Profession

	Alcoholic	Nonalcoholic	Sample Size
Clergy	32 (58.25)	268 (241.75)	300
Educators	51 (48.54)	199 (201.46)	250
Executives	67 (58.25)	233 (241.75)	300
Merchants	83 (67.96)	267 (282.04)	350
Total	233	967	1200

Carry out a chi-square test to determine whether the incidence rate of alcoholism appears to be the same in all four groups. Use $\alpha = .05$.

SOLUTION AND DISCUSSION

1. *Hypotheses.* Let us denote the proportions of alcoholics in the populations of the clergy, educators, executives, and merchants by $p_1, p_2, p_3,$ and p_4, respectively. Based on independent random samples from four binomial populations, we want to test the null hypothesis

$$H_0: \quad p_1 = p_2 = p_3 = p_4$$

2. *Level of significance.* We are given $\alpha = .05$.

3. *Test statistic.*

$$\chi^2 = \sum_{\text{cells}} \frac{(O - E)^2}{E}$$

4. *Rejection region.* We reject H_0 for large values of χ^2. With $\alpha = .05$ and d.f. = 3, the tabulated upper 5% of χ^2 is 7.81 so $R: \chi^2 \geq 7.81$.

5. *Calculation.* The expected cell frequencies, shown in parentheses in Table 8, are computed by multiplying the row and column totals and dividing the results by 1200. The χ^2 statistic is computed in Table 9 (page 480).

6. *Conclusion.* The observed value $\chi^2 = 20.59$ is in the rejection region. The null hypothesis is rejected at level $\alpha = .05$.

Since $\chi^2_{.01} = 11.34$, the P-value is much smaller than .01. (The accompanying MINITAB computer calculation gives P-value $= .000$.) We strongly reject the null hypothesis of equal proportions of alcoholics among the different professions.

TABLE 9 The Values of $(O - E)^2/E$ for the Data in Table 8

	Alcoholic	Nonalcoholic
Clergy	11.83	2.85
Educators	.12	.03
Executives	1.31	.32
Merchants	3.33	.80

$$20.59 = \chi^2$$
$$\text{d.f. of } \chi^2 : (4 - 1)(2 - 1) = 3$$

Examining Table 9, we notice that a large contribution to the χ^2 statistic has come from the first row. This is because the relative frequency of alcoholics among the clergy is quite low in comparison to the others, as one can see from Table 8.

MINITAB Output for Example 11
See Exercise 4.11 for the commands.

Chi-Square Test

```
        Expected counts are printed below observed counts

                   Alcohol Non-alco    Total

          1             32      268      300
                     58.25   241.75

          2             51      199      250
                     48.54   201.46

          3             67      233      300
                     58.25   241.75

          4             83      267      350
                     67.96   282.04

       Total           233      967     1200

    Chi-Sq = 11.829   + 2.850 +
              0.124   + 0.030 +
              1.314   + 0.317 +
              3.329   + 0.802  = 20.597
    DF = 3,   P-Value = 0.000
```

Reject H_0 if the P-value is smaller than the specified level α.

THE CHI-SQUARE TEST FOR 2 × 2 TABLES AND THE Z-TEST

Example 10 contains an application of the Z-test for testing the equality of two proportions p_1 and p_2 in the context of remembering the key point of a commercial. The same test can be performed by the chi-square test. The accompanying computer output gives the chi-square statistic = 6.780.

```
       Expected counts are printed below observed counts

               C1        C2      Total
         1     43        37        80
              50.67     29.33

         2     52        18        70
              44.33     25.67

     Total     95        55       150

   ChiSq =    1.160 +  2.004   +
              1.326 +  2.290   = 6.780

   df = 1
```

The two tests, Z and chi-square, will always agree and the P-values will always be the same for two-sided alternatives since $\chi^2 = Z^2$. In this application, $(2.61)^2 = 6.81$ and the difference is due to roundoff error. (More accurately $Z = 2.604$.)

For comparing two proportions, the Z-test is preferred because it gives the choice of one- or two-sided tests.

Exercises

4.1 Soccer has become a popular sport, especially among grade school children. Eighty second-grade boys and seventy-five second-grade girls were asked whether they play on a youth soccer team. The results are given in the table.

	Play	Don't Play	Total
Boys	34	41	75
Girls	59	21	80
Total	93	62	155

Determine whether there are different participation rates for second grade boys and girls.

(a)   Formulate the null and alternative hypotheses.

(b)   Test the null hypothesis at level $\alpha = .05$.

4.2   Many industrial air pollutants adversely affect plants. Sulfur dioxide causes leaf damage in the form of intraveinal bleaching in many sensitive plants. In a study of the effect of a given concentration of sulfur dioxide in the air on three types of garden vegetables, 40 plants of each type are exposed to the pollutant under controlled greenhouse conditions. The frequencies of severe leaf damage are recorded in the following table.

	Leaf Damage		
	Severe	Moderate or None	Total
Lettuce	32	8	40
Spinach	28	12	40
Tomato	19	21	40
Total	79	41	120

Analyze these data to determine whether the incidence of severe leaf damage is alike for the three types of plants. In particular,

(a)   Formulate the null hypothesis.

(b)   Test the null hypothesis with $\alpha = .10$.

(c)   Construct three individual 95% confidence intervals and plot them on graph paper.

4.3   When a new product is introduced in a market, it is important for the manufacturer to evaluate its performance during the critical months after its distribution. A study of market penetration, as it is called, involves sampling consumers and assessing their exposure to the product. Suppose that the marketing division of a company selects random samples of 200, 150, and 300 consumers from three cities and obtains the following data from them.

	Never Heard of the Product	Heard About It but Did Not Buy	Bought It at Least Once	Total
City 1	36	55	109	200
City 2	45	56	49	150
City 3	54	78	168	300
Total	135	189	326	650

Do these data indicate that the extent of market penetration differs in the three cities?

4.4　Nausea from air sickness affects some travelers. In a comparative study of the effectiveness of two brands of motion sickness pills, brand A pills were given to 45 persons randomly selected from a group of 90 air travelers, while the other 45 persons were given brand B pills. The following results were obtained.

	Degree of Nausea				Total
	None	Slight	Moderate	Severe	
Brand A	18	17	6	4	45
Brand B	11	14	14	6	45

Do these observations demonstrate that the two brands of pills are significantly different in quality? Test at $\alpha = .05$.

4.5　A study on the possible effect of hearing impairment in children on the development of a secure attachment to their mothers was reported in A. R. Lederberg and C. E. Mobley, "The Effect of Hearing Impairment on the Quality of Attachment and Mother–Toddler Interaction," *Child Development*, **61** (1990), pp. 1596–1604. The following results were obtained from an observation of 41 hearing-impaired children and 41 hearing children along with their mothers.

Group	Nature of Attachment		Total
	Secure	Insecure	
Hearing-impaired	23	18	41
Hearing	25	16	41

Do these results demonstrate a significant difference between the two groups of children with regard to the nature of attachment? Answer by using

(a)   The χ^2 test at $\alpha = .05$

(b)   The Z-test and calculating the significance probability

4.6   The toxicity of four combinations (designated as treatments A, B, C, and D) of an insecticide and a herbicide was studied in an experiment. Four batches of fruit flies were randomly assigned to these treatments and the following results were recorded.

Treatments	Number of Flies		Batch Size
	Dead	Alive	
A	58	57	115
B	43	77	120
C	56	42	98
D	45	75	120

(a)   Are there significant differences in the mortality rates of fruit flies under the four treatments? Test at $\alpha = .10$.

(b)   Let p_A, p_B, p_C, and p_D denote the probabilities of death of fruit flies under the application of treatments A, B, C, and D, respectively. Construct a 95% confidence interval for each of these probabilities individually.

4.7   Using the data for treatments A and C in Exercise 4.6, make a 2 × 2 table and test $H_0: p_A = p_C$ versus $H_1: p_A \neq p_C$ at $\alpha = .05$ with

(a)   The χ^2 test

(b)   The Z-test

4.8   Refer to the data for treatments A and D in Exercise 4.6.

(a)   Is there strong evidence for a higher mortality rate under treatment A than under treatment D? Answer by calculating the P-value.

(b)   Construct a 95% confidence interval for the difference $p_A - p_D$.

4.9   Researchers studying sleep disorders collected data from the general population to serve as a reference set. Randomly selected state employees were asked the question,

"According to what others have told you, please estimate how often you snore."

The responses were categorized by whether or not the person snored 3 or more times per week and by gender.[3]

	Weekly Snoring		
Gender	Less than 3 times	3 or more times	Total
Female	32	23	55
Male	31	65	96
Total	63	88	151

Test the equality of proportions for males and females using the χ^2 test with $\alpha = .05$.

4.10   Osteoporosis or a loss of bone minerals is a common cause of broken bones in the elderly. A researcher on aging conjectures that bone mineral loss can be reduced by regular physical therapy or certain kinds of physical activity. A study is conducted on 200 elderly subjects of approximately the same age divided into control, physical therapy, and physical activity groups. After a suitable period of time, the nature of change in bone mineral content is observed.

	Change in Bone Mineral			
	Appreciable Loss	Little Change	Appreciable Increase	Total
Control	38	15	7	60
Therapy	22	32	16	70
Activity	15	30	25	70
Total	75	77	48	200

Do these data indicate that the change in bone mineral varies for the different groups? Use the output given on page 486 to answer.

[3]Courtesy of T. Young.

Chi-Square Test

Expected counts are printed below observed counts

	C1	C2	C3	Total
1	38	15	7	60
	22.50	23.10	14.40	
2	22	32	16	70
	26.25	26.95	16.80	
3	15	30	25	70
	26.25	26.95	16.80	
Total	75	77	48	200

Chi-Sq = 10.678 + 2.840 + 3.803 +
          0.688 + 0.946 + 0.038 +
          4.821 + 0.345 + 4.002 = 28.162
DF = 4,   P-Value = 0.000

(a) Would you reject the null hypothesis that the populations are the same, when $\alpha = .05$? Comment on the strength of evidence against the null hypothesis.

(b) Which cells make a large contribution to the chi-square statistic? Is the corresponding observed frequency too high or too low?

USING A COMPUTER

4.11 The analysis of a contingency table can be conveniently done on a computer. For an illustration, we present here a MINITAB analysis of the data on commercials in Table 5.

Data:

C1: 19 24
C2: 24 28
C3: 37 18

Dialog box:	**Session command**
Stat > Tables > Chisquare Test	MTB > CHISQUARE C1-C3
Type *C1*-C3 in **Columns containing** the table.	
Click **OK.**	

Output:

Chi-Square Test

```
        Expected counts are printed below observed counts

                    C1        C2       C3     Total
             1       19        24       37       80
                  22.93     27.73    29.33

             2       24        28       18       70
                  20.07     24.27    25.67

        Total       43        52       55      150

        Chi-Sq = 0.675 + 0.503 + 2.004 +
                 0.771 + 0.574 + 2.290 = 6.816
        DF = 2, P-Value = 0.033
```

(a) Compare this output with the calculations presented in Table 7.

(b) Do Exercise 4.3 on a computer.

5. CONTINGENCY TABLE WITH NEITHER MARGIN FIXED (TEST OF INDEPENDENCE)

When two traits are observed for each element or unit of a random sample, the data can be simultaneously classified with respect to these traits. We then obtain a two-way contingency table in which neither set of marginal totals is fixed so both are random.

Example 12 One sample simultaneously classified according to two characteristics
A random sample of 500 persons is questioned regarding political affiliation and attitude toward a tax reform program. From the observed frequency table given in Table 10, we wish to answer the following question: Do the data indicate that the pattern of opinion is different between the two political groups?

TABLE 10 Political Affiliation and Opinion

	Favor	Indifferent	Opposed	Total
Democrat	138	83	64	285
Republican	64	67	84	215
Total	202	150	148	500

Unlike Example 2, here we have a single random sample but each sampled individual elicits two types of responses: political affiliation and attitude. In the present context, the null hypothesis of "no difference" amounts to saying that the two types of responses are independent. In other words, attitude toward the program is unrelated to or independent of a person's political affiliation. ∎

To cite a few other examples: A random sample of employed persons may be classified according to educational attainment and type of occupation; college students may be classified according to the year in college and attitude toward a dormitory regulation; flowering plants may be classified with respect to type of foliage and size of flower.

A typical inferential aspect of cross-tabulation is the study of whether the two characteristics appear to be manifested independently or certain levels of one characteristic tend to be associated or contingent with some levels of another.

Example 13   A χ^2 test of independence—Are political affiliation and opinion related? Analyze the data in Example 12.

SOLUTION AND DISCUSSION   The 2 × 3 contingency table is given in Table 10. Here a single random sample of 500 persons is classified into six cells.

Dividing the cell frequencies by the sample size 500, we obtain the relative frequencies shown in Table 11. Its row marginal totals, .570 and .430, represent the sample proportions of Democrats and Republicans, respectively. Likewise, the column marginal totals show the sample proportions in the three categories of attitude.

TABLE 11   Proportion of Observations in Each Cell

	Favor	Indifferent	Opposed	Total
Democrat	.276	.166	.128	.570
Republican	.128	.134	.168	.430
Total	.404	.300	.296	1

Imagine a classification of the entire population. The unknown population proportions (i.e., the probabilities of the cells) are represented by the entries in Table 12, where the subscripts D and R stand for Democrat and Republican, and 1, 2, and 3 refer to the "favor," "indifferent," and "opposed" categories. Table 12 is the population analogue of Table 11, which shows the sample proportions. For instance,

TABLE 12   Cell Probabilities

	Favor	Indifferent	Opposed	Row Marginal Probability
Democrat	p_{D1}	p_{D2}	p_{D3}	p_D
Republican	p_{R1}	p_{R2}	p_{R3}	p_R
Column marginal probability	p_1	p_2	p_3	1

Cell probability $p_{D1} = P(\text{Democrat } \textbf{and} \text{ in favor})$
Row marginal probability $p_D = P(\text{Democrat})$
Column marginal probability $p_1 = P(\text{in favor})$

We are concerned with testing the null hypothesis that the two classifications are independent. Recall from Chapter 4 that the probability of the intersection of independent events is the product of their probabilities. Thus, the independence of the two classifications means that $p_{D1} = p_D p_1$, $p_{D2} = p_D p_2$, and so on. The hypothesis of independence can be formalized as

H_0: Each cell probability is the product of the corresponding pair of marginal probabilities.

To construct a χ^2 test, we need to determine the expected frequencies. Under H_0, the expected cell frequencies are

$$500 p_D p_1 \qquad 500 p_D p_2 \qquad 500 p_D p_3$$
$$500 p_R p_1 \qquad 500 p_R p_2 \qquad 500 p_R p_3$$

These involve the unknown marginal probabilities that must be estimated from the data. Referring to Table 10, we calculate the estimates as

$$\hat{p}_D = \frac{285}{500}, \qquad \hat{p}_R = \frac{215}{500}$$

$$\hat{p}_1 = \frac{202}{500}, \qquad \hat{p}_2 = \frac{150}{500}, \qquad \hat{p}_3 = \frac{148}{500}$$

Then, the expected frequency in the first cell is estimated as

$$500 \hat{p}_D \hat{p}_1 = 500 \times \frac{285}{500} \times \frac{202}{500}$$

$$= \frac{285 \times 202}{500} = 115.14$$

Notice that the expected frequency for each cell of Table 10 is of the form

$$\frac{\text{Row total} \times \text{Column total}}{\text{Grand total}}$$

Table 13(a) on page 490 presents the observed cell frequencies along with the expected frequencies shown in parentheses. The quantities $(O - E)^2/E$ and the χ^2 statistic are then calculated in Table 13(b).

Having calculated the χ^2 statistic, it now remains to determine its d.f. by

$$\text{d.f. of } \chi^2 = (\text{No. of rows} - 1) \times (\text{No. of columns} - 1)$$
$$= (2 - 1) \times (3 - 1) = 2$$

We choose level of significance $\alpha = .05$ and the tabulated upper 5% point of χ^2 with d.f. $= 2$ is 5.99. Because the observed χ^2 is larger than the tabulated value, the null hypothesis of independence is rejected at $\alpha = .05$. In fact, it would be rejected even for $\alpha = .01$.

TABLE 13(a)   The Observed and Expected Cell Frequencies for the Data in Table 10

	Favor	Indifferent	Opposed
Democrat	138 (115.14)	83 (85.50)	64 (84.36)
Republican	64 (86.86)	67 (64.50)	84 (63.64)

TABLE 13(b)   The Values of $(O - E)^2/E$

	Favor	Indifferent	Opposed	Total
Democrat	4.539	.073	4.914	
Republican	6.016	.097	6.514	
				$22.153 = \chi^2$

An inspection of Table 13(b) reveals that large contributions to the value of χ^2 have come from the corner cells. Moreover, comparing the observed and expected frequencies in Table 13(a), we see that the support for the program draws more from the Democrats than Republicans.   ∎

From our analysis of the contingency table in Example 13, the procedure for testing independence in a general $r \times c$ contingency table is readily apparent.

Chi-Square Test of Independence

$$\text{Expected cell frequency} = \frac{\text{Row total} \times \text{Column total}}{\text{Grand total}}$$

The test statistic is

$$\chi^2 = \sum_{\text{cells}} \frac{(O - E)^2}{E}$$

$$\begin{aligned}\text{d.f. of } \chi^2 &= (r - 1)(c - 1) \\ &= (\text{No. of rows} - 1) \times (\text{No. of columns} - 1)\end{aligned}$$

which is identical to the d.f. of χ^2 for testing homogeneity.

The Null Hypothesis of Independence

H_0:   Each cell probability equals the product of the corresponding row and column marginal probabilities.

In summary, the χ^2 test statistic, its d.f., and the rejection region for testing independence are the same as when testing homogeneity. The expected cell frequencies are determined in the same way. Only the statement of the null hypothesis is different.

SPURIOUS DEPENDENCE

When the χ^2 test leads to a rejection of the null hypothesis of independence, we conclude that the data provide evidence of a **statistical association** between the two characteristics. However, we must refrain from making the hasty interpretation that these characteristics are directly related. A claim of causal relationship must draw from common sense, which statistical evidence must not be allowed to supersede.

Two characteristics may appear to be strongly related due to the common influence of a third factor that is not included in the study. In such cases, the dependence is called a **spurious dependence.** For instance, if a sample of individuals is classified in a 2 × 2 contingency table according to whether or not they are heavy drinkers and whether or not they suffer from respiratory trouble, we would probably find a high value for χ^2 and conclude that a strong statistical association exists between drinking habit and lung condition. But the reason for the association may be that most heavy drinkers are also heavy smokers and the smoking habit is a direct cause of respiratory trouble. This discussion should remind the reader of a similar warning given in Chapter 3 regarding the interpretation of a correlation coefficient between two sets of measurements. In the context of contingency tables, examples of spurious dependence are sometimes called **Simpson's paradox,** which is discussed in Chapter 3, Section 2.

Exercises

5.1 Applicants for public assistance are allowed an appeals process when they feel unfairly treated. At such a hearing, the applicant may choose self-representation or representation by an attorney. The appeal may result in an increase, decrease, or no change of the aid recommendation. Court records of 320 appeals cases provided the following data.

Type of Representation	Amount of Aid		
	Increased	Unchanged	Decreased
Self	59	108	17
Attorney	70	63	3

Are the patterns of the appeals decision significantly different between the two types of representation? Test at $\alpha = .05$.

5.2   Interviews with 185 persons engaged in a stressful occupation revealed the following classification.

	Alcoholic	Nonalcoholic	Total
Depressed	54	27	81
Not depressed	22	82	104
Total	76	109	185

Do these observations demonstrate an association between alcoholism and mental depression? Test at $\alpha = .01$.

5.3   In a study of handedness,[4] a sample of 408 children who were adopted in early infancy was classified according to their own handedness and the handedness of their adoptive parents.

Adoptive Parents	Handedness of Adopted Offspring		Total
	Right	Left	
Both right handed	307	48	355
At least one left handed	47	6	53
Total	354	54	408

Test the null hypothesis of independence. Take $\alpha = .05$.

5.4   A survey was conducted to study people's attitude toward television programs that show violence. A random sample of 1250 adults was selected and classified according to gender and response to the question: Do you think there is a link between violence on TV and crime?

	Response		
	Yes	No	Not sure
Male	378	237	26
Female	438	146	25

[4]*Source:*   L. Carter-Saltzman, "Biological and Sociocultural Effects on Handedness: Comparison Between Biological and Adoptive Families," *Science, 209* (Sept. 1980), pp. 1263–1265, Table 3. Copyright ©1980 by AAAS.

Do the survey data show that attitude and gender are associated? Test the null hypothesis of independence at $\alpha = .05$.

5.5   A survey was conducted by sampling 400 persons who were questioned regarding union membership and attitude toward decreased national spending on social welfare programs. The cross-tabulated frequency counts are presented.

	Support	Indifferent	Opposed	Total
Union	112	36	28	176
Nonunion	84	68	72	224
Total	196	104	100	400

Can these observed differences be explained by chance or are attitude and membership status dependent? Test with $\alpha = .05$.

5.6   The food supplies department circulates a questionnaire among U.S. armed forces personnel to gather information about preferences for various types of foods. One question is, "Do you prefer black olives to green olives on the lunch menu?" A random sample of 435 respondents is classified according to both olive preference and geographical area of home, and the following data are recorded.

Region	Number Preferring Black Olives	Total No. of Respondents
West	65	118
East	59	135
South	48	90
Midwest	43	92

Are preference and geographical area dependent? Test with $\alpha = .05$.

5.7   In a study of factors that regulate behavior, three kinds of subjects are identified: overcontrollers, average controllers, and undercontrollers, with the first group being most inhibited. Each subject is given the routine task of filling a box with buttons and all subjects are told they can stop whenever they wish. Whenever a subject indicates he or she wishes to stop, the experimenter asks, "Don't you really want to continue? The number of subjects in each group who stop and the number who continue are given in the following table.

Controller	Continue	Stop	Total
Over	9	9	18
Average	8	12	20
Under	3	14	17
Total	20	35	55

Are controller group and experimenter's influence associated?

5.8   According to a survey,[5] 1256 persons responded to the question "have you ever been a designated driver?" The yes/no response was cross-classified with respect to gender.

	Yes	No	Total
Male	444	117	561
Female	588	107	695
Total	1032	224	1256

Test the null hypothesis of independence between gender and designated driver experience. Take $\alpha = .05$.

KEY IDEAS AND FORMULAS

Inference About a Population Proportion When n Is Large

Parameter of interest

p = the population proportion of individuals possessing a stated characteristic

Inferences are based on $\hat{p} = \dfrac{X}{n}$, the sample proportion.

[5]Wisconsin Department of Transportation, Traffic/Safety Study 1997.

1. A point estimator of p is $\hat{p}$.

$$\text{Estimated standard error} = \sqrt{\frac{\hat{p}\hat{q}}{n}}, \quad \text{where } \hat{q} = 1 - \hat{p}$$

$$100(1 - \alpha)\% \text{ error margin} = z_{\alpha/2}\sqrt{\frac{\hat{p}\hat{q}}{n}}$$

2. A $100(1 - \alpha)\%$ confidence interval for p is

$$\left(\hat{p} - z_{\alpha/2}\sqrt{\frac{\hat{p}\hat{q}}{n}}, \quad \hat{p} + z_{\alpha/2}\sqrt{\frac{\hat{p}\hat{q}}{n}} \right)$$

3. To test hypotheses about p, the test statistic is

$$Z = \frac{\hat{p} - p_0}{\sqrt{p_0 q_0 / n}}$$

where p_0 is the value of p under H_0. The rejection region is right-sided, left-sided, or two-sided according to $H_1: p > p_0$, $H_1: p < p_0$, or $H_1: p \neq p_0$, respectively.

Comparing Two Binomial Proportions—Large Samples

Data:

X = No. of successes in n_1 trials with success probability $P(S) = p_1$

Y = No. of successes in n_2 trials with success probability $P(S) = p_2$

To test $H_0: p_1 = p_2$ versus $H_1: p_1 \neq p_2$, use the Z-test:

$$Z = \frac{\hat{p}_1 - \hat{p}_2}{\sqrt{\hat{p}(1 - \hat{p})}\sqrt{\frac{1}{n_1} + \frac{1}{n_2}}} \quad \text{with} \quad R: \; Z \geq z_{\alpha/2} \;\; \text{or} \;\; Z \leq -z_{\alpha/2}$$

where

$$\hat{p}_1 = \frac{X}{n_1}, \quad \hat{p}_2 = \frac{Y}{n_2}, \quad \hat{p} = \frac{X + Y}{n_1 + n_2}$$

To test $H_0: p_1 = p_2$ versus $H_1: p_1 > p_2$, use the Z-test with R: $Z \geq z_\alpha$.

A $100(1 - \alpha)\%$ confidence interval for $p_1 - p_2$ is

$$(\hat{p}_1 - \hat{p}_2) \pm z_{\alpha/2}\sqrt{\frac{\hat{p}_1(1 - \hat{p}_1)}{n_1} + \frac{\hat{p}_2(1 - \hat{p}_2)}{n_2}}$$

Testing Homogeneity in an $r \times c$ Contingency Table

Data: Independent random samples from r populations, each sample classified in c response categories.

Null hypothesis: In each response category, the probabilities are equal for all the populations.

Test statistic

$$\chi^2 = \sum_{\text{cells}} \frac{(O - E)^2}{E}, \qquad \text{d.f.} = (r - 1)(c - 1)$$

where for each cell

$$O = \text{Observed cell frequency}$$

$$E = \frac{\text{Row total} \times \text{Column total}}{\text{Grand total}}$$

Rejection region: $\chi^2 \geq \chi_\alpha^2$

Testing Independence in an $r \times c$ Contingency Table

Data: A random sample of size n is simultaneously classified with respect to two characteristics; one has r categories and the other has c categories.

Null hypothesis: The two classifications are independent; that is, each cell probability is the product of the row and column marginal probabilities.

Test statistic and rejection region: Same as when testing homogeneity.

Limitation

All inference procedures of this chapter require large samples. The χ^2 tests are appropriate if no expected cell frequency is too small (≥ 5 is normally required).

6. REVIEW EXERCISES

6.1   Calculate a $100(1 - \alpha)\%$ confidence interval for the population proportion in each of the following cases.

(a) $n = 1600$, $\hat{p} = .36$, $1 - \alpha = .98$

(b) $n = 920$, $X = 567$, $1 - \alpha = .95$

6.2   A random sample of 2000 persons from the labor force of a large city are interviewed, and 165 of them are found to be unemployed.

(a)   Estimate the rate of unemployment based on the data.

(b)   Establish a 95% error margin for your estimate.

6.3   Referring to Exercise 6.2, compute a 98% confidence interval for the rate of unemployment.

6.4   Let p = proportion of adults in a city who required a lawyer in the past year.

(a)   Determine the rejection region for an $\alpha = .05$ level test of $H_0: p = .25$ against $H_1: p > .25$.

(b)   If 65 persons in a random sample of 200 required lawyer services, what does the test conclude?

6.5   Referring to Exercise 6.4, obtain a 90% confidence interval for p.

6.6   A survey of 130 high school students shows that 38 of them work over 20 hours per week during the school year. Does it demonstrate the conjecture that more than 25% of the high school students work over 20 hours per week during the school year? (Answer by calculating the P-value.)

6.7   A marketing manager wishes to determine whether lemon-scented and almond-scented dishwashing liquids are equally liked by consumers. Out of 250 consumers interviewed, 145 expressed their preference for the lemon-scented and the remaining 105 preferred the almond-scented.
  (a)   Do these data provide strong evidence that there is a difference in popularity between the two scented liquids? (Test with $\alpha = .05$.)
  (b)   Construct a 95% confidence interval for the population proportion of consumers who prefer the almond-scented liquid.

6.8   A genetic model suggests that 80% of the plants grown from a cross between two given strains of seeds will be of the dwarf variety. After breeding 200 of these plants, 136 were observed to be of the dwarf variety.
  (a)   Does this observation strongly contradict the genetic model?
  (b)   Construct a 95% confidence interval for the true proportion of dwarf plants obtained from the given cross.

6.9   A researcher in a heart-and-lung center wishes to estimate the rate of incidence of respiratory disorders among middle-aged males who have been smoking more than two packs of cigarettes per day during the last five years. How large a sample should the researcher select to be 95% confident that the error of estimation of the proportion of the population afflicted with respiratory disorders does not exceed .03? (The true value of p is expected to be near .15.)

6.10   One wishes to estimate the proportion of car owners who purchase more than $500,000 of liability coverage in their automobile insurance policies.
  (a)   How large a sample should be chosen to estimate the proportion with a 95% error margin of .008? (Use $p^* = .15$.)
  (b)   A random sample of 400 car owners is taken, and 56 of them are found to have chosen this extent of coverage. Construct a 95% confidence interval for the population proportion.

6.11   An experiment was conducted to study whether cloud seeding reduces the occurrences of hail. At a hail-prone geographical area, seeding was done on 50 stormy days and another 165 stormy days were also observed without seeding. The following counts were obtained.

	Days	
	Seeded	Not Seeded
Hail	7	43
No hail	43	122
Total	50	165

Do these data substantiate the conjecture that seeding reduces the chance of hail? (Answer by determining the P-value.)

6.12 Referring to the data of Exercise 6.11, calculate a 90% confidence interval for the difference between the probabilities of hail with and without seeding.

6.13 Refer to Exercise 7.4, Chapter 3, and the data concerning a vaccine for type B hepatitis.

	Hepatitis	No Hepatitis	Total
Vaccinated	11	538	549
Not vaccinated	70	464	534
Total	81	1002	1083

Do these data indicate that there is a different rate of incidence of hepatitis between the vaccinated and nonvaccinated participants? Use the χ^2 test for homogeneity in a contingency table. Use $\alpha = .05$.

6.14 Refer to the data in Exercise 6.13.

(a) Use the Z-test for testing the equality of two population proportions with a two-sided alternative. Verify the relation $\chi^2 = Z^2$ by comparing their numerical values.

(b) If the alternative is that the incidence rate is lower for the vaccinated group, which of the two tests should be used?

6.15 To compare the effectiveness of four drugs in relieving postoperative pain, an experiment was done by randomly assigning 195 surgical patients to the drugs under study. Recorded here are the number of patients assigned to each drug and the number of patients who were free of pain for a period of five hours.

	Free of Pain	No. of Patients Assigned
Drug 1	23	52
Drug 2	30	48
Drug 3	19	50
Drug 4	29	45

(a)   Make a 4 × 2 contingency table showing the counts of patients who were free of pain and those who had pain, and test the null hypothesis that all four drugs are equally effective. (Use $\alpha = .05$.)

(b)   Let p_1, p_2, p_3, and p_4 denote the population proportions of patients who would be free of pain under the use of drugs 1, 2, 3, and 4, respectively. Calculate a 90% confidence interval for each of these probabilities individually.

6.16   Using the data for drugs 1 and 3 in Exercise 6.15, make a 2 × 2 contingency table and test $H_0: p_1 = p_3$ versus $H_1: p_1 \neq p_3$ at $\alpha = .05$ employing

(a)   The χ^2 test

(b)   The Z-test

6.17   Based on interviews of couples seeking divorces, a social worker compiles the following data related to the period of acquaintanceship before marriage and the duration of marriage.

Acquaintanceship Before Marriage	Duration of Marriage		Total
	≤4 years	>4 years	
Under $\frac{1}{2}$ year	11	8	19
$\frac{1}{2}$–$1\frac{1}{2}$ years	28	24	52
Over $1\frac{1}{2}$ years	21	19	40
Total	60	51	111

Perform a test to determine whether the data substantiate an association between the stability of a marriage and the period of acquaintanceship prior to marriage. Use $\alpha = .05$.

6.18   By polling a random sample of 350 undergraduate students, a campus press obtains the following frequency counts regarding student attitude toward a proposed change in dormitory regulations.

	Favor	Indifferent	Oppose	Total
Male	93	72	21	186
Female	55	79	30	164
Total	148	151	51	350

Does the proposal seem to appeal differently to male and female students?

6.19   In a study of possible genetic influence of parental hand preference (refer to Exercise 5.3), a sample of 400 children was classified according to each child's handedness and the handedness of the biological parents.

Parents' Handedness (Father × Mother)	Handedness of Biological Offspring		Total
	Right	Left	
Right × Right	303	37	340
Right × Left	29	9	38
Left × Right	16	6	22
Total	348	52	400

Do these findings demonstrate an association between the handedness of parents and their biological offspring?

6.20   ***Pooling contingency tables can produce spurious association.***   A large organization is being investigated to determine whether its recruitment is sex-biased. Tables 14 and 15, respectively, show the classification of applicants for secretarial and for sales positions according to gender and result of interview. Table 16 is an aggregation of the corresponding entries of Table 14 and Table 15.

TABLE 14   Secretarial Positions

	Offered	Denied	Total
Male	25	50	75
Female	75	150	225
Total	100	200	300

TABLE 15   Sales Positions

	Offered	Denied	Total
Male	150	50	200
Female	75	25	100
Total	225	75	300

TABLE 16   Secretarial and Sales Positions

	Offered	Denied	Total
Male	175	100	275
Female	150	175	325
Total	325	275	600

(a) Verify that the χ^2 statistic for testing independence is zero for each of the data sets given in Tables 14 and 15.

(b) For the pooled data given in Table 16, compute the value of the χ^2 statistic and test the null hypothesis of independence.

(c) Explain the paradoxical result that there is no sex bias in any job category, but the combined data indicate sex discrimination.

6.21 Refer to the admissions–gender data in Table 4, page 100, Chapter 3. Test for gender bias by testing the null hypothesis of independence of gender and admission decision, with $\alpha = .05$.

ACTIVITIES TO IMPROVE UNDERSTANDING: STUDENT PROJECTS

Formulate a question of interest concerning two treatments where the response is either a *Success* or a *Failure*. Conduct a comparative experiment and report your activities.

Statement of Purpose

Plan for Data Collection

Treatments

Treatment 1 ⸺

Treatment 2 ⸺

Populations

Population 1 _____

Population 2 _____

Response variable

Randomization   Specify your randomization and how the sample is to be collected.

Calculations
Determine confidence intervals and conduct any appropriate test of hypotheses.

Summary of Findings
Interpret your tests and confidence intervals. Relate findings to the purpose of the study. Include a graphical display.

CHAPTER 12

Regression Analysis— Simple Linear Regression

Chapter Objectives

After reading this chapter, you should be able to

▶ Discuss a straight-line regression model.
▶ Fit a straight line to pairs of observations.
▶ Construct confidence intervals and conduct tests for the slope of a line.
▶ Use the fitted line to make inferences about the mean.
▶ Predict a new observation using the fitted line.
▶ Understand the relation between correlation and straight-line regression.

1. INTRODUCTION

Regression analysis concerns the study of relationships between variables with the object of identifying, estimating, and validating the relationship. The estimated relationship can then be used to predict one variable from the value of the other variable(s). In this chapter, we introduce the subject with specific reference to the straight-line model.

There is a natural relationship between the price, y, and the age, x, of a used car. We expect that the price will drop as the age increases. For an arbitrarily chosen used car of a given age x, however, the price also depends on many other factors. Does the car have high mileage, rust spots, or accident damage? Besides (1) mileage, (2) rust, and (3) damage, there are also other factors such as (4) options like air conditioning, automatic transmission, and power steering, and (5) whether the car was serviced according to the maintenance schedule. Price may also depend on the area of the state or country where it is located.

The price of a car can be approximated by a linear relationship with age x, at least over a limited range of ages. Given the age x of an arbitrary car, we would approximate its price by the linear relation $y = \beta_0 + \beta_1 x$ for some constants β_0 and β_1.

The interpretation of the linear, or straight-line, relation is reviewed in Figure 1. The line is determined by two constants: the **intercept** β_0 is the value where the line cuts the y-axis, and the **slope** specifies the amount that y increases whenever x increases by one unit. If the linear relation specifies that y decreases as x increases, the slope is negative as in Figure 1(b).

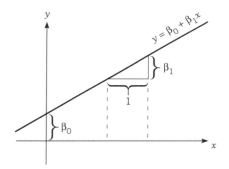

(a) An increasing relation:
Slope β_1 is positive.

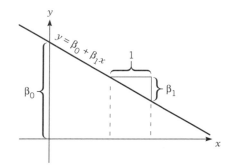

(b) A decreasing relation:
Slope β_1 is negative.

Figure 1. Graph of straight line $y = \beta_0 + \beta_1 x$

Usually, a strict linear relation will not be satisfactory. A more reasonable relation allows for some adjustment e to the linear approximation:

$$Y = \beta_0 + \beta_1 x + e$$

In the car example, e can be positive or negative depending on factors other than age of car, including those mentioned above. We will model the adjustment e, called the **error**, as a random variable with mean zero so the mean price of cars with age x is $\beta_0 + \beta_1 x$.

A straight-line regression model can be formulated and tentatively entertained but data need to be collected and plotted to check conformance with the model.

Example 1   **A scatter diagram reveals the form of relation between price and age**
A potential purchaser of a used Saturn automobile decided to collect data from the newspaper on advertised price and age.[1] To simplify the discussion, price is reported only to the nearest thousand dollars.

As expected, the older cars have lower prices than the newer ones, but it is difficult to determine an overall pattern from the tabled data. Make a scatter diagram to see whether a pattern exists.

TABLE 1   Advertised Price (in Thousands of Dollars) and Age (in Years)

Age (years) x	Price (thousands) y
6	6
5	9
4	8
3	10
2	11
2	12
1	11
1	13

SOLUTION AND DISCUSSION   The construction of scatter diagrams is discussed in Section 4 of Chapter 3. Recall that, for each pair of values (age, price), age is located on the horizontal axis and price on the vertical axis. In the scatter plot shown in Figure 2 (page 506), the points seem to cluster about a straight line with a negative slope. The points do not lie exactly on a straight line but the pattern is consistent with a straight line masked by error.

[1]From Minneapolis and St. Paul newspapers. Exercise 6.8 gives the data to more digits.

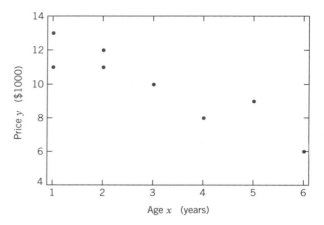

Figure 2. A scatter diagram of the age–price data in Table 1

We could draw a line, by eye, on the scatter diagram and read off the estimates of the parameters β_0 and β_1. In Section 3, we present an objective method of estimation called **least squares.** We will obtain the estimated straight line

$$\hat{y} = 13.4 - 1.13x$$

Once estimates are obtained, four types of inference problems can be addressed. Later, we assume normal distributions for price that result in sampling distributions for the estimators. These lead to point estimates, confidence intervals, and tests of hypotheses for:

1. The value of β_0:   The average new car price, after it is driven off the lot

2. The value of β_1:   The average price change per year for the population of current used cars

3. The value of $\beta_0 + \beta_1 x$ for a fixed age x: The population mean price of used cars of age x

4. The price of a particular used car of age x

To address these inference problems, we select a random sample of n used cars and record the price y_i and age x_i, for the ith car, $i = 1, 2, \ldots, n,$ as in Example 1. These data can be used to provide answers for each of the four types of inference problems.

We return to these problems in Section 4 after we review and summarize the steps thus far.

2. FORMULATING A MODEL FOR A STRAIGHT-LINE REGRESSION

In our used car example, age can be used to predict the price. Whenever a tentative straight-line relation is studied for the purpose of predicting one variable y from another variable x, it is called **simple regression** or **simple linear regression.**

The variable x acts as an independent variable whose values are often set by the experimenter. The observed response variable y depends on x but is also subject to unaccountable variation or error.

Notation

$x =$ independent variable, also called predictor variable, causal variable, or input variable

$y =$ dependent or response variable

Pairs of (x, y) values need to be collected to understand the relation between x and y. At each instance, x is the fixed value of the independent variable and y is the corresponding response. For a generic experiment, we use n to denote the total number of pairs and a subscript to denote the case number. For instance, in the used car example, there are $n = 8$ cars and $(x_2, y_2) = (5, 9)$ are the age and price of the second car. Table 2 gives this general structure for simple linear regression.

TABLE 2   Date Structure for a Simple Regression

Setting of the Independent Variable	Response
x_1	y_1
x_2	y_2
x_3	y_3
.	.
.	.
.	.
x_n	y_n

The data must be plotted in a scatter diagram, as illustrated in Figure 1, to see whether a linear or even some other type of relation exists. If you cannot see a clustering of points along a line or curve, masked by error, it is unlikely that x and y are related.

First Step in the Analysis

Plotting a scatter diagram is an important preliminary step prior to undertaking a formal statistical analysis of the relationship between two variables.

Statistical ideas must be introduced when the points in the scatter diagram do not lie exactly on a straight line. The price of a used car depends not only on age but on condition, mileage, and many other known and unknown variables. Consequently, we include another term to capture the variation due to other factors or variables. That is, we have the following model

Response = [Straight line with input x] + [Effects due to other sources]

The first term on the right-hand side is the mean response of Y for the given input x. The second term on the right-hand side is called the error term. This last term is modeled as a random variable. In other words, we formulate the model as a straight-line relation masked by random error.

Statistical Model for a Straight-Line Regression

We assume that the response Y is a random variable that is related to the input variable x by

$$Y_i = \beta_0 + \beta_1 x_i + e_i \qquad i = 1, \ldots, n$$

where

1. Y_i denotes the response corresponding to the ith experimental run in which the input variable x is set at the value x_i.

2. $e_1, \ldots, e_n$ are the unknown error components that are superimposed on the true linear relation. These are unobservable random variables, which we assume are independently and normally distributed with mean zero and an unknown standard deviation σ.

3. The parameters β_0 and β_1, which together determine the straight line, are unknown.

According to this model, the observation Y_i corresponding to level x_i of the controlled variable is one observation from the normal distribution with mean = $\beta_0 + \beta_1 x_i$ and standard deviation = σ. This statistical model is illustrated in Figure 3, which shows a few normal distributions for the response variable Y for different values of the input variable x. All these distributions have the same standard deviation and their means lie on the unknown true straight line $\beta_0 + \beta_1 x$. Aside from the fact that σ is unknown, the line on which the means of these normal distributions are located is also unknown. In fact, an important objective of the statistical analysis is to estimate this line.

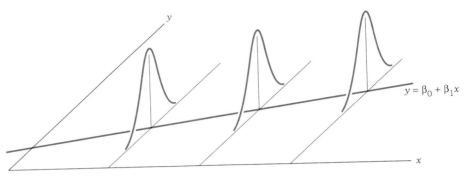

Figure 3. Normal distributions of Y with means on a straight line

Exercises

2.1   Identify the predictor variable x and the response variable y in each of the following situations.

(a)   The carbon monoxide level in the blood is to be related to the number of cigarettes a person smokes per day.

(b)   A market analyst wishes to relate local advertising expenses with product sales in several areas of the state.

2.2   Identify the predictor variable x and the response variable y in each of the following situations.

(a)   Noise level, measured at a point along a main city street, is to be related to the number of cars per minute that pass this point.

(b)   A computer scientist collected data from students on the number of hours they studied for the midterm examination and their grade on the exam.

2.3   Plot the line $y = 2 + 3x$ on graph paper by locating the points for $x = 1$ and $x = 4$. What is its intercept? What is its slope?

2.4   Identify the values of the parameters β_0, β_1, and σ in the statistical model

$$Y = 3 + 5x + e$$

where e is a normal random variable with mean 0 and standard deviation 4.

2.5   Identify the values of the parameters β_0, β_1, and σ in the statistical model

$$Y = 8 - 6x + e$$

where e is a normal random variable with mean 0 and standard deviation 4.

2.6   Under the linear regression model,
   (a)   Determine the mean and standard deviation of Y, for $x = 3$, when $\beta_0 = 1$, $\beta_1 = 3$, and $\sigma = 2$.
   (b)   Repeat part (a) with $x = 1$.

2.7   Under the linear regression model,
   (a)   Determine the mean and standard deviation of Y, for $x = 1$, when $\beta_0 = 2$, $\beta_1 = 4$, and $\sigma = 3$.
   (b)   Repeat part (a) with $x = 2$.

2.8   Graph the straight line for the means of the linear regression model

$$Y = \beta_0 + \beta_1 x + e$$

   having $\beta_0 = -3$, $\beta_1 = 4$, and the normal random variable e has standard deviation 3.

2.9   Graph the straight line for the means of the linear regression model $Y = \beta_0 + \beta_1 x + e$ having $\beta_0 = 7$ and $\beta_1 = 2$.

2.10   Consider the linear regression model

$$Y = \beta_0 + \beta_1 x + e$$

   where $\beta_0 = -1$, $\beta_1 = -2$, and the normal random variable e has standard deviation 2.
   (a)   What is the mean of the response Y when $x = 3$? When $x = 6$?
   (b)   Will the response at $x = 2$ always be larger than that at $x = 4$? Explain.

2.11   Consider the linear regression model $Y = \beta_0 + \beta_1 x + e$, where $\beta_0 = 3$, $\beta_1 = 5$, and the normal random variable e has standard deviation 3.
   (a)   What is the mean of the response Y when $x = 4$? When $x = 5$?
   (b)   Will the response at $x = 5$ always be larger than that at $x = 4$? Explain.

2.12   Data on the monthly rent and number of bedrooms are to be collected for apartments that are available in one area of a city. The goal is to relate the monthly rent to the number of bedrooms.
   (a)   Specify the independent variable and the dependent or response variable.
   (b)   Name two other variables of interest that would be part of the error term in the simple regression model.
   (c)   Explain the meaning of the regression parameter β_1 in terms of the mean rent for one additional bedroom.

2.13   Data on the monthly rent and number of square feet are to be collected for apartments that are available in one area of a city. The goal is to relate the monthly rent to the number of square feet.

(a) Specify the independent variable and the dependent or response variable.

(b) Name two other variables of interest that would be part of the error term in the simple regression model.

(c) Explain the meaning of the regression parameter β_1 in terms of the mean rent for one additional square foot of space.

2.14 A person seeking a permanent partner often evaluates potential candidates on the basis of looks, intelligence, and other variables. Experience teaches that few persons, among the many who are thought to be good candidates, turn out to be suitable as a permanent partner. Still, the use of predictor variables helps to narrow the field.

Suppose you can quantify both intelligence and suitability after you get to know a person. Then, you can develop a regression equation to predict the response—suitability as a permanent partner.

(a) Select two variables that would be good predictor variables.

(b) For each variable specified in part (a), would your relation be appropriate if you were just looking for someone to go with to a Saturday night party? Comment on the choice of the predictor variables and so on.

(c) Would your choice of predictor variables, and the resulting predictions, be suitable for use by others? Why or why not?

3. FITTING A LINE BY THE METHOD OF LEAST SQUARES

Let us tentatively assume that the formulation of a straight-line model is correct. The **method of least squares** provides an objective and efficient method for estimating the two coefficients that determine the line and for estimating the variance of the error term.

Suppose that an arbitrary line $y = b_0 + b_1 x$ is drawn on the scatter diagram as it is in Figure 4. At the value x_i of the independent variable, the y value predicted by this line is $b_0 + b_1 x_i$ whereas the observed value is y_i. The discrepancy between the observed and predicted y values is then $(y_i - b_0 - b_1 x_i) = d_i$, which is the **vertical** distance of the point from the line.

Considering such discrepancies at all the n points, we take

$$D = \sum_{i=1}^{n} d_i^2 = \sum_{i=1}^{n} (y_i - b_0 - b_1 x_i)^2$$

as an overall measure of the discrepancy of the observed points from the trial line $y = b_0 + b_1 x$. The magnitude of D obviously depends on the line that is drawn. In other words, it depends on b_0 and b_1, the two quantities that determine the trial line. A good fit will make D as small as possible.

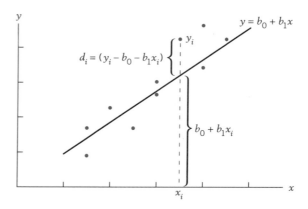

Figure 4. Deviations of the observations from a line
$y = b_0 + b_1 x$

The **principle of least squares** specifies that b_0 and b_1 are to be chosen to minimize the sum of squares

$$D = \sum_{i=1}^{n} (y_i - b_0 - b_i x_i)^2$$

The quantities b_0 and b_1 thus determined are denoted by $\hat{\beta}_0$ and $\hat{\beta}_1$, respectively, and called the **least squares estimates** of the regression parameters β_0 and β_1. The best fitting straight line is then given by the equation

$$\hat{y} = \hat{\beta}_0 + \hat{\beta}_1 x$$

Most often, computer software is used to perform the calculations that give the least squares estimators of the intercept, slope, and fitted line, and the estimated standard deviation of the error. Table 3 illustrates some typical output. The more complete output, given in Table 7, will be examined later after we introduce statistical procedures for estimation and testing.

To perform these calculations with a hand-held calculator, we follow the procedure outlined in Table 4. The basic quantities $\bar{x}$, $\bar{y}$, S_{xx}, S_{xy}, and S_{yy} are

TABLE 3   MINITAB Output for Fitting a Straight Line
Using the Data in Table 1, Example 1

The commands are given in Exercise 4.11.

```
Data:

Age     6 5 4  3  2  2  1  1
Price   6 9 8 10 11 12 11 13
```

Regression Analysis

```
The regression equation is
Price = 13.4 - 1.13 Age
S = 0.9682
```

TABLE 4   Steps in a Simple Regression Analysis

1. Data

x	x_1	x_2	$\cdots$	x_n
y	y_1	y_2	$\cdots$	y_n

2. Plot scatter diagram.

3. Compute the two sums Σx, Σy. Then find the three sums Σx^2, Σy^2, Σxy to use the second forms

$$S_{xx} = \Sigma(x - \bar{x})^2 \qquad\qquad S_{xx} = \Sigma x^2 - (\Sigma x)^2/n$$
$$S_{xy} = \Sigma(x - \bar{x})(y - \bar{y}) \quad \text{or} \quad S_{xy} = \Sigma xy - (\Sigma x)(\Sigma y)/n$$
$$S_{yy} = \Sigma(y - \bar{y})^2 \qquad\qquad S_{yy} = \Sigma y^2 - (\Sigma y)^2/n$$

4. Compute least squares estimates: $\hat{\beta}_0 = \bar{y} - \hat{\beta}_1 \bar{x}$
$$\hat{\beta}_1 = S_{xy}/S_{xx}$$

5. Form the least squares line: $\hat{y} = \hat{\beta}_0 + \hat{\beta}_1 x$

6. Find the predicted y_i at x_i using (6): $\hat{y}_i = \hat{\beta}_0 + \hat{\beta}_1 x_i$, $\quad i = 1, \ldots, n$

7. Find the estimated error (residual):
$$\hat{e}_i = y_i - \hat{y}_i = (\text{Observed } y - \text{Predicted } y)$$

8. Check whether the estimated residuals in (7) come from a normal population. Use a normal-scores plot, dot diagram, or histogram.

9. Compute SSE $= S_{yy} - S_{xy}^2/S_{xx}$

10. Compute the estimated standard deviation of the errors: $S = \sqrt{\dfrac{\text{SSE}}{n-2}}$

11. Compute the correlation coefficient: $r = \dfrac{S_{xy}}{\sqrt{S_{xx}S_{yy}}}$

determined first. These last three quantities were introduced in Section 4 of Chapter 3. We next calculate the least squares estimates.

> Least squares estimate of β_0
> $$\hat{\beta}_0 = \bar{y} - \hat{\beta}_1 \bar{x}$$
> Least squares estimate of β_1
> $$\hat{\beta}_1 = \frac{S_{xy}}{S_{xx}}$$

The estimates $\hat{\beta}_0$ and $\hat{\beta}_1$ can then be used to locate the best fitting line:

Fitted (or estimated) regression line

$$\hat{y} = \hat{\beta}_0 + \hat{\beta}_1 x$$

As we have already explained, this line provides the best fit to the data in the sense that the sum of squares of the deviations or

$$\sum_{i=1}^{n} (y_i - \hat{\beta}_0 - \hat{\beta}_1 x_i)^2$$

is the smallest.

The individual deviations of the observations y_i from the fitted values $\hat{y}_i = \hat{\beta}_0 + \hat{\beta}_1 x_i$ are called the **residuals**, and we denote these by $\hat{e}_i$.

Residuals

$$\hat{e}_i = y_i - \hat{\beta}_0 - \hat{\beta}_1 x_i \qquad i = 1, \ldots, n$$

Although some residuals are positive and some negative, a property of the least squares fit is that the **sum of the residuals is always zero.**

The **residual sum of squares** or the **sum of squares due to error** is

$$\mathrm{SSE} = \sum_{i=1}^{n} \hat{e}_i^2$$

$$= S_{yy} - \frac{S_{xy}^2}{S_{xx}}$$

An estimate of σ^2 is obtained by dividing SSE by $(n - 2)$. The reduction by 2 is because two degrees of freedom are lost from estimating the two parameters β_0 and β_1.

Estimate of Variance

The estimator of the error variance σ^2 is

$$S^2 = \frac{SSE}{n - 2}$$

The least squares calculations following the steps in Table 4, for the used car data in Table 1, are illustrated in Table 5.

TABLE 5 Computations for the Least Squares Line, SSE, and Residuals Using the Data of Table 1

x	y	x^2	xy	y^2	$\hat{\beta}_0 + \hat{\beta}_1 x$	e_i
6	6	36	36	36	6.625	−.625
5	9	25	45	81	7.750	1.250
4	8	16	32	64	8.875	−.875
3	10	9	30	100	10.000	0
2	11	4	22	121	11.125	−.125
2	12	4	24	144	11.125	.875
1	11	1	11	121	12.250	−1.250
1	13	1	13	169	12.250	.750
Total 24	80	96	213	836		0

$$\bar{x} = 3 \quad \bar{y} = 10$$

$$S_{xx} = 96 - \frac{(24)^2}{8} = 24 \qquad \hat{\beta}_1 = \frac{-27}{24} = -1.125$$

$$S_{xy} = 213 - \frac{24 \times 80}{8} = -27 \qquad \hat{\beta}_0 = 10 - (-1.125)3 = 13.375$$

$$S_{yy} = 836 - \frac{(80)^2}{8} = 36 \qquad SSE = 36 - \frac{(-27)^2}{24} = 5.625$$

The equation of the line fitted by the least squares method is then

$$\hat{y} = 13.375 - 1.125 x$$

The residuals $\hat{e}_i = y_i - 13.375 + 1.125 x_i$ are computed in the last column of Table 4. The sum of squares of these residuals is

$$\Sigma \hat{e}_i^2 = (-.625)^2 + (1.250)^2 + \cdots + (.750)^2 = 5.625$$

This agrees with our previous calculation of SSE.

Finally, the estimate of σ^2, the variance of the error, is

$$s^2 = \frac{SSE}{n-2} = \frac{5.625}{8-2} = .9375$$

The estimate of the standard deviation of the error, σ, is $s = \sqrt{.9375} = .9682$.

By drawing the least squares line $\hat{y} = 13.375 - 1.125\,x$ on the scatter plot, as shown in Figure 5, we can visually judge the fit. Typically some points will lie above the line and some below.

Notice that the sum of the residuals $\hat{e}_i$ is zero. The **sum of the residuals is always zero.** That is, whatever the data $(x_1, y_1), \ldots, (x_n, y_n)$, the least squares line is centered in the scatter of data in the sense that the sample mean of the residuals is always zero.

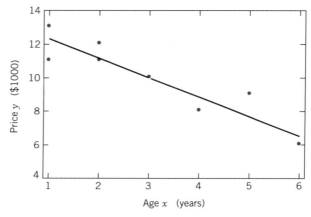

Figure 5. Least squares line and scatter diagram for the used car data

Exercises

3.1   Given the five pairs of (x, y) values,

x	0	1	6	3	5
y	4	3	0	2	1

(a)   Construct a scatter diagram.
(b)   Calculate $\bar{x}, \bar{y}, S_{xx}, S_{xy}$, and S_{yy}.
(c)   Calculate the least squares estimates $\hat{\beta}_0$ and $\hat{\beta}_1$.
(d)   Determine the fitted line and draw the line on the scatter diagram.

3.2   Given these six pairs of (x, y) values,

x	1	2	3	3	4	5
y	9	5	6	3	3	1

(a)   Plot the scatter diagram.
(b)   Calculate $\bar{x}, \bar{y}, S_{xx}, S_{xy}$, and S_{yy}.
(c)   Calculate the least squares estimates $\hat{\beta}_0$ and $\hat{\beta}_1$.
(d)   Determine the fitted line and draw the line on the scatter diagram.

3.3   Refer to Exercise 3.1.
(a)   Find the residuals and verify that they sum to zero.
(b)   Calculate the residual sum of squares SSE by
(i)   Adding the squares of the residuals.
(ii)   Using the formula SSE $= S_{yy} - S_{xy}^2/S_{xx}$.
(c)   Obtain the estimate of σ^2.

3.4   Refer to Exercise 3.2.
(a)   Find the residuals and verify that they sum to zero.
(b)   Calculate the residual sums of squares SSE by
(i)   Adding the squares of the residuals.
(ii)   Using the formula SSE $= S_{yy} - S_{xy}^2/S_{xx}$.
(c)   Obtain the estimate of σ^2.

3.5   Given the five pairs of (x, y) values,

x	0	1	2	3	4
y	2	1	4	5	8

(a)   Calculate $\bar{x}, \bar{y}, S_{xx}, S_{xy}$, and S_{yy}.
(b)   Calculate the least squares estimates $\hat{\beta}_0$ and $\hat{\beta}_1$.
(c)   Determine the fitted line.

3.6   Given the five pairs of (x, y) values,

x	1	3	5	7	9
y	4	3	6	8	9

(a) Calculate $\bar{x}$, $\bar{y}$, S_{xx}, S_{xy}, and S_{yy}.

(b) Calculate the least squares estimates $\hat{\beta}_0$ and $\hat{\beta}_1$.

(c) Determine the fitted line.

3.7 Computing from a data set of (x, y) values, we obtained the following summary statistics:

$$n = 14, \qquad \bar{x} = 3.5, \qquad \bar{y} = 2.32$$
$$S_{xx} = 10.82, \qquad S_{xy} = 2.677, \qquad S_{yy} = 1.035$$

(a) Obtain the equation of the best fitting straight line.

(b) Calculate the residual sum of squares.

(c) Estimate σ^2.

3.8 Computing from a data set of (x, y) values, we obtained the following summary statistics:

$$n = 18, \qquad \bar{x} = 1.2, \qquad \bar{y} = 5.1$$
$$S_{xx} = 14.10, \qquad S_{xy} = 2.31, \qquad S_{yy} = 2.01$$

(a) Obtain the equation of the best fitting straight line.

(b) Calculate the residual sum of squares.

(c) Estimate σ^2.

4. INFERENCES FOR A STRAIGHT-LINE MODEL

We are now prepared to test hypotheses, construct confidence intervals, and make predictions in the context of straight-line regression. These inferences are based on the point estimators and estimated standard errors presented in Table 6.

TABLE 6 Point Estimator and Its Estimated Standard Error for Each Type of Problem in Regression

Type of Problem	Estimator	Estimated Standard Error
1. Intercept $\hat{\beta}_0$	$\hat{\beta}_0 = \bar{y} - \hat{\beta}_1 \bar{x}$	$S\sqrt{\dfrac{1}{n} + \dfrac{\bar{x}^2}{S_{xx}}}$
2. Slope $\hat{\beta}_1$	$\hat{\beta}_1 = S_{xy}/S_{xx}$	$S/\sqrt{S_{xx}}$
3. Mean response for a given input x^*	$\hat{\beta}_0 + \hat{\beta}_1 x^*$	$S\sqrt{\dfrac{1}{n} + \dfrac{(x^* - \bar{x})^2}{S_{xx}}}$
4. Single response y for a given input x^*	$\hat{\beta}_0 + \hat{\beta}_1 x^*$	$S\sqrt{1 + \dfrac{1}{n} - \dfrac{(x^* - \bar{x})^2}{S_{xx}}}$

where $\hat{\beta}_0$, $\hat{\beta}_1$, S, S_{xx}, and $\bar{x}$ are found in the list of the steps in simple regression.

Confidence intervals take the form

Point estimate $\pm\ t_{\alpha/2}$ (Estimated standard error of point estimate)

where $t_{\alpha/2}$ is the upper $\alpha/2$ point of the t distribution with $n - 2$ degrees of freedom. For the prediction of a new case, the fourth type of problem, the interval is called a **prediction interval.**

Tests of hypotheses are based on test statistics of the form

$$T = \frac{\text{Point estimator} - \text{Null hypothesis parameter value}}{\text{Estimated standard error of estimator}}$$

Under the null hypothesis, these statistics have a t distribution with $n - 2$ degrees of freedom.

For example, consider testing whether the slope is positive, when the sample size is 20. To test $H_0: \beta_1 = 0$ versus $H_1: \beta_1 > 0$, at level of significance .05, with $n = 20$, we take

$$T = \frac{\hat{\beta}_1}{S/\sqrt{S_{xx}}}$$

and the rejection region is $R: T \geq t_{.05}$ where $t_{.05}$ is based on $20 - 2 = 18$ degrees of freedom. To test the null hypothesis $H_0: \beta_1 = \beta_{10}$, where β_{10} is some nonzero value, just replace the numerator $\hat{\beta}_1$ by $\hat{\beta}_1 - \beta_{10}$.

INFERENCE CONCERNING THE SLOPE β_1

In a regression analysis problem, it is of special interest to determine whether the expected response does or does not vary with the magnitude of the input variable x. According to the linear regression model,

$$\text{Expected response} = \beta_0 + \beta_1 x$$

This does not change with a change in x if and only if $\beta_1 = 0$. We can therefore test the null hypothesis $H_0: \beta_1 = 0$ against a one- or a two-sided alternative, depending on the nature of the relation that is anticipated.

Example 2   A test concerning slope—The mean price decrease per year
Do the data in Table 1 provide strong evidence that the mean price of a used car decreases with age? Test with $\alpha = .05$.

SOLUTION AND DISCUSSION

1. *Hypotheses.*   The slope β_1 is the mean change in price over one year. For a decreasing relation, the slope β_1 must be negative. Therefore we test the null hypothesis $H_0: \beta_1 = 0$ against the one-sided alternative hypothesis $H_1: \beta_1 < 0$.

2. *Level of significance.*   We are given $\alpha = .05$.

3. *Test statistic.*   From Table 6, the test statistic is

$$T = \frac{\hat{\beta}_1}{S/\sqrt{S_{xx}}}$$

4. *Rejection region.*   Since $t_{.05} = 1.943$ for $n - 2 = 8 - 2 = 6$ d.f., and the alternative hypothesis is left-sided, we set the rejection region as $R: T \leq -1.943$.

5. *Calculations.*   Using the calculations in Table 5, we have

$$\hat{\beta}_1 = -1.125$$

$$s^2 = \frac{SSE}{n-2} = \frac{5.625}{6} = .9375 \qquad s = .9682$$

$$\text{Estimated S.E.}(\hat{\beta}_1) = \frac{s}{\sqrt{S_{xx}}} = \frac{.9682}{\sqrt{24}} = .1976$$

so the test statistic takes the value

$$t = \frac{-1.125}{.1976} = -5.692$$

6. *Conclusion.*   The observed value of the test statistic lies in the rejection region, so we reject $H_0: \beta_1 = 0$ in favor of $H_1: \beta_1 < 0$, at level $\alpha = .05$.

   Moreover, -5.692 is smaller than -3.707 where $t_{.005} = 3.707$. The *P*-value is smaller than .005. A computer program gives

$$P[T \leq -5.692] = .0006 \quad (\text{see Figure 6})$$

   The evidence is very strong that the mean price decreases with age.

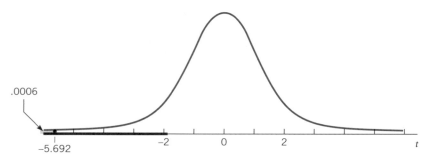

Figure 6. *P*-value = .0006

Warning.   If the test fails to reject $H_0: \beta_1 = 0$, we must not conclude that y does not depend on x. First, the absence of a linear relation has been established only over the limited range of x values in the experiment. There may be dependence of y on x if x is varied over a wider range. Second, this is a test of a linear relation. There may be a relation but it might follow a curve. This case would be revealed by inspection of the scatter diagram.

   We can also obtain confidence intervals for the slope β_1.

Example 3

A confidence interval for the slope—Mean price change per year
Construct a 95% confidence interval for the slope of the regression line in reference to the data in Table 1.

SOLUTION AND DISCUSSION

From Table 6, the 95% confidence interval for β_1 has the form

$$\left(\hat{\beta}_1 - t_{.025} \frac{S}{\sqrt{S_{xx}}}, \qquad \hat{\beta}_1 + t_{.025} \frac{S}{\sqrt{S_{xx}}} \right)$$

From Example 2, $\hat{\beta}_1 = -1.125$ and $s/\sqrt{S_{xx}} = .1976$. Since $t_{.025} = 2.447$ for $n - 2 = 8 - 2 = 6$ d.f., the 95% confidence interval becomes

$$(-1.125 - 2.447 \times .1976, \qquad -1.125 + 2.447 \times .1976)$$
$$\text{or} \qquad (-1.609, \quad -.641)$$

We are 95% confident that the mean price decrease per year of age will be between .641 and 1.609 thousand dollars. ∎

STATISTICAL REASONING

Reducing Your Time in Line

Grocery shopping in a busy store is a familiar activity to most of us. After filling your shopping cart with the desired items, you next want to check out as soon as possible. With several lines to choose from, you naturally join the shortest line. This strategy cannot guarantee that you will always complete your checkout before others who later join some other line. However, over many different trips, this strategy saves time.

Bob Peterson/FPG International

This strategy, learned by experience, is an example of an informal regression analysis. Consider the time to wait in line, before reaching the checkout counter, as the response variable Y. This time in line is linked to the number of persons ahead of us. When the number of persons is large, the wait will typically be long. A small number of persons typically means a short wait. That is, the number of persons in line is a good predictor variable. Without performing any formal calculations, we have concluded that the expected waiting time is smallest for the shortest line.

Even though time in line is linked to number of persons ahead of us, the relation really connects the *expected value* of waiting time to the number of persons. Other variables that contribute to the error, or variation about the expected relationship, include the speed of the checkout person, the number of manufacturers' coupons that are redeemed, and the number of personal checks that require verification. Because of these and other variables, entering the shortest line will not always get you to the checkout counter fastest. It will, however, do so on average because the expected times are shortest with this strategy.

The regression relation in expected value can provide useful predictions in spite of the presence of errors or variation about the relation.

Another predictor variable, the number of items in the carts of persons ahead of us, can often provide even better prediction of the time waiting in line. Professor J often joins a slightly longer line if he notices one very full cart in the shortest line. This, too, has been found to shorten the average time in line.

When a good predictor variable can be identified, the value of the response variable can be estimated or forecasted with improved accuracy. Better strategies can then be developed.

INFERENCE ABOUT THE INTERCEPT β_0

Although somewhat less important in practice, inferences similar to those above for β_1 can be provided for the parameter β_0. The procedures are again based on the t distribution with d.f. $= n - 2$.

Example 4   A confidence interval for the intercept

Construct a 95% confidence interval for the intercept of the regression line in reference to the data in Table 1.

SOLUTION AND DISCUSSION   From Table 6, the 95% confidence interval for β_0 has the form

$$\left(\hat{\beta}_0 - t_{.025}\, S \sqrt{\frac{1}{n} + \frac{\bar{x}^2}{S_{xx}}}, \quad \hat{\beta}_0 + t_{.025}\, S \sqrt{\frac{1}{n} + \frac{\bar{x}^2}{S_{xx}}} \right)$$

From the calculations below Table 5, we found $\hat{\beta}_0 = 13.375$, $\bar{x} = 3$, and $S_{xx} = 24$. Also $s = .9682$. Since $t_{.025} = 2.447$ for 6 d.f., the 95% confidence interval for β_0 becomes

$$\left(13.375 - 2.447 \times .9682 \sqrt{\frac{1}{8} + \frac{3^2}{24}}, \quad 13.375 + 2.447 \times .9682 \sqrt{\frac{1}{8} + \frac{3^2}{24}} \right)$$

$$= (13.375 - 1.675, \quad 13.375 + 1.675) \quad \text{or} \quad (11.70, \quad 15.05)$$

Note that β_0 represents the mean price for a car of age 0. No new cars were included in the data so this is an extrapolation for one year beyond the data. It may be representative of the value of new cars, at least after they are a few weeks old. ∎

ESTIMATION OF THE MEAN RESPONSE FOR A SPECIFIED x VALUE

The least squares fitted line can be used to predict the mean response at a specified value x^* of the input variable. According to the regression model introduced in Section 2, the expected response at a value x^* of the input variable is $\beta_0 + \beta_1 x^*$. This expected response is estimated by $\hat{\beta}_0 + \hat{\beta}_1 x^*$, the ordinate of the regression line at $x = x^*$.

Example 5   Confidence interval for mean price
Refer to Example 1 and Table 1. Construct a 95% confidence interval for the mean price at age (a) $x^* = 3.5$ and (b) $x^* = 5.5$ years.

SOLUTION AND DISCUSSION   From Table 6, the 95% confidence interval for the mean response $\beta_0 + \beta_1 x^*$ has the form

$$\hat{\beta}_0 + \hat{\beta}_1 x^* \pm t_{.025} \, S \sqrt{\frac{1}{n} + \frac{(x^* - \bar{x})^2}{S_{xx}}}$$

The fitted regression line is

$$\hat{y} = \hat{\beta}_0 + \hat{\beta}_1 x = 13.375 - 1.125x$$

(a)   The expected price for a 3.5-year-old car is

$$\hat{\beta}_0 + \hat{\beta}_1 x = 13.375 - 1.125 \times 3.5 = 9.438 \text{ thousand dollars}$$

$$\text{Estimated standard error} = s \sqrt{\frac{1}{n} + \frac{(x^* - \bar{x})^2}{S_{xx}}}$$

$$= .9682 \times \sqrt{\frac{1}{8} + \frac{(3.5 - 3)^2}{24}} = .3563$$

A 95% confidence interval for the mean price of a 3.5-year-old car is

$$(9.438 - 2.447 \times .3563, \quad 9.438 + 2.447 \times .3563)$$
$$\text{or} \quad (8.57, \quad 10.31)$$

We are 95% confident that the mean price of a 3.5-year-old car is between 8.57 and 10.31 thousand dollars.

(b)   The expected price for a 5.5-year-old car is

$$\hat{\beta}_0 + \hat{\beta}_1 x = 13.375 - 1.125 \times 5.5 = 7.188 \text{ thousand dollars}$$

$$\text{Estimated standard error} = s \sqrt{\frac{1}{n} + \frac{(x^* - \bar{x})^2}{S_{xx}}}$$

$$= .9682 \times \sqrt{\frac{1}{8} + \frac{(5.5 - 3)^2}{24}} = .6011$$

A 95% confidence interval for the mean price of a 5.5-year-old car is

$$(7.188 - 2.447 \times .6011, \qquad 7.188 + 2.447 \times .6011)$$
$$\text{or} \qquad (5.72, \quad 8.66)$$

We are 95% confident that the mean price of a 5.5-year-old car is between 5.72 and 8.66 thousand dollars.

This second interval is more than 50% longer than the first. The formula for standard error contains the term $(x^* - \bar{x})^2$, which increases as x^* increases in distance from $\bar{x}$. In general, estimation is more precise for values of x near the mean $\bar{x}$ than for values far from the mean. ∎

PREDICTION OF A SINGLE RESPONSE FOR A SPECIFIED x VALUE

Suppose we want to predict the price of a specific used car of age x^*. This problem is different from estimating the mean price of cars of age x^*. The prediction is still based on the fitted regression line but the interval is larger. It must allow for the fact that a single car's price is not on the regression line but deviates from it by an error that has variance σ^2.

Example 6   Prediction interval for price
Refer to Example 1 and Table 1. Construct a 95% prediction interval for the price of a used car having age $x^* = 3.5$ years.

SOLUTION AND DISCUSSION   From Table 6, the 95% prediction interval for the price of a car of age 3.5 years has the form

$$\hat{\beta}_0 + \hat{\beta}_1 x^* \pm t_{.025} \, S \sqrt{1 + \frac{1}{n} + \frac{(x^* - \bar{x})^2}{S_{xx}}}$$

From Example 5, the expected price for a 3.5-year-old car is

$$\hat{\beta}_0 + \hat{\beta}_1 x = 13.375 - 1.125 \times 3.5 = 9.438 \text{ thousand dollars}$$

$$\text{Estimated standard error} = s \sqrt{1 + \frac{1}{n} + \frac{(x^* - \bar{x})^2}{S_{xx}}}$$

$$= .9682 \times \sqrt{1 + \frac{1}{8} + \frac{(3.5 - 3)^2}{24}} = 1.032$$

A 95% prediction interval for the price of a specific 3.5-year-old car is

$$(9.438 - 2.447 \times 1.032, \qquad 9.438 + 2.447 \times 1.032) \qquad \text{or} \qquad (6.91, \quad 11.96)$$

We are 95% confident that the price of a 3.5-year-old car will be between 6.91 and 11.96 thousand dollars.

Notice that this interval is much larger than the corresponding interval in Example 5 that pertains to the mean price of all cars of age 3.5 years. ∎

Regression analyses are most conveniently done on a computer. A more complete selection of the output from the computer software package MINITAB, for the data in Example 1, is given in Table 7.

TABLE 7   MINITAB Computer Output for the Data in Example 1

Regression Analysis

```
The regression equation is
Price = 13.4 − 1.13 Age
```

Predictor	Coef	StDev	T	P
Constant	13.3750	0.6847	19.54	0.000
Age	−1.1250	0.1976	−5.69	0.001

```
S = 0.9682   R-Sq = 84.4%
```

Analysis of Variance

Source	DF	SS	MS	F	P
Regression	1	30.375	30.375	32.40	0.001
Error	6	5.625	0.938		
Total	7	36.000			

The output shows that $\hat{\beta}_1 = -1.125$ with estimated standard error .1976 and $s = .9682$, which agrees with Example 2. The **F-test,** based on the **F-distribution,** is equivalent to the t-test for $\beta_1 = 0$.

In the preceding discussion, we have used the data of Example 1 to illustrate the various inferences associated with a straight-line regression model. Example 7 gives applications to a different data set.

Example 7   Statistical analysis of straight-line relation—Skill related to training
In a study to determine how the skill in doing a complex assembly job is influenced by the amount of training, 15 new recruits were given varying amounts of training ranging between 3 and 12 hours. After the training, their times to perform the job were recorded. After denoting x = duration of training (in hours) and y = time to do the job (in minutes), the following summary statistics were calculated.

$$\bar{x} = 7.2, \qquad S_{xx} = 33.6, \qquad S_{xy} = -57.2$$
$$\bar{y} = 45.6, \qquad S_{yy} = 160.2$$

(a) Determine the equation of the best fitting straight line.

(b) Do the data substantiate the claim that the mean job time depends on the hours of training? Test with $\alpha = .02$.

(c) Estimate the mean job time for 9 hours of training and construct a 95% confidence interval.

(d) Find the predicted y for $x = 35$ hours and comment on the result.

SOLUTION AND DISCUSSION

Using the summary statistics, we find:

(a) The least squares estimates are

$$\hat{\beta}_1 = \frac{S_{xy}}{S_{xx}} = \frac{-57.2}{33.6} = -1.702$$

$$\hat{\beta}_0 = \bar{y} - \hat{\beta}_1 \bar{x} = 45.6 - (-1.702) \times 7.2 = 57.85$$

So, the equation of the fitted line is

$$\hat{y} = 57.85 - 1.702x$$

(b) 1. *Hypotheses.* The slope β_1 is the change in mean job time for each additional hour of training. The claim that mean job time depends on training, that the slope β_1 is not zero, becomes the alternative hypothesis. We test $H_0: \beta_1 = 0$ versus $H_1: \beta_1 \neq 0$.

2. *Level of significance.* We are given $\alpha = .02$.

3. *Test statistic.*

$$T = \frac{\hat{\beta}_1}{S/\sqrt{S_{xx}}}$$

4. *Rejection region.* Since $\alpha = .02$ and $t_{.01} = 2.650$ for $n - 2 = 15 - 2 = 13$ d.f., and the alternative hypothesis is two-sided, we set the rejection region as $R: T \leq -2.650$ or $T \geq 2.650$.

5. *Calculations.*

$$\text{SSE} = S_{yy} - \frac{S_{xy}^2}{S_{xx}} = 160.2 - \frac{(-57.2)^2}{33.6} = 62.824$$

$$s = \sqrt{\frac{\text{SSE}}{n - 2}} = \sqrt{\frac{62.824}{13}} = 2.198$$

$$\text{Estimated S.E.}(\hat{\beta}_1) = \frac{s}{\sqrt{S_{xx}}} = \frac{2.198}{\sqrt{33.6}} = .379$$

The t statistic has the value

$$t = \frac{-1.702}{.379} = -4.49$$

6. *Conclusion.*   The observed value -4.49 is less than -2.650 so we reject $H_0: \beta_1 = 0$ in favor of $H_1: \beta_1 \neq 0$, at level $\alpha = .02$. Moreover, -4.49 is smaller than -3.012 where $t_{.005} = 3.012$. The P-value is smaller than .010. (See Figure 7.)

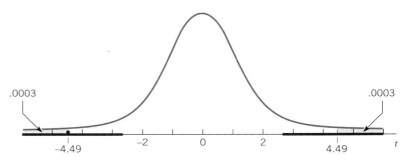

Figure 7. P-value $= .0006$ for two-sided test

A computer calculation gives

$$P[T \leq -4.49] + P[T \geq 4.49] = 2(.0003) = .0006$$

The evidence is very strong that the mean job time changes with hours of training, within the range covered in the experiment.

(c) The expected job time corresponding to $x^* = 9$ hours is estimated as

$$\hat{\beta}_0 + \hat{\beta}_1 x^* = 57.85 + (-1.702) \times 9$$
$$= 42.53 \text{ minutes}$$

and its

$$\text{Estimated S.E.} = s\sqrt{\frac{1}{15} + \frac{(9 - 7.2)^2}{33.6}} = .888$$

Since $t_{.025} = 2.160$ with d.f. $= 13$, the required confidence interval is

$$42.53 \pm 2.160 \times .888 = 42.53 \pm 1.92$$
$$\text{or} \quad (40.6, \quad 44.5) \text{ minutes}$$

(d) Since $x = 35$ hours is far beyond the experimental range of 3 to 12 hours, it is not sensible to predict y at $x = 35$ using the fitted regression line. Here, a formal calculation gives

$$\text{Predicted job time} = 57.85 - 1.702 \times 35$$
$$= -1.72 \text{ minutes}$$

which is a nonsensical result.   ■

Caution.   Extreme caution should be exercised in extending a fitted regression line to make **long-range predictions** far away from the range of x values
covered in the experiment. Not only does the confidence interval become so
wide that predictions based on it can be extremely unreliable, but an even
greater danger exists. If the pattern of the relationship between the variables
changes drastically at a distant value of x, the data provide no information
with which to detect such a change. Figure 8 illustrates this situation. We
would observe a good linear relationship if we experimented with x values in
the 5–10 range, but if the fitted line were extended to estimate the response
at $x^* = 20$, then our estimate would drastically miss the mark.

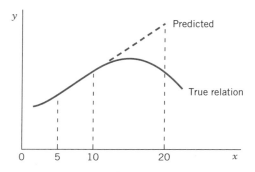

Figure 8. Danger in long-range prediction

Exercises

4.1   Given the five pairs of (x, y) values,

x	0	1	6	3	5
y	4	3	0	2	1

(a)   Calculate the least squares estimates $\hat{\beta}_0$ and $\hat{\beta}_1$. Also estimate the
error variance σ^2.

(b)   Test $H_0: \beta_1 = 0$ versus $H_1: \beta_1 \neq 0$ with $\alpha = .05$.

(c)   Estimate the expected y value corresponding to $x = 2.5$ and give a
90% confidence interval.

(d)   Construct a 90% confidence interval for the intercept β_0.

4.2   Refer to Exercise 4.1. Obtain a 95% confidence interval for β_1.

4.3   Given these five pairs of (x, y) values,

x	1	2	3	4	5
y	.9	2.1	2.4	3.3	3.8

(a)   Calculate the least squares estimates $\hat{\beta}_0$ and $\hat{\beta}_1$. Also estimate the error variance σ^2.

(b)   Test $H_0: \beta_1 = 1$ versus $H_1: \beta_1 \neq 1$ with $\alpha = .05$.

(c)   Estimate the expected y value corresponding to $x = 3.5$ and give a 95% confidence interval.

(d)   Construct a 90% confidence interval for the intercept β_0.

4.4   For a random sample of seven homes that were recently sold in a city suburb, the assessed values x and the selling prices y are

x (thousand dollars)	83.5	90.0	70.5	100.8	110.2	94.6	120.0
y (thousand dollars)	88.0	91.2	76.2	107.0	111.0	99.0	118.0

(a)   Plot the scatter diagram.

(b)   Determine the equation of the least squares regression line and draw this line on the scatter diagram.

(c)   Construct a 95% confidence interval for the slope of the regression line.

4.5   Refer to the data in Exercise 4.4.

(a)   Estimate the expected selling price of homes that were assessed at $90,000 and construct a 95% confidence interval.

(b)   For a single home that was assessed at $90,000, give a 95% confidence interval for the selling price.

4.6   According to the computer output in Table 8 (page 530),

(a)   What model is fitted?

(b)   Test, with $\alpha = .05$, whether the x term is needed in the model.

4.7   According to the computer output in Table 8 (page 530),

(a)   Predict the mean response when $x = 5000$.

(b)   Find a 90% confidence interval for the mean response when $x = 5000$. You will need the additional information: $n = 30$, $\bar{x} = 8354$, and $\Sigma (x_i - \bar{x})^2 = 97,599,296$.

TABLE 8   Computer Output for Exercises 4.6 and 4.7

```
THE REGRESSION EQUATION IS
Y = 994 + 0.104X
```

PREDICTOR	COEF	STDEV	T-RATIO	P
CONSTANT	994.0	254.7	3.90	0.001
X	0.10373	0.02978	3.48	0.002

```
S = 299.4   R-SQ = 30.2%
```

ANALYSIS OF VARIANCE

SOURCE	DF	SS	MS	F	P
REGRESSION	1	1087765	1087765	12.14	0.002
ERROR	28	2509820	89636		
TOTAL	29	3597585			

4.8   According to the computer output in Table 9,
  (a)   What model is fitted?
  (b)   Test, with $\alpha = .05$, whether the x term is needed in the model.

4.9   According to the computer output in Table 9,
  (a)   Predict the mean response when $x = 3$.
  (b)   Find a 90% confidence interval for the mean response when $x = 3$. You will need the additional information: $n = 25$, $\bar{x} = 1.793$, and $\Sigma (x_i - \bar{x})^2 = 1.848$.
  (c)   Find a 90% confidence interval for the mean response when $x = 2$.

TABLE 9   Computer Output for Exercises 4.8 and 4.9

```
THE REGRESSION EQUATION IS
Y = 0.338 + 0.831X
```

PREDICTOR	COEF	STDEV	T-RATIO	P
CONSTANT	0.3381	0.1579	2.14	0.043
X	0.83099	0.08702	9.55	0.000

```
S = 0.1208   R-SQ = 79.9%
```

ANALYSIS OF VARIANCE

SOURCE	DF	SS	MS	F	P
REGRESSION	1	1.3318	1.3318	91.20	0.000
ERROR	23	0.3359	0.0146		
TOTAL	24	1.6676			

4.10 A car dealer specializing in Corvettes enlarged his facilities and offered a number of models for sale during the open house. His data on price and age of Corvettes are listed in the table.

Age (years)	Price ($1000)
1	39.9
2	32.0
4	25.0
5	20.0
6	16.0
6	20.0
10	13.0
11	13.7
11	11.0
12	12.0
12	20.0
12	9.0
12	9.0
13	12.5
15	7.0

(a) Construct a scatter diagram of price versus age.

(b) Calculate the least squares estimates and the fitted line.

(c) Predict the price of a 19-year-old model. What danger is there in making this prediction?

USING A COMPUTER

4.11 We illustrate the MINITAB commands for fitting a straight line, using the age–price car data in Table 1.

```
Data C12T1.DAT:
Age:          6   5   4    3    2    2    1    1
Price:        6   9   8   10   11   12   11   13
Dialog box:                     Session command
Stat > Regression > Regression   MTB > REGRESS C2 ON 1 PRED C1
Type C2 in Response.
Type C1 in Predictors. Click OK.
```

Output:

Regression Analysis

```
The regression equation is
Price = 13.4 - 1.13 Age
```

Predictor	Coef	StDev	T	P
Constant	13.3750	0.6847	19.54	0.000
Age	-1.1250	0.1976	-5.69	0.001

```
S = 0.9682          R-Sq = 84.4%
```

Use MINITAB or some other package program to regress the marine growth on freshwater growth for the fish growth data in Table D.7 of the Data Bank. Do separate regression analyses for

(a) All fish

(b) Males

(c) Females

Your analysis should include (i) a scatter diagram, (ii) a fitted line, (iii) a determination if β_1 differs from zero. Also find a 95% confidence interval for the population mean when the freshwater growth is 100.

4.12 Use MINITAB or some other package program to obtain the linear regression of the Computer Anxiety Score on the Computer Attitude Score as given in Table D.4 of the Data Bank. Do separate regression analyses for

(a) All students

(b) Males

(c) Females

Your analysis should include (i) a scatter diagram, (ii) a fitted line, (iii) a determination if β_1 differs from zero. Also find a 95% confidence interval for the population mean when the computer attitude score is 2.5.

4.13 Use MINITAB or some other package to obtain the scatter diagram and the regression line of the final number of situps on the initial number of situps given in Table D.5 of the Data Bank.

5. CHECKING THE ADEQUACY OF THE STRAIGHT-LINE MODEL

Until now, we have tentatively assumed that the straight-line model is correct. The inference procedures can be misleading if there are serious violations of the model. How do we check the adequacy of the model?

To examine this question, we view an observation y_i as consisting of two components. One part is explained by the fitted regression line and the other is the unexplained or residual.

$$y_i = (\hat{\beta}_0 + \hat{\beta}_1 x_i) + (y_i - \hat{\beta}_0 - \hat{\beta}_1 x_i)$$

Observed	Explained by	Residual or
y value	linear relation	deviation from
		linear relation

Below we show how to use the residuals to check departures from the model assumptions, but first we study the first component due to regression.

THE STRENGTH OF A LINEAR RELATION

From the discussion in Chapter 3, we know that when the sample correlation coefficient $r = S_{xy}/\sqrt{S_{xx}S_{yy}}$ is near 1 or -1, the points in the scatter diagram lie almost on a straight line. In these cases, y can be predicted accurately by the least squares line because almost all of the variation in y is **explained** by the linear dependence on x.

More generally, we wish to determine the proportion of variation in the observations that is explained by the regression. The total variation in the observed responses

$$S_{yy} = \sum_{i=1}^{n}(y_i - \bar{y})^2$$

and the unexplained part of the variation is the sum of squares of the residuals or $\text{SSE} = S_{yy} - S_{xy}^2/S_{xx}$. This leaves the difference $S_{yy} - \text{SSE}$ as the sum of squares due to regression. This decomposes the sum of squares into the two components corresponding to the decomposition of the observation: linear relation and residual.

Decomposition of Variability

$$S_{yy} = \frac{S_{xy}^2}{S_{xx}} + \text{SSE}$$

Total	Variability explained	Residual or
variability of y	by the linear relation	unexplained
		variability

The first term on the right-hand side of this equality is called the sum of squares (SS) due to linear regression. Likewise, the total variability S_{yy} is also called the total SS of y. For the straight-line model to provide a good fit to the data, the SS due to the linear regression should comprise a major portion of S_{yy}.

As an index of how well the straight-line model fits, it is then reasonable to consider the proportion of the y-variability explained by the linear relation:

$$\frac{\text{SS due to linear regression}}{\text{Total SS of } y} = \frac{S_{xy}^2 / S_{xx}}{S_{yy}} = \frac{S_{xy}^2}{S_{xx} S_{yy}} = r^2$$

where r is the sample correlation coefficient. Thus, the square of the sample correlation coefficient represents the proportion of the y-variability explained by the linear relation.

The strength of a linear relation is measured by

$$r^2 = \frac{S_{xy}^2}{S_{xx} S_{yy}}$$

which is the square of the sample correlation coefficient r.

The quantity r^2 is called the **coefficient of determination**.

From the expressions for r and $\hat{\beta}_1$, we have

$$r = \frac{S_{xy}}{\sqrt{S_{xx} S_{yy}}} = \frac{\sqrt{S_{xx}}}{\sqrt{S_{yy}}} \frac{S_{xy}}{S_{xx}} = \frac{\sqrt{S_{xx}}}{\sqrt{S_{yy}}} \hat{\beta}_1$$

The sample correlation coefficient, r, and the least squares estimate of slope, $\hat{\beta}_1$, have the same sign.

Example 8   The proportion of variation in price explained by age
Refer to the price data in Example 1.

(a) Determine the decomposition of variation and the proportion of variation explained by the linear regression. Also obtain the correlation r.

(b) Locate the decomposition of variance and proportion of variance explained in the computer output in Table 7, page 525.

SOLUTION AND DISCUSSION   (a) In Table 5, page 515, we obtained $S_{yy} = 36$ and SSE $= 5.625$. By subtraction, the sum of squares due to regression is $S_{yy} - $ SSE $= 36 - 5.625 = 30.375$. The decomposition of variability is

$$\begin{array}{ccccc} 36 & = & 30.375 & + & 5.625 \\ \text{Total} & & \text{Regression} & & \text{Residual} \\ \text{SS} & & \text{SS} & & \text{SS} \end{array}$$

and the proportion of variation in y explained by the regression is

$$r^2 = \frac{30.375}{36} = .844$$

This is quite high and we consider x = age to be a useful predictor of price.

Since the estimate of slope $\hat{\beta}_1 = -1.125$ is negative, r must be negative so $r = -\sqrt{.844} = -.919$.

(b)   The output includes the **analysis of variance** table

```
s = 0.9682   R-sq = 84.4%
```

```
Analysis of Variance
```

SOURCE	DF	SS	MS	F	P
Regression	1	30.375	30.375	32.40	0.001
Error	6	5.625	0.938		
Total	7	36.000			

The column labeled SS, Sums of Squares, gives the decomposition of variability. The proportion of variation explained by regression is given in percent: 84.4% = R-sq. ∎

CHECKING THE RESIDUALS

We now look at the second component of the observations, the residuals, $\hat{e}_i = y_i - \hat{\beta}_0 - \hat{\beta}_1 x_i$, to see what they can tell us. The residuals contain all of the sample information regarding departures from the model assumptions. They should be plotted for visual inspection.

The assumption of normal errors can be checked by making a histogram, dot diagram, and normal-scores plot. Figure 9 gives the normal-scores plot for the used car data. The pattern is nearly a straight line but there are really too few residuals to get accurate information.

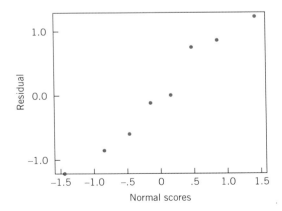

Figure 9. A normal-scores plot of the residuals for the used car data in Example 1

Other plots can reveal other discrepancies. Dot diagrams can reveal outliers among the residuals, although these can usually be seen in the scatter diagram. Plots of residuals versus the predicted values can reveal situations where the variance of observations is not constant but varies systematically with the input variable. Some typical cases are shown in Figure 10.

If the points form a horizontal band around zero, as in Figure 10(a), then no abnormality is indicated. In Figure 10(b), the width of the band increases

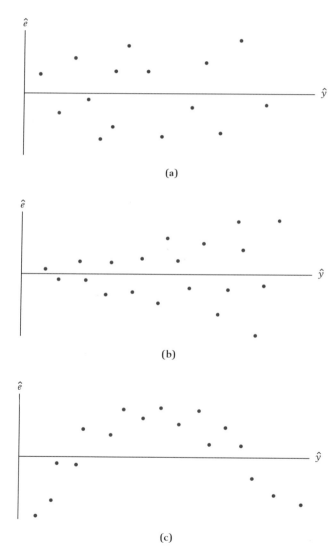

Figure 10. Plot of residual versus predicted value: **(a)** Constant spread; **(b)** Increasing spread; **(c)** Curved pattern

noticeably with increasing values of $\hat{y}$. This indicates that the error variance σ^2 tends to increase with an increasing level of response. We would then suspect the validity of the assumption of constant variance in the model. Figure 10(c) shows residuals that form a systematic pattern. Instead of being randomly distributed around the $\hat{y}$-axis, they tend first to increase steadily and then decrease. This would lead us to suspect that the model is inadequate and a squared term or some other nonlinear x term should be considered.

Exercises

5.1   Computing from a data set of (x, y) values, the following summary statistics are obtained.

$$n = 14, \qquad \bar{x} = 1.2, \qquad \bar{y} = 5.1$$
$$S_{xx} = 14.10, \qquad S_{xy} = 2.31, \qquad S_{yy} = 2.01$$

Determine the proportion of variation in y that is explained by linear regression.

5.2   Given $S_{xx} = 10.1$, $S_{yy} = 16.5$, and $S_{xy} = 9.3$, determine
(a)   the proportion of variation in y that is explained by linear regression
(b) r

5.3   A calculation shows that $S_{xx} = 92$, $S_{yy} = 457$, and $S_{xy} = 160$. Determine
(a)   the proportion of variation in y that is explained by linear regression
(b) r

5.4   Refer to Exercise 4.1.
(a)   What proportion of the y-variability is explained by the linear regression on x?
(b)   Find the sample correlation coefficient.
(c)   Calculate the residual sum of squares.
(d)   Estimate σ^2.

5.5   Referring to Exercise 4.4, determine the proportion of variation in the y values that is explained by the linear regression of y on x. What is the sample correlation coefficient between the x and y values?

5.6   Refer to Exercise 4.6. According to the computer output in Table 8, find the proportion of y-variability explained by x.

5.7   Refer to Exercise 4.8. According to the computer output in Table 9, find the proportion of y-variability explained by x.

KEY IDEAS AND FORMULAS

In its simplest form, **regression analysis** deals with studying the manner in which the **response variable** y depends on a **predictor variable** x.

The first important step in studying the relation between the variables y and x is to plot the **scatter diagram** of the data (x_i, y_i), $i = 1, \ldots, n$. If this plot indicates an approximate linear relation, a **straight-line regression model** is formulated:

$$
\begin{array}{ccccc}
\text{Response} & & (\text{A straight line in } x) & & (\text{Random error}) \\
Y_i & = & \beta_0 + \beta_1 x_i & + & e_i
\end{array}
$$

The random errors are assumed to be independent, normally distributed, and have mean 0 and equal standard deviations σ.

The regression parameters β_0 and β_1 are estimated by the **method of least squares**, which minimizes the sum of squared deviations $\Sigma (y_i - \beta_0 - \beta_1 x_i)^2$. The least squares estimates $\hat{\beta}_0$ and $\hat{\beta}_1$ determine the **best fitting regression line** $\hat{y} = \hat{\beta}_0 + \hat{\beta}_1 x$ which serves to predict y from x.

The differences $(y_i - \hat{y}_i) = (\text{Observed response} - \text{Predicted response})$ are called the **residuals**.

The adequacy of a straight line fit is measured by r^2, which represents the proportion of y-variability that is explained by the linear relation between y and x. A low value of r^2 indicates only that a linear relation is not appropriate—there may still be a relation on a curve.

Least squares estimators

$$
\hat{\beta}_1 = \frac{S_{xy}}{S_{xx}} \qquad \hat{\beta}_0 = \bar{y} - \hat{\beta}_1 \bar{x}
$$

Fitted regression line

$$
\hat{y} = \hat{\beta}_0 + \hat{\beta}_1 x
$$

Residuals

$$
\hat{e}_i = y_i - \hat{y}_i = y_i - \hat{\beta}_0 - \hat{\beta}_1 x_i
$$

Residual sum of squares

$$
\text{SSE} = \sum_{i=1}^{n} \hat{e}_i^2 = S_{yy} - \frac{S_{xy}^2}{S_{xx}}
$$

Estimate of σ^2

$$
S^2 = \frac{\text{SSE}}{n - 2}
$$

Inferences

1. Inferences concerning the **slope** β_1 are based on the

    estimator $\hat{\beta}_1$

    $$\text{estimated S.E.} = \frac{S}{\sqrt{S_{xx}}}$$

    and the sampling distribution

    $$T = \frac{\hat{\beta}_1 - \beta_1}{S/\sqrt{S_{xx}}}, \qquad \text{d.f.} = n - 2$$

    A $100(1 - \alpha)\%$ confidence interval for β_1 is

    $$\hat{\beta}_1 \pm t_{\alpha/2} \frac{S}{\sqrt{S_{xx}}}$$

    To test $H_0: \beta_1 = \beta_{10}$, the test statistic is

    $$T = \frac{\hat{\beta}_1 - \beta_{10}}{S/\sqrt{S_{xx}}}, \qquad \text{d.f.} = n - 2$$

2. Inferences concerning the **intercept** β_0 are based on the

    estimator $\hat{\beta}_0$

    $$\text{estimated S.E.} = S\sqrt{\frac{1}{n} + \frac{\bar{x}^2}{S_{xx}}}$$

    and the sampling distribution

    $$T = \frac{\hat{\beta}_0 - \beta_0}{S\sqrt{\dfrac{1}{n} + \dfrac{\bar{x}^2}{S_{xx}}}}, \qquad \text{d.f.} = n - 2$$

    A $100(1 - \alpha)\%$ confidence interval for β_0 is

    $$\hat{\beta}_0 \pm t_{\alpha/2}\, S\sqrt{\frac{1}{n} + \frac{\bar{x}^2}{S_{xx}}}$$

3. At a specified $x = x^*$, the expected response is $\beta_0 + \beta_1 x^*$. Inferences about the **expected response** are based on the

    estimator $\hat{\beta}_0 + \hat{\beta}_1 x^*$

    $$\text{estimated S.E.} = S\sqrt{\frac{1}{n} + \frac{(x^* - \bar{x})^2}{S_{xx}}}$$

A $100(1 - \alpha)\%$ confidence interval for the expected response at x^* is given by

$$\hat{\beta}_0 + \hat{\beta}_1 x^* \pm t_{\alpha/2} \, S \sqrt{\frac{1}{n} + \frac{(x^* - \bar{x})^2}{S_{xx}}}$$

4. A single response at a specified $x = x^*$ is predicted by $\hat{\beta}_0 + \hat{\beta}_1 x^*$ with

$$\text{estimated S.E.} = S \sqrt{1 + \frac{1}{n} + \frac{(x^* - \bar{x})^2}{S_{xx}}}$$

A $100(1 - \alpha)\%$ confidence interval for predicting a single response is

$$\hat{\beta}_0 + \hat{\beta}_1 x^* \pm t_{\alpha/2} \, S \sqrt{1 + \frac{1}{n} + \frac{(x^* - \bar{x})^2}{S_{xx}}}$$

Decomposition of Variability

Variability explained by the linear relation $= \dfrac{S_{xy}^2}{S_{xx}} = \hat{\beta}_1^2 S_{xx}$

Residual or unexplained variability $=$ SSE
Total y-variability $= S_{yy}$

Proportion of y-variability explained by linear regression

$$r^2 = \frac{S_{xy}^2}{S_{xx} S_{yy}}$$

Sample correlation coefficient

$$r = \frac{S_{xy}}{\sqrt{S_{xx} S_{yy}}}$$

6. REVIEW EXERCISES

6.1 Given the nine pairs of (x, y) values:

x	1	1	1	2	3	3	4	5	5
y	9	7	8	10	15	12	19	24	21

(a) Plot the scatter diagram.
(b) Calculate $\bar{x}, \bar{y}, S_{xx}, S_{yy}$, and S_{xy}.
(c) Determine the equation of the least squares fitted line and draw the line on the scatter diagram.
(d) Find the predicted y corresponding to $x = 3$.

6.2   Refer to Exercise 6.1.

(a)   Find the residuals.

(b)   Calculate the SSE by (i) summing the squares of the residuals and also (ii) using the formula $SSE = S_{yy} - S_{xy}^2/S_{xx}$.

(c)   Estimate the error variance.

6.3   Refer to Exercise 6.1.

(a)   Construct a 95% confidence interval for the slope of the regression line.

(b)   Obtain a 90% confidence interval for the expected y value corresponding to $x = 4$.

6.4   Given

$$n = 20, \qquad \Sigma x = 160, \qquad \Sigma y = 240$$
$$\Sigma x^2 = 1536, \qquad \Sigma xy = 1832, \qquad \Sigma y^2 = 2965$$

(a)   Find the equation of the least squares regression line.

(b)   Calculate the sample correlation coefficient between x and y.

(c)   Comment on the adequacy of the straight-line fit.

USING A COMPUTER

6.5   The calculations involved in a regression analysis become increasingly tedious with larger data sets. Access to a computer proves to be of considerable advantage. Illustrated here is a computer-based analysis of linear regression using the MINITAB package. See Exercise 4.11 for the commands.

Data									
X:	3	3	4	5	6	6	7	8	8	9
Y:	9	5	12	9	14	16	22	18	24	22

MINITAB output includes all the results that are basic to a linear regression analysis. The important pieces in the output are shown in Table 10 on page 542. Identify

(a)   The least squares estimates

(b)   SSE

(c)   Estimated standard errors of $\hat{\beta}_0$ and $\hat{\beta}_1$

(d)   The t statistics for testing $H_0: \beta_0 = 0$ and $H_0: \beta_1 = 0$

(e) r^2

(f)   The decomposition of the total sum of squares into the sum of squares explained by the linear regression and the residual sum of squares

TABLE 10   MINITAB Regression Analysis

Regression Analysis

The regression equation is
$Y = -1.07 + 2.74\,X$

Predictor	Coef	StDev	T	P
Constant	-1.071	2.751	-0.39	0.707
X	2.7408	0.4411	6.21	0.000

$S = 2.821$ R-Sq = 82.8%

Analysis of Variance

Source	DF	SS	MS	F	P
Regression	1	307.25	307.25	38.62	0.000
Error	8	63.65	7.96		
Total	9	370.90			

6.6   Many college students obtain college degree credits by demonstrating their proficiency on exams developed as part of the College Level Examination Program (CLEP). Based on their scores on the College Qualification Test (CQT), it would be helpful if students could predict their scores on a corresponding portion of the CLEP exam. The following data (courtesy of R. W. Johnson) are for x = Total CQT score and y = Mathematical CLEP score.

x	170	147	166	125	182	133	146	125	136	179
y	698	518	725	485	745	538	485	625	471	798

x	174	128	152	157	174	185	171	102	150	192
y	645	578	625	558	698	745	611	458	538	778

(a)   Find the least squares fit of a straight line.
(b)   Construct a 95% confidence interval for the slope.
(c)   Construct a 95% confidence interval for the CLEP score of a student who obtains a CQT score of 150.
(d)   Repeat (c) with $x = 175$ and $x = 195$.

6.7   Use MINITAB or some other package to obtain the scatter diagram, correlation coefficient, and the regression line of the final time to run 1.5 miles on the initial times given in Table D.5 of the Data Bank.

6.8   The more detailed prices of the used Saturn automobiles in Example 1 are given in the table.[2]

Age (years)	Price (thousands)
6	5,950
5	8,950
4	8,195
3	9,600
2	11,400
2	11,995
1	10,900
1	13,500

Repeat the regression analysis with these more accurate data. Your analysis should include (a) a scatter diagram, (b) fitted line, (c) determination whether β_1 differs from zero, (d) the proportion of variance explained by regression, and (e) comments on the adequacy of the fit.

ACTIVITIES TO IMPROVE UNDERSTANDING: STUDENT PROJECTS

Formulate a question concerning a response variable of interest and a possible predictor variable. Report your activities.

Statement of Purpose

Plan for Data Collection

Identify the variables

Response or dependent variable _____

Predictor variable _____

Population _____

Randomization   Specify your randomization of the order of runs if that is under your control.

Calculations

Summary of Findings
Interpret your tests and confidence intervals. Relate findings to the purpose of the study.

[2]In Example 1, we rounded 13,500 to 13 thousand.

Summation Notation

A.1 SUMMATION AND ITS PROPERTIES

The addition of numbers is basic to our study of statistics. To avoid a detailed and repeated writing of this operation, the symbol Σ (the Greek capital letter *sigma*) is used as mathematical shorthand for the operation of addition.

Summation Notation Σ

The notation $\sum_{i=1}^{n} x_i$ represents the sum of n numbers $x_1, x_2, \ldots, x_n$ and is read as the sum of all x_i with i ranging from 1 to n.

$$\sum_{i=1}^{n} x_i = x_1 + x_2 + \cdots + x_n$$

The term following the sign Σ indicates the quantities that are being summed, and the notations on the bottom and the top of the Σ specify the range of the terms being added. For instance,

$$\sum_{i=1}^{3} x_i = x_1 + x_2 + x_3$$

$$\sum_{i=1}^{4} (x_i - 3) = (x_1 - 3) + (x_2 - 3) + (x_3 - 3) + (x_4 - 3)$$

Example   Suppose that the four measurements in a data set are given as $x_1 = 2$, $x_2 = 5$, $x_3 = 3$, $x_4 = 4$. Compute the numerical values of

(a) $\sum_{i=1}^{4} x_i$ (b) $\sum_{i=1}^{4} 6$

(c) $\sum_{i=1}^{4} 2x_i$ (d) $\sum_{i=1}^{4} (x_i - 3)$

(e) $\sum_{i=1}^{4} x_i^2$ (f) $\sum_{i=1}^{4} (x_i - 3)^2$

SOLUTION AND DISCUSSION

(a) $\sum_{i=1}^{4} x_i = x_1 + x_2 + x_3 + x_4 = 2 + 5 + 3 + 4 = 14$

(b) $\sum_{i=1}^{4} 6 = 6 + 6 + 6 + 6 = 4(6) = 24$

(c) $\sum_{i=1}^{4} 2x_i = 2x_1 + 2x_2 + 2x_3 + 2x_4 = 2\left(\sum_{i=1}^{4} x_i\right) = 2 \times 14 = 28$

(d) $\sum_{i=1}^{4} (x_i - 3) = (x_1 - 3) + (x_2 - 3) + (x_3 - 3) + (x_4 - 3)$

$$= \sum_{i=1}^{4} x_i - 4(3) = 14 - 12 = 2$$

(e) $\sum_{i=1}^{4} x_i^2 = x_1^2 + x_2^2 + x_3^2 + x_4^2 = 2^2 + 5^2 + 3^2 + 4^2 = 54$

(f) $\sum_{i=1}^{4} (x_i - 3)^2 = (x_1 - 3)^2 + (x_2 - 3)^2 + (x_3 - 3)^2 + (x_4 - 3)^2$

$$= (2 - 3)^2 + (5 - 3)^2 + (3 - 3)^2 + (4 - 3)^2$$
$$= 1 + 4 + 0 + 1 = 6$$

Alternatively, noting that $(x_i - 3)^2 = x_i^2 - 6x_i + 9$, we can write

$$\sum_{i=1}^{4} (x_i - 3)^2 = \sum_{i=1}^{4} (x_i^2 - 6x_i + 9)$$

$$= (x_1^2 - 6x_1 + 9) + (x_2^2 - 6x_2 + 9) + (x_3^2 - 6x_3 + 9)$$
$$+ (x_4^2 - 6x_4 + 9)$$

$$= \sum_{i=1}^{4} x_i^2 - 6\left(\sum_{i=1}^{4} x_i\right) + 4(9)$$

$$= 54 - 6(14) + 36 = 6 \qquad \blacksquare$$

A few basic properties of the summation operation are apparent from the numerical demonstration in the example.

Some Basic Properties of Summation

If a and b are fixed numbers,

$$\sum_{i=1}^{n} bx_i = b\sum_{i=1}^{n} x_i$$

$$\sum_{i=1}^{n} (bx_i + a) = b\sum_{i=1}^{n} x_i + na$$

$$\sum_{i=1}^{n} (x_i - a)^2 = \sum_{i=1}^{n} x_i^2 - 2a\sum_{i=1}^{n} x_i + na^2$$

Exercises

1. Demonstrate your familiarity with the summation notation by evaluating the following expressions when $x_1 = 4$, $x_2 = -2$, $x_3 = 1$.

    (a) $\displaystyle\sum_{i=1}^{3} x_i$ (b) $\displaystyle\sum_{i=1}^{3} 7$ (c) $\displaystyle\sum_{i=1}^{3} 5x_i$

    (d) $\displaystyle\sum_{i=1}^{3} (x_i - 2)$ (e) $\displaystyle\sum_{i=1}^{3} (x_i - 3)$ (f) $\displaystyle\sum_{i=1}^{3} (x_i - 2)^2$

    (g) $\displaystyle\sum_{i=1}^{3} x_i^2$ (h) $\displaystyle\sum_{i=1}^{3} (x_i - 3)^2$ (i) $\displaystyle\sum_{i=1}^{3} (x_i^2 - 6x_i + 9)$

2. Five measurements in a data set are $x_1 = 4$, $x_2 = 3$, $x_3 = 6$, $x_4 = 5$, $x_5 = 7$. Determine

    (a) $\displaystyle\sum_{i=1}^{5} x_i$ (b) $\displaystyle\sum_{i=2}^{3} x_i$ (c) $\displaystyle\sum_{i=1}^{5} 2$

    (d) $\displaystyle\sum_{i=1}^{5} (x_i - 6)$ (e) $\displaystyle\sum_{i=1}^{5} (x_i - 6)^2$ (f) $\displaystyle\sum_{i=1}^{4} (x_i - 5)^2$

A.2 SOME BASIC USES OF Σ IN STATISTICS

Let us use the summation notation and its properties to verify some computational facts about the sample mean and variance.

$\Sigma (x_i - \bar{x}) = 0$

The total of the deviations about the sample mean is always zero. Since $\bar{x} = (x_1 + x_2 + \cdots + x_n)/n$, we can write

$$\sum_{i=1}^{n} x_i = x_1 + x_2 + \cdots + x_n = n\bar{x}$$

Consequently, whatever the observations,

$$\sum_{i=1}^{n} (x_i - \bar{x}) = (x_1 - \bar{x}) + (x_2 - \bar{x}) + \cdots + (x_n - \bar{x})$$

$$= x_1 + x_2 + \cdots + x_n - n\bar{x}$$

$$= n\bar{x} - n\bar{x} = 0$$

We could also verify this directly with the second property for summation in A.1, when $b = 1$ and $a = -\bar{x}$.

ALTERNATIVE FORMULA FOR s^2

By the quadratic rule of algebra,

$$(x_i - \bar{x})^2 = x_i^2 - 2\bar{x}x_i + \bar{x}^2$$

Therefore,

$$\Sigma (x_i - \bar{x})^2 = \Sigma x_i^2 - \Sigma 2\bar{x}x_i + \Sigma \bar{x}^2$$
$$= \Sigma x_i^2 - 2\bar{x}\Sigma x_i + n\bar{x}^2$$

Using $(\Sigma x_i)/n$ in place of $\bar{x}$, we get

$$\Sigma (x_i - \bar{x})^2 = \Sigma x_i^2 - \frac{2(\Sigma x_i)^2}{n} + \frac{n(\Sigma x_i)^2}{n^2}$$

$$= \Sigma x_i^2 - \frac{2(\Sigma x_i)^2}{n} + \frac{(\Sigma x_i)^2}{n}$$

$$= \Sigma x_i^2 - \frac{(\Sigma x_i)^2}{n}$$

We could also verify this directly from the third property for summation, in A.1, with $a = \bar{x}$.

This result establishes that

$$s^2 = \frac{\sum_{i=1}^{n} (x_i - \bar{x})^2}{n - 1} = \frac{\Sigma x_i^2 - (\Sigma x_i)^2/n}{n - 1}$$

so the two forms of s^2 are equivalent.

SAMPLE CORRELATION COEFFICIENT

The sample correlation coefficient and slope of the fitted regression line contain a term

$$S_{xy} = \sum_{i=1}^{n} (x_i - \bar{x})(y_i - \bar{y})$$

which is a sum of the products of the deviations. To obtain the alternative form, first note that

$$(x_i - \bar{x})(y_i - \bar{y}) = x_i y_i - x_i \bar{y} - \bar{x}y_i + \bar{x}\bar{y}$$

We treat $x_i y_i$ as a single number, with index i, and conclude that

$$\Sigma (x_i - \bar{x})(y_i - \bar{y}) = \Sigma x_i y_i - \Sigma x_i \bar{y} - \Sigma \bar{x}y_i + \Sigma \bar{x}\bar{y}$$
$$= \Sigma x_i y_i - \bar{y}\Sigma x_i - \bar{x}\Sigma y_i + n\bar{x}\bar{y}$$

Since $\bar{x} = (\Sigma x_i)/n$ and $\bar{y} = (\Sigma y_i)/n$,

$$\Sigma (x_i - \bar{x})(y_i - \bar{y}) = \Sigma x_i y_i - \frac{(\Sigma y_i)}{n} \Sigma x_i - \frac{(\Sigma x_i)}{n} \Sigma y_i + \frac{n(\Sigma x_i)}{n} \frac{(\Sigma y_i)}{n}$$

$$= \Sigma x_i y_i - \frac{(\Sigma x_i)(\Sigma y_i)}{n}$$

Consequently, either $\Sigma (x_i - \bar{x})(y_i - \bar{y})$ or $\Sigma x_i y_i - (\Sigma x_i)(\Sigma y_i)/n$ can be used for the calculation of S_{xy} with similar choices for S_{xx} and S_{yy} in the calculation of r.

Tables

TABLE 1   Random Digits

Row										
1	0695	7741	8254	4297	0000	5277	6563	9265	1023	5925
2	0437	5434	8503	3928	6979	9393	8936	9088	5744	4790
3	6242	2998	0205	5469	3365	7950	7256	3716	8385	0253
4	7090	4074	1257	7175	3310	0712	4748	4226	0604	3804
5	0683	6999	4828	7888	0087	9288	7855	2678	3315	6718
6	7013	4300	3768	2572	6473	2411	6285	0069	5422	6175
7	8808	2786	5369	9571	3412	2465	6419	3990	0294	0896
8	9876	3602	5812	0124	1997	6445	3176	2682	1259	1728
9	1873	1065	8976	1295	9434	3178	0602	0732	6616	7972
10	2581	3075	4622	2974	7069	5605	0420	2949	4387	7679
11	3785	6401	0540	5077	7132	4135	4646	3834	6753	1593
12	8626	4017	1544	4202	8986	1432	2810	2418	8052	2710
13	6253	0726	9483	6753	4732	2284	0421	3010	7885	8436
14	0113	4546	2212	9829	2351	1370	2707	3329	6574	7002
15	4646	6474	9983	8738	1603	8671	0489	9588	3309	5860
16	7873	7343	4432	2866	7973	3765	2888	5154	2250	4339
17	3756	9204	2590	6577	2409	8234	8656	2336	7948	7478
18	2673	7115	5526	0747	3952	6804	3671	7486	3024	9858
19	0187	7045	2711	0349	7734	4396	0988	4887	7682	8990
20	7976	3862	8323	5997	6904	4977	1056	6638	6398	4552
21	5605	1819	8926	9557	2905	0802	7749	0845	1710	4125
22	2225	5556	2545	7480	8804	4161	0084	0787	2561	5113
23	2549	4166	1609	7570	4223	0032	4236	0169	4673	8034
24	6113	1312	5777	7058	2413	3932	5144	5998	7183	5210
25	2028	2537	9819	9215	9327	6640	5986	7935	2750	2981
26	7818	3655	5771	4026	5757	3171	6435	2990	1860	1796
27	9629	3383	1931	2631	5903	9372	1307	4061	5443	8663
28	6657	5967	3277	7141	3628	2588	9320	1972	7683	7544
29	4344	7388	2978	3945	0471	4882	1619	0093	2282	7024
30	3145	8720	2131	1614	1575	5239	0766	0404	4873	7986
31	1848	4094	9168	0903	6451	2823	7566	6644	1157	8889
32	0915	5578	0822	5887	5354	3632	4617	6016	8989	9482
33	1430	4755	7551	9019	8233	9625	6361	2589	2496	7268
34	3473	7966	7249	0555	6307	9524	4888	4939	1641	1573
35	3312	0773	6296	1348	5483	5824	3353	4587	1019	9677
36	6255	4204	5890	9273	0634	9992	3834	2283	1202	4849
37	0562	2546	8559	0480	9379	9282	8257	3054	4272	9311
38	1957	6783	4105	8976	8035	0883	8971	0017	6476	2895
39	7333	1083	0398	8841	0017	4135	4043	8157	4672	2424
40	4601	8908	1781	4287	2681	6223	0814	4477	3798	4437

TABLE 1   *(Continued)*

Row									
41	2628	2233	0708	0900	1698	2818	3931	6930	9273	6749
42	5318	8865	6057	8422	6992	9697	0508	3370	5522	9250
43	6335	0852	8657	8374	0311	6012	9477	0112	8976	3312
44	0301	8333	0327	0467	6186	1770	4099	9588	5382	8958
45	1719	9775	1566	7020	4535	2850	0207	4792	6405	1472
46	8907	8226	4249	6340	9062	3572	7655	6707	3685	1282
47	6129	5927	3731	1125	0081	1241	2772	6458	9157	4543
48	7376	3150	8985	8318	8003	6106	4952	8492	2804	3867
49	9093	3407	4127	9258	3687	5631	5102	1546	2659	0831
50	1133	3086	9380	5431	8647	0910	6948	2257	0946	1245
51	4567	0910	8495	2410	1088	7067	8505	9083	4339	2440
52	6141	8380	2302	4608	7209	5738	9765	3435	9657	6061
53	1514	8309	8743	3096	0682	7902	8204	7508	8330	1681
54	7277	1634	7866	9883	0916	6363	5391	6184	8040	3135
55	4568	4758	0166	1509	2105	0976	0269	0278	7443	2431
56	9200	7599	7754	4534	4532	3102	6831	2387	4147	2455
57	3971	8149	4431	2345	6436	0627	0410	1348	6599	1296
58	2672	9661	2359	8477	3425	8150	6918	8883	1518	4708
59	1524	3268	3798	3360	2255	0371	7610	9114	9466	0901
60	6817	9007	5959	0767	1166	7317	7502	0274	6340	0427
61	6762	3502	9559	4279	9271	9595	3053	4918	7503	5169
62	5264	0075	6655	4563	7112	7264	3240	2150	8180	1361
63	5070	8428	5149	2137	8728	9110	2334	9709	8134	3925
64	1664	3379	5273	9367	6950	6828	1711	7082	4783	0147
65	6962	7141	1904	6648	7328	2901	6396	9949	6274	1672
66	7541	4289	4970	2922	6670	8540	9053	3219	8881	1897
67	5244	4651	2934	6700	8869	0926	4191	1364	0926	2874
68	2939	3890	0745	2577	7931	3913	7877	2837	2500	8774
69	4266	6207	8083	6564	5336	5303	7503	6627	6055	3606
70	7848	5477	5588	3490	0294	3609	1632	5684	1719	6162
71	3009	1879	0440	7916	6643	9723	5933	0574	2480	6893
72	9865	7813	7468	8493	3293	1071	7183	9462	2363	6529
73	1196	1251	2368	1262	5769	9450	7485	4039	4985	6612
74	1067	3716	8897	1970	8799	5718	4792	7292	4589	4554
75	5160	5563	6527	7861	3477	6735	7748	4913	6370	2258
76	4560	0094	8284	7604	1667	9286	2228	9507	1838	4646
77	7697	2151	4860	0739	4370	3992	8121	2502	7670	4470
78	8675	2997	9783	7306	4116	6432	7233	4611	7121	9412
79	3597	3520	5995	0892	3470	4581	1068	8801	1254	8607
80	4281	8802	5880	6212	6818	8162	0052	1755	7107	5197

TABLE 1   *(Continued)*

Row										
81	0101	0907	9057	2263	0059	8553	7855	7758	1020	1264
82	8179	0109	4412	6044	7167	4209	5250	4570	1984	8276
83	8980	9662	9333	6598	2990	8173	1753	1135	1409	2042
84	3050	2450	9252	6724	2697	7933	9540	3700	6561	2790
85	4465	1307	8782	6763	9202	5594	7166	7050	4462	0426
86	1925	5402	1379	3556	5109	4846	9827	2881	5574	9027
87	8753	4602	1838	4624	4632	2512	2652	4804	1624	5116
88	2645	9197	4541	4822	7883	3352	3202	0906	3676	8141
89	4287	5473	4493	7086	4271	9140	3315	7073	4533	0653
90	5280	5426	7240	2154	7952	3804	8097	9328	8069	6894
91	9553	3136	2112	1369	5562	7360	5530	8074	6488	3682
92	2975	7924	0253	3503	9383	9454	3320	3234	9255	3527
93	2596	7274	8967	8138	6868	0385	4467	3792	3844	8700
94	4192	7440	6410	6064	4561	0411	9187	9940	2866	3345
95	3980	8594	9935	8560	0229	8778	2386	7852	4031	0627
96	1822	1177	6846	3997	5822	9188	2479	7951	3051	0110
97	8415	2623	2358	8895	5125	0173	3182	4151	4419	9049
98	2123	5798	5444	3282	8022	3931	4429	6028	5385	6845
99	1754	4076	3507	3705	7459	7544	6127	4820	3760	6476
100	3967	9997	0695	3562	9997	2934	8469	9706	4763	7132
101	7604	6645	6633	6288	5488	8355	9295	9637	5410	0452
102	6357	0216	1685	4308	0391	1517	1952	0108	1258	5498
103	5241	0554	6072	2412	1915	4451	0633	0449	9059	6873
104	9683	0618	2433	0154	0816	9885	3562	7392	4406	2994
105	8073	7718	9374	0965	8861	0018	2152	1736	5187	9347
106	3685	5901	6296	7748	6815	8033	5646	8691	3885	1550
107	9354	1854	1914	2592	9939	2468	0190	5882	3964	6938
108	2604	3040	9664	3962	4600	1314	8163	7869	2059	8203
109	9371	8390	6971	4931	1142	8588	2240	9256	7805	0153
110	5463	5569	1657	2797	9026	7754	8501	1953	1364	7787
111	5832	6510	1728	0531	9770	5790	8294	2702	4318	2494
112	6977	1478	4053	5836	5773	5706	8840	6575	6984	0196
113	6653	3177	7173	1053	8117	5818	2177	7524	3839	2438
114	2043	3329	3149	8591	8213	7941	0324	0275	2808	5787
115	1892	6495	7363	8840	6126	5749	5841	5564	3296	8176
116	4279	6686	2795	2572	6915	5770	0723	5003	6124	0041
117	9018	3226	1024	4455	4743	8634	7086	9462	5603	4961
118	6588	0445	5301	0442	7270	4287	9827	7666	4020	6061
119	3258	2829	5949	6280	9178	3614	8680	6705	1311	2408
120	9213	0161	4449	9084	8199	7330	4284	5061	1971	1008

TABLE 2 Cumulative Binomial Probabilities

$$P[X \le c] = \sum_{x=0}^{c} \binom{n}{x} p^x (1 - p)^{n-x}$$

	c	.05	.10	.20	.30	.40	.50	.60	.70	.80	.90	.95
n = 1	0	.950	.900	.800	.700	.600	.500	.400	.300	.200	.100	.050
	1	1.000	1.000	1.000	1.000	1.000	1.000	1.000	1.000	1.000	1.000	1.000
n = 2	0	.902	.810	.640	.490	.360	.250	.160	.090	.040	.010	.002
	1	.997	.990	.960	.910	.840	.750	.640	.510	.360	.190	.097
	2	1.000	1.000	1.000	1.000	1.000	1.000	1.000	1.000	1.000	1.000	1.000
n = 3	0	.857	.729	.512	.343	.216	.125	.064	.027	.008	.001	.000
	1	.993	.972	.896	.784	.648	.500	.352	.216	.104	.028	.007
	2	1.000	.999	.992	.973	.936	.875	.784	.657	.488	.271	.143
	3	1.000	1.000	1.000	1.000	1.000	1.000	1.000	1.000	1.000	1.000	1.000
n = 4	0	.815	.656	.410	.240	.130	.063	.026	.008	.002	.000	.000
	1	.986	.948	.819	.652	.475	.313	.179	.084	.027	.004	.000
	2	1.000	.996	.973	.916	.821	.688	.525	.348	.181	.052	.014
	3	1.000	1.000	.998	.992	.974	.938	.870	.760	.590	.344	.185
	4	1.000	1.000	1.000	1.000	1.000	1.000	1.000	1.000	1.000	1.000	1.000
n = 5	0	.774	.590	.328	.168	.078	.031	.010	.002	.000	.000	.000
	1	.977	.919	.737	.528	.337	.188	.087	.031	.007	.000	.000
	2	.999	.991	.942	.837	.683	.500	.317	.163	.058	.009	.001
	3	1.000	1.000	.993	.969	.913	.813	.663	.472	.263	.081	.023
	4	1.000	1.000	1.000	.998	.990	.969	.922	.832	.672	.410	.226
	5	1.000	1.000	1.000	1.000	1.000	1.000	1.000	1.000	1.000	1.000	1.000
n = 6	0	.735	.531	.262	.118	.047	.016	.004	.001	.000	.000	.000
	1	.967	.886	.655	.420	.233	.109	.041	.011	.002	.000	.000
	2	.998	.984	.901	.744	.544	.344	.179	.070	.017	.001	.000
	3	1.000	.999	.983	.930	.821	.656	.456	.256	.099	.016	.002
	4	1.000	1.000	.998	.989	.959	.891	.767	.580	.345	.114	.033
	5	1.000	1.000	1.000	.999	.996	.984	.953	.882	.738	.469	.265
	6	1.000	1.000	1.000	1.000	1.000	1.000	1.000	1.000	1.000	1.000	1.000
n = 7	0	.698	.478	.210	.082	.028	.008	.002	.000	.000	.000	.000
	1	.956	.850	.577	.329	.159	.063	.019	.004	.000	.000	.000
	2	.996	.974	.852	.647	.420	.227	.096	.029	.005	.000	.000
	3	1.000	.997	.967	.874	.710	.500	.290	.126	.033	.003	.000
	4	1.000	1.000	.995	.971	.904	.773	.580	.353	.148	.026	.004
	5	1.000	1.000	1.000	.996	.981	.938	.841	.671	.423	.150	.044
	6	1.000	1.000	1.000	1.000	.998	.992	.972	.918	.790	.522	.302
	7	1.000	1.000	1.000	1.000	1.000	1.000	1.000	1.000	1.000	1.000	1.000

TABLE 2 *(Continued)*

	c	.05	.10	.20	.30	.40	.50	.60	.70	.80	.90	.95
n = 8	0	.663	.430	.168	.058	.017	.004	.001	.000	.000	.000	.000
	1	.943	.813	.503	.255	.106	.035	.009	.001	.000	.000	.000
	2	.994	.962	.797	.552	.315	.145	.050	.011	.001	.000	.000
	3	1.000	.995	.944	.806	.594	.363	.174	.058	.010	.000	.000
	4	1.000	1.000	.990	.942	.826	.637	.406	.194	.056	.005	.000
	5	1.000	1.000	.999	.989	.950	.855	.685	.448	.203	.038	.006
	6	1.000	1.000	1.000	.999	.991	.965	.894	.745	.497	.187	.057
	7	1.000	1.000	1.000	1.000	.999	.996	.983	.942	.832	.570	.337
	8	1.000	1.000	1.000	1.000	1.000	1.000	1.000	1.000	1.000	1.000	1.000
n = 9	0	.630	.387	.134	.040	.010	.002	.000	.000	.000	.000	.000
	1	.929	.775	.436	.196	.071	.020	.004	.000	.000	.000	.000
	2	.992	.947	.738	.463	.232	.090	.025	.004	.000	.000	.000
	3	.999	.992	.914	.730	.483	.254	.099	.025	.003	.000	.000
	4	1.000	.999	.980	.901	.733	.500	.267	.099	.020	.001	.000
	5	1.000	1.000	.997	.975	.901	.746	.517	.270	.086	.008	.001
	6	1.000	1.000	1.000	.996	.975	.910	.768	.537	.262	.053	.008
	7	1.000	1.000	1.000	1.000	.996	.980	.929	.804	.564	.225	.071
	8	1.000	1.000	1.000	1.000	1.000	.998	.990	.960	.866	.613	.370
	9	1.000	1.000	1.000	1.000	1.000	1.000	1.000	1.000	1.000	1.000	1.000
n = 10	0	.599	.349	.107	.028	.006	.001	.000	.000	.000	.000	.000
	1	.914	.736	.376	.149	.046	.011	.002	.000	.000	.000	.000
	2	.988	.930	.678	.383	.167	.055	.012	.002	.000	.000	.000
	3	.999	.987	.879	.650	.382	.172	.055	.011	.001	.000	.000
	4	1.000	.998	.967	.850	.633	.377	.166	.047	.006	.000	.000
	5	1.000	1.000	.994	.953	.834	.623	.367	.150	.033	.002	.000
	6	1.000	1.000	.999	.989	.945	.828	.618	.350	.121	.013	.001
	7	1.000	1.000	1.000	.998	.988	.945	.833	.617	.322	.070	.012
	8	1.000	1.000	1.000	1.000	.998	.989	.954	.851	.624	.264	.086
	9	1.000	1.000	1.000	1.000	1.000	.999	.994	.972	.893	.651	.401
	10	1.000	1.000	1.000	1.000	1.000	1.000	1.000	1.000	1.000	1.000	1.000
n = 11	0	.569	.314	.086	.020	.004	.000	.000	.000	.000	.000	.000
	1	.898	.697	.322	.113	.030	.006	.001	.000	.000	.000	.000
	2	.985	.910	.617	.313	.119	.033	.006	.001	.000	.000	.000
	3	.998	.981	.839	.570	.296	.113	.029	.004	.000	.000	.000
	4	1.000	.997	.950	.790	.533	.274	.099	.022	.002	.000	.000
	5	1.000	1.000	.988	.922	.753	.500	.247	.078	.012	.000	.000
	6	1.000	1.000	.998	.978	.901	.726	.467	.210	.050	.003	.000
	7	1.000	1.000	1.000	.996	.971	.887	.704	.430	.161	.019	.002
	8	1.000	1.000	1.000	.999	.994	.967	.881	.687	.383	.090	.015
	9	1.000	1.000	1.000	1.000	.999	.994	.970	.887	.678	.303	.102
	10	1.000	1.000	1.000	1.000	1.000	1.000	.996	.980	.914	.686	.431
	11	1.000	1.000	1.000	1.000	1.000	1.000	1.000	1.000	1.000	1.000	1.000

TABLE 2 *(Continued)*

		.05	.10	.20	.30	.40	p .50	.60	.70	.80	.90	.95
	c											
$n = 12$	0	.540	.282	.069	.014	.002	.000	.000	.000	.000	.000	.000
	1	.882	.659	.275	.085	.020	.003	.000	.000	.000	.000	.000
	2	.980	.889	.558	.253	.083	.019	.003	.000	.000	.000	.000
	3	.998	.974	.795	.493	.225	.073	.015	.002	.000	.000	.000
	4	1.000	.996	.927	.724	.438	.194	.057	.009	.001	.000	.000
	5	1.000	.999	.981	.882	.665	.387	.158	.039	.004	.000	.000
	6	1.000	1.000	.996	.961	.842	.613	.335	.118	.019	.001	.000
	7	1.000	1.000	.999	.991	.943	.806	.562	.276	.073	.004	.000
	8	1.000	1.000	1.000	.998	.985	.927	.775	.507	.205	.026	.002
	9	1.000	1.000	1.000	1.000	.997	.981	.917	.747	.442	.111	.020
	10	1.000	1.000	1.000	1.000	1.000	.997	.980	.915	.725	.341	.118
	11	1.000	1.000	1.000	1.000	1.000	1.000	.998	.986	.931	.718	.460
	12	1.000	1.000	1.000	1.000	1.000	1.000	1.000	1.000	1.000	1.000	1.000
$n = 13$	0	.513	.254	.055	.010	.001	.000	.000	.000	.000	.000	.000
	1	.865	.621	.234	.064	.013	.002	.000	.000	.000	.000	.000
	2	.975	.866	.502	.202	.058	.011	.001	.000	.000	.000	.000
	3	.997	.966	.747	.421	.169	.046	.008	.001	.000	.000	.000
	4	1.000	.994	.901	.654	.353	.133	.032	.004	.000	.000	.000
	5	1.000	.999	.970	.835	.574	.291	.098	.018	.001	.000	.000
	6	1.000	1.000	.993	.938	.771	.500	.229	.062	.007	.000	.000
	7	1.000	1.000	.999	.982	.902	.709	.426	.165	.030	.001	.000
	8	1.000	1.000	1.000	.996	.968	.867	.647	.346	.099	.006	.000
	9	1.000	1.000	1.000	.999	.992	.954	.831	.579	.253	.034	.003
	10	1.000	1.000	1.000	1.000	.999	.989	.942	.798	.498	.134	.025
	11	1.000	1.000	1.000	1.000	1.000	.998	.987	.936	.766	.379	.135
	12	1.000	1.000	1.000	1.000	1.000	1.000	.999	.990	.945	.746	.487
	13	1.000	1.000	1.000	1.000	1.000	1.000	1.000	1.000	1.000	1.000	1.000
$n = 14$	0	.488	.229	.044	.007	.001	.000	.000	.000	.000	.000	.000
	1	.847	.585	.198	.047	.008	.001	.000	.000	.000	.000	.000
	2	.970	.842	.448	.161	.040	.006	.001	.000	.000	.000	.000
	3	.996	.956	.698	.355	.124	.029	.004	.000	.000	.000	.000
	4	1.000	.991	.870	.584	.279	.090	.018	.002	.000	.000	.000
	5	1.000	.999	.956	.781	.486	.212	.058	.008	.000	.000	.000
	6	1.000	1.000	.988	.907	.692	.395	.150	.031	.002	.000	.000
	7	1.000	1.000	.998	.969	.850	.605	.308	.093	.012	.000	.000
	8	1.000	1.000	1.000	.992	.942	.788	.514	.219	.044	.001	.000
	9	1.000	1.000	1.000	.998	.982	.910	.721	.416	.130	.009	.000
	10	1.000	1.000	1.000	1.000	.996	.971	.876	.645	.302	.044	.004
	11	1.000	1.000	1.000	1.000	.999	.994	.960	.839	.552	.158	.030
	12	1.000	1.000	1.000	1.000	1.000	.999	.992	.953	.802	.415	.153
	13	1.000	1.000	1.000	1.000	1.000	1.000	.999	.993	.956	.771	.512
	14	1.000	1.000	1.000	1.000	1.000	1.000	1.000	1.000	1.000	1.000	1.000

TABLE 2   *(Continued)*

		.05	.10	.20	.30	.40	.50	.60	.70	.80	.90	.95
	c											
n = 15	0	.463	.206	.035	.005	.000	.000	.000	.000	.000	.000	.000
	1	.829	.549	.167	.035	.005	.000	.000	.000	.000	.000	.000
	2	.964	.816	.398	.127	.027	.004	.000	.000	.000	.000	.000
	3	.995	.944	.648	.297	.091	.018	.002	.000	.000	.000	.000
	4	.999	.987	.836	.515	.217	.059	.009	.001	.000	.000	.000
	5	1.000	.998	.939	.722	.403	.151	.034	.004	.000	.000	.000
	6	1.000	1.000	.982	.869	.610	.304	.095	.015	.001	.000	.000
	7	1.000	1.000	.996	.950	.787	.500	.213	.050	.004	.000	.000
	8	1.000	1.000	.999	.985	.905	.696	.390	.131	.018	.000	.000
	9	1.000	1.000	1.000	.996	.966	.849	.597	.278	.061	.002	.000
	10	1.000	1.000	1.000	.999	.991	.941	.783	.485	.164	.013	.001
	11	1.000	1.000	1.000	1.000	.998	.982	.909	.703	.352	.056	.005
	12	1.000	1.000	1.000	1.000	1.000	.996	.973	.873	.602	.184	.036
	13	1.000	1.000	1.000	1.000	1.000	1.000	.995	.965	.833	.451	.171
	14	1.000	1.000	1.000	1.000	1.000	1.000	1.000	.995	.965	.794	.537
	15	1.000	1.000	1.000	1.000	1.000	1.000	1.000	1.000	1.000	1.000	1.000
n = 16	0	.440	.185	.028	.003	.000	.000	.000	.000	.000	.000	.000
	1	.811	.515	.141	.026	.003	.000	.000	.000	.000	.000	.000
	2	.957	.789	.352	.099	.018	.002	.000	.000	.000	.000	.000
	3	.993	.932	.598	.246	.065	.011	.001	.000	.000	.000	.000
	4	.999	.983	.798	.450	.167	.038	.005	.000	.000	.000	.000
	5	1.000	.997	.918	.660	.329	.105	.019	.002	.000	.000	.000
	6	1.000	.999	.973	.825	.527	.227	.058	.007	.000	.000	.000
	7	1.000	1.000	.993	.926	.716	.402	.142	.026	.001	.000	.000
	8	1.000	1.000	.999	.974	.858	.598	.284	.074	.007	.000	.000
	9	1.000	1.000	1.000	.993	.942	.773	.473	.175	.027	.001	.000
	10	1.000	1.000	1.000	.998	.981	.895	.671	.340	.082	.003	.000
	11	1.000	1.000	1.000	1.000	.995	.962	.833	.550	.202	.017	.001
	12	1.000	1.000	1.000	1.000	.999	.989	.935	.754	.402	.068	.007
	13	1.000	1.000	1.000	1.000	1.000	.998	.982	.901	.648	.211	.043
	14	1.000	1.000	1.000	1.000	1.000	1.000	.997	.974	.859	.485	.189
	15	1.000	1.000	1.000	1.000	1.000	1.000	1.000	.997	.972	.815	.560
	16	1.000	1.000	1.000	1.000	1.000	1.000	1.000	1.000	1.000	1.000	1.000

TABLE 2   *(Continued)*

						p					
		.05	.10	.20	.30	.40	.50	.60	.70	.80	.90	.95
	c											
$n = 17$	0	.418	.167	.023	.002	.000	.000	.000	.000	.000	.000	.000
	1	.792	.482	.118	.019	.002	.000	.000	.000	.000	.000	.000
	2	.950	.762	.310	.077	.012	.001	.000	.000	.000	.000	.000
	3	.991	.917	.549	.202	.046	.006	.000	.000	.000	.000	.000
	4	.999	.978	.758	.389	.126	.025	.003	.000	.000	.000	.000
	5	1.000	.995	.894	.597	.264	.072	.011	.001	.000	.000	.000
	6	1.000	.999	.962	.775	.448	.166	.035	.003	.000	.000	.000
	7	1.000	1.000	.989	.895	.641	.315	.092	.013	.000	.000	.000
	8	1.000	1.000	.997	.960	.801	.500	.199	.040	.003	.000	.000
	9	1.000	1.000	1.000	.987	.908	.685	.359	.105	.011	.000	.000
	10	1.000	1.000	1.000	.997	.965	.834	.552	.225	.038	.001	.000
	11	1.000	1.000	1.000	.999	.989	.928	.736	.403	.106	.005	.000
	12	1.000	1.000	1.000	1.000	.997	.975	.874	.611	.242	.022	.001
	13	1.000	1.000	1.000	1.000	1.000	.994	.954	.798	.451	.083	.009
	14	1.000	1.000	1.000	1.000	1.000	.999	.988	.923	.690	.238	.050
	15	1.000	1.000	1.000	1.000	1.000	1.000	.998	.981	.882	.518	.208
	16	1.000	1.000	1.000	1.000	1.000	1.000	1.000	.998	.977	.833	.582
	17	1.000	1.000	1.000	1.000	1.000	1.000	1.000	1.000	1.000	1.000	1.000
$n = 18$	0	.397	.150	.018	.002	.000	.000	.000	.000	.000	.000	.000
	1	.774	.450	.099	.014	.001	.000	.000	.000	.000	.000	.000
	2	.942	.734	.271	.060	.008	.001	.000	.000	.000	.000	.000
	3	.989	.902	.501	.165	.033	.004	.000	.000	.000	.000	.000
	4	.998	.972	.716	.333	.094	.015	.001	.000	.000	.000	.000
	5	1.000	.994	.867	.534	.209	.048	.006	.000	.000	.000	.000
	6	1.000	.999	.949	.722	.374	.119	.020	.001	.000	.000	.000
	7	1.000	1.000	.984	.859	.563	.240	.058	.006	.000	.000	.000
	8	1.000	1.000	.996	.940	.737	.407	.135	.021	.001	.000	.000
	9	1.000	1.000	.999	.979	.865	.593	.263	.060	.004	.000	.000
	10	1.000	1.000	1.000	.994	.942	.760	.437	.141	.016	.000	.000
	11	1.000	1.000	1.000	.999	.980	.881	.626	.278	.051	.001	.000
	12	1.000	1.000	1.000	1.000	.994	.952	.791	.466	.133	.006	.000
	13	1.000	1.000	1.000	1.000	.999	.985	.906	.667	.284	.028	.002
	14	1.000	1.000	1.000	1.000	1.000	.996	.967	.835	.499	.098	.011
	15	1.000	1.000	1.000	1.000	1.000	.999	.992	.940	.729	.266	.058
	16	1.000	1.000	1.000	1.000	1.000	1.000	.999	.986	.901	.550	.226
	17	1.000	1.000	1.000	1.000	1.000	1.000	1.000	.998	.982	.850	.603
	18	1.000	1.000	1.000	1.000	1.000	1.000	1.000	1.000	1.000	1.000	1.000

TABLE 2   *(Continued)*

						p						
		.05	.10	.20	.30	.40	.50	.60	.70	.80	.90	.95
	c											
n = 19	0	.377	.135	.014	.001	.000	.000	.000	.000	.000	.000	.000
	1	.755	.420	.083	.010	.001	.000	.000	.000	.000	.000	.000
	2	.933	.705	.237	.046	.005	.000	.000	.000	.000	.000	.000
	3	.987	.885	.455	.133	.023	.002	.000	.000	.000	.000	.000
	4	.998	.965	.673	.282	.070	.010	.001	.000	.000	.000	.000
	5	1.000	.991	.837	.474	.163	.032	.003	.000	.000	.000	.000
	6	1.000	.998	.932	.666	.308	.084	.012	.001	.000	.000	.000
	7	1.000	1.000	.977	.818	.488	.180	.035	.003	.000	.000	.000
	8	1.000	1.000	.993	.916	.667	.324	.088	.011	.000	.000	.000
	9	1.000	1.000	.998	.967	.814	.500	.186	.033	.002	.000	.000
	10	1.000	1.000	1.000	.989	.912	.676	.333	.084	.007	.000	.000
	11	1.000	1.000	1.000	.997	.965	.820	.512	.182	.023	.000	.000
	12	1.000	1.000	1.000	.999	.988	.916	.692	.334	.068	.002	.000
	13	1.000	1.000	1.000	1.000	.997	.968	.837	.526	.163	.009	.000
	14	1.000	1.000	1.000	1.000	.999	.990	.930	.718	.327	.035	.002
	15	1.000	1.000	1.000	1.000	1.000	.998	.977	.867	.545	.115	.013
	16	1.000	1.000	1.000	1.000	1.000	1.000	.995	.954	.763	.295	.067
	17	1.000	1.000	1.000	1.000	1.000	1.000	.999	.990	.917	.580	.245
	18	1.000	1.000	1.000	1.000	1.000	1.000	1.000	.999	.986	.865	.623
	19	1.000	1.000	1.000	1.000	1.000	1.000	1.000	1.000	1.000	1.000	1.000
n = 20	0	.358	.122	.012	.001	.000	.000	.000	.000	.000	.000	.000
	1	.736	.392	.069	.008	.001	.000	.000	.000	.000	.000	.000
	2	.925	.677	.206	.035	.004	.000	.000	.000	.000	.000	.000
	3	.984	.867	.411	.107	.016	.001	.000	.000	.000	.000	.000
	4	.997	.957	.630	.238	.051	.006	.000	.000	.000	.000	.000
	5	1.000	.989	.804	.416	.126	.021	.002	.000	.000	.000	.000
	6	1.000	.998	.913	.608	.250	.058	.006	.000	.000	.000	.000
	7	1.000	1.000	.968	.772	.416	.132	.021	.001	.000	.000	.000
	8	1.000	1.000	.990	.887	.596	.252	.057	.005	.000	.000	.000
	9	1.000	1.000	.997	.952	.755	.412	.128	.017	.001	.000	.000
	10	1.000	1.000	.999	.983	.872	.588	.245	.048	.003	.000	.000
	11	1.000	1.000	1.000	.995	.943	.748	.404	.113	.010	.000	.000
	12	1.000	1.000	1.000	.999	.979	.868	.584	.228	.032	.000	.000
	13	1.000	1.000	1.000	1.000	.994	.942	.750	.392	.087	.002	.000
	14	1.000	1.000	1.000	1.000	.998	.979	.874	.584	.196	.011	.000
	15	1.000	1.000	1.000	1.000	1.000	.994	.949	.762	.370	.043	.003
	16	1.000	1.000	1.000	1.000	1.000	.999	.984	.893	.589	.133	.016
	17	1.000	1.000	1.000	1.000	1.000	1.000	.996	.965	.794	.323	.075
	18	1.000	1.000	1.000	1.000	1.000	1.000	.999	.992	.931	.608	.264
	19	1.000	1.000	1.000	1.000	1.000	1.000	1.000	.999	.988	.878	.642
	20	1.000	1.000	1.000	1.000	1.000	1.000	1.000	1.000	1.000	1.000	1.000

TABLE 2 *(Continued)*

	c	.05	.10	.20	.30	.40	.50	.60	.70	.80	.90	.95
							p					
$n = 25$	0	.277	.072	.004	.000	.000	.000	.000	.000	.000	.000	.000
	1	.642	.271	.027	.002	.000	.000	.000	.000	.000	.000	.000
	2	.873	.537	.098	.009	.000	.000	.000	.000	.000	.000	.000
	3	.966	.764	.234	.033	.002	.000	.000	.000	.000	.000	.000
	4	.993	.902	.421	.090	.009	.000	.000	.000	.000	.000	.000
	5	.999	.967	.617	.193	.029	.002	.000	.000	.000	.000	.000
	6	1.000	.991	.780	.341	.074	.007	.000	.000	.000	.000	.000
	7	1.000	.998	.891	.512	.154	.022	.001	.000	.000	.000	.000
	8	1.000	1.000	.953	.677	.274	.054	.004	.000	.000	.000	.000
	9	1.000	1.000	.983	.811	.425	.115	.013	.000	.000	.000	.000
	10	1.000	1.000	.994	.902	.586	.212	.034	.002	.000	.000	.000
	11	1.000	1.000	.998	.956	.732	.345	.078	.006	.000	.000	.000
	12	1.000	1.000	1.000	.983	.846	.500	.154	.017	.000	.000	.000
	13	1.000	1.000	1.000	.994	.922	.655	.268	.044	.002	.000	.000
	14	1.000	1.000	1.000	.998	.966	.788	.414	.098	.006	.000	.000
	15	1.000	1.000	1.000	1.000	.987	.885	.575	.189	.017	.000	.000
	16	1.000	1.000	1.000	1.000	.996	.946	.726	.323	.047	.000	.000
	17	1.000	1.000	1.000	1.000	.999	.978	.846	.488	.109	.002	.000
	18	1.000	1.000	1.000	1.000	1.000	.993	.926	.659	.220	.009	.000
	19	1.000	1.000	1.000	1.000	1.000	.998	.971	.807	.383	.033	.001
	20	1.000	1.000	1.000	1.000	1.000	1.000	.991	.910	.579	.098	.007
	21	1.000	1.000	1.000	1.000	1.000	1.000	.998	.967	.766	.236	.034
	22	1.000	1.000	1.000	1.000	1.000	1.000	1.000	.991	.902	.463	.127
	23	1.000	1.000	1.000	1.000	1.000	1.000	1.000	.998	.973	.729	.358
	24	1.000	1.000	1.000	1.000	1.000	1.000	1.000	1.000	.996	.928	.723
	25	1.000	1.000	1.000	1.000	1.000	1.000	1.000	1.000	1.000	1.000	1.000

TABLE 3   Standard Normal Probabilities

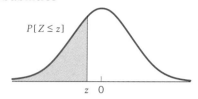

z	.00	.01	.02	.03	.04	.05	.06	.07	.08	.09
−3.5	.0002	.0002	.0002	.0002	.0002	.0002	.0002	.0002	.0002	.0002
−3.4	.0003	.0003	.0003	.0003	.0003	.0003	.0003	.0003	.0003	.0002
−3.3	.0005	.0005	.0005	.0004	.0004	.0004	.0004	.0004	.0004	.0003
−3.2	.0007	.0007	.0006	.0006	.0006	.0006	.0006	.0005	.0005	.0005
−3.1	.0010	.0009	.0009	.0009	.0008	.0008	.0008	.0008	.0007	.0007
−3.0	.0013	.0013	.0013	.0012	.0012	.0011	.0011	.0011	.0010	.0010
−2.9	.0019	.0018	.0018	.0017	.0016	.0016	.0015	.0015	.0014	.0014
−2.8	.0026	.0025	.0024	.0023	.0023	.0022	.0021	.0021	.0020	.0019
−2.7	.0035	.0034	.0033	.0032	.0031	.0030	.0029	.0028	.0027	.0026
−2.6	.0047	.0045	.0044	.0043	.0041	.0040	.0039	.0038	.0037	.0036
−2.5	.0062	.0060	.0059	.0057	.0055	.0054	.0052	.0051	.0049	.0048
−2.4	.0082	.0080	.0078	.0075	.0073	.0071	.0069	.0068	.0066	.0064
−2.3	.0107	.0104	.0102	.0099	.0096	.0094	.0091	.0089	.0087	.0084
−2.2	.0139	.0136	.0132	.0129	.0125	.0122	.0119	.0116	.0113	.0110
−2.1	.0179	.0174	.0170	.0166	.0162	.0158	.0154	.0150	.0146	.0143
−2.0	.0228	.0222	0217	.0212	.0207	.0202	.0197	.0192	.0188	.0183
−1.9	.0287	.0281	.0274	.0268	.0262	.0256	.0250	.0244	.0239	.0233
−1.8	.0359	.0351	.0344	.0336	.0329	.0322	.0314	.0307	.0301	.0294
−1.7	.0446	.0436	.0427	.0418	.0409	.0401	.0392	.0384	.0375	.0367
−1.6	.0548	.0537	.0526	.0516	.0505	.0495	.0485	.0475	.0465	.0455
−1.5	.0668	.0655	.0643	.0630	.0618	.0606	.0594	.0582	.0571	.0559
−1.4	.0808	.0793	.0778	.0764	.0749	.0735	.0721	.0708	.0694	.0681
−1.3	.0968	.0951	.0934	.0918	.0901	.0885	.0869	.0853	.0838	.0823
−1.2	.1151	.1131	.1112	.1093	.1075	.1056	.1038	.1020	.1003	.0985
−1.1	.1357	.1335	.1314	.1292	.1271	.1251	.1230	.1210	.1190	.1170
−1.0	.1587	.1562	.1539	.1515	.1492	.1469	.1446	.1423	.1401	.1379
−.9	.1841	.1814	.1788	.1762	.1736	.1711	.1685	.1660	.1635	.1611
−.8	.2119	.2090	.2061	.2033	.2005	.1977	.1949	.1922	.1894	.1867
−.7	.2420	.2389	.2358	.2327	.2297	.2266	.2236	.2206	.2177	.2148
−.6	.2743	.2709	.2676	.2643	.2611	.2578	.2546	.2514	.2483	.2451
−.5	.3085	.3050	.3015	.2981	.2946	.2912	.2877	.2843	.2810	.2776
−.4	.3446	.3409	.3372	.3336	.3300	.3264	.3228	.3192	.3156	.3121
−.3	.3821	.3783	.3745	.3707	.3669	.3632	.3594	.3557	.3520	.3483
−.2	.4207	.4168	.4129	.4090	.4052	.4013	.3974	.3936	.3897	.3859
−.1	.4602	.4562	.4522	.4483	.4443	.4404	.4364	.4325	.4286	.4247
−.0	.5000	.4960	.4920	.4880	.4840	.4801	.4761	.4721	.4681	.4641

TABLE 3   *(Continued)* Standard Normal Probabilities

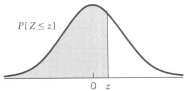

$P[Z \le z]$

z	.00	.01	.02	.03	.04	.05	.06	.07	.08	.09
.0	.5000	.5040	.5080	.5120	.5160	.5199	.5239	.5279	.5319	.5359
.1	.5398	.5438	.5478	.5517	.5557	.5596	.5636	.5675	.5714	.5753
.2	.5793	.5832	.5871	.5910	.5948	.5987	.6026	.6064	.6103	.6141
.3	.6179	.6217	.6255	.6293	.6331	.6368	.6406	.6443	.6480	.6517
.4	.6554	.6591	.6628	.6664	.6700	.6736	.6772	.6808	.6844	.6879
.5	.6915	.6950	.6985	.7019	.7054	.7088	.7123	.7157	.7190	.7224
.6	.7257	.7291	.7324	.7357	.7389	.7422	.7454	.7486	.7517	.7549
.7	.7580	.7611	.7642	.7673	.7703	.7734	.7764	.7794	.7823	.7852
.8	.7881	.7910	.7939	.7967	.7995	.8023	.8051	.8078	.8106	.8133
.9	.8159	.8186	.8212	.8238	.8264	.8289	.8315	.8340	.8365	.8389
1.0	.8413	.8438	.8461	.8485	.8508	.8531	.8554	.8577	.8599	.8621
1.1	.8643	.8665	.8686	.8708	.8729	.8749	.8770	.8790	.8810	.8830
1.2	.8849	.8869	.8888	.8907	.8925	.8944	.8962	.8980	.8997	.9015
1.3	.9032	.9049	.9066	.9082	.9099	.9115	.9131	.9147	.9162	.9177
1.4	.9192	.9207	.9222	.9236	.9251	.9265	.9279	.9292	.9306	.9319
1.5	.9332	.9345	.9357	.9370	.9382	.9394	.9406	.9418	.9429	.9441
1.6	.9452	.9463	.9474	.9484	.9495	.9505	.9515	.9525	.9535	.9545
1.7	.9554	.9564	.9573	.9582	.9591	.9599	.9608	.9616	.9625	.9633
1.8	.9641	.9649	.9656	.9664	.9671	.9678	.9686	.9693	.9699	.9706
1.9	.9713	.9719	.9726	.9732	.9738	.9744	.9750	.9756	.9761	.9767
2.0	.9772	.9778	.9783	.9788	.9793	.9798	.9803	.9808	.9812	.9817
2.1	.9821	.9826	.9830	.9834	.9838	.9842	.9846	.9850	.9854	.9857
2.2	.9861	.9864	.9868	.9871	.9875	.9878	.9881	.9884	.9887	.9890
2.3	.9893	.9896	.9898	.9901	.9904	.9906	.9909	.9911	.9913	.9916
2.4	.9918	.9920	.9922	.9925	.9927	.9929	.9931	.9932	.9934	.9936
2.5	.9938	.9940	.9941	.9943	.9945	.9946	.9948	.9949	.9951	.9952
2.6	.9953	.9955	.9956	.9957	.9959	.9960	.9961	.9962	.9963	.9964
2.7	.9965	.9966	.9967	.9968	.9969	.9970	.9971	.9972	.9973	.9974
2.8	.9974	.9975	.9976	.9977	.9977	.9978	.9979	.9979	.9980	.9981
2.9	.9981	.9982	.9982	.9983	.9984	.9984	.9985	.9985	.9986	.9986
3.0	.9987	.9987	.9987	.9988	.9988	.9989	.9989	.9989	.9990	.9990
3.1	.9990	.9991	.9991	.9991	.9992	.9992	.9992	.9992	.9993	.9993
3.2	.9993	.9993	.9994	.9994	.9994	.9994	.9994	.9995	.9995	.9995
3.3	.9995	.9995	.9995	.9996	.9996	.9996	.9996	.9996	.9996	.9997
3.4	.9997	.9997	.9997	.9997	.9997	.9997	.9997	.9997	.9997	.9998
3.5	.9998	.9998	.9998	.9998	.9998	.9998	.9998	.9998	.9998	.9998

TABLE 4 Percentage Points of t Distributions

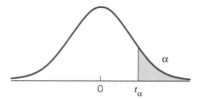

d.f. \\ α	.25	.10	.05	.025	.01	.00833	.00625	.005
1	1.000	3.078	6.314	12.706	31.821	38.190	50.923	63.657
2	.816	1.886	2.920	4.303	6.965	7.649	8.860	9.925
3	.765	1.638	2.353	3.182	4.541	4.857	5.392	5.841
4	.741	1.533	2.132	2.776	3.747	3.961	4.315	4.604
5	.727	1.476	2.015	2.571	3.365	3.534	3.810	4.032
6	.718	1.440	1.943	2.447	3.143	3.287	3.521	3.707
7	.711	1.415	1.895	2.365	2.998	3.128	3.335	3.499
8	.706	1.397	1.860	2.306	2.896	3.016	3.206	3.355
9	.703	1.383	1.833	2.262	2.821	2.933	3.111	3.250
10	.700	1.372	1.812	2.228	2.764	2.870	3.038	3.169
11	.697	1.363	1.796	2.201	2.718	2.820	2.981	3.106
12	.695	1.356	1.782	2.179	2.681	2.779	2.934	3.055
13	.694	1.350	1.771	2.160	2.650	2.746	2.896	3.012
14	.692	1.345	1.761	2.145	2.624	2.718	2.864	2.977
15	.691	1.341	1.753	2.131	2.602	2.694	2.837	2.947
16	.690	1.337	1.746	2.120	2.583	2.673	2.813	2.921
17	.689	1.333	1.740	2.110	2.567	2.655	2.793	2.898
18	.688	1.330	1.734	2.101	2.552	2.639	2.775	2.878
19	.688	1.328	1.729	2.093	2.539	2.625	2.759	2.861
20	.687	1.325	1.725	2.086	2.528	2.613	2.744	2.845
21	.686	1.323	1.721	2.080	2.518	2.601	2.732	2.831
22	.686	1.321	1.717	2.074	2.508	2.591	2.720	2.819
23	.685	1.319	1.714	2.069	2.500	2.582	2.710	2.807
24	.685	1.318	1.711	2.064	2.492	2.574	2.700	2.797
25	.684	1.316	1.708	2.060	2.485	2.566	2.692	2.787
26	.684	1.315	1.706	2.056	2.479	2.559	2.684	2.779
27	.684	1.314	1.703	2.052	2.473	2.552	2.676	2.771
28	.683	1.313	1.701	2.048	2.467	2.546	2.669	2.763
29	.683	1.311	1.699	2.045	2.462	2.541	2.663	2.756
30	.683	1.310	1.697	2.042	2.457	2.536	2.657	2.750
40	.681	1.303	1.684	2.021	2.423	2.499	2.616	2.704
60	.679	1.296	1.671	2.000	2.390	2.463	2.575	2.660
120	.677	1.289	1.658	1.980	2.358	2.428	2.536	2.617
∞	.674	1.282	1.645	1.960	2.326	2.394	2.498	2.576

TABLE 5   Percentage Points of χ^2 Distributions

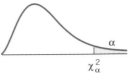

d.f. α	.99	.975	.95	.90	.50	.10	.05	.025	.01
1	.0002	.001	.004	.02	.45	2.71	3.84	5.02	6.63
2	.02	.05	.10	.21	1.39	4.61	5.99	7.38	9.21
3	.11	.22	.35	.58	2.37	6.25	7.81	9.35	11.34
4	.30	.48	.71	1.06	3.36	7.78	9.49	11.14	13.28
5	.55	.83	1.15	1.61	4.35	9.24	11.07	12.83	15.09
6	.87	1.24	1.64	2.20	5.35	10.64	12.59	14.45	16.81
7	1.24	1.69	2.17	2.83	6.35	12.02	14.07	16.01	18.48
8	1.65	2.18	2.73	3.49	7.34	13.36	15.51	17.53	20.09
9	2.09	2.70	3.33	4.17	8.34	14.68	16.92	19.02	21.67
10	2.56	3.24	3.94	4.87	9.34	15.99	18.31	20.48	23.21
11	3.05	3.81	4.57	5.58	10.34	17.28	19.68	21.92	24.72
12	3.57	4.40	5.23	6.30	11.34	18.55	21.03	23.34	26.22
13	4.11	5.01	5.89	7.04	12.34	19.81	22.36	24.74	27.69
14	4.66	5.62	6.57	7.79	13.34	21.06	23.68	26.12	29.14
15	5.23	6.26	7.26	8.55	14.34	22.31	25.00	27.49	30.58
16	5.81	6.90	7.96	9.31	15.34	23.54	26.30	28.85	32.00
17	6.41	7.56	8.67	10.09	16.34	24.77	27.59	30.19	33.41
18	7.01	8.23	9.39	10.86	17.34	25.99	28.87	31.53	34.81
19	7.63	8.90	10.12	11.65	18.34	27.20	30.14	32.85	36.19
20	8.26	9.59	10.85	12.44	19.34	28.41	31.41	34.17	37.57
21	8.90	10.28	11.59	13.24	20.34	29.62	32.67	35.48	38.93
22	9.54	10.98	12.34	14.04	21.34	30.81	33.92	36.78	40.29
23	10.20	11.69	13.09	14.85	22.34	32.01	35.17	38.08	41.64
24	10.86	12.40	13.85	15.66	23.34	33.20	36.42	39.36	42.98
25	11.52	13.11	14.61	16.47	24.34	34.38	37.65	40.65	44.31
26	12.20	13.84	15.38	17.29	25.34	35.56	38.89	41.92	45.64
27	12.88	14.57	16.15	18.11	26.34	36.74	40.11	43.19	46.96
28	13.56	15.30	16.93	18.94	27.34	37.92	41.34	44.46	48.28
29	14.26	16.04	17.71	19.77	28.34	39.09	42.56	45.72	49.59
30	14.95	16.78	18.49	20.60	29.34	40.26	43.77	46.98	50.89
40	22.16	24.42	26.51	29.05	39.34	51.81	55.76	59.34	63.69
50	29.71	32.35	34.76	37.69	49.33	63.17	67.50	71.42	76.15
60	37.48	40.47	43.19	46.46	59.33	74.40	79.08	83.30	88.38
70	45.44	48.75	51.74	55.33	69.33	85.53	90.53	95.02	100.43
80	53.54	57.15	60.39	64.28	79.33	96.58	101.88	106.63	112.33
90	61.75	65.64	69.13	73.29	89.33	107.57	113.15	118.14	124.12
100	70.06	74.22	77.93	82.36	99.33	118.50	124.34	129.56	135.81

TABLE 6 Percentage Points of $F(\nu_1, \nu_2)$ Distributions

$\boxed{\alpha = .10}$

$F_\alpha(\nu_1, \nu_2)$

ν_2 \ ν_1	1	2	3	4	5	6	7	8	9	10	12	15	20	25	30	40	60
1	39.86	49.50	53.59	55.83	57.24	58.20	58.91	59.44	59.86	60.19	60.71	61.22	61.74	62.05	62.26	62.53	62.79
2	8.53	9.00	9.16	9.24	9.29	9.33	9.35	9.37	9.38	9.39	9.41	9.42	9.44	9.45	9.46	9.47	9.47
3	5.54	5.46	5.39	5.34	5.31	5.28	5.27	5.25	5.24	5.23	5.22	5.20	5.18	5.17	5.17	5.16	5.15
4	4.54	4.32	4.19	4.11	4.05	4.01	3.98	3.95	3.94	3.92	3.90	3.87	3.84	3.83	3.82	3.80	3.79
5	4.06	3.78	3.62	3.52	3.45	3.40	3.37	3.34	3.32	3.30	3.27	3.24	3.21	3.19	3.17	3.16	3.14
6	3.78	3.46	3.29	3.18	3.11	3.05	3.01	2.98	2.96	2.94	2.90	2.87	2.84	2.81	2.80	2.78	2.76
7	3.59	3.26	3.07	2.96	2.88	2.83	2.78	2.75	2.72	2.70	2.67	2.63	2.59	2.57	2.56	2.54	2.51
8	3.46	3.11	2.92	2.81	2.73	2.67	2.62	2.59	2.56	2.54	2.50	2.46	2.42	2.40	2.38	2.36	2.34
9	3.36	3.01	2.81	2.69	2.61	2.55	2.51	2.47	2.44	2.42	2.38	2.34	2.30	2.27	2.25	2.23	2.21
10	3.29	2.92	2.73	2.61	2.52	2.46	2.41	2.38	2.35	2.32	2.28	2.24	2.20	2.17	2.16	2.13	2.11
11	3.23	2.86	2.66	2.54	2.45	2.39	2.34	2.30	2.27	2.25	2.21	2.17	2.12	2.10	2.08	2.05	2.03
12	3.18	2.81	2.61	2.48	2.39	2.33	2.28	2.24	2.21	2.19	2.15	2.10	2.06	2.03	2.01	1.99	1.96
13	3.14	2.76	2.56	2.43	2.35	2.28	2.23	2.20	2.16	2.14	2.10	2.05	2.01	1.98	1.96	1.93	1.90
14	3.10	2.73	2.52	2.39	2.31	2.24	2.19	2.15	2.12	2.10	2.05	2.01	1.96	1.93	1.91	1.89	1.86
15	3.07	2.70	2.49	2.36	2.27	2.21	2.16	2.12	2.09	2.06	2.02	1.97	1.92	1.89	1.87	1.85	1.82
16	3.05	2.67	2.46	2.33	2.24	2.18	2.13	2.09	2.06	2.03	1.99	1.94	1.89	1.86	1.84	1.81	1.78
17	3.03	2.64	2.44	2.31	2.22	2.15	2.10	2.06	2.03	2.00	1.96	1.91	1.86	1.83	1.81	1.78	1.75
18	3.01	2.62	2.42	2.29	2.20	2.13	2.08	2.04	2.00	1.98	1.93	1.89	1.84	1.80	1.78	1.75	1.72
19	2.99	2.61	2.40	2.27	2.18	2.11	2.06	2.02	1.98	1.96	1.91	1.86	1.81	1.78	1.76	1.73	1.70
20	2.97	2.59	2.38	2.25	2.16	2.09	2.04	2.00	1.96	1.94	1.89	1.84	1.79	1.76	1.74	1.71	1.68
21	2.96	2.57	2.36	2.23	2.14	2.08	2.02	1.98	1.95	1.92	1.87	1.83	1.78	1.74	1.72	1.69	1.66
22	2.95	2.56	2.35	2.22	2.13	2.06	2.01	1.97	1.93	1.90	1.86	1.81	1.76	1.73	1.70	1.67	1.64
23	2.94	2.55	2.34	2.21	2.11	2.05	1.99	1.95	1.92	1.89	1.84	1.80	1.74	1.71	1.69	1.66	1.62
24	2.93	2.54	2.33	2.19	2.10	2.04	1.98	1.94	1.91	1.88	1.83	1.78	1.73	1.70	1.67	1.64	1.61
25	2.92	2.53	2.32	2.18	2.09	2.02	1.97	1.93	1.89	1.87	1.82	1.77	1.72	1.68	1.66	1.63	1.59
26	2.91	2.52	2.31	2.17	2.08	2.01	1.96	1.92	1.88	1.86	1.81	1.76	1.71	1.67	1.65	1.61	1.58
27	2.90	2.51	2.30	2.17	2.07	2.00	1.95	1.91	1.87	1.85	1.80	1.75	1.70	1.66	1.64	1.60	1.57
28	2.89	2.50	2.29	2.16	2.06	2.00	1.94	1.90	1.87	1.84	1.79	1.74	1.69	1.65	1.63	1.59	1.56
29	2.89	2.50	2.28	2.15	2.06	1.99	1.93	1.89	1.86	1.83	1.78	1.73	1.68	1.64	1.62	1.58	1.55
30	2.88	2.49	2.28	2.14	2.05	1.98	1.93	1.88	1.85	1.82	1.77	1.72	1.67	1.63	1.61	1.57	1.54
40	2.84	2.44	2.23	2.09	2.00	1.93	1.87	1.83	1.79	1.76	1.71	1.66	1.61	1.57	1.54	1.51	1.47
60	2.79	2.39	2.18	2.04	1.95	1.87	1.82	1.77	1.74	1.71	1.66	1.60	1.54	1.50	1.48	1.44	1.40
120	2.75	2.35	2.13	1.99	1.90	1.82	1.77	1.72	1.68	1.65	1.60	1.55	1.48	1.45	1.41	1.37	1.32
∞	2.71	2.30	2.08	1.94	1.85	1.77	1.72	1.67	1.63	1.60	1.55	1.49	1.42	1.38			

TABLE 6 (Continued) Percentage Points of $F(\nu_1, \nu_2)$ Distributions

$$\alpha = .05$$

$F_\alpha(\nu_1, \nu_2)$

ν_2 \ ν_1	1	2	3	4	5	6	7	8	9	10	12	15	20	25	30	40	60
1	161.5	199.5	215.7	224.6	230.2	234.0	236.8	238.9	240.5	241.9	243.9	246.0	248.0	249.3	250.1	251.1	252.2
2	18.51	19.00	19.16	19.25	19.30	19.33	19.35	19.37	19.38	19.40	19.41	19.43	19.45	19.46	19.46	19.47	19.48
3	10.13	9.55	9.28	9.12	9.01	8.94	8.89	8.85	8.81	8.79	8.74	8.70	8.66	8.63	8.62	8.59	8.57
4	7.71	6.94	6.59	6.39	6.26	6.16	6.09	6.04	6.00	5.96	5.91	5.86	5.80	5.77	5.75	5.72	5.69
5	6.61	5.79	5.41	5.19	5.05	4.95	4.88	4.82	4.77	4.74	4.68	4.62	4.56	4.52	4.50	4.46	4.43
6	5.99	5.14	4.76	4.53	4.39	4.28	4.21	4.15	4.10	4.06	4.00	3.94	3.87	3.83	3.81	3.77	3.74
7	5.59	4.74	4.35	4.12	3.97	3.87	3.79	3.73	3.68	3.64	3.57	3.51	3.44	3.40	3.38	3.34	3.30
8	5.32	4.46	4.07	3.84	3.69	3.58	3.50	3.44	3.39	3.35	3.28	3.22	3.15	3.11	3.08	3.04	3.01
9	5.12	4.26	3.86	3.63	3.48	3.37	3.29	3.23	3.18	3.14	3.07	3.01	2.94	2.89	2.86	2.83	2.79
10	4.96	4.10	3.71	3.48	3.33	3.22	3.14	3.07	3.02	2.98	2.91	2.85	2.77	2.73	2.70	2.66	2.62
11	4.84	3.98	3.59	3.36	3.20	3.09	3.01	2.95	2.90	2.85	2.79	2.72	2.65	2.60	2.57	2.53	2.49
12	4.75	3.89	3.49	3.26	3.11	3.00	2.91	2.85	2.80	2.75	2.69	2.62	2.54	2.50	2.47	2.43	2.38
13	4.67	3.81	3.41	3.18	3.03	2.92	2.83	2.77	2.71	2.67	2.60	2.53	2.46	2.41	2.38	2.34	2.30
14	4.60	3.74	3.34	3.11	2.96	2.85	2.76	2.70	2.65	2.60	2.53	2.46	2.39	2.34	2.31	2.27	2.22
15	4.54	3.68	3.29	3.06	2.90	2.79	2.71	2.64	2.59	2.54	2.48	2.40	2.33	2.28	2.25	2.20	2.16
16	4.49	3.63	3.24	3.01	2.85	2.74	2.66	2.59	2.54	2.49	2.42	2.35	2.28	2.23	2.19	2.15	2.11
17	4.45	3.59	3.20	2.96	2.81	2.70	2.61	2.55	2.49	2.45	2.38	2.31	2.23	2.18	2.15	2.10	2.06
18	4.41	3.55	3.16	2.93	2.77	2.66	2.58	2.51	2.46	2.41	2.34	2.27	2.19	2.14	2.11	2.06	2.02
19	4.38	3.52	3.13	2.90	2.74	2.63	2.54	2.48	2.42	2.38	2.31	2.23	2.16	2.11	2.07	2.03	1.98
20	4.35	3.49	3.10	2.87	2.71	2.60	2.51	2.45	2.39	2.35	2.28	2.20	2.12	2.07	2.04	1.99	1.95
21	4.32	3.47	3.07	2.84	2.68	2.57	2.49	2.42	2.37	2.32	2.25	2.18	2.10	2.05	2.01	1.96	1.92
22	4.30	3.44	3.05	2.82	2.66	2.55	2.46	2.40	2.34	2.30	2.23	2.15	2.07	2.02	1.98	1.94	1.89
23	4.28	3.42	3.03	2.80	2.64	2.53	2.44	2.37	2.32	2.27	2.20	2.13	2.05	2.00	1.96	1.91	1.86
24	4.26	3.40	3.01	2.78	2.62	2.51	2.42	2.36	2.30	2.25	2.18	2.11	2.03	1.97	1.94	1.89	1.84
25	4.24	3.39	2.99	2.76	2.60	2.49	2.40	2.34	2.28	2.24	2.16	2.09	2.01	1.96	1.92	1.87	1.82
26	4.23	3.37	2.98	2.74	2.59	2.47	2.39	2.32	2.27	2.22	2.15	2.07	1.99	1.94	1.90	1.85	1.80
27	4.21	3.35	2.96	2.73	2.57	2.46	2.37	2.31	2.25	2.20	2.13	2.06	1.97	1.92	1.88	1.84	1.79
28	4.20	3.34	2.95	2.71	2.56	2.45	2.36	2.29	2.24	2.19	2.12	2.04	1.96	1.91	1.87	1.82	1.77
29	4.18	3.33	2.93	2.70	2.55	2.43	2.35	2.28	2.22	2.18	2.10	2.03	1.94	1.89	1.85	1.81	1.75
30	4.17	3.32	2.92	2.69	2.53	2.42	2.33	2.27	2.21	2.16	2.09	2.01	1.93	1.88	1.84	1.79	1.74
40	4.08	3.23	2.84	2.61	2.45	2.34	2.25	2.18	2.12	2.08	2.00	1.92	1.84	1.78	1.74	1.69	1.64
60	4.00	3.15	2.76	2.53	2.37	2.25	2.17	2.10	2.04	1.99	1.92	1.84	1.75	1.69	1.65	1.59	1.53
120	3.92	3.07	2.68	2.45	2.29	2.18	2.09	2.02	1.96	1.91	1.83	1.75	1.66	1.60	1.55	1.50	1.43
∞	3.84	3.00	2.61	2.37	2.21	2.10	2.01	1.94	1.88	1.83	1.75	1.67	1.57	1.51	1.46	1.39	1.32

Data Bank

The Jump River Electric Company serves several counties in northern Wisconsin. Much of the area is forest and lakes. The data on power outages from one recent summer include date, time, duration of outage (hours), and cause.

TABLE D.1 Power Outages

Date	Time	Duration	Cause
6/11	4:00 PM	5.50	Trees and limbs
6/12	8:00 PM	1.50	Trees and limbs
6/16	8:30 AM	2.00	Trees and limbs
6/17	5:30 AM	2.00	Trees and limbs
6/17	5:00 PM	8.00	Windstorm
6/21	4:30 PM	2.00	Trees and limbs
6/26	3:00 AM	3.00	Trees and limbs
6/26	2:00 PM	1.75	Unknown
6/26	12:00 AM	2.00	Lightning blew up transformer
7/03	6:00 PM	1.50	Trees and limbs
7/04	9:00 AM	2.50	Unknown
7/04	5:00 PM	1.50	Trees and limbs
7/05	5:00 AM	1.50	Trees and limbs
7/08	7:00 AM	3.50	Lightning
7/20	12:00 AM	1.50	Unknown
7/21	9:00 AM	1.50	Animal
7/28	12:00 PM	1.00	Animal
7/30	7:00 PM	1.00	Squirrel on transformer
7/31	7:30 AM	0.50	Squirrel on cutout
8/04	6:00 AM	2.50	Trees and limbs
8/06	8:00 PM	2.00	Beaver
8/09	5:30 AM	1.50	Fuse-flying squirrel
8/11	12:00 AM	3.00	Beaver-cut trees
8/13	1:00 AM	1.00	Unknown
8/15	12:30 AM	1.50	Animal
8/18	8:00 AM	1.50	Trees and limbs
8/20	4:00 AM	2.00	Transformer fuse
8/25	9:00 AM	1.00	Animal
8/26	2:00 AM	10.00	Lightning
8/27	3:00 AM	1.00	Trees and limbs

Madison recruits for the fire department must complete a timed test that simulates working conditions. It includes placing a ladder against a building, pulling out a section of fire hose, dragging a weighted object, and crawling in a simulated attic environment. The times, in seconds, for recruits to complete the test for a Madison firefighter job are shown in Table D.2.

TABLE D.2 Time to Complete Firefighters Physical Test (seconds)

425	389	380	421	438	331	368	417	403	416	385	315
427	417	386	386	378	300	321	286	269	225	268	317
287	256	334	342	269	226	291	280	221	283	302	308
296	266	238	286	317	276	254	278	247	336	296	259
270	302	281	228	317	312	327	288	395	240	264	246
294	254	222	285	254	264	277	266	228	347	322	232
365	356	261	293	354	236	285	303	275	403	268	250
279	400	370	399	438	287	363	350	278	278	234	266
319	276	291	352	313	262	289	273	317	328	292	279
289	312	334	294	297	304	240	303	255	305	252	286
297	353	350	276	333	285	317	296	276	247	339	328
267	305	291	269	386	264	299	261	284	302	342	304
336	291	294	323	320	289	339	292	373	410	257	406
374	268										

Natural resource managers have attempted to use the Satellite Landsat Multispectral Scanner data for improved land-cover classification. The intensities of reflected light recorded on the near-infrared band of a thermatic mapper are given in Table D.3. Table D.3a gives readings from a country known to consist of forest and Table D.3b readings from urban areas.

TABLE D.3a   Near Infrared Light Reflected from Forest Areas

77	77	78	78	81	81	82	82	82	82	82	83	83	84	84	84	84	85	
86	86	86	86	86	87	87	87	87	87	87	87	89	89	89	89	89	89	89
90	90	90	91	91	91	91	91	91	91	91	91	91	93	93	93	93	93	93
94	94	94	94	94	94	94	94	94	94	94	94	95	95	95	95	95	96	96
96	96	96	96	97	97	97	97	97	97	97	97	97	98	99	100	100	100	100
100	100	100	100	100	101	101	101	101	101	101	102	102	102	102	102			
102	103	103	104	104	104	105	107											

TABLE D.3b   Near Infrared Light Reflected from Urban Areas

71	72	73	74	75	77	78	79	79	79	79	80	80	80	81	81	81	82	82	82	82
84	84	84	84	84	84	85	85	85	85	85	85	86	86	87	88	90	91	94		

Beginning accounting students must learn to audit in a computerized environment. A sample of beginning accounting students took a test that is summarized by two scores shown in Table D.4: the Computer Attitude Scale (CAS), based on 20 questions, and the Computer Anxiety Rating Scale (CARS), based on 19 questions. (Courtesy of Douglas Stein.) Males are coded as 1 and females as 0.

TABLE D.4   Computer Attitude and Anxiety Scores

Gender	CAS	CARS	Gender	CAS	CARS
0	2.85	2.90	1	3.30	3.47
1	2.60	2.32	1	2.90	3.05
0	2.20	1.00	1	2.60	2.68
1	2.65	2.58	0	2.25	1.90
1	2.60	2.58	0	1.90	1.84
1	3.20	3.05	1	2.20	1.74
1	3.65	3.74	0	2.30	2.58
0	2.55	1.90	0	1.80	1.58
1	3.15	3.32	1	3.05	2.47
1	2.80	2.74	1	3.15	3.32
0	2.40	2.37	0	2.80	2.90
1	3.20	3.11	0	2.35	2.42
0	3.05	3.32	1	3.70	3.47
1	2.60	2.79	1	2.60	4.00
1	3.35	2.95	0	3.50	3.42
0	3.75	3.79	0	2.95	2.53
0	3.00	3.26	1	2.80	2.68
1	2.80	3.21			

Data were collected on students taking the course Conditioning 1, designed to introduce students to a variety of training techniques to improve cardiorespiratory fitness, muscular strength, and flexibility (Table D.5). (Courtesy of K. Baldridge.)

c1   Gender (1 = male, 2 = female)
c2   Pretest percent body fat
c3   Posttest percent body fat
c4   Pretest time to run 1.5 miles (seconds)
c5   Posttest time to run 1.5 miles (seconds)
c6   Pretest time to row 2.5 kilometers (seconds)
c7   Posttest time to row 2.5 kilometers (seconds)
c8   Pretest number of situps completed in 1 minute
c9   Posttest number of situps completed in 1 minute

TABLE D.5   Physical Fitness Improvement

Gender	Pretest % Fat	Posttest % Fat	Pretest Run	Posttest Run	Pretest Row	Posttest Row	Pretest Situps	Posttest Situps
1	15.1	12.4	575	480	621	559	60	67
1	17.1	18.5	766	672	698	595	42	45
2	25.5	15.0	900	750	840	725	32	36
2	19.5	17.0	715	610	855	753	28	38
2	21.7	20.6	705	585	846	738	46	54
1	17.7	18.5	820	670	630	648	18	41
2	22.7	17.2	880	745	860	788	22	33
2	26.6	22.4	840	725	785	745	29	39
2	36.4	32.5	1065	960	780	749	27	40
1	9.7	8.0	630	565	673	588	32	49
2	31.0	25.0	870	780	746	689	39	54
2	6.0	6.0	580	494	756	714	37	49
2	25.1	22.8	1080	806	852	838	43	38
1	15.1	13.4	720	596	674	576	48	62
2	23.0	21.1	780	718	846	783	30	38
2	9.7	9.7	945	700	890	823	38	44
1	7.0	6.1	706	657	652	521	52	63
1	16.6	15.1	650	567	740	615	38	47
2	21.7	19.5	686	662	762	732	29	47
2	23.7	21.7	758	718	830	719	35	42
2	24.7	21.7	870	705	754	734	24	37
1	11.6	8.9	480	460	640	587	55	60
2	19.5	16.0	715	655	703	731	36	44
1	4.8	4.2	545	530	625	571	40	45
2	31.1	21.6	840	790	745	728	30	36

(continued)

TABLE D.5 *(Continued)*

Gender	Pretest % Fat	Posttest % Fat	Pretest Run	Posttest Run	Pretest Row	Posttest Row	Pretest Situps	Posttest Situps
1	23.2	19.9	617	622	637	620	35	32
1	12.5	12.2	635	600	805	736	37	52
2	29.3	21.7	790	715	821	704	41	42
2	19.5	16.0	750	702	1043	989	42	45
1	10.7	9.8	622	567	706	645	40	46
2	26.6	21.7	722	725	741	734	49	61
2	27.6	23.7	641	598	694	682	40	47
1	25.9	18.8	708	609	685	593	20	21
1	27.5	16.0	675	637	694	682	35	37
1	9.4	7.6	618	566	610	579	36	50
1	12.5	9.8	613	552	610	575	46	50
2	27.5	30.9	705	660	746	691	31	36
2	18.3	15.0	853	720	748	694	31	34
1	6.6	4.7	496	476	623	569	45	45
2	23.7	21.7	860	750	758	711	42	55
2	23.2	22.1	905	636	759	726	26	30
2	18.0	14.8	900	805	823	759	31	29
2	22.7	18.3	767	741	808	753	28	32
1	19.9	14.8	830	620	632	586	39	45
1	6.6	5.7	559	513	647	602	44	49
1	14.3	9.8	699	652	638	602	34	35
1	21.7	18.5	765	735	674	615	41	48
1	13.9	9.4	590	570	599	571	40	42
1	20.1	19.3	770	672	675	611	37	48
1	15.1	8.9	602	560	656	578	47	58
2	14.8	11.0	741	610	768	687	34	41
2	23.4	19.1	723	641	711	695	36	41
2	23.7	16.3	648	601	802	740	44	51
1	28.0	17.7	842	702	790	765	26	29
1	4.7	4.7	558	540	660	600	40	46
2	26.6	23.7	750	565	720	670	36	52
2	21.7	18.3	608	592	707	697	42	46
1	11.6	8.9	537	495	610	572	55	60
2	22.7	18.3	855	694	800	712	41	50
2	21.7	18.3	630	614	785	743	50	55
2	25.9	21.9	902	820	771	717	34	38
1	28.4	21.7	780	664	756	703	57	64
2	21.9	20.8	665	670	673	667	41	44
2	27.5	22.7	675	646	689	674	49	50
1	5.7	3.8	473	472	551	546	53	53
2	22.7	22.0	715	682	678	672	40	43

TABLE D.5   *(Continued)*

Gender	Pretest % Fat	Posttest % Fat	Pretest Run	Posttest Run	Pretest Row	Posttest Row	Pretest Situps	Posttest Situps
2	33.2	25.7	795	740	817	721	30	31
2	16.0	13.6	688	615	811	705	45	49
1	14.3	11.6	530	497	589	570	39	50
1	22.7	19.5	840	705	788	780	29	32
2	25.0	18.3	690	618	816	701	42	50
1	4.2	3.2	545	527	577	543	55	59
2	25.7	21.7	760	727	849	724	39	44
1	13.9	9.4	620	515	689	580	32	41
1	12.2	7.6	605	564	661	614	35	38
1	21.7	20.6	688	625	750	686	39	32
1	7.0	5.1	590	529	631	619	60	65
2	34.6	27.5	720	694	690	698	34	45
1	3.2	3.2	500	459	644	599	35	37
1	6.1	5.1	540	492	579	546	56	60
2	28.4	21.7	885	825	804	733	33	36

Salmon fisheries support a primary industry in Alaska and their management is of high priority. Salmon are born in freshwater rivers and streams, but then swim out into the ocean for a few years before returning to spawn and die. To identify the origins of mature fish and equably divide the catch of returning salmon between Alaska and the Canadian provinces, researchers have studied the growth of their scales. The growth the first year in freshwater is measured by the width of the growth rings for that period of life, and marine growth is measured by the width of the growth rings for the first year in the ocean environment. A set of these measurements, collected by the Alaska Department of Fish and Game, are given in Table D.6. (Courtesy of K. Jensen and B. Van Alen.)

TABLE D.6 Radius of Growth Zones for Freshwater and First Marine Year

Males		Females	
Freshwater Growth	First Year Marine Growth	Freshwater Growth	First Year Marine Growth
147	444	131	405
139	446	113	422
160	438	137	428
99	437	121	469
120	405	139	424
151	435	144	402
115	394	161	440
121	406	107	410
109	440	129	366
119	414	123	422
130	444	148	410
110	465	129	352
127	457	119	414
100	498	134	396
115	452	139	473
117	418	140	398
112	502	126	434
116	478	116	395
98	500	112	334
98	589	117	455
83	480	97	439
85	424	134	511
88	455	88	432
98	439	99	381
74	423	105	418
58	411	112	475
114	484	98	436
88	447	80	431

TABLE D.6 *(Continued)*

Males		Females	
Freshwater Growth	First Year Marine Growth	Freshwater Growth	First Year Marine Growth
77	448	139	515
86	450	97	508
86	493	103	429
65	495	93	420
127	470	85	424
91	454	60	456
76	430	115	491
44	448	113	474
42	512	91	421
50	417	109	451
57	466	122	442
42	496	68	363

Insecticides, including the long banned DDT, which imitate the human reproductive hormone estrogen, may cause serious health problems in humans and animals. Researchers examined the reproductive development of young alligators hatched from eggs taken from (1) Lake Apopka, which is adjacent to an EPA Superfund site, and (2) Lake Woodruff, which acted as a control. The contaminants at the first lake, including DDT, were thought to have caused reproductive disorders in animals living in and around the lake. The concentrations of the sex steroids estradiol and testosterone in the blood of alligators were determined by radioimmunoassay both at about six months of age and then again after the alligators were stimulated with LH, a pituitary hormone. (Courtesy of L. Guillette.)

The data are coded as (* indicates missing):

x_1 = 1, Lake Apopka, and = 0, Lake Woodruff

x_2 = 1 male and = 0 female

x_3 = E2 = estradiol concentration (pg/ml)

x_4 = T = testosterone concentration (pg/ml)

x_5 = LHE2 = estradiol concentration after receiving LH (pg/ml)

x_6 = LHT = testosterone concentration after receiving LH (pg/ml)

TABLE D.7 Alligator Data

Lake Apopka						Lake Woodruff					
Lake	Sex	E2	T	LHE2	LHT	Lake	Sex	E2	T	LHE2	LHT
x_1	x_2	x_3	x_4	x_5	x_6	x_1	x_2	x_3	x_4	x_5	x_6
1	1	38	22	134	15	0	1	29	47	46	10
1	1	23	24	109	28	0	1	64	20	82	76
1	1	53	8	16	12	0	1	19	60	*	*
1	1	37	6	220	13	0	1	36	75	19	72
1	1	30	7	114	11	0	1	27	12	118	95
1	0	60	19	184	7	0	1	16	54	33	64
1	0	73	23	143	13	0	1	15	33	99	19
1	0	64	16	228	13	0	1	72	53	29	20
1	0	101	8	163	10	0	1	85	100	72	0
1	0	137	9	83	7	0	0	139	20	82	2
1	0	88	7	200	12	0	0	74	4	170	75
1	0	73	19	220	21	0	0	83	18	125	45
1	0	257	8	194	37	0	0	35	43	19	76
1	0	138	10	221	3	0	0	106	9	142	5
1	0	220	15	101	5	0	0	47	52	24	62
1	0	88	10	141	7	0	0	38	8	68	20
						0	0	65	15	32	50
						0	0	21	7	140	4
						0	0	68	16	110	3
						0	0	70	16	58	18
						0	0	112	14	78	5

Answers to Selected Odd–Numbered Exercises

CHAPTER 1

8.1 Population: entire set of responses from all U.S. teens 13 to 17. Sample: responses from the 1055 teens contacted.

8.3 Population: semen concentrations of all adult males. Sample: concentrations of the 14,847 men in the study.

8.5 Population: preferences of all adults in the city. Sample: preferences of persons who send in votes. Not representative since self-selected.

8.7 (b) Sample is collection of use/non-use of credit card by the 102 customers.

8.9 "soft" not well defined. Could rate on five-point scale (1) paper towel to (5) cotton.

8.11 Population: entire collection of yes/no responses for all readers. Sample not representative.

8.15 Mice or you and me?

8.17 We selected buses 24, 23, 49, and 45 on list.

CHAPTER 2

3.1 The frequency table for blood type:

Blood Type	Frequency	Relative Frequency
O	16	$.40 = \frac{16}{40}$
A	18	$.45 = \frac{18}{40}$
B	4	$.10 = \frac{4}{40}$
AB	2	$.05 = \frac{2}{40}$
Total	40	1.00

3.3 (a) The relative frequencies are Drive alone .625, Car pool .075, Ride bus .175, and Other .125.

3.5 No place for lengths between 7 and 8 inches.

3.7 120 should be in the last class.

3.9 (a) Yes   (d) No   (e) No

3.23 237, 268, 291, . . . , 504, 507, 621

3.25 18, 19, 20, . . . , 28, 29, 30

3.29 Misleading. Sales of Company C appear at least 7 times larger.

4.1 (a) Median = 8, $\bar{x}$ = 9
(b) Median = 2, $\bar{x}$ = 2

4.3 $\bar{x}$ = 26.75, median = 25

4.5 (a) $\bar{x}$ = 1160.7, median = 950
(b) The median is best.

4.7 10.5

4.9 (a) $\bar{x}$ = 9.28 days
(b) Sample median = 4.0 days, $\bar{x}$ is higher because of a few large observations.

4.11 (a) Sample median = 244
(b) $\bar{x}$ = 206.5

4.13 (a) Sample mean = 6.775
(b) Sample median = 6.5

4.15 181.5

4.17 (b) $\bar{x}$ = 26.62; sample median = 24
(c) Q_1 = 22, Q_2 = 24, and Q_3 = 32

4.19 (a) Sample median = 153
(b) Q_1 = 135.5, Q_3 = 166.5

4.21 Sample median = 94, Q_1 = 73 and Q_3 = 105

4.23 (a) Median = 6.5; Q_1 = 3, Q_3 = 9.5
(b) 90th percentile = 13

4.25 (b) Mean = 25.44°C; median = 25.56°C

5.1 (a) Weather, traffic, and running/walking
(b) 9.857   (c) 1.356

5.3 (b) s^2 = 7, s = 2.646

5.5 (b) s^2 = 18.67, s = 4.32

5.9 34.857

5.11 (a) 6.673 (b) 2.583

5.13 (a) 25.51

5.17 32

5.19 8.9 days

5.21 (a) .95 and 1

5.23 (a) $\bar{x} \pm s$ = (.97, 8.97) has proportion = .962, guidelines = .95

5.31 (a) $\bar{x}$ = 7.51, s = 12.10

5.33 (b) Maximum = 55 for Truman, minimum = 4 for Kennedy

6.1 No trend but 50 calls for worker 20 may be low.

6.3 Downward trend is dominant feature.

7.1 (a) Relative frequencies for government employees are .179 and .172.

7.5 (a) Yes   (c) No

7.7 (a) Median = 68   (c) .526, .263

7.9 Median = 55, Q_1 = 51, Q_3 = 58
7.11 (a) $\bar{x}$ = 10, s = 2
    (c) Mean = -30, standard deviation = 6
7.13 (a) $\bar{x}$ = 69.61, s = 2.97
    (c) Median = 70, Q_1 = 68, Q_3 = 72
7.19 (a & b) (188.61, 290.59) has proportion = .698;
    guidelines = .68
7.21 (a) Median = 4.505, Q_1 = 4.30, Q_3 = 4.70
    (b) 4.935
    (c) $\bar{x}$ = 4.5074, s = .368
7.23 (a) Median = .14, Q_1 = .10, Q_3 = .35
    (b) $\bar{x}$ = .331, s = .514
7.25 (b) Not reasonable with time trend
7.27 (b) .78   (c) No speeding tickets 32–35 mph
7.29 (b) Nonsense. The ordering of the numbers is
    meaningless for nominal data.

CHAPTER 3

2.1 (c) The pill seems to reduce the proportions of
    severe and moderate nausea.
2.3 Proportion is highest for B7.
2.7 (b)

	Major				
	B	H	P	S	Total
Male	.245	.082	.102	.286	.715
Female	.122	0	.082	.082	.286
Total	.367	.082	.184	.368	1.001

2.11 (b) Research hospital best for either condition.
3.1 Flip coin; if H then first gets control.
5.5 (b) High negative correlation
    (c) r = $-.992$
5.7 (b) r = .80
    (c) Moderate negative correlation, r = $-.75$
5.9 r is positive and moderate, r = .5
5.11 r = .818
5.17 .994
6.1 (b) Would like conclusion to apply to a typical
    day.
7.1 Attorney improves chances of an increase.
7.5 $-.757$
7.7 .902
7.9 (b) .988
7.11 (a) Negative; typically, the lighter the car the
    more miles per gallon.
    (c) Negative; an older plane would probably
    require more work to keep it running properly.

    (e) Positive; higher temperature tends to make
people more thirsty.

CHAPTER 4

2.1 (a) (ii), (v)   (d) (vi)   (f) (iii), (v)
2.5 (a) {0, 1}   (b) {0, 1, . . . , 344}
    (c) {t: 90 < t < 425.4}
2.7 (a) S = {BJ, BL, JB, JL, LB, LJ}
    (b) A = {LB, LJ}
        B = {JL, LJ}
2.11 (c) $\frac{3}{8}$
2.13 (a) S = {1, 2, 3}
        $P(1)$ = $\frac{3}{8}$, $P(2)$ = $\frac{2}{8}$, $P(3)$ = $\frac{3}{8}$
    (b) $\frac{3}{4}$
2.15 (c) $P(A)$ = $\frac{5}{36}$, $P(B)$ = $\frac{1}{6}$, $P(C)$ = $\frac{1}{2}$, $P(D)$ = $\frac{1}{6}$
2.17 (b) .52
2.19 (b) $\frac{2}{3}$
2.21 (a) .167
3.1 (c) (i) AB = {o, b} (iii) $A\overline{B}$ = {g, w}
3.3 (b) AB = {d2, d6, d7}, $P(AB)$ = .38
    (d) $\overline{A}\,\overline{C}$ = {d3, d4}, $P(\overline{A}\,\overline{C})$ = .16
3.5 (b) A or B = {2, 3, 4}, AB = {3}
3.7 (a) $P(A)$ = .16, $P(AB)$ = .08
3.9 (b) $P(A\overline{B})$ = .32, $P(\overline{A}B)$ = .16, $P(\overline{A}\,\overline{B})$ = .32
3.11 (b) $P(A\overline{B})$ = .12
    (c) $P(A$ or $B)$ = .52
    (d) $P(A\overline{B}$ or $\overline{A}B)$ = .27
3.13 (a) $P(A)$ = .65
    (b) $P(A\overline{B})$ = .15
3.15 .4
3.17 (b) .4 (i) $P(\overline{A}\,\overline{B})$ = .35 (iii) $P(\overline{A}\,\overline{B}\,\overline{C})$ = .10
5.1 (a) .471   (b) .529   (c) .719
5.3 $P(B|A)$ = .0099, not independent
5.7 (a) $P(A)$ = .4, $P(B)$ = .4
    (b) Not independent because
    $P(AB)$ = .1 ≠ $P(A)P(B)$
5.9 (b) 5.4%   (c) .334
5.11 (a) .61   (c) .641
5.13 (a) $\frac{2}{7}$ (b) $P(\overline{A}|B)$ = $\frac{5}{7}$
5.17 (a) $\frac{2}{9}$ (b) $\frac{7}{12}$
5.19 (a) BC, $P(BC)$ = 0
    (c) $\overline{B}$, $P(\overline{B})$ = .9
5.21 .000 000 000 016
5.23 (b) .784
5.25 (a) .65   (b) No
6.1 (a) 28   (c) 1140   (e) 27405
6.3 (a) 126   (b) 60
6.5 (a) 66   (b) 63

6.7 (a) .446   (b) .103
6.9 No
6.11 (a) No   (c) No
7.3 (a) A = {win, tie}
    (c) A = {0, 1, . . . , 55}
7.5 A = {p: .10 < p < 1}
7.7 (a) .5   (b) .75   (c) .167
7.9 .5
7.13 (b) 2/3
7.15 (a) Either a faulty transmission or faulty brakes
    (b) Transmission, brakes, and exhaust system all faulty
7.17 $P(A)$ = .47, $P(AB)$ = .13, $P(A$ or $B)$ = .68
7.21 (a) .18   (b) .73
7.23 (a) .25   (b) .34
7.27 (a) $P(A)P(C)$ = .15, $P(AC)$ = .15, independent
    (b) $P(A\bar{B})P(C)$ = .1, $P(A\bar{B}C)$ = .15, not independent
7.29 (a) .509   (b) .294   (c) Not independent
7.31 (c) .64   (d) .49
7.33 (a) .092   (b) .538
7.35 (a) .82

CHAPTER 5

2.1 (a) Discrete   (b) Continuous   (c) Discrete
2.3 (a)

Possible choices	x	Possible choices	x
{2, 4}	2	{4, 7}	3
{2, 6}	4	{4, 8}	4
{2, 7}	5	{6, 7}	1
{2, 8}	6	{6, 8}	2
{4, 6}	2	{7, 8}	1

(b)

x	1	2	3	4	5	6
$f(x)$	.2	.3	.1	.2	.1	.1

2.7

x	0	2	4
$f(x)$	.61	.23	.16

2.11 (a) .4   (b) .533

2.15 (a)

x	1	2	3	4	5	6
$f(x)$	$\frac{1}{13}$	$\frac{3}{13}$	$\frac{5}{13}$	$\frac{2}{13}$	$\frac{1}{13}$	$\frac{1}{13}$

2.17 $f(2)$ = .1, $f(4)$ = .1, $f(6)$ = .1

2.21 (a) .92   (b) .63   (c) .80
2.23 (a) .45
    (b) .25
    (c) The capacity must be increased by 2 to a total of 4.
3.1 (b) $E(X)$ = 1, σ = 1
3.3 $1208
3.5 μ = 1.2, σ = .917
3.9 (a) $E(X)$ = 1.424
    (b) σ = 1.351
3.13 (b)

x	0	1	2	3	4	Total
$f(x)$	.4096	.4096	.1536	.0256	.0016	1.0000

(c) $E(X)$ = .8
3.15 Median = 1
3.17 (c) .64
4.1 (a) Bernoulli model plausible
    (b) Trials are dependent
4.3 Identify S = yellow, F = other colors
    (a) Yes, p = .4
    (b) No
    (c) No
4.5 (a) Yes, $P(S)$ = .5
    (b) Not independent
4.7 (a) Bernoulli trials model is not appropriate.
    (b) Appropriate
4.9 (a) .316   (b) .004 = $\frac{1}{256}$

4.13 (c)

x	0	1	2
$f(x)$	$\frac{16}{81}$	$\frac{40}{81}$	$\frac{25}{81}$

5.1 (a) Yes, n = 9, p = $\frac{1}{6}$
    (b) No, because the number of trials is not fixed.
5.3 (b) .132
5.5 (c) .3852
5.7 (a) .051   (b) .996   (c) .949
5.11 (a) .231   (b) .007   (c) .007
5.13 (a) .163   (b) .537
5.17 .411
5.19 (a) .215   (c) .283
5.23 (b) Mean = 12.308, sd = 2.919
6.1 (a) .072   (b) .073
6.5 .917 outside of control limits
7.1 (a) $-3, -1, 1, 3$

(b) $X = -3$: *TTT*
$X = -1$: *HTT, THT, TTH*
$X = 1$: *HHT, HTH, THH*
$X = 3$: *HHH*

7.3

w	0	1	2
$f(w)$	.605	.350	.045

7.5

x	0	1	2	3
$f(x)$	.212	.509	.255	.024

7.7 (a) 3.9   (b) 1.14
7.9 (a) .002   (b) $-$0.68
7.11 (a) $4500   (b) $2000
7.13 (a)

x	0	5	100	1000	4000
$f(x)$	.8952	.0850	.0190	.0006	.0002

(b) $3.725   (c) .9802
7.15 (a)

x	-15	-5	5	15
$f(x)$	.1458	.3936	.3543	.1036

7.19 (a) $F(3) = .44$, $F(4) = .72$, $F(5) = .90$,
$F(6) = 1.00$
7.21 (a) Not plausible, because the assumption of
independence is questionable.
(b) Plausible
7.23 .125
7.25 (a) .36   (b) .216   (c) .432
7.29 (a) Binomial distribution with $n = 5$, $p = .4$
(b) .913, .078, 2
7.31 .984
7.33 (b) Not binomial. p not the same all trials.
7.39 $P[X \leq 5] = .416$, $P[X = 6] = .192$
7.41 $E(X) = 3.2$, sd$(X) = 1.6$

CHAPTER 6

2.1 (a) Test several cars in lab.
2.3 (a) .25   (d) 0
2.5 Median = 1
2.7 (a) Not symmetric—long tail to right.
4.1 (a) .8790
(c) .0336

4.3 (a) .1210
(c) .8708
4.5 (b) .7016
(c) .3705
4.7 (b) .84
(e) $z = .94$
(f) $-.32$
4.9 (a) .6628
(c) .0455
(e) .8670
(g) .2524
4.11 (a) .5557   (d) 1.08
5.1 (a) .1056
(c) .0668
(e) .9477
5.3 (a) $b = 159.8$
5.7 .7338
5.9 .1817
5.11 Shortest person is 65.4 inches, tallest is 72.6
inches.
5.13 (b) $Z = \dfrac{X - 350}{25}$
6.1 No. Results for persons in same family are
dependent.
6.3 (a) $z = -4/3$ or -1.33, probability = .0918
6.5 (b) No, since $np = 2$, p is too small.
6.9 (a) $z = (18 - 23.7)/4.672 = -1.22$,
probability = .1112
6.11 $z = (20 - 17.5)/3.623 = .69$, probability =
.2451
6.13 The z-values are $z = (16 - 11.783)/3.019 =$
1.40 and $z = (10 - 11.783)/3.019 = -.59$.
Probability = .6416
6.15 481
8.1 .5
8.5 (a) $z = -1.36$
(c) $z = .65$
8.7 (a) .7642
(c) .0456
8.9 (a) .8106   (c) .1056   (e) .5091
8.11 (a) .1949
(c) .2090
8.13 .4013
8.15 (a) .0228
8.17 (a) $z = -1.24$, .1075
(b) Smaller than .0000
8.19 (b) Not appropriate

CHAPTER 7

2.1 (a) Statistic
(c) Parameter

2.5 (b)

$\bar{x}$	Probability $f(\bar{x})$
1	$\frac{1}{9}$
2	$\frac{2}{9}$
3	$\frac{3}{9}$
4	$\frac{2}{9}$
5	$\frac{1}{9}$
Total	1

2.7 No, larger size berries are often on top.

2.9

Value	0	3	9	12
Probability	.160	.090	.360	.390

3.1 (a) $E(\overline{X}) = \mu = 99$, sd$(\overline{X}) = 3.5$
3.3 (a) sd $(\overline{X}) = 2$
(c) sd$(\overline{X}) = .5$
3.5 $E(\overline{X}) = 3$, sd$(\overline{X}) = \sqrt{\frac{4}{3}}$
3.9 .2709
3.11 1.26
3.13 (a) $\overline{X}$ is approximately $N(20, .5)$.
(b) .0668
3.15 .145
3.17 .885
4.1 (b) Sampling distribution of $\overline{X}$

$\bar{x}$	Probability $f(\bar{x})$
2	$\frac{1}{16}$
3	$\frac{2}{16}$
4	$\frac{3}{16}$
5	$\frac{4}{16}$
6	$\frac{3}{16}$
7	$\frac{2}{16}$
8	$\frac{1}{16}$
Total	1

(d) $E(\overline{X}) = 5$, sd$(\overline{X}) = \sqrt{2.5}$
4.3 (a) sd$(\overline{X}) = 17.5$
4.5 (b) Exactly a normal distribution
(c) .7698
4.7 (a) $N(12.1 (3.2)/3)$
(c) About 25% smaller than 10

4.11 (a) Approximately .632
(b) (52.83, 57.17)
4.13 .869
4.15

$\bar{x}$	0	1	2	3	4	5	6	8	9	12
$f(\bar{x})$	.027	.135	.225	.125	.054	.180	.150	.036	.060	.008

CHAPTER 8

2.1 (a) Point estimate   (b) Not point estimate
2.3 (a) 8 ounces could be considered as a point estimate of the population mean weight of all bags manufactured.
(b) Expected weight
2.5 (a) S.E. = 1.873, 95% error margin = 3.67
(c) S.E. = 3.130, 92% error margin = 5.48
2.7 (a) $\bar{x} = 12.17$, estimated S.E. = .211
2.11 (a) 99   (c) 74
2.13 133
3.1 (84.62, 88.38)
3.3 About 180
3.5 (28.94, 31.46) grams
3.7 (420, 446)
3.9 (9.26, 10.34) miles
3.11 (3.96, 4.78) pounds
3.13 (354, 841) stings
3.15 .8502 or about 85%
4.1 (a) $H_0: \mu = 17.2$ versus $H_1: \mu > 17.2$
(b) $H_0: \mu = 210$ seconds versus $H_1: \mu < 210$ seconds
4.3 (a) Correct decision if $\mu = 17.2$; wrong decision if $\mu > 17.2$, type II error.
(b) Correct decision if $\mu = 210$ seconds, wrong decision if $\mu < 210$ seconds, type II error.
4.7 (a) $\alpha = .025$
(b) $c = 21.10$
4.9 Observed $z = -2.48$ is in $R: Z \geq 1.96$ or $Z \leq -1.96$; H_0 is rejected at $\alpha = .05$.
4.11 H_0 would not be rejected.
4.13 (a) .0089, strong rejection
(c) .1032, not a strong rejection
4.15 (a) $H_0: \mu = 15$, $H_1: \mu < 15$, $R: Z \leq -1.96$, observed $z = -1.12$, H_0 is not rejected at $\alpha = .025$.
(b) .1314
4.17 $H_0: \mu = 3000$, $H_1: \mu \neq 3000$, $R: Z \geq 1.96$ or $Z \leq -1.96$; observed $z = 2.13$, H_0 is rejected at $\alpha = .05$.

5.1 (a) *S.E.* = .126, 98% error margin .292
5.3 (a) 47 minutes
(b) 1.58 minutes
5.5 (a) $n = 68$
(b) $n = 161$
5.9 (a) $\bar{x} = 126.9$, 95.4% error margin = 2.8
(b) (124.6, 129.2)
5.11 (a) (8.38, 9.14)
(b) (8.17, 9.35)
5.13 (b) μ = mean concentration. $H_0: \mu = .008$,
$H_1: \mu > .008$
5.15 (b) $c = 2.0776$
5.17 (a) $Z = \dfrac{\bar{X} - 20}{S/\sqrt{n}}$, $R: Z \leq -1.96$ or $Z \geq 1.96$
5.19 (b) *P*-value = $2P[Z \leq -1.48]$ about .14
5.21 (b) *P*-value = $2P[Z \leq -1.34]$ about .18

CHAPTER 9

2.1 (a) 2.015
(b) -2.101
2.3 (b) 4.541
(c) -1.706
2.5 (a) 1.943
(b) 2.131
2.7 (b) Between .025 and .05
(c) Between .05 and .10
3.1 (a) (135.5, 144.5)
(b) Length = 9.0, center = 140
(c) Usually different since S is random.
3.3 (25.7, 38.5)
3.7 (a) $\bar{x} = 22.4$, $s = 5.98$
(b) (17.7, 27.1)
3.9 $t = 1.94$, reject H_0
3.11 *P*-value = $P[T > 2.5]$, about .01
3.13 $t = 3.09$, reject H_0
3.15 (4.10, 4.78)
3.17 (b) (162.82, 197.18) acres
3.19 (b) $t = 1.38$, fail to reject H_0 even with $\alpha = .10$
3.21 (a) Fail to reject H_0
4.1 (a) H_0 is not rejected.
(b) H_0 is rejected.
4.3 (a) No
(d) Can't tell
4.5 (a) (2.37, 11.19)
(b) H_0 would not be rejected.
5.1 (a) 15.51   (c) 3.81
5.3 (a) .05   (c) .875

5.5 $H_0: \sigma = 1.5$, $H_1: \sigma > 1.5$,
$\chi^2 = \dfrac{(n-1)S^2}{(1.5)^2}$, d.f. = 9, $R: \chi^2 \geq \chi^2_{.05} = 16.92$.
Observed $\chi^2 = 20.6$; H_0 is rejected at $\alpha = .05$.
Assumption: normal population.
5.7 (a) 1.178
(b) (.86, 1.94)
5.9 (4.70, 10.83) feet
5.11 (a) (6.67, 26.21) mg/kg
6.1 (a) 2.132   (c) -2.132
6.5 (39.8, 54.2)
6.7 (c) $t = -.82$, fail to reject H_0
6.9 (b) *P*-value = $P[T \leq -2.67]$ about .01
6.11 $t = -3.33$. Reject H_0 in favor of $H_1: \mu < 40$
6.13 $t = 3.69$. *P*-value smaller than .005
6.15 (a) 9.49   (c) .711
6.17 (a) $\chi^2 = \dfrac{(n-1)S^2}{(10)^2}$, d.f. = 24, $\chi^2_{.05} =$
36.42, $R: \chi^2 \geq 36.42$.
Observed $\chi^2 = 40.16$; H_0 is rejected.
6.19 (a) $s = 10.174$

CHAPTER 10

1.3 (a) {(T, E), (C, R), (S, G)}
{(T, C), (E, R), (S, G)}
{(T, R), (C, E), (S, G)}
(b) There are 9 sets, each consisting of two pairs.
2.1 (a) $\bar{x} - \bar{y} = 9$, estimated *S.E.* = 2.52
(b) (4.1, 13.9)
2.3 $H_0: \mu_1 - \mu_2 = 0$, $H_1: \mu_1 - \mu_2 \neq 0$,
$R: Z \leq -1.96$ or $Z \geq 1.96$.
Observed $z = -4.17$; H_0 is rejected at $\alpha = .05$.
2.5 (a) $H_0: \mu_1 - \mu_2 = 0$, $H_1: \mu_1 - \mu_2 < 0$
(b) $Z = \dfrac{\bar{X} - \bar{Y}}{\sqrt{\dfrac{S_1^2}{n_1} + \dfrac{S_2^2}{n_2}}}$, $R: Z \leq -1.645$
(c) Observed $z = -2.22$, *P*-value = .0132
2.7 *P*-value = $P[Z \geq 4.55]$, smaller than .0002
2.9 Assume normal populations with equal σ's.
(a) (10.28, 13.64)
(b) $R: T \geq 2.650$.
Observed $t = 2.52$; H_0 is not rejected.
2.11 (a) $(-6.4, 29.4)$ (using the conservative T^*,
d.f. = 6)
2.13 $H_0: \mu_1 - \mu_2 = 0$, $H_1: \mu_1 - \mu_2 > 0$,

$$Z = \frac{\overline{X} - \overline{Y}}{\sqrt{\dfrac{S_1^2}{n_1} + \dfrac{S_2^2}{n_2}}}$$

Observed $z = 3.01$, P-value $= .0013$. There is strong evidence in support of H_1.

2.17 (a) $H_0: \mu_1 - \mu_2 = 0$, $H_1: \mu_1 - \mu_2 < 0$, $t_{.05} = 1.734$, d.f. $= 18$, $R: T \le -1.734$. Observed $t = -1.81$; H_0 is rejected at $\alpha = .05$. The mean job time is significantly less with method 1.
(b) Normal populations with equal σ's.
(c) $(-9.09, .69)$

4.1 (a) $t = -.23$ (b) d.f. $= 3$

4.3 (a) $H_0: \delta = 0$, $H_1: \delta > 0$, $T = \dfrac{\overline{D}}{S_D / \sqrt{n}}$,

d.f. $= 11$, $R: T \ge 1.796$.
Observed $t = 1.86$; H_0 is rejected at $\alpha = .05$. Treatment A has a significantly higher mean response.
(b) $(-.98, 11.48)$

4.7 $(1.52, 7.90)$

4.9 $H_0: \delta = 0$, $H_1: \delta \ne 0$, $T = \dfrac{\overline{D}}{S_D / \sqrt{n}}$,

d.f. $= 8$, $R: T \le -2.306$ or $T \ge 2.306$
Observed $t = 1.48$; H_0 is not rejected. The difference is not significant at $\alpha = .05$.

4.11 (b) $(.43, 6.77)$

5.1 (a) $(24.6, 33.4)$
(c) $R: Z \le -1.645$, observed $z = -3.16$; H_0 is rejected at $\alpha = .05$.

5.3 $(-13.14, -10.12)$

5.5 (a) $H_0: \mu_A - \mu_B = 0$, $H_1: \mu_A - \mu_B > 0$

(b) $Z = \dfrac{\overline{X} - \overline{Y}}{\sqrt{\dfrac{S_1^2}{n_1} + \dfrac{S_2^2}{n_2}}}$, $R: Z \ge 1.28$

(c) Observed $z = 2.09$, H_0 is rejected.
P-value $= .0183$.

5.7 (a) $s_{\text{pooled}}^2 = 4.857$ (b) $t = .68$ with d.f. $= 7$

5.9 $(74.87, 81.73)$ minutes for μ_1

5.11 $(-12.3, 44.3)$ pounds

5.15 $H_0: \delta = 0$, $H_1: \delta \ne 0$, $T = \dfrac{\overline{D}}{S_D / \sqrt{n}}$, d.f. $= 8$;

$R: T \le -2.896$ or $T \ge 2.896$

Observed $t = 1.85$; H_0 is not rejected. The difference is not significant.
(b) $(-.005, 2.403)$

5.17 $(.45, 3.55)$ centimeters

5.19 (a) $t = -1.10$, fail to reject H_0.

5.21 $(-7.4, 8.0)$. Both very much alike.

CHAPTER 11

2.1 (a) $\hat{p} = .62$, 95% error margin $= .135$
(b) $\hat{p} = .113$, 95% error margin $= .029$

2.3 (a) $(.51, .73)$
(b) $(.09, .14)$

2.5 $\hat{p} = .622$, 80% error margin $= .09$

2.7 (b) $\hat{p} = .512$, 98% confidence interval $(.49, .53)$

2.11 637

2.13 (b) $H_0: p = .12$ versus $H_1: p < .12$

2.15 (c) (i) $H_0: p = .60$ versus $H_1: p \ne .60$, (ii) $Z = (\hat{p} - .60)/.0522$, (iii) $R: Z \ge 2.33$ or $Z \le -2.33$

2.17 $H_0: p = .19$, $H_1: p < .19$, $R: Z \le -1.645$, observed $z = -2.76$; H_0 is rejected at $\alpha = .05$.

2.19 P-value $= .0150$, strong support for conjecture.

3.1 (a) $(-.08, .32)$
(b) Observed $z = -3.39$, P-value $= .0003$. H_0 is strongly rejected.

3.3 (a) $H_0: p_A = p_B$, $H_1: p_A < p_B$, $R: Z \le -1.645$. Observed $z = -2.52$; H_0 is rejected at $\alpha = .05$.
(b) $(.03, .27)$

3.7 (b) $(.023, .072)$

3.9 (b) $(.11, .51)$

3.11 $(.12, .29)$

3.13 (b) $(.08, .14)$

3.15 (a) $(.17, .38)$

4.1 (b) $\chi^2 = .281$. Fail to reject equality.

4.3 $\chi^2 = 24.608$. Reject equality at $\alpha = .05$.

4.5 (b) $Z = -.45$. Fail to reject equal proportions.

4.7 (b) $Z = -.997$. Fail to reject equality.

4.9 $\chi^2 = 9.640$. Reject equality, evidence is strong.

5.1 Observed $\chi^2 = 15.734$, d.f. $= 2$, $\chi_{.05}^2 = 5.99$. The appeals decision depends on type of representation.

5.5 Observed $\chi^2 = 27.847$, d.f. $= 2$, $\chi_{.01}^2 = 9.21$. The null hypothesis of independence is rejected at $\alpha = .01$. Attitude depends on membership.

5.7 Observed $\chi^2 = 4.134$, d.f. $= 2$, $\chi_{.05}^2 = 5.991$, we fail to reject independence of group and experimenter's influence.

6.1 (a) $(.33, .39)$

6.3 (a) $(.068, .097)$

6.7 (b) (.36, .48)

6.9 545

6.11 P-value = .0384. Quite strong evidence for conjecture.

6.13 (a) Observed χ^2 = 48.24, d.f. = 1, $\chi^2_{.01}$ = 6.63. The null hypothesis of homogeneity even is rejected at α = .01.

6.19 Observed χ^2 = 9.146, d.f. = 2, $\chi^2_{.05}$ = 5.99. H_0 is rejected at α = .05.

CHAPTER 12

2.1 (a) x = number of cigarettes per day, y = carbon monoxide level

2.3 intercept = 2, slope = 3

2.5 β_0 = 8, β_1 = −6, σ = 4

2.11 (a) 23, 28

2.13 (c) mean price for 1 square foot

3.1 (d) $\hat{y}$ = 3.845 − .615x

3.3 (c) s^2 = .0513

3.7 (a) $\hat{y}$ = 1.454 + .247x
       (c) s^2 = .031

4.1 (a) $\hat{\beta}_0$ = 3.845, $\hat{\beta}_1$ = −.615, s^2 = .0513
       (b) T = −13.8, reject H_0: β_1 = 0
       (d) (3.45, 4.24)

4.5 (a) (91.31, 96.11)
       (b) (87.30, 100.12)

4.7 (b) (1316.4, 1709.0)

4.11 (b) $\hat{y}$ = 469.6 − .2579 x

4.13 $\hat{y}$ = 10.33 + .8990 x

5.1 r^2 = .188

5.5 r^2 = .979, r = .989

5.7 .799

6.1 (c) $\hat{y}$ = 3.737 + 3.655x

6.3 (a) (2.86, 4.45)
       (b) (17.10, 19.61)

6.5 (b) 63.65
       (e) r^2 = .828

6.7 $\hat{y}$ = 146.4 + .6888 x

Index